Cytokines and Colony Stimulating Factors

METHODS IN MOLECULAR BIOLOGY™

John M. Walker, SERIES EDITOR

METHODS IN MOLECULAR BIOLOGY™

Cytokines and Colony Stimulating Factors

Methods and Protocols

Edited by

Dieter Körholz

and

Wieland Kiess

Department of Pediatrics, University of Leipzig, Leipzig, Germany

Humana Press **Totowa, New Jersey**

Softcover reprint of the hardcover 1st edition 2003
999 Riverview Drive, Suite 208
Totowa, New Jersey 07512

www.humanapress.com

This publication is printed on acid-free paper. ∞
ANSI Z39.48-1984 (American Standards Institute) Permanence of Paper for Printed Library Materials.

Cover illustration: Figure 1A (background) and Figure 2 (inset) from Chapter 9, "Angiogenic Cytokines: Quantitative and Functional Analysis," by Angelika Eggert and John H. Maris.

Production Editor: Jessica Jannicelli.

Cover design by Patricia F. Cleary.

For additional copies, pricing for bulk purchases, and/or information about other Humana titles, contact Humana at the above address or at any of the following numbers: Tel.: 973-256-1699; Fax: 973-256-8341; E-mail: humana@humanapr.com; or visit our Website: www.humanapress.com

10 9 8 7 6 5 4 3 2 1

Library of Congress Cataloging in Publication Data

Cytokines and colony stimulating factors : methods and protocols / edited by Dieter Körholz and Wieland Kiess.
p. cm. -- (Methods in molecular biology ; v. 215)
Includes bibliographical references and index.
ISBN 978-1-61737-318-3 e-ISBN 978-1-59259-345-3
1. Cytokines--Laboratory manuals. 2. Colony-stimulating factors (Physiology)--Laboratory manuals. I. Körholz, Dieter. II. Kiess, W. (Wieland) III. Methods in molecular biology (Totowa, N.J.) ; v. 215.

QR185.8.C95 C959 2003
616.07'9--dc21

2002027346

Preface

The immune system is a complex network in which different cell types and soluble factors interact to efficiently eliminate various kinds of microorganisms as well as aberrant cell clones. The roots of immunologic investigations reach far into the past. In 430 BC, Thucydides reported that survivors of the plague did not present a second time with similar symptoms. The first report of a successful immunotherapy was made by Edward Jenner in 1798 who found a protective effect of cowpox vaccination against human pox. Since then, much knowledge has been accumulated; today, investigations of the molecular mechanisms of immune regulation are of central research interest. The novel insights into gene polymorphisms and gene regulation gathered from this work has improved our knowledge of individual immune reactions and risk factors in overcoming infections. Strategies to use the immune system for cancer treatment have been propelled by the discovery of divergent immunoregulatory cytokines and the introduction of new gene therapy strategies to modify immune responses. Recently, the discovery of various dendritic cells has focused attention on these cell types as central elements of the immune response and to the possibility of dendritic cell expansion, maturation, and consecutive stimulation with immunoreactive tumor-specific peptides. Similarly, methods for ex vivo expansion of various stem cell-derived cell types have led to an improved therapeutic management of various benign and malignant diseases. However, hope for greater therapeutic success in the clinical setting, and therefore patient benefits, has been quite limited with these methods. In the field of immunoregulation, enormous effort is still required to increase fundamental understanding and the therapeutic benefit. Only basic science and further research will provide for means to achieve this goal.

Cytokines and Colony Stimulating Factors: Methods and Protocols is intended to promote all the present knowledge and understanding generated by the many novel technologies described here in the hope of spreading awareness of its useful applications. Our book aims to address the needs of novice investigators hoping to work in this field who need profound and practical information to get started, as well as seasoned investigators seeking to extend their technical range. The book not only provides detailed descriptions of methods, but also includes a section on troubleshooting in each chapter indicated in the Notes section. With the spreading use of these methods and the growing activity in the field, many qualified scientists can work on the therapeutic benefit of these studies for future patients suffering from these diseases.

Dieter Körholz
Wieland Kiess

Contents

Contributors

NICHOLAS ACHILLE • *Department of Pathology, Loyola University Stritch School of Medicine, Maywood, IL*

BEATRIZ ALBELLA • *Hematopoiesis Project, CIEMAT, Madrid, Spain*

URSULA BANNING • *Division of Pediatric Hematology and Oncology, Department of Pediatrics, University of Leipzig, Germany*

OLIVER B. BETZ • *Center for Molecular Orthopaedics, Harvard Medical School, Boston, MA*

MARTIN BORNHÄUSER • *Medical Clinic I, Carl Gustav Carus University, Dresden, Germany*

MINKA BRELOER • *Division of Medical Microbiology and Immunology, Bernhard-Nocht-Institute for Tropical Medicine, Hamburg, Germany*

JUAN ANTONIO BUEREN • *Hematopoiesis Project, CIEMAT, Madrid, Spain*

QUAN CAI • *Department of Surgery, University of Pittsburgh, PA*

BRUNO CANQUE • *Laboratoire d'Immunologie, Institut National de la Santé et de la Recherche Médicale, Paris, France*

MARINA CAVAZZANA-CALVO • *Gene and Cell Therapy Unit, Institut National de la Santé et de la Recherche Médicale, Hôpital Necker Enfants Malades, Paris, France*

STEPHEN J. CHANOCK • *Immunocompromised Host Section, Pediatric Oncology Branch, National Cancer Institute, National Institutes of Health, Gaithersburg, MD*

YUTI CHERNAJOVSKY • *Bone and Joint Research Unit, William Harvey Research Institute, Barts and Royal London Medical School, Queen Mary, University of London, UK*

TIMOTHY M. CLAY • *Program in Molecular Therapeutics, Department of Surgery, Duke University Medical Center, Durham, NC*

ISAAC COHEN • *Department of Cell and Molecular Biology, Northwestern University, Chicago, IL*

GENEVIEVE DE SAINT BASILE • *Institut National de la Santé et de la Recherche Médicale, Hôpital Necker Enfants Malades, Paris, France*

DAGMAR DILLOO • *Department of Pediatric Hematology and Oncology, University Hospital, Heinrich-Heine-University, Düsseldorf, Germany*

HANNA DREJA • *Bone and Joint Research Unit, William Harvey Research Institute, Barts and Royal London Medical School, Queen Mary, University of London, UK*

YASUHIRO EBIHARA • *Division of Cellular Therapy, Advanced Clinical Research Center, Institute of Medical Science, University of Tokyo, Japan*

ANGELIKA EGGERT • *Department of Hematology/Oncology, University Children's Hospital of Essen, Essen, Germany*

PETER D. EMANUEL • *Department of Medicine, Comprehensive Cancer Center, University of Alabama, Birmingham, AL*

CHRISTOPHER H. EVANS • *Center for Molecular Orthopaedics, Harvard Medical School, Boston, MA*

ROBERTO FURLAN • *Neuroimmunology Unit, Department of Neuroscience, San Raffaele Scientific Institute - DIBIT, Milan, Italy*

ALESSANDRA GALETTO • *Department of Oncological Surgery, University of Turin, Italy*

BÉATRICE GAUGLER • *Laboratoire d'Immunologie des Tumeurs, Institut Paoli-Calmettes, Université de la Méditerranée, Institut National de la Santé et de la Recherche Médicale, Marseille, France*

STEVEN C. GHIVIZZANI • *Center for Molecular Orthopaedics, Harvard Medical School, Boston, MA*

JEAN CLAUDE GLUCKMAN • *Laboratoire d'Immunologie, Institut National de la Santé et de la Recherche Médicale, Paris, France*

JOHN R. GORDON • *Department of Veterinary Microbiology, Western College of Veterinary Medicine, University of Saskatchewan, Saskatoon, Canada*

DAVID GOULD • *Bone and Joint Research Unit, William Harvey Research Institute, Barts and Royal London Medical School, Queen Mary, University of London, UK*

ELVIRE GOUZE • *Molecular Orthopedics, Harvard Medical School, Boston, MA*

JEAN-NOEL GOUZE • *Center for Molecular Orthopaedics, Harvard Medical School, Boston, MA, and Electrical Engineering and Computer Sciences Laboratory, Massachusetts Institute of Technology, Cambridge, MA*

SUSANNE GRETHER-BECK • *Institut für Umweltmedizinische Forschung (IUF), Heinrich-Heine-University, Düsseldorf, Germany*

GUILLERMO GUENECHEA • *Hematopoiesis Project, CIEMAT, Madrid, Spain*

SALIMA HACEIN-BEY-ABINA • *Gene and Cell Therapy Unit, Institut National de la Santé et de la Recherche Médicale, Hôpital Necker Enfants Malades, Paris, France*

Kunisuke Himeno • *Department of Parasitology and Immunology, Kyushu University Graduate School of Medicine, Fukuoka, Japan*
Bodo Hoffmeister • *Institute of Medical Immunology, Charité, Humboldt-University, Berlin, Germany*
Xin-Sheng Ju • *Max-Delbrück-Center for Molecular Medicine Berlin, Germany*
Pawel Kalinski • *Department of Surgery, University of Pittsburgh, PA*
Martien Kapsenberg • *Department of Cell Biology and Histology, Department of Dermatology, Academic Medical Center, University of Amsterdam, Netherlands*
Florian Kern • *Institute of Medical Immunology, Charité, Humboldt-University, Berlin, Germany*
Elham Khatamzas • *Institute of Medical Immunology, Charité, Humboldt-University, Berlin, Germany*
Felix Kiecker • *Institute of Medical Immunology, Charité, Humboldt-University, Berlin, Germany*
Wieland Kiess • *Department of Pediatrics, University of Leipzig, Germany*
Dieter Körholz • *Division of Pediatric Hematology and Oncology, Department of Pediatrics, University of Leipzig, Germany*
Christof M. Kramm • *Department of Pediatric Hematology and Oncology, University Hospital, Heinrich-Heine-University, Düsseldorf, Germany*
Jean Krutmann • *Institut für Umweltmedizinische Forschung (IUF), Heinrich-Heine-University, Düsseldorf, Germany*
Deanne M. R. Lathers • *Research Service, Hines Veterans Affairs Hospital, Hines, IL, and Departments of Pathology and Otolaryngology, Loyola University Strich School of Medicine, Maywood, IL*
Phil Lefebvre • *Department of Medicine, Northwestern University, Chicago, IL*
Thomas Lehrnbecher • *Department for Pediatric Hematology and Oncology, Children's Hospital III, Johann Wolfgang Goethe University, Frankfurt/Main, Germany*
Renz Mang • *Clinical and Experimental Photodermatology, Department of Dermatology, Heinrich-Heine-University, Düsseldorf, Germany*
Peggy C. Marconi • *Section of Microbiology, Department of Clinical and Experimental Medicine, University of Ferrara, Italy*
John H. Maris • *Division of Oncology, Children's Hospital of Philadelphia, Philadelphia, PA*
Gianvito Martino • *Neuroimmunology Unit, Department of Neuroscience, San Raffaele Scientific Institute - DIBIT, Milan, Italy*

LINA MATERA • *Laboratory of Tumor Immunology, Department of Internal Medicine, University of Turin, Italy*
MOHAMAD MOHTY • *Laboratoire d'Immunologie des Tumeurs, Institut Paoli-Calmettes, Université de la Méditerranée, Marseille, France*
SOLVEIG H. MORÉ • *Division of Medical Microbiology and Immunology, Bernhard-Nocht-Institute for Tropical Medicine, Hamburg, Germany*
EVELIEN NAESSENS • *Department of Clinical Chemistry, Microbiology and Immunology, Ghent University, Belgium*
TIM NIEHUES • *Hematology and Oncology, University Hospital, Heinrich-Heine-University, Düsseldorf, Germany*
CHARLOTTE M. NIEMEYER • *Pediatric Hematology and Oncology, Department of Pediatrics, University of Freiburg, Germany*
MICHAEL I. NISHIMURA • *Department of Surgery, University of Chicago, IL*
DANIEL OLIVE • *Laboratoire d'Immunologie des Tumeurs, Institut Paoli-Calmettes, Université de la Méditerranée, Institut National de la Santé et de la Recherche Médicale U119, Marseille, France*
GLYN D. PALMER • *Center for Molecular Orthopaedics, Harvard Medical School, Boston, MA*
CIRIACO A. PICCIRILLO • *Laboratory of Immunology, National Institute of Allergy and Infectious Diseases, National Institutes of Health, Bethesda, MD*
STEFANO PLUCHINO • *Neuroimmunology Unit, Department of Neuroscience, San Raffaele Scientific Institute - DIBIT, Milan, Italy*
GÉRALD J. PRUD'HOMME • *Department of Pathology, McGill University, Montreal, Canada*
SERGIO QUEROL • *Barcelona Cord Blood Bank, Centre de Transfusió i Banc de Teixits, Hospital Durán i Reynals, Barcelona, Spain*
NIKOLAI G. RAINOV • *Walton Centre for Neurology and Neurosurgery, University of Liverpool, UK*
MICHELE ROSENZWAJG • *Laboratoire d'Immunologie, Institut National de la Santé et de la Recherche Médicale, Paris, France*
TOHRU SAKAI • *Department of Nutrition, University of Tokushima School of Medicine, Tokushima, Japan*
JOSÉ CARLOS SEGOVIA • *Hematopoiesis Project, CIEMAT, Madrid, Spain*
JOOST H. N. SCHUITEMAKER • *Department of Cell Biology and Histology, Academic Medical Center, University of Amsterdam, Netherlands*
CHRISTIAN SCHULTZ • *Department of Paediatrics, Medical University Lübeck, Germany*
VOLKER SCHUSTER • *Division of Pediatric Immunology, Department of Pediatrics, University of Leipzig, Germany*

MICHAEL R. SHURIN • *Clinical Immunopathology, University of Pittsburgh Medical Center, Cancer Institute, Pittsburgh, PA*
HAUKE SIEVERTS • *Department of Hematology/Oncology, University Children's Hospital of Heidelberg, Heidelberg, Germany*
VERONIQUE STOVE • *Department of Clinical Chemistry, Microbiology and Immunology, Ghent University, Belgium*
INGOLF SUREL • *Institute of Medical Immunology, Charité, Humboldt-University, Berlin, Germany*
KOSTAS TRIANTAPHYLLOPOULOS • *Bone and Joint Research Unit, William Harvey Research Institute, Barts and Royal London Medical School, Queen Mary, University of London, UK*
KOHICHIRO TSUJI • *Division of Cellular Therapy, Advanced Clinical Research Center, Institute of Medical Science, University of Tokyo, Japan*
TAKAHIRO UEDA • *Division of Cellular Therapy, Advanced Clinical Research Center, Institute of Medical Science, University of Tokyo, Japan*
BRUNO VERHASSELT • *Department of Clinical Chemistry, Microbiology and Immunology, Ghent University, Belgium*
PEDRO VIEIRA • *Department of Cell Biology and Histology, Academic Medical Center, University of Amsterdam, Netherlands*
HANS-DIETER VOLK • *Institute of Medical Immunology, Charité, Humboldt-University, Berlin, Germany*
ARNE VON BONIN • *Division of Medical Microbiology and Immunology, Bernhard-Nocht-Institute for Tropical Medicine, Hamburg, Germany*
THOMAS VRAETZ • *Pediatric Hematology and Oncology, Department of Pediatrics, University of Freiburg, Germany*
VOLKER WAHN • *Children's Clinic, Uckermark Hospital, Schwedt/Oder, Germany*
HENNING WALCZAK • *Division of Apoptosis Regulation, German Cancer Research Center, Heidelberg, Germany*
KERSTIN WILLMANN • *BD Biosciences, San Jose, CA*
JOSH WU • *Department of Dermatology, University of Colorado Health Science Center, Denver, CO*
JIM XIANG • *Saskatoon Cancer Center, Departments of Microbiology and Oncology, University of Saskatchewan, Saskatoon, Canada*
M. RITA I. YOUNG • *Research Service, Hines Veterans Affairs Hospital, Hines, IL, and Departments of Pathology and Otolaryngology, Loyola University Strich School of Medicine, Maywood, IL*
MARTIN ZENKE • *Max-Delbrück-Center for Molecular Medicine Berlin, Germany*
ANDREE ZIBERT • *Department of Pediatric Hematology and Oncology, University Hospital, Heinrich-Heine-University, Düsseldorf, Germany*

I

INTRODUCTION

1

Primary Immunodeficiencies Caused by Defects of Cytokines and Cytokine Receptors

Volker Wahn

1. Introduction

The immune system responds to antigenic stimulation with a complex array of molecular events involving antigen-presenting cells, B-cells, T-cells, and phagocytes. Cytokines and their respective receptors are intimately involved in regulating such immune responses. Their pivotal role can be illustrated in animal models for which certain cytokines or their receptors have been deleted.

It is beyond the scope of this review to discuss our knowledge on growth factor or receptor deficiencies derived from animal models. I would rather like to focus on observations in children with selective molecular defects. For example, pulmonary alveolar proteinosis in some patients has been found to be associated with mutations in the genes for granulocyte macrophage–colony-stimulating factor (GM-CSF) receptor β-chain, which is shared with receptors for interleukin (IL)-3 and IL-5. Certainly, in the future we should be aware of further pathogenic mutations in humans that, to date, have been demonstrated in experimental animals only.

The majority of mutations in cytokine or cytokine receptor genes result in inherited immunodeficiencies. Our current knowledge on such disorders, therefore, will be summarized.

2. Interleukin-1

There is one report on defective IL-1 production associated with immunodeficiency with some evidence of familiality *(1)*. The siblings described

From: *Methods in Molecular Biology, vol. 215: Cytokines and Colony Stimulating Factors: Methods and Protocols*
Edited by: D. Körholz and W. Kiess © Humana Press Inc., Totowa, NJ

suffered from fungal infections similar to patients with chronic mucocutaneous candidiasis and from bacterial infections. Monocytes of the two patients once stimulated in culture with either lipopolysaccharide (LPS) or silica produced <1% of IL-1 compared with either normal cells or cells from patients with other immunodeficiencies. Appropriate studies on the DNA level were not performed to date and the molecular basis of this deficiency remains to be elucidated.

2.1. IL-1 Receptor

The first report on a possible abnormality of the IL-1 receptor probably comes from Chu et al. *(2)*. A 10-yr-old boy with recurrent infections had a normal B- and T-cell phenotype and normal immunoglobulins. His cells in vitro had a diminished response to mitogens and an absent response to antigens. IL-1 production was normal. T-Cell blasts from the patient were unable to absorb out IL-1 activity from an IL-1 preparation that may possibly be explained by the absence of the receptor or a molecular alteration within the receptor.

Another report in the literature describes a 15-yr-old girl with recurrent bacterial infections whose cells did not produce tumor necrosis factor (TNF) upon stimulation with several stimuli including IL-1 *(3)*. The molecular defect was assumed to be localized early in the signal transduction pathway, not necessarily in the receptor; however, appropriate studies to identify the defect on the molecular level were not performed. So, to date, information on IL-1 receptor deficiencies are restricted to those derived from IL-1 receptor knockout mice and have no real correlate in humans yet.

2.2. Interleukin-2

Two types of primary IL-2 deficiency should be distinguished: selective IL-2 deficiency and IL-2 deficiency as part of multiple cytokine deficiency. Both will be discussed separately.

2.2.1. Selective IL-2 Deficiency

Weinberg and Parkman *(4)* reported on a male infant with T+B+ SCID. IL-2 production was markedly deficient. mRNA encoding interferon (IFN)-γ was present at normal amounts, whereas that encoding IL-2 was completely absent. On the DNA level, no defects related to structural genes or the promotor or enhancer region could be identified. Bone marrow transplantation (BMT) was attempted, but the child died as a result of hemorrhagic pancreatitis. A similar case has been published with mainly similar findings *(5)*.

Further cases with analogous findings have been published thereafter. Remarkable may be the case of Sorensen et al. *(6)* because the boy they

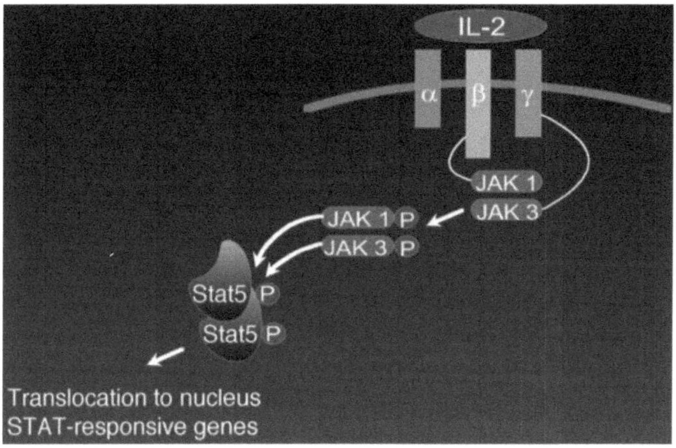

Fig. 1. After binding of IL-2 to IL-2R JAK1 associates with the β-chain and JAK3 is constitutively associated with γc. Next, tyrosine phosphorylation occurs in both JAKs and IL-2Rβ before STAT5 is activated (phosphorylated), dimerized, and transferred to the nucleus, where it can bind to DNA at the appropriate site.

described was 15 yr of age at diagnosis. The molecular basis for absent IL-2 synthesis in all these cases is still unclear.

2.2.2. Multiple Lymphokine Deficiency

Pahwa et al. *(7)* described a with T+B+ SCID whose cells were unable to secrete appropriate amounts of IL-2 in response to cellular stimulation. Other cytokines were initially not measured. Two attempts to correct SCID by bone marrow transplantation failed, but the child responded well to in vivo administration of recombinant IL-2. This child was further studied by Chatila et al. *(8)*. The authors found that in addition to IL-2, the synthesis of IL-3, IL-4, and IL-5 was grossly deficient, and IFN-γ was secreted at moderately decreased amounts. GM-CSF was produced at normal levels. In 1993, the same group presented the molecular basis of this defect *(9)*: Nuclear factor of activated T-cells (NF-AT) did not appropriately bind to both its response element in the IL-2 enhancer and an NF-AT-like response element in the IL-4 enhancer. To date, however, no mutation has been reported explaining this phenomenon.

2.3. IL-2 Receptor

The structure and function of the receptor is briefly summarized in **Fig. 1**. It is composed of an α-chain (CD25), a β-chain (CD122), and a γ-chain

(CD132) shared by many interleukin receptors (IL-4, IL-7, IL-9, IL-15) and is thus called common γ-chain. Deficiencies of all three receptor chains have been identified in humans.

2.3.1. IL-2 Receptor α-Chain Deficiency

This disorder was first reported by Sharfe et al. ***(10)***. The affected patient presented with cytomegalovirus (CMV) pneumonitis, oral thrush and candida esophagitis. He had low normal numbers of T-cells but normal B-cell function. Mitogen stimulation was low and could not be increased by the addition of IL-2. There was an excessive T-cell infiltration of several organs accompanied by tissue atrophy and inflammation. The child was successfully treated by BMT. The molecular basis was a truncation mutation of CD25. As a consequence of this mutation, apoptosis in the thymus was decreased, which might impair negative selection ***(11)***. Autoreactive T-cell clones may leave the thymus and infiltrate tissues as a sort of an autoimmune process.

2.3.2. IL-2 Receptor/IL-15 Receptor β-Chain Deficiency

Interleukin-15 is essential for natural killer (NK) cell development. Gilmour et al. ***(12)*** reported a male infant born to nonconsanguinous parents with a typical history for SCID. Surface marker analysis revealed the phenotype: T low, B+, and NK–. γc and JAK3 were normal, indicating a new molecular defect. Almost no JAK3 phosphorylation could be observed upon IL-2 stimulation. The explanation came from studies of the β-chain because almost no expression of β-chain mRNA and protein could be found. Familiality of the defect, however, was not established.

2.3.3. IL-2 Receptor γ-Chain Deficiency (X-Linked SCID)

The disease must be suspected in all boys with T–B+ SCID ***(13)***. In the meantime, far more than 100 children have been analyzed on the molecular level. As a result, a large database is now available that gives an overview of the type and frequency of known mutations ***(14)***. The disease is of special interest because it is the first human disease that seems to have been cured by gene therapy (*see* Chapter 19).

2.3.4. Combined IL-2 and IL-2 Receptor Deficiency

So far, there is only one report on this deficiency, that by Doi et al. ***(15)***. The authors described a child with combined immunodeficiency (ID) whose cells in vitro did not respond to T-cell mitogens, did not produce IL-2, and did not express the IL-2R. The brother of the child seemed to have died of the same disease, giving some evidence of familiality. DNA coding regions for IL-2 and IL-2R were found to be normal but no mRNA could be found, indicating a transcriptional problem.

2.4. IL-4 Receptor

Mutations within genes encoding this receptor have been associated with hyper-IgE syndrome in some patients. However, because these observations are still a matter of controversial debate, they will not be discussed here.

2.5. IL-7 Receptor

The receptor is composed of two chains, one of which is the γc chain mentioned earlier, the other is the IL-7Rα chain. Like the IL-2R, they are linked to JAK3 and JAK1, respectively, and transmit signals to the nucleus using STAT5 and other pathways ***(16)***. Signals contribute to recombination events of T-cell receptor genes.

Once Puel et al. studied the IL-7 receptor in Epstein–Barr-virus (EBV)-transformed cell lines of children with T–B+NK+ SCID ***(17)***, they identified two subjects with specific mutations: One had a splice-junction acceptor mutation on one strand (intron 4) and the other strand had a nonsense mutation with a premature stop codon (exon 5). Acting in concert these mutations prevented production of a functional IL-7Rα protein. The second patient had a two-amino-acid change in the extracellular domain of the IL-7Rα chain, both of which turned out to be a common polymorphism. A further splice mutation, however, was responsible for absence of the IL-7Rα chain. Treatment consisted of BMT.

2.6. Interleukin-12

Interleukin-12 is one of the major cytokines stimulating production of IFN-γ. It is produced by antigen-presenting cells upon activation and stimulates NK cells and T-cells (**Fig. 2**). Absence of IL-12 results in diminished or absent production of IFN-γ, which, in turn, is responsible for increased susceptibility to mycobacterial and other infections.

The first report on defective IL-12 production in familial disseminated mycobacterial infection (*Myobacterium avium*) comes from Frucht and Holland ***(18)***. The authors showed in three affected family members that patient monocytes did not appropriately produce IL-12, which uses responsible for deficient production of IFN-γ synthesis. In vitro IFN-γ synthesis could be initiated by addition of exogenous IL-12, indicating integrity of the IL-12 receptor. In vivo patients were successfully treated with IFN-γ. The genetic basis of abnormal IL-12 synthesis in this family has not been studied so far.

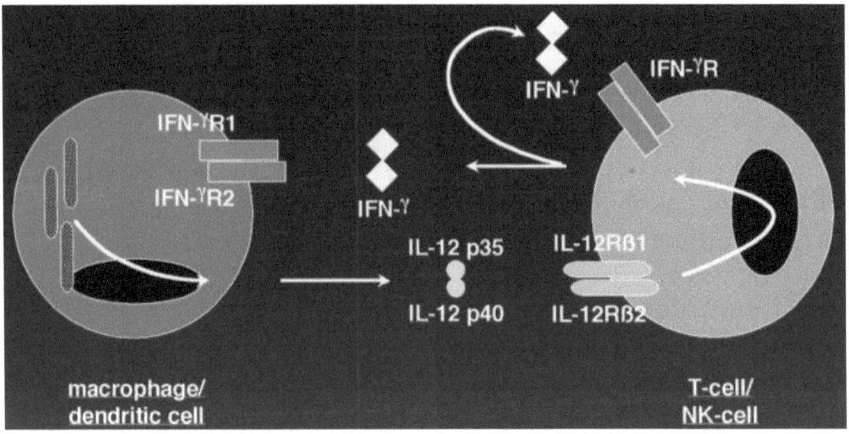

Fig. 2. Induction of IFN-γ by mycobacteria. (From **ref. *20***.)

Such studies were accomplished by Altare et al. ***(19)*** in a child with Bacille–Calmette–Guérin (BCG) and *Salmonella* enteritidis-disseminated infection. The authors found a large homocygous deletion within the p40-subunit gene precluding expression of functional IL-12 p70 cytokine expression. In vitro complementation of the cells with a wild-type IL-12 p40 gene restored IL-12 and, in turn, IFN-γ production.

2.6.1. IL-12 Receptor

The receptor is a heterodimer consisting of a β1-chain and a β2-chain. In 1998, complete IL-12Rβ1 deficiency was described by two groups in several children with disseminated other mycobacterial (BCG, non-tuberculous [NTM]) and Salmonella infections. All patients were homocygous for recessive mutations ***(20)***. As the receptor is expressed on the cell surface, the diagnosis can be established by flow cytometry. Treatment may consist of parenteral administration of IFN-γ. An international registry was established to collect all information on the spectrum of these abnormalities.

2.7. Interferon-γ

There are some reports in the literature on cases and families with impaired production of IFN-γ ***(21)*** in association with atypical mycobacterial infections or cryptosporidiosis. Unfortunately, no data on the IL-12 system were presented and it seems likely that these cases, upon reinvestigation, will turn out as being IL-12 or IL-12R deficient.

2.7.1. Interferon-γ Receptor

Interferon-γ can exert its effects only if its receptor is functional. This receptor is a heterodimer that is tetramerized following binding of IFN-γ. In patients with inappropriate receptor function, the most impressive clinical finding is the increased susceptibility to mainly mycobacterial (BCG, NTM) and, to a lesser extent, Salmonella infections. With regard to congenital immunodeficiencies, five subtypes must be differentiated ***(22)***.

2.7.2. Complete Recessive IFN-γR1 Deficiency

This disease must be suspected in very young children with severe BCG or NTM infections, especially if the granulomas formed are lepromatous rather than tuberculous. The diagnosis can be suspected by flow cytometry (absence of CD119 on monocytes). Definitive diagnosis is established on the genetic level by studies of the *IFNGR1* gene. Treatment consists of bone marrow transplantation in the absence of active mycobacterial disease.

2.7.3. Complete Recessive IFN-γR2 Deficiency

Clinical aspects in the only decribed patient did not differ from those in other deficiencies. The disease must be suspected if patients cells in vitro do not respond to high concentrations of IFN-γ. A reliable diagnosis can only be made if the IFNGR2 exons and flanking intron regions are sequenced.

2.7.4. Partial Recessive IFN-γR1 Deficiency

There are no specific clinical aspects in this disease. In vitro cells from children do respond in vitro, in contrast to complete deficiency, to high concentrations of IFN-γ, and the authors suspected that the missense mutation found in the *IFNGR1* gene reduces the affinity of the encoded receptor to its ligand. IFN-γ binds to the R1 chain. Another finding specific for this variant was mature granulomas, in contrast to the complete deficiency, where mature granulomas are never observed. Definitive diagnosis can only be established by sequencing of the gene. High doses of IFN-γ in vivo may provide improved control of mycobacteria.

2.7.5. Partial Recessive IFN-γR2 Deficiency

In the single described 20-yr-old patient, the receptor seemed to be weakly expressed on the cell surface by flow-cytometric studies. A missense mutation was responsible for the functional receptor defect. In vitro STAT1 translocation and HLA-DR expression following stimulation was impaired but could be detected. In vivo, the patient seemed to benefit from IFN-γ therapy.

2.7.6. Partial Dominant IFN-γR1 Deficiency

The basis for this disease, which has been found in currently 18 patients, is a heterozygous frame-shift small deletion in exon 6. The mutant alleles encode a truncated receptor with no more than five intracellular amino acids. The receptor can bind IFN-γ normally and dimerize normally, but it is unable to transduce intracellular signals. The diagnosis is made by appropriate genetic studies. In vitro, high doses of IFN-γ can trigger cellular responses. In vivo, some patients respond to administration of IFN-γ.

3. Notes

Inherited deficiencies of cytokines and their respective receptors are the human equivalent to corresponding knockout mice. Several such deficiencies have already been found and experiments of nature improved our understanding of molecular events. With now 23 interleukins and many other cytokines already defined we have to expect that more disorders can be found where genetic disturbances of these systems are involved.

4. Note Added in Proof

Updated information may be relevant and should be mentioned. First, a second cause for multiple cytokine deficiency could be identified in two siblings *(23)*. Its molecular basis seems to be defective nuclear NF-AT translocation, leading to the same biological effect as did defective NF-AT binding to its enhancer. Second, more molecular variants leading to increased susceptibility to mycobacterial infections have been described *(24)*. For those interested in this field, reading these two publications is highly recommended.

References

1. Komiyama, A., Ichikawa, M., Kanda, H., Aoyama K., Yasui, K., Yamazaki, M., Kawai, H., et al. (1988) Defective interleukin-1 production in a familial monocyte disorder with acombined abnormality of mobility and phagocytosis-killing. *Clin. Exp. Immunol.* **73,** 500–504.
2. Chu, E. T., Rosenwasser, L. J., Dinarello, C. A., Rosen, F. S., and Geha, R. S. (1984) Immunodeficiency with defective T-cell response to interleukin 1. *Proc. Natl. Acad. Sci. USA* **81,** 4945–4949.
3. Kuhns, D. B., Long Priel, D. A., and Gallin, J. I. (1997) Endotoxin and IL-1 hyporesponsiveness in a patient with recurrent bacterial infections. *J. Immunol.* **158,** 3959–3964.
4. Weinberg, K. and Parkman, R. (1990) Severe combined immunodeficeincy due to a specific defect in the production of interleukin-2. *N. Engl. J. Med.* **322,** 1718–1723.

5. Disanto, J. P., Keever, C. A., Small, T. N., Nichols, G. L., O'Reilly, R. J., and Flomenberg, N. (1990) Absence of interleukin 2 production in a severe combined immunodeficiency disease syndrome with T cells. *J. Exp. Med.* **171,** 1697–1704.
6. Sorensen, R. U., Boehm, K. D., Kaplan, D., and Berger, M. (1992) Cryptococcal osteomyelitis and cellular immunodeficiency associated with interleukin-2 deficiency. *J. Pediatr.* **121,** 873–879.
7. Pahwa, R., Chatila, T., Pahwa, S., Paradise, C., Day, N. K., Geha, R., et al. (1989) Recombinant interleukin 2 therapy in severe combined immunodeficiency disease. *Proc. Natl. Acad. Sci. USA* **86,** 5069–5073.
8. Chatila, T., Castigli, E., Pahwa, R., Pahwa, S., Chirmule, N., Oyaizu, N., et al. (1990) Primary combined immunodeficiency resulting from defective transcription of multiple T-cell lymphokine genes. *Proc. Natl. Acad. Sci. USA* **87,** 10,033–10,037.
9. Castigli, E., Pahwa, R., Good, R. A., Geha, R. S., and Chatila, T. A. (1993) Molecular basis of a multiple lymphokine deficiency in a patient with severe combined immunodeficiency. *Proc. Natl. Acad. Sci. USA* **90,** 4728–4732.
10. Sharfe, N., Dadi, H. K., Shahar, M., and Roifman, C. M. (1997) Human disorder arising from mutation of the alpha chain of the interleukin-2 receptor. *Proc. Natl. Acad. Sci. USA* **94,** 3168–3171.
11. Roifman, C. M. (2000) Human IL-2 receptor α chain deficiency. *Pediat. Res.* **48,** 6–11.
12. Gilmour, K. C., Fujii, H., Cranston, T., Davies, E. G., Kinnon, C., and Gaspar, H. B. (2001) Defective expression of the interleukin-2/interleukin-15 receptor beta subunit leads to a natural killer cell-deficient form of severe combined immunodeficiency. *Blood* **98,** 877–879.
13. Uribe, L. and Weinberg, K. I. (1998) X-linked SCID and other defects of cytokine pathways. *Semin. Hematol.* **35,** 299–309.
14. Puck, J. M. (1996) IL2Rgbase: a database of γc-chain defects causing human X-SCID. *Immunol. Today* **17,** 507–511.
15. Doi, S., Saiki, O., Tanaka, T., Ha-Kawa, K., Igarashi, T., Fujita, T., et al. (1988) Cellular and genetic analyses of IL-2 production and IL-2 receptor expression in a patient with familial T-cell-dominant immunodeficiency. *Clin. Immunol. Immunopathol.* **46,** 24–36.
16. Puel, A. and Leonard, W. J. (2000) Interleukin-7 receptor α chain-dependent signalling is required for T-cell development. *Immunol. Allergy Clin. North Am.* **20,** 51–63.
17. Puel, A., Ziegler, S. F., Buckley, R. H., and Leonard, W. J. (1998) Defective IL-7R expression in T–B+NK+ severe combined immunodeficiency. *Nat. Genet.* **20,** 394–397.
18. Frucht, D. M. and Holland, S. M. (1996) Defective monocyte costimulation for IFN-γ production in familial disseminated *Mycobacterium avium* complex infection. *J. Immunol.* **157,** 411–416.
19. Altare, F., Lamms, D., Revy, P., Jouanguy, E., Döffinger, R., Lamhamedi, S., et al. (1998) Inherited interleukin 12 deficiency in a child with Bacille–Calmette–

Guérin and Salmonella enteritidis disseminated infection. *J. Clin. Invest.* **102,** 2035–2040.

20. Altare, F., Jouanguy, E., Lamhamedi, S., Döffinger, R., Fischer, A., and Casanova, J. L. (1998) Mendelian susceptibility to mycobacterial infection in man. *Curr. Opin. Immunol.* **10,** 413–417.
21. Levin, M., Newport, M. J., D'Souza, S., Kalabalikis, P., Brown, I. N., Lenicker, H. M., et al. (1995) Familial disseminated atypical mycobacterial infection in childhood: a human mycobacterial sysceptibility gene? *Lancet* **345,** 79–83.
22. Picard, C., Baud, O., Fieschi, C., and Casanova, J. L. (2000) Diagnosis and management of inheritable disorders of Interferon-γ-mediated immunity. *Immunol. Allergy Clin. North Am.* **20,** 65–75.
23. Feske, S., Draeger, R., Peter, H. H., and Rao, A. (2000) Impaired NFAT regulation and its role in a severe combined immunodeficiency. *Immunobiology* **202,** 134–150.
24. Casanova, J. L. and Abel, L. (2002) Genetic dissection of immunity to mycobacteria: the human model. *Annu. Rev. Immunol.* **20,** 581–620.

II

Detection Assays for Cytokines and Growth Factors

2

Intracellular Detection of T-Cell Cytokines

Differentiation of TH1 and TH2 Cells

Ursula Banning and Dieter Körholz

1. Introduction

The use of cytometry has become very important in different fields of modern biology and medicine *(1–4)*. In addition to the simple measurement of cell surface marker expression to define certain cellular subsets in basic research as well as in clinical settings, more complex assays have been developed in recent years to analyze, for example, physiological responses, apoptosis, or cell cycling *(5–9)*. The expression of intracellular proteins is another important feature for characterizing cellular function. The development of multicolor cytometers as well as the growing number of fluorochromes and dyes makes it possible to perform more complex analyses.

Cytokines play an important role in the interaction of different cells *(10)*. Evaluation of cytokine content together with cell surface marker expression makes it possible to understand the relationship between different cell types and elucidate their special role in the hematopoietic system. The use of a cytometer makes it possible to analyze cells on a single-cell level in a comparatively short time. Staining of cell surface markers is currently a routinely used method in many laboratories. However, the detection of intracellular proteins is more complicated. Staining methods depend on the permeabilization of the membrane of the cell. Different solutions, such as formaldehyde or alcohols, are used to fixate the cells before incubation with detergents, which permeabilize the membrane reversible *(11,12)*. To date, ready-to-use solutions are available that have been optimized for the use in flow cytometry. Primarily, cells have to be stimulated to express a detectable amount of suspective protein. In addition,

From: *Methods in Molecular Biology, vol. 215:*
Cytokines and Colony Stimulating Factors: Methods and Protocols
Edited by: D. Körholz and W. Kiess © Humana Press Inc., Totowa, NJ

stimulated cells have to be treated with Monensin or Brefeldin A to prevent the secretion of the produced proteins ***(13–15)***.

T-cells play an important role in the immune system ***(16)***. T-cell-derived cytokines control pro-inflammatory and anti-inflammatory processes. The differentiation of T-cells according to the types of cytokines produced by these cells lead to the concept of TH-1 and TH-2 cells, which might be important for the evaluation of several diseases, such as autoimmune diseases, human immunodeficiency virus (HIV)-associated pathology, or graft-versus-host disease, one of the severest side effects of allogeneic transplantation ***(17–21)***. The TH-1 and TH-2 cell subset can be distinguished by analysis of cytoplasmic interferon (IFN)-γ (TH-1) or interleukin (IL)-4 (TH-2).

2. Materials

2.1. Isolation of Mononuclear Cells from Whole Blood

1. Blood, drawn in 9 mL EDTA or lithium–heparin tubes (Sarstedt, Nurenberg, Germany).
2. Phosphate-buffered saline (PBS): 8 g NaCl, 1.42 g $Na_2HPO_4 \bullet H_2O$, 0.2 g KH_2PO_4, 0.2 g KCl, add 1 L distilled water, pH 7.3, sterilize, and store at room temperature.
3. Bicoll (Biochrome, Berlin, Germany), sterile, should be stored at 4–25°C and protected from light (storage in cold will increase the shelf life).
4. RPMI 1640 medium supplemented with the following:
   10% heat-inactivated fetal calf serum (FCS)
   10 U/mL penicillin
   10 µg/mL streptomycin
   2 m*M* L-glutamine
   This can be stored at 4°C up to 3 wk.
5. 0.83% Ammonium chloride: 8.29 g NH_4Cl, 1 g $KHCO_3$, 0.0371 g EDTA; add 1 L with distilled water, sterilize, and store at room temperature.

2.2. Stimulation

1. 96 Flat-bottom 96-well cell culture plates.
2. Phorbol 12-myristate 13-acetate (PMA).
3. Phytohemagglutinine.
4. Ionomycin (Sigma, St. Louis, MO); dissolved in 96% ethanol, aliquoted, and stored at –80°C until use.
5. GolgiStop (BD Pharmingen, Heidelberg, Germany); should be stored at 4°C. **Caution:** contains Monensin (toxic!) and is highly flammable.

2.3. Staining of Cells

1. 6-mL Polypropylene tubes.
2. Fix/Perm solution (BD Pharmingen, Heidelberg, Germany); stored in cold. **Caution:** contains formaldehyde and saponin.

Table 1
Staining Panel

Tube	FITC	PE	ECD	PC5
1	IgG1 surface	IgG1 cytoplasmic	IgG1 surface	IgG1 surface IgG2a surface
2	IgG1 cytoplasmic	IgG1 surface	IgG1 surface	IgG1 surface IgG2a surface
3	CD4 surface	IL-4 cytoplasmic	CD3 surface	CD14 surface
4	IFN cytoplasmic	CD4 surface	CD3 surface	CD14 surface
5	CD8 surface	IL-4 cytoplasmic	CD3 surface	CD56 surface
6	IFN cytoplasmic	CD8 surface	CD3 surface	CD56 surface

3. Wash buffer (BD Pharmingen, Heidelberg, Germany); stored in cold. **Caution:** contains sodium azide and saponin.
4. Fluorescence-labeled antibodies (staining panel is described in **Table 1**):
   Anti-CD3-ECD (clone: UCHT1; isotype: mouse IgG1)
   Anti-CD4-FITC (clone: 13B8.2; isoytype: mouse IgG1)
   Anti-CD4-PE (clone: 13B8.2; isoytype: mouse IgG1)
   Anti-CD8 FITC (clone: B9.11; isoytype: mouse IgG1)
   Anti-CD8 PE (clone: B9.11; isoytype: mouse IgG1)
   Anti-CD14 PC5 (clone: RMO52; isoytype: mouse IgG2a)
   Anti-CD56 PC5 (clone: N901-NKH1; isoytype: mouse IgG1)
   Anti-IFN-γ FITC (clone: B27; isoytype: mouse IgG1)
   Anti-IL-4 PE (clone: 8D4-8; isoytype: mouse IgG1)

2.4. Flow Cytometric Analysis

The analysis was done on a Beckman Coulter Epics XL flow cytometer (Beckman Coulter, Krefeld, Germany). For measurement and analysis, EXPO 32 software was used. The daily routine involves use of FlowCheck Fluorespheres (No. 6605359, Beckman Coulter, Krefeld, Germany) for check up of the cytometer, use of Coulter Isoton II (no. 8448011) for running, and Coulter Clenz Solution (no. 8456930) for cleaning.

3. Methods

3.1. Isolation of Mononuclear Cells from Whole Blood

1. Heparinized whole-blood samples are 1:2 diluted with PBS, pH 7.3 (*see* **Note 1**).
2. 5 mL of Ficoll are overlayed by up to 10 mL of the whole blood–PBS solution.
3. Centrifugate for 15 min at 800*g* without break.
4. Transfer mononuclear cell fraction to a new tube and wash with PBS to remove residual Ficoll solution (600*g*, 5 min).

5. Lysis of remaining erythrocytes may be done by incubation of the cell pellet with 1–2 mL 0.83% ammonium chloride for 5 min.
6. Incubation should be followed by intensive washing to remove ammonium chloride solution completely (at minimum, three times with 20 mL PBS).
7. After washing steps, the cell number is determined by counting.
8. Cells are suspended in RPMI 1640 supplemented with 10% heat-inactivated FCS and penicillin/streptomycine at a final density of 10^6/mL (*see* **Note 2**).

3.2. Stimulation

1. Transfer 200 μL of cell suspension (=2×10^5 cells) per well to a 96-well flat-bottom cell culture plate.
2. Stimulate the cells with PMA (10 ng/mL) and Ionomycin (1 μg/mL) for 4 h.
3. Add GolgiStop (0.6 μL of stock solution 1:5 diluted with PBS) to the cultures to inhibit secretion of the produced cytokines.
4. Incubate the cells for 4 h at 37°C, 5% CO_2 (*see* **Note 3**).

3.3. Staining of Cells

1. After stimulation, transfer cells from two wells (approx 4×10^5 cells) into a 6-mL polypropylene-tube suitable for cytometer use.
2. To guarantee transfer of possibly all cells, wash the wells with PBS twice.
3. After the addition of 1 mL PBS, centrifugate the cells at 600*g* for 5 min.
4. Discard the supernatant and resupend the cells in 100 μL PBS.
5. For cell surface analysis, stain the cells with 10 μL of each assigned fluorescence-labeled antibody and incubate cells for 10 min at room temperature in the dark. For analysis of different lymphocyte subsets, staining may be carried out according to the panel listed in **Table 1**.
6. Wash with 1 mL PBS and discard supernatant
7. Add 250 μL fixing and permeabilizing solution (Fix/Perm, BD Pharmingen).
8. Vortex cells and incubate for 25 min at 4°C.
9. After fixation and permeabilization, wash cells two times with 1 mL Wash-Buffer (BD Pharmingen, diluted 1:10 with distilled water).
10. Resuspend cells in 100 μL PBS.
11. After the addition of 10 μL cytokine-specific fluorescence-labeled antibody, vortex cells very gently and incubate for 30 min at 4°C (*see* **Note 4**).
12. Wash cells twice with 1 mL wash buffer.
13. Repeat washing with 1 mL PBS.
14. Resuspend cells in 500 μL PBS and measure with the cytometer (*see* **Note 5**).

A typical experiment is shown in **Fig. 1**.

4. Notes

1. To combine intracellular staining with immunophenotyping and detection of plasma cytokine concentration, portions of whole blood and plasma should be taken before dilution with PBS!

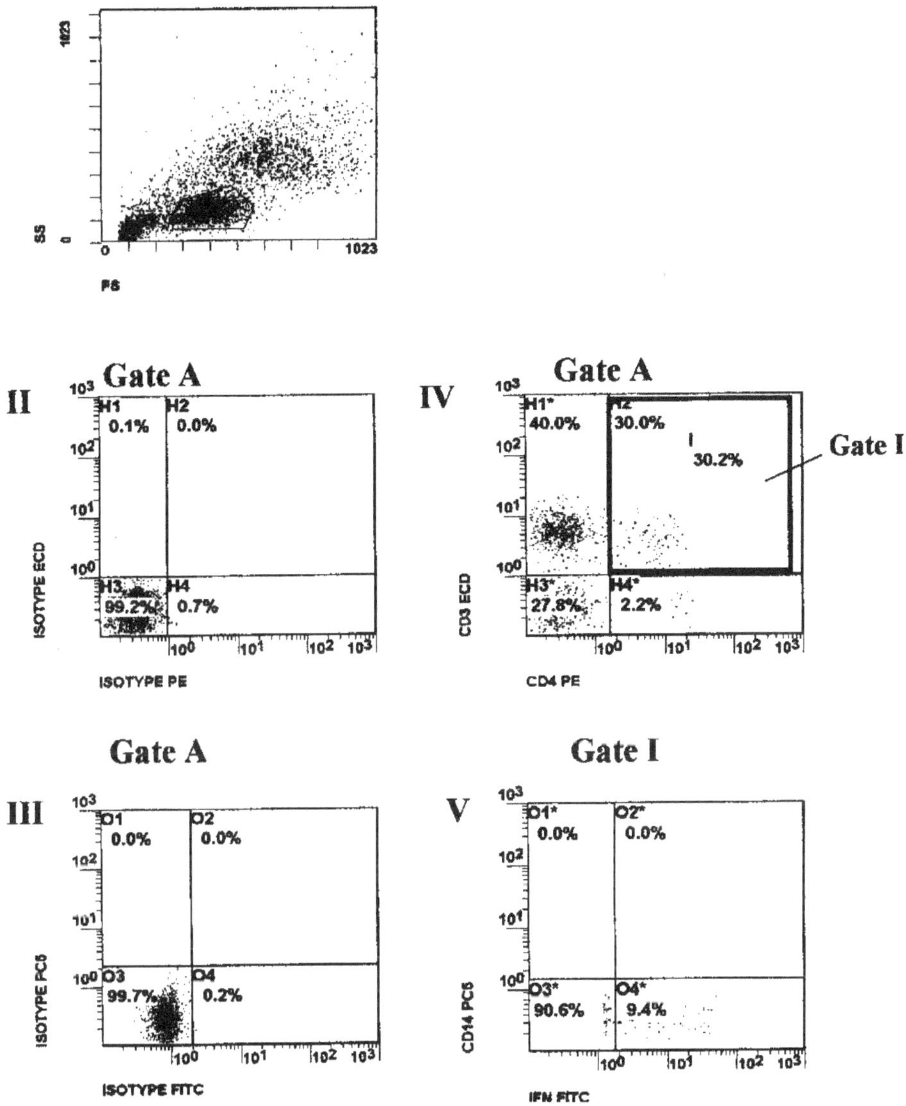

Fig. 1. IFN-γ content of CD3-positive lymphocytes. Shown is a cytoplasmic staining of cells from a patient after allogeneic transplantation. Lymphocytes are gated in gate A after FSC/SSC dot plot. (**I**). Left panel shows isotype control (**II–III**); right panel shows specific staining: (**IV**) CD4 PE vs CD3 ECD. (**V**) shows the IFN-γ positive portion of CD3–CD4 double positive cells (gate I in **IV**). No CD14-positive monocytes can be found in the gate.

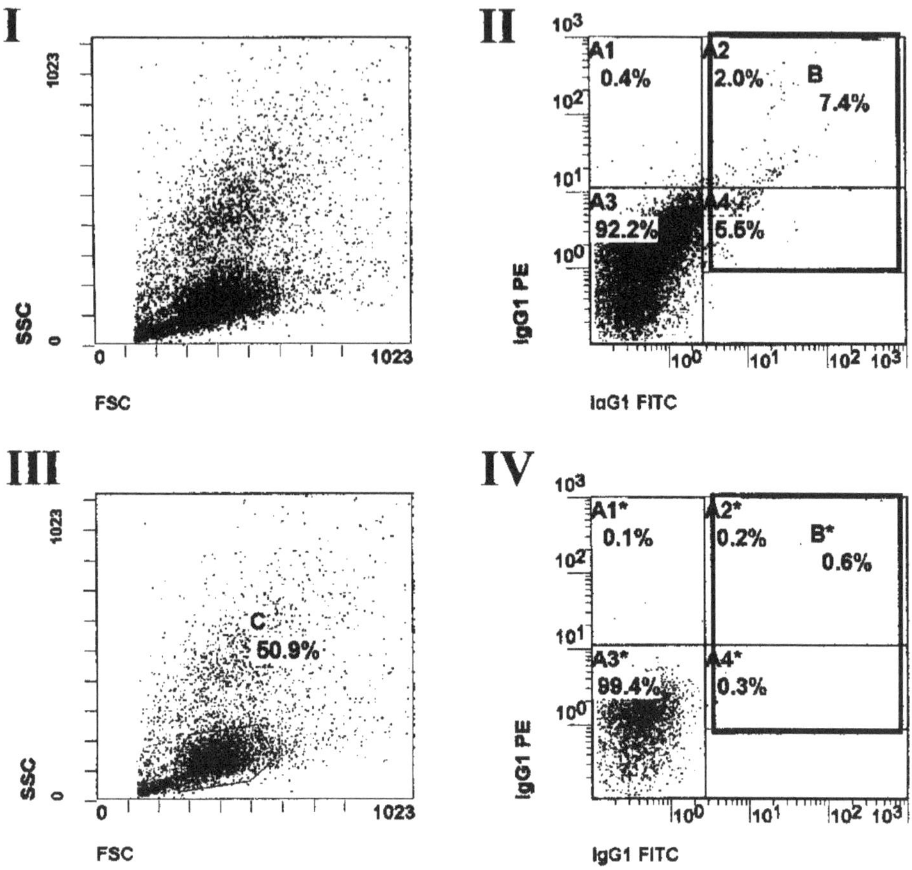

Fig. 2. Reduction of nonspecific staining by gating. Shown is a cytoplasmic isotype control staining of cells from a patient after allogeneic transplantation. Left panel shows FSC/SSC dot plot (**I** + **III**), right panel shows fluorescence signals in FL1 and FL2 (**II** + **IV**). A great portion of unspecific staining occurred (**II**, gate B). Nonspecific stained cells in gate B can be reduced by gating lymphocytes, as shown in (**III**), and analysis of gate C, shown in **IV**.

2. It is possible to keep the isolated mononuclear cells at 4°C overnight in medium and proceed on the next day. It is also possible to interupt the staining procedure after fixation of the cells. However, best results are obtained without interruptions.
3. The stimulation time depends on the observed cells and the cytokines of interest. Therefore, kinetic experiments might be necessary to determine the optimal time for stimulation.
4. It is important to control the cells that are in the antibody containing solution after vortexing (sometimes cell pellets move while vortexing)!

5. Unfortunately, nonspecific staining by isotype control antibodies may occur. This might be reduced by blocking of the cells with Fc block (e.g., from BD Pharmingen) prior to staining. However, background staining can also be reduced by gating strategies. Most of the background results from cell debris or cell aggregates. Gating the population of interest by FSC/SSC characteristics can reduce the nonspecific staining dramatically (e.g., *see* **Fig. 2**).

References

1. Griffioen, A. W., Rijkers, G. T., and Cambier, J. C. (1991) Flow cytometric analysis of intracellular calcium: the polyclonal and antigen-specific response inhuman B lymphocytes. *Methods* **2,** 219.
2. Reddy, S., Rayburn, H., and Vonmelcher, H. (1992) Fluorescence-activated sorting of totipotent embryonic stem-cells expressing deveolpmentally regulated lac z fusion genes. *PNAS* **89,** 6721.
3. Leary, J. F. (1994) Strategies in rare cell detection and isolation. *Methods Cell Biol.* **42,** 331.
4. Young, J. S., Varma. A., and Diguisto, D. (1996) Retention of quiescent hematopoietic cells with high proliferative potential. *Blood* **87,** 545.
5. Wieder, E. D., Hang, H., and Fox, M. H. (1993) Measurement of intracellular pH using flow cytometry with carboxy-SNARF-1. *Cytometry* **14,** 916.
6. Vermes, I., Haanen, C., Steffens-Nakken, H., and Reutelingsperger, C. (1995) A novel assay for apoptosis. Flow cytometric detection of phosphatidylserine expression on early apoptotic cells using fluorescein-labelled Annexin V. *J. Immunol. Methods* **184,** 29.
7. Vermes, I., Haanen, C., and Reutelingsperger, C. (2000) Flow cytometry of apoptotic cell death. *J. Immunol. Methods* **243,** 167.
8. Lecoeur, H., Fevrier, M., Garcia, S., Riviere, Y., and Gougeon, M. (2001) A novel flow cytometric assay for quantitation and multiparametric characterization of cell-mediated cytotoxicity. *J. Immunol. Methods* **253,** 177.
9. Heine, J., Jaeger, K., Osthaus, A., Weingaertner, N., Munte, S., Piepenbrock, S., et al. (2000) Anaesthesia with propofol decreases FMLP-induced neutrophil respiratory burst but not phagocytosis compared with isoflurane. *Br. J. Anaesth.* **85,** 424.
10. Thomson, A. W. (1998) *The Cytokine Handbook*, 3rd ed. Academic, New York.
11. Schroff, R. W., Bucana, C. D., Klein, R. A., Farrell, M. M., and Morgan, A. C., Jr. (1984) Detection of intracytoplasmic antigens by flow cytometry. *J. Immunol. Methods* **70,** 167.
12. Jacob, M. C., Favre, M., and Bensa, J. C. (1991) Membrane cell permeabilization with saponin and multiparametric analysis by flow cytometry. *Cytometry* **12,** 550.
13. Sander, B., Andersson, J., and Andersson, U. (1991) Assessment of cytokines by immunofluorescence and paraformaldehyde-saponin procedure. *Immunol. Rev.* **119,** 65.
14. Verheyen, J., Bönig, H., Kim, Y. M., Banning, U., Mauz-Körholz, C., Kramm, C., et al. (2000) Regulation of interleukin-2 induced interleukin-5 and interleukin-13

production in human peripheral blood mononuclear cells. *Scand. J. Immunol.* **51,** 45.
15. Chalmers, I. M. H., Janossy, G., Contreras, M., and Navarrete, C. (1998) Intracellular cytokine profile of cord and adult blood lymphocytes. *Blood* **92,** 11.
16. Mosmann, T. R. and Sad, S. (1996) The expanding universe of T-cell subsets: Th1, Th2 and more. *Immunol. Today* **17,** 138–145.
17. Jung, T., Lack, G., Schauer, U., Uberruck, W., Renz, H., and Gelfand, E. W. (1995) Allergens, IgE, mediators, inflammatory mechanisms. Decreased frequency of interferon γ and interleukin-2-producing cells in patients with atopic diseases measured at the single cell level. *J. Allergy Clin. Immunol.* **96,** 515.
18. Meyaard, L., Hovenkamp, E., Keet, I. P. M., Hooibrink, B., Jong, I. C. H., Otto, S. A., et al. (1996) Single-cell analysis of IL-4 and IFN-γ production by T cells from HIV-infected individuals. *J. Immunol.* **157,** 2712.
19. Carayol, G., Bourhis, J. H., Guillard, M., Bosq, J., Pailler, C., Castagna, L., et al. (1997) Quantitative analysis of T Helper 1, T Helper 2, and inflammatory cytokine expression in patients after allogeneic bone marrow transplantation. *Transplantation* **63,** 1307.
20. Murphy, W. J. and Blazar, B. R. (1999) New strategies for preventing graft-versus-host disease. *Curr. Opin. Immunol.* **11,** 509.
21. Pan, L., Delmonte, J., Jalonen, C. K., and Ferrara, J. L. (1995) Pretreatment of donor mice with granulocyte colony stimulatory factor polarizes donor T-lymphocytes towards type 2 cytokine production and reduces severity of experimental graft versus host disease. *Blood* **86,** 4422.

3

Analysis of Cytokine Profiles in Human Skin

Jean Krutmann, Renz Mang, and Susanne Grether-Beck

1. Introduction

Human skin cells as well as skin-infiltrating cells such as T-lymphocytes are endowed with the capacity to produce a multitude of cytokines. This cytokine expression is of central importance to preserve the normal skin homeostasis; any alteration has profound effects on skin morphology and can give rise to a variety of inflammatory and immunological skin disorders (i.e., atopic dermatitis) *(1,2)*. Within recent years, cytokine profiles in the skin of patients suffering from several diseases has been observed. These studies have been made possible through the availability of *in situ* techniques that allow the detection and monitoring of immunomodulatory consequences in cells of lesional skin of patients. In this chapter, the highly sensitive reverse transcription–polymerase chain reaction (RT-PCR)-based assessment of cytokine expression in human skin will be described.

1.1. RT-PCR

The polymerase chain reaction (PCR) is an in vitro nucleic acid amplification method. The reaction comprises repeated thermal cycling of the reaction mixture. Each cycle consists a set of time- and temperature-controlled incubations. The function of each incubation is to denature the target nucleic acid at a temperature in the region of 94°C, anneal at a temperature dependent on their calculated annealing temperature, and extend the primers by the thermostabile DNA polymerase catalyzed addition of nucleotides to the 3′ end of each primer at a temperature of about 72°C. If each cycle were 100% efficient, each cycle would double the number of copies of the original target sequence. PCR may

From: *Methods in Molecular Biology, vol. 215: Cytokines and Colony Stimulating Factors: Methods and Protocols*
Edited by: D. Körholz and W. Kiess © Humana Press Inc., Totowa, NJ

be used to analyze or determine the presence of specific DNA targets or, after a reverse transcription step, to amplifying RNA sequences as DNA copies (RT-PCR). RT-PCR is a sensitive analytical method and is a suitable approach for the analysis of mRNA expression in cells. In this method a housekeeping gene (i.e., β-actin), which is transcribed constitutively in most cell types and tissues, is used as an invariant internal control for variations in RT-PCR and product detection steps.

1.2. High-Performance Liquid Chromatography

High-performance liquid chromatography (HPLC) is a form of liquid chromatography to separate compounds that are dissolved in solution. HPLC instruments consist of a reservoir of mobile phase, a pump, an injector, a separation column, and a detector. Compounds are separated by injecting a plug of the sample mixture onto the column. The different components in the mixture pass through the column at different rates because of differences in their partitioning behavior between the mobile liquid phase and the stationary phase. For quantification, these fractions are then evaluated by an on-line ultraviolet (UV)-spectrophotometer (260 nm). Data can be obtained as histograms of absorption (DNA concentration) on the *y*-axis and elution time (DNA retention time) on the *x*-axis for each PCR product. Values for areas under the curve of each PCR product can be used for further calculations *(3)*.

2. Materials

2.1. Tissues and RNA Extraction

1. 4-mm (Minimum) punch biopsies (*see* **Notes 1** and **2**).
2. RNeasy Kit (Qiagen, Hilden, Germany).
3. Ethanol, 70% and 100%.
4. β-Mercaptoethanol.

2.2. RT-PCR

1. Primers (sequences for some important examples; *see* **Table 1**).
2. GeneAMP® RNA PCR Core Kit (Perkin-Elmer, Applied Biosystems, Weiterstadt, Germany).
3. PlatinumTaq® DNA Polymerase (Life Technologies, Karlsruhe, Germany).
4. Thermal cycler (i.e., PTC-200; MJ Research, Watertown, MA).

2.3. High-Performance Liquid Chromatography

1. HPLC device (Dionex, Idstein, Germany; formerly Gynkotek, Germering, Germany) consisting of a pump (M-480), an injector (GINA 50), an UV detector (UV-160S), and a reservoir of mobile phases (buffer A: 1 *M* NaCl and 25 m*M* Tris-HCl [pH 9.0]; buffer B: 25 m*M* Tris/HCl [pH 9.0]).

Table 1
Primer Pairs Specific for Cytokines and β-Actin (Cytokinesequence 5′–3′)

Cytokine primer	Primer sequence	Product size (bp)	Annealing temp. (°C)
Interferon-γ	AGTTATATCTTGGCTTTTCA ACCGAATAATTAGTCAGCTT	501	54
Interleukin-4	CTTCCCCCTCTGTTCTTCCT TTCCTGCCGAGCCGTTTCAG	423	57
Interleukin-12 p35	ACCCAGGAATGTTTCCCATGC TCTGTCAATAGTCACTGCCCG	409	57
Interleukin-12 p40	AAAGGAGGCGAGGTTCTAAGCC TTTGCGGCAGATGACCGTGG	437	57
β-Actin	GTGGGGCGCCCCAGGCACCA CTCCTTAATGTCACGCACGATTTC	661	60

2. Separation column: PE TSK® DEAE-NPR Column (Perkin-Elmer Europa, Vaterstetten, Germany).
3. Personal computer system and integration software.

3. Methods

3.1. RNA Extraction and RT Reaction

For analysis of cytokine profiles in human skin, a RT-PCR-based semiquantitative method has been employed. This technique allows one to monitor cytokine mRNA expression in 4-mm punch biopsies that are taken from skin of patients *(2,4)*. The specimens are washed twice in 0.9% NaCl for 20 s to remove blood and then snap-frozen in liquid nitrogen. Frozen biopsy specimens are grounded to powder and total RNA is extracted. Historically, we have used a modified chloroform/phenol method *(5)*. The method of choice is now the use of the Quiagen RNeasy Kit (*see* the instruction pamphlet).

In order to semiquantitatively assess cytokine mRNA expression in these samples, a differential RT-PCR method has to be employed *(2–4)*. For this purpose, total RNA is reverse-transcribed using MuLV (murine leukemia virus) reverse transcriptase and random hexamers (GeneAMP® RNA PCR Core Kit).

1. Prepare a master mix for reverse transcription by adding the reagents in the following order and proportions. 25 m*M* $MgCl_2$ solution: 4 µL, 10X PCR buffer II, dGTP, dATP, dTTP, dCTP: 2 µL; RNase inhibitor, MuLV reverse transcriptase, random hexamers: 1 µL, RNA-containing sample: 3 µL.
2. Reverse-transcribe at 42°C for 60 min (i.e., in the thermocycler or a water bath).
3. Terminate the reaction at 95°C for 5 min and cool down.

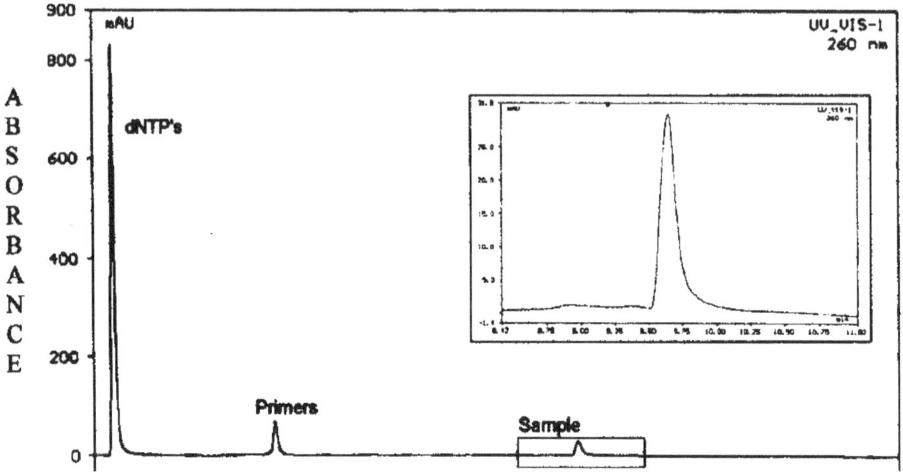

Fig. 1. Example of an HPLC diagram. The first peak represents the deoxynucleotidtriphosphates (dNTP), the second peak represents the primers, and the third peak represents the amplification product.

For further information, *see* the instruction pamphlet of the GeneAMP® RNA PCR Core Kit.

3.2. PCR Reaction

1. PCR is carried out in a final volume of 50 µL containing 1 µ*M* of each primer, 0.5 U PlatinumTaq® DNA polymerase, template cDNA (5 µL [for β-actin] and 30 µL [for genes], in a reaction buffer consisting of 50 m*M* KCl, 10 m*M* Tris-HCl (pH 8.3), and 2 m*M* $MgCl_2$.
2. Perform amplication in a thermocycler with approriate annealing temperature. For our cytokine primers, amplification was found to be linear for up to 38–40 cycles and up to 32 cycles for the β-actin.
3. To ensure the identity of products, it is recommended to collect their chromatogram peaks and to digest them with an appropriate restriction endonuclease. Fragments are then subsequently visualized on agarose gel by ethidium bromide staining. Lengths of the restriction fragments are compared to those deduced from published mRNA sequences of the respective cytokines (e.g., by using PC/GENE software [Intelligenetics Inc., Mountain View, CA]).

3.3. High-Performance Liquid Chromatography (HPLC)

For the exact information about the use of HPLC, *see* the instruction pamphlets. After data collection, the peaks will be integrated by the software. For an example, *see* **Fig. 1**.

4. Notes

1. As controls you need biopsies of healthy skin.
2. This will confirm that the total system is clean.

References

1. Grewe, M., Bruijnzeel-Koomen, C. A. F. M., Schöpf, E., Thepen, T., Langeveld-Wildschut, A. G., Ruzicka, T., et al. (1998) A role for Th1 and Th2 cells in the immunopathogenesis of atopic dermatitis. *Imunol. Today* **19,** 359–361.
2. Grewe, M., Gyufko, K., Schöpf, E., and Krutmann, J. (1994) Lesional expression of interferon-gamma in atopic eczema. *Lancet* **343,** 25–26.
3. Henninger, H. P., Hoffmann, R., Grewe, M., Schulze-Specking, A., and Decker, K. (1993) Purification and quantitative analysis of nucelic acids by anion-exchange–high-performance liquid chromatogrpahy. *Biochem. Biophys. Hoppe–Seyler* **374,** 625–632.
4. Grewe, M., Walter, S., Gyufko, K., Czech, W., Schöpf, E., and Krutmann, J. (1997) Analysis of the cytokine pattern expressed in situ in inhalant allergen patch test reactions of atopic dermatitis patients. *J. Invest. Dermatol.* **105,** 407–410.
5. Chomzynski, P. and Sacchi, N. (1987) Single step method of RNA isolation by acid guanidium thiocyabate–phenol–chloroform extraction. *Anal. Biochem.* **162,** 156–161.
6. Thepen, T., Langeveld-Wildschut, E. G., Bihari, I. C., and Bruynzeel-Koomen, C. A. F. M. (1996) Biphasic response against aeroallergen in atopic dermatitis showing a switch from an intial Th2 response in situ: an immunocytochemical study. *J. Allergy Clin. Immunol.* **97,** 828–838.

4

Intracytoplasmic Detection of Proinflammatory Cytokines and Chemokines in Monocytes by Flow Cytometry

Christian Schultz

1. Introduction

Intracytoplasmic detection of cytokines by flow cytometry has revolutionized the area of cell biology in the past few years *(1–3)*. It represents a powerful analytical technique in which individual cells can be simultaneously analyzed for several parameters, including cell size and granularity, as well as the coexpressed levels of surface and intracellular markers defined by fluorescent antibodies. After stimulation of cells in whole-blood culture, cells are stabilized and permeabilized to allow monoclonal antibodies penetrating through cell membranes, the cytosol, and the membranes of the endoplasmatic reticulum and Golgi complex. Paraformaldehyde fixation preserves cell morphology and intracellular antigenicity, also enabling the cells to withstand permeabilization by detergent. Reversible permeabilization of cells is done with saponin, a plant glycoside, which mainly interacts with cholesterol in cell membranes. Both substances maintain the light-scattering properties of leukocytes which are important for subsequent flow-cytometric analysis. A considerable increase in the fluorescence intensity is achieved by blocking intracellular transport processes with monensin, resulting in the accumulation of cytokine proteins in the Golgi complex *(4)*. Cells are then incubated with cytokine- and chemokine-specific as well as surface antigen-specific fluorochrome-conjugated antibodies that allow cytokine detection directly at the single-cell level by flow cytometry (*see* **Note 1**). Up to now, studies investigating intracytoplasmic cytokines focused primarily on T-lymphocytes. The method introduced allows the

From: *Methods in Molecular Biology, vol. 215: Cytokines and Colony Stimulating Factors: Methods and Protocols*
Edited by: D. Körholz and W. Kiess © Humana Press Inc., Totowa, NJ

determination of monocyte-derived cytokines and chemokines, which represent important mediators of inflammation (interleukin-1α [IL-1α], IL-6, tumor necrosis factor-α [TNF-α]), cell differentiation, and chemotaxis (IL-8, monocyte chemoattractant protein-1 [MCP-1], macrophage inflammatory protein-1α [MIP-1α]).

2. Materials

2.1. Cell Culture

1. Blood is drawn in 4.5-mL lithium–heparin tubes (Sarstedt, Nürnbrecht).
2. Roswell Park Memorial Institute (RPMI) 1640 medium with 2 g/L $NaHCO_3$ and without L-glutamine should be stored at 4°C (Seromed Biochrome, Berlin, Germany; F 1215).
3. 200 µL of each of the supplements penicillin/streptomycin (Seromed Biochrome A 2213), 200 m*M* N-acetyl-L-alanyl-L-glutamine (Seromed Biochrome K 0202), nonessential amino acids (Seromed Biochrome K0293), and 100 m*M* Na–pyruvate (Seromed Biochrome L0473) are stored at –20°C in Eppendorf tubes.
4. A solution containing 19.2 mL of RPMI 1640 and 800 µL of supplements must be prepared fresh as required.
5. Cultivation of blood cells is done in sterile multiwell plates (Greiner Labortechnik, Frickenhausen, Germany).

2.2. Stimulation of Cells, Protein Transport Inhibition

1. Hanks' balanced salt solution (HBSS; Sigma, Deisenhofen, Germany; H-4891) and 350 mg $NaHCO_3$ are mixed with 1 L of distilled water. This solution should be stored at 4°C in the dark. Deterioration of the liquid medium may be recognized by pH change, precipitate matter throughout the solution, cloudy appearance, or color change.
2. 10 mg of lipopolysaccharide (LPS) (Sigma, L3129) is added to 10 mL of HBSS solution and stored as 100-µL aliquots at –20°C. This stock solution is stable for roughly 1 yr. For further processing, 1 µL of stock solution is combined with 1 mL of freshly prepared RPMI medium (*see* **Subheading 2.1.**), resulting in a concentration of 0.1% LPS.
3. To obtain a concentration of 10 m*M* monensin stock solution (Sigma, M 5273), 69.3 mg are filled up to 10 mL with ethanol. This solution should be stored at –20°C. For further processing, 10 µL of stock solution is combined with 990 µL of freshly prepared RPMI medium (*see* **Subheading 2.1.**), containing 100 µ*M* monensin. Monensin must be handled with care because it is particularly toxic.

2.3. Fixation Reagents

1. Paraformaldehyde (PFA) powder is stored at 4°C (Riedel de Haen, Seelze, Germany). To obtain a 4% solution, 1 g of PFA powder is added to 25 mL HBSS, heated to 57°C for 60 min, and, subsequently, cooled down to room temperature

for 3 h. This solution can be stored at 4°C for 3 d. PFA is toxic, mutagenic, and cancerogenic.
2. Nonfat, dry-milk powder is commercially available. To obtain a 5% solution, 2.5 g are added to 50 mL of phosphate buffer solution.

2.4. Permeabilization Reagents

1. 1 *M* *N*-2-hydroxyethylpiperazine-*N'*-2-ethanesufonic acid solution (HEPES, Seromed Biochrome) and saponin powder (Riedel de Haen) are stored at room temperature. To obtain a 10% stock solution, 1 g of saponin is added to 10 mL HBSS, heated to 57°C for 60 min, and, subsequently, cooled down to room temperature for 3 h. This solution can be stored at 4°C for 2 wk.
2. The 0.1% saponin buffer should be made fresh as required. It consists of 1 mL 10% stock solution of saponin, 1 mL HEPES, and 98 mL HBSS.

2.5. Fluorochrome-Conjugated Monoclonal Antibodies

Anti-human CD14-PE (clone: M5E2, isotype: mouse IgG2a), CD3-CyChrome (17A2, rat IgG2b), anti-human fluorescein isothiocyanate (FITC)-conjugated and purified IL-1α (364-3B3-14, mouse IgG1), IL-6 (MQ2-13A5, rat IgG1), IL-8 (G265-8, mouse IgG2b), TNF-α (Mab11, mouse IgG1), MCP-1 (5D3-F7, mouse IgG1) and MIP-1α (11A3, mouse IgG2a) were all purchased from Pharmingen (Heidelberg, Germany).

2.6. Flow-Cytometric Analysis

The following reagents were obtained from Beckmann Coulter (Krefeld, Germany): Clenz Clearing solution (no. 8546930), Isoton II (no. 8448011), Flow-Check Fluorospheres (no. 6605359).

3. Methods

3.1. Cell Culture

1. Blood is drawn in 4.5-mL lithium–heparin tubes (Sarstedt, Nürnbrecht, Germany) and must be processed within 20 h (*see* **Note 2**).
2. Whole blood is suspended in RPMI 1640 supplemented with 1% penicillin/streptomycin, 2 m*M* glutamine, 1 m*M* pyruvate, and nonessential amino acids (Seromed Biochrome, Berlin, Germany) at a concentration of 5×10^6 leukocytes/mL (*see* **Note 3**).

3.2. Stimulation of Cells and Protein Transport Inhibition

1. 1.5-mL aliquots of cell culture containing 5×10^6 leukocytes/mL are incubated in multiwell plates with LPS at 37°C, 5% CO_2. Optimal conditions are 5 h of stimulation, with LPS at a final concentration of 30 ng/mL (*see* **Note 4**).
2. Simultaneously monensin is added at a final concentration of 3 μ*M* to inhibit cytokine secretion.

3.3. Fixation of Cells

1. Cell cultures are transferred to plastic tubes, diluted with 5 mL HBSS, and vortexed briefly.
2. After spinning at 360*g* for 10 min at room temperature, supernatants are aspirated and discarded.
3. 1 mL of PFA 4% is added and vortexed briefly.
4. After 10 min of incubation at 4°C, tubes are filled again with HBSS, vortexed briefly, and spinned for 10 min at 360*g*.
5. Supernatants are aspirated and discarded.
6. The cell suspension is resuspended in 1 mL of nonfat dry milk (5%) for 16 h at 4°C in the dark to reduce nonspecific binding *(5)*.

3.4. Permeabilization of Cells and Antigen Staining

1. After spinning at 360*g* for 10 min at room temperature, supernatants are aspirated and discarded.
2. 1 mL of 0.1% saponin buffer is added and vortexed briefly.
3. 200-μL Aliquots of this cell suspension are then transferred to tubes containing 5 μg of anti-human CD14, CD3 and 0.5 μg of anti-human IL-1α, IL-6, IL-8, TNF-α, MCP-1, or MIP-1α fluorochrome-conjugated antibodies (*see* **Note 4**).
4. After incubation for 20 min at 4°C in the dark, 1 mL of 0.1% saponin buffer is added and vortexed briefly.
5. Cell suspensions are spun at 270*g* for 5 min at room temperature and supernatants are discarded.
6. 0.5 mL of HBSS are added to all tubes, vortexed briefly, and stored at 4°C in the dark until analysis by flow cytometry.

3.5. Negative Control

1. In parallel, a surplus of purified anti-human IL-1α, IL-6, IL-8, TNF-α, MCP-1, and MIP-1α antibodies served as negative controls (*see* **Note 5**). 200 μL Aliquots of cell suspension are transferred to tubes containing 5 μg of anti-human CD14, CD3 and 5 μg of purified anti-human IL-1α, IL-6, IL-8, TNF-α, MCP-1 or MIP-1α.
2. After incubation for 20 min at 4°C in the dark, 1 mL of 0.1% saponin buffer is added and vortexed briefly.
3. Cell suspensions are spun at 270*g* for 5 min at room temperature and supernatants are discarded.
4. Anti-human IL-1α, IL-6, IL-8, TNF-α, MCP-1, or MIP-1α fluorochrome-conjugated antibodies are added at a concentration of 0.5 μg each.
5. After incubation for 20 min at 4°C in the dark, 1 mL of 0.1% saponin buffer is added and vortexed briefly.
6. Cell suspensions are spun at 270*g* for 5 min at room temperature and supernatants are discarded.

7. 0.5 mL of HBSS are added to all tubes, vortexed briefly, and stored at 4°C in the dark until analysis by flow cytometry.

3.6. Flow-Cytometric Analysis

1. Flow-cytometric analyis is performed on an EPICS XL flow cytometer with System II™, Version 1.0 software that has been calibrated daily with Flow-Check Fluorospheres (Coulter Electronics, Krefeld, Germany).
2. A total of 2000 CD14-positive monocytes are acquired from each sample, gating on CD14-positive cells (*see* **Fig. 1**). The purity of the gated cells is controlled by use of CD3/CD14 double staining. It is regarded to be correct if the gate included at least 95% of CD14-positive cells. Granulocytes and cell debris are excluded by forward- and side-scattered light signals.
3. Preincubation with unlabeled anticytokine monoclonal antibodies before the addition of fluorochrome-labeled antibodies of an identical specificity completely blocks specific binding (*see* **Subheading 3.5.** and **Note 5**). The quadrant markers for the bivariate dot plots are set according to this negative control allowing less than 2% of positive cells in the right upper quadrant (*see* **Fig. 2**).
4. Stimulated monocytes stained with fluorochrome-labeled antibodies against cytokines (*see* **Subheading 3.4.**) represent the percentage of cytokine-producing CD14-positive monocytes. All events in the right upper quadrant according to the negative control are defined as cytokine positive (*see* **Fig. 2**).

4. Notes

1. The time-scale of the technique is as follows. Preparation of solutions and cell culture takes approx 2 h. After that, 5 h of stimulation follows. Subsequently, fixation of cells in PFA lasts 1 h, followed by a resuspension in nonfat dry milk overnight for 16 h. On the second day, the distribution of fluorochrome-conjugated surface and anticytokine antibodies to tubes before the addition of cell suspensions will take nearly 1 h. Permeabilization, surface antigen staining, and cytokine staining will take approx 2 h. Measurement by flow cytometry will take nearly 2–4 h depending on the number of cytokines and samples that will be investigated simultaneously.
2. It is important to process the blood within 20 h of storage. After this period of time, a profound variation of cytokine positive cells is seen, especially for IL-6-positive monocytes. In contrast, the amount of IL-8-positive monocytes remains quite stable even after 48 h of specimen age.
3. After performing cell separation techniques, a profound ex vivo induction of cytokines could be demonstrated at the protein and mRNA levels *(6,7)*. Therefore, the whole-blood culture assay should be preferred. It preserves cellular interactions and approximates physiological reactions occurring in vivo *(8)*. The intracytoplasmic detection of cytokines by flow cytometry enables a high-resolution analysis of particular cell types within heterogeneous cell populations without the need for artifactual cell separation procedures (*see* **Note 6**).

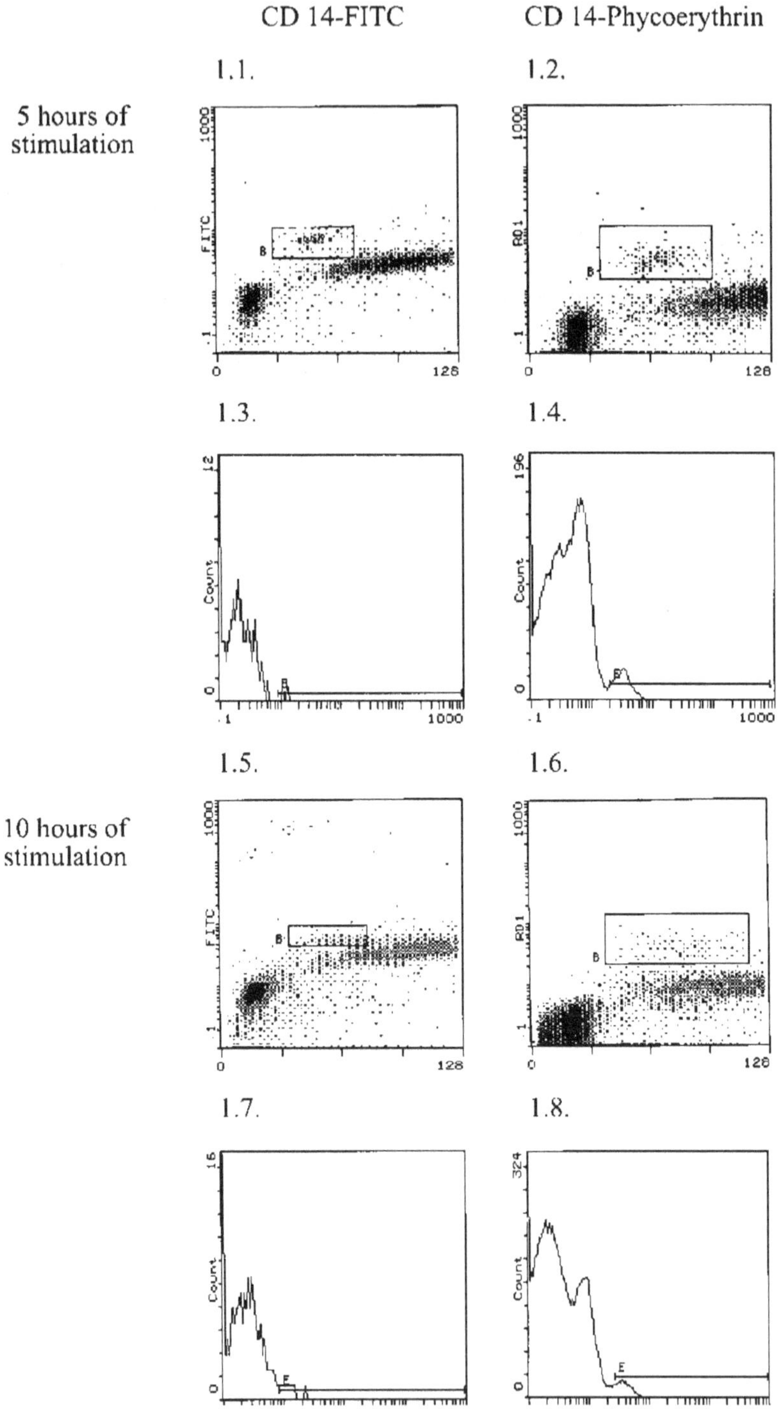
CD 14-FITC
CD 14-Phycoerythrin
5 hours of stimulation
10 hours of stimulation
1.1.
1.2.
1.3.
1.4.
1.5.
1.6.
1.7.
1.8.
FITC
RD1
Count
1000
128
196
324

Fig. 1.

4. During stimulation with LPS, the expression of CD14 on monocytes decreases, especially after 5 h of stimulation. Activated neutrophils in whole blood produce mediators that can induce a CD14 proteolysis *(9)*. This phenomenon is possibly responsible for the impaired differentiation between CD14 positively and negatively stained cells. However, using phycoerythrin-conjugated monoclonal antibodies against CD14 instead of FITC-conjugated monoclonal antibodies, a strikingly improved discrimination is possible between CD14 positively and negatively stained cells (**Fig. 1**).
5. An intense enhancement of nonspecific binding during the fixation and stimulation procedure *(5)* makes a correct interpretation of data without the use of adequate negative controls difficult. It has been demonstrated that a surplus of purified anticytokine antibodies blocks specific binding and allows an excellent differentiation between positively and negatively stained cells *(5)*. In comparison with the most often used isotype-matched antibodies as negative controls, a higher amount of cytokines up to 2.1-fold is seen *(10)*. Isotype-matched antibodies display the inherent staining background for a given fluorescent antibody, but they fail to cover the enhancement of nonspecific staining induced by the procedure of stimulation, permeabilization, and fixation (**Fig. 3**). Therefore, the purified antibody blocking control should be preferred to each sample. Nonstimulated cells are also an unreliable negative control because cytokine positive cells can be found without in vitro stimulation. Moreover, the augmentation of nonspecific staining during stimulation cannot be detected in nonstimulated cells (**Fig. 3**).
6. Related techniques: The enzyme-linked immunospot (ELISPOT) assay is an enzyme-linked immuosorbent assay (ELISA)-based technique that enables detection and enumeration of cytokine-secreting cells at the single-cell level *(11)*. In contrast to the introduced technique, colored spots represent secreted but not intracytoplasmic cytokines of a single cell and are counted by microscope. This analysis is even more time-consuming than flow-cytometric analysis. Secreted cytokines can also be investigated by flow cytometry if cells are microencapsulated in an agarose matrix, which is used to capture biotin-labeled

Fig. 1. *(see opposite page)* Comparison between FITC- and PE-conjugated monoclonal antibodies in the detection of CD14-positive cells. Dot plots show staining with FITC- and PE-conjugated CD14 on a logarithmic scale (*y*-axes) vs side-scatter characteristics on a linear scale (*x*-axes). The rectangle (B) demonstrates the gate of CD14 positive cells (1.1, 1.2, 1.5, 1.6). Single-parameter histograms display the relative cell number on a linear scale (*y*-axes) vs CD14 staining on a logarithmic scale (*x*-axes). The peak of CD14-positive monocytes is shown to the right of the solid lines (E) (1.3, 1.4, 1.7, 1.8). CD14 expression decreases especially after 10 h compared with 5 h of stimulation with LPS. The use of PE-conjugated monoclonal antibodies (right-hand panels) allows a better differentiation between CD14 positively and negatively stained cells.

2.1: Interleukin-1α

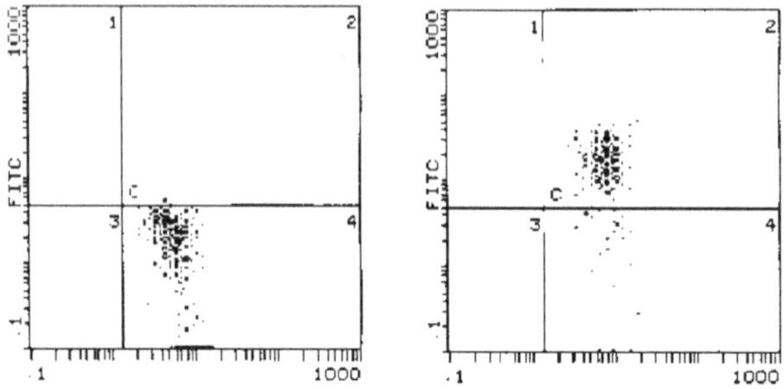

2.2: Interleukin-6

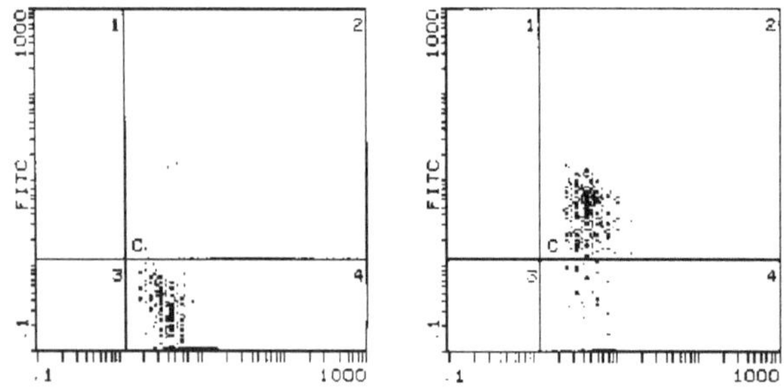

2.3: Tumor necrosis factor-α

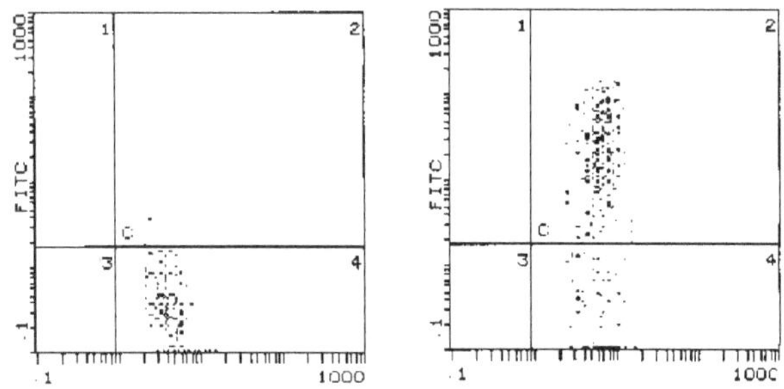

2.4: Macrophage inflammatory protein-1

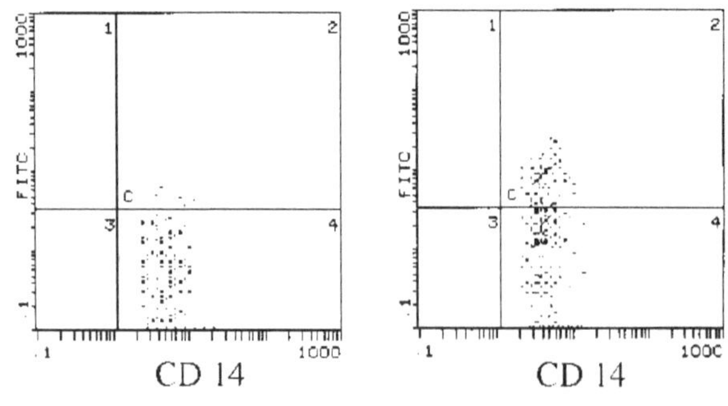

Fig. 2.

cytokine-specific antibodies using avidin as a bridge. Secreted cytokines are captured by this antibody and are detected by a second fluorochrome-labeled cytokine-specific antibody by flow cytometry ***(12)***. This technique is preferable if cells should stay viable for subsequent assays. A good correlation with minor kinetic differences has been demonstrated between the determination of cytokines in the supernatant by ELISA and intracytoplasmic cytokine detection by flow cytometry (***1,13***, personal observations), as well as intracytoplasmic and surface cytokine expression ***(14)***.

Fig. 2. *(see opposite page)* Intracytoplasmic staining of IL-1α, IL-6, TNF-α, and MIP-1α in monocytes. Representative dot plots show staining with FITC-conjugated monoclonal antibodies against cytokines (*y*-axes) vs PE-conjugated monoclonal antibodies against CD14 (*x*-axes) on a logarithmic scale. Positive events are completely blocked if stimulated cells are preincubated with unlabeled anticytokine monoclonal antibodies before the addition of fluorochrome-labeled antibodies of an identical specificity (left-hand panels). The quadrant markers (C) for the bivariate dot plots are set according to this negative control, allowing less than 2% of positive cells in the right upper quadrants. Stimulated monocytes stained with fluorochrome-labeled antibodies against IL-1α (2.1), IL-6 (2.2), TNF-α (2.3), and MIP-1 (2.4) are shown in the right upper quadrants (right-hand panels). They represent cytokine-producing monocytes of the CD14-positive cell population. IL-8 staining in monocytes is shown in **Fig. 3**.

Fig. 3. *(see following page)* The ability of different negative controls to discriminate specific from nonspecific staining is illustrated in a representative case. Dot plots (left-hand panels) show staining with FITC-conjugated monoclonal antibodies against IL-8 (*y*-axes) vs PE-conjugated monoclonal antibodies against CD14 (*x*-axes) on a logarithmic scale. Single-parameter histograms (right-hand panels) show the relative cell number on a linear scale (*y*-axes) vs IL-8 staining on a logarithmic scale (*x*-axes). The quadrants (C) and solid lines (D) are set according to the purified antibody-blocking control (*see* **Subheading 3.6.**). IL-8-positive monocytes are represented in dot plots in the right upper quadrants and in single-parameter histograms to the right of the solid lines (3.4). The incubation with purified anti-IL-8 antibodies prior to staining abrogates selectively specific staining and exposes the peak of nonspecific staining (3.1) compared with the positive control (3.4). Stimulated cells stained with isotype-matched antibodies (3.2) failed to detect the enhancement of nonspecific staining after stimulation with LPS, illustrated by the profound distance between the peaks in 3.1 and 3.2. Furthermore, the use of nonstimulated cells did not exactly differentiate specific from nonspecific staining compared with the purified antibody-blocking control (3.1 and 3.3).

3.1: Purified antibody-blocking control

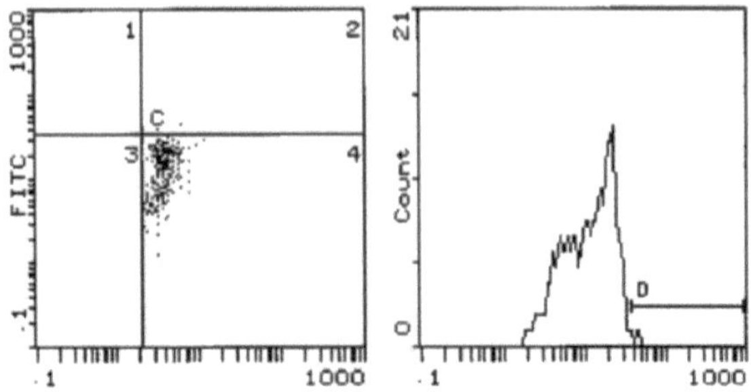

3.2: Isotype control

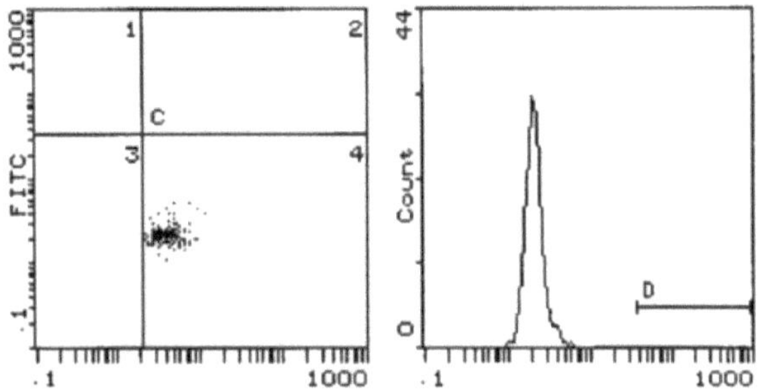

3.3: Non-stimulated cells

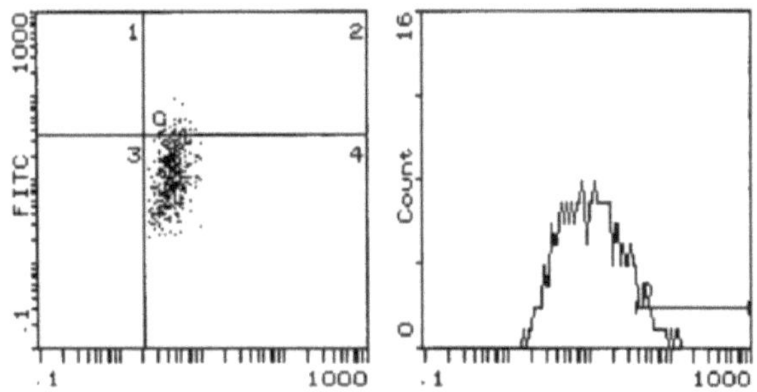

3.4: Positive control

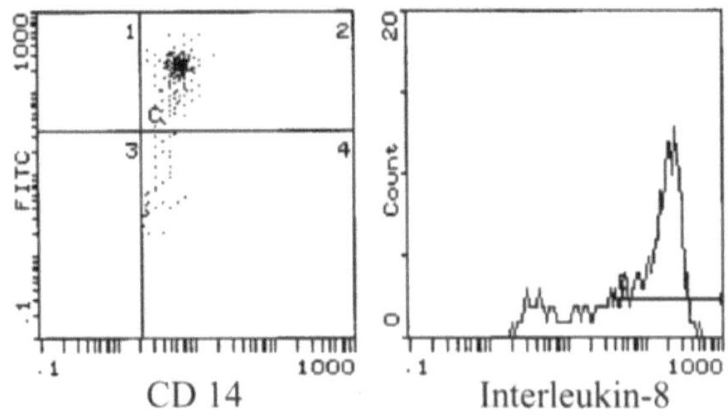

Fig. 3.

References

1. Jung, T., Schauer, U., Heusser, C., Neumann, C., and Rieger, C. (1993) Detection of intracellular cytokines by flow cytometry. *J. Immunol. Methods* **159,** 197–207.
2. Sander, B., Andersson, J., and Andersson, U. (1993) Assessment of cytokines by immunofluorescence and the paraformaldehyde-saponin procedure. *Immunol. Rev.* **119,** 65–93.
3. Schauer, U., Jung, T., Krug, N., and Frew, A. (1996) Measurement of intracellular cytokines. *Immunol. Today* **17,** 305–306.
4. Tartakoff, A. M. (1983) Perturbation of vesicular traffic with the carboxylic ionophore monensin. *Cell* **32,** 1026–1028.
5. Prussin, C. and Metcalfe, D. D. (1995) Detection of intracytoplasmic cytokine using flow cytometry and directly conjugated anti-cytokine antibodies. *J. Immunol. Methods* **188,** 117–128.
6. De Groote, D., Zangerle, P. F., Gevaert, Y., Fassotte, M. F., Beguin, Y., Noizat-Pirenne, F., et al. (1992) Direct stimulation of cytokines (IL-1β, TNF-a, IL-6, IL-2, IFN-γ, and GM-CSF) in whole blood. I. Comparison with isolated PBMC stimulation. *Cytokine* **4,** 239–248.
7. Härtel, C., Bein, G., Müller-Steinhardt, M., and Klüter, H. (2001) Ex vivo induction of cytokine mRNA expression in human blood samples. *J. Immunol. Methods* **249,** 63–71.
8. Maino, V. C., Ruitenberg, J. J., and Suni, M. A. (1996) Flow cytometric method for analysis of cytokine expression in clinical samples. *Clin. Immunol.* **16,** 95–98.
9. Le-Barillec, K., Pidard, D., Balloy, V., and Chignard, M. (2000) Human neutrophil cathepsin G down-regulates LPS-mediated monocyte activation through CD14 proteolysis. *J. Leuk. Biol.* **68,** 209–215.
10. Schultz, C., Rott, C., Temming, P., von Puttkamer, J., and Bucsky, P. (2002) Influence of specimen age and use of different negative controls in determination of intracytoplasmic cytokine levels after whole-blood culture assay. *Clin. Diagn. Lab Immunol.* **9,** 295–298.
11. Czerkinsky, C., Nilsson, L. A., Nygren, H., Ouchterloney, Ö., and Tarkowski, A. (1983) A solid-phase enzyme-linked immunospot (ELISPOT) assay for enumeration of specific antibody-secreting cells. *J. Immunol. Methods* **65,** 109–121.
12. Turcanu, V. and Williams, N. A. (2001) Cell identification and isolation on the basis of cytokine secretion: a novel tool for investigating immune responses. *Nat. Med.* **7,** 373–376.
13. Sander, B., Höiden, I., Andersson, U., Möller, E., and Abrams, J. S. (1993) Similar frequencies and kinetics of cytokine producing cells in murine peripheral blood and spleen. Cytokine detection by immunoassay and intracellular immunostaining. *J. Immunol. Methods* **166,** 201–214.
14. Scheffold, A., Assenmacher, M., Reiners-Schramm, L., Lauster, R., and Radbruch, A. (2000) High-sensitivity immunofluorescence for detection of the pro- and anti-inflammatory cytokines gamma interferon and interleukin-10 on the surface of cytokine-secreting cells. *Nat. Med.* **6,** 107–110.

5

Flow-Cytometric Immune Function Methodology for Human Peripheral Blood Dendritic Cells

Kerstin Willmann

1. Introduction

The following describes the use of flow cytometry to detect the immune response of circulating CD11c+ and CD11c– (CD123 high, pre-DC2) dendritic cells on a single-cell level *(1–4)*. Flow cytometry is an analytical and preparative tool for immunology *(5,6)*. Its use has rapidly advanced the accurate quantification of peripheral blood cell populations. Today, it facilitates the assessment of patient diagnosis and treatment *(7,8)*. Recent studies have demonstrated the importance of investigating not only the presence but also the functionality of specific leukocyte subsets *(9–12)*. Several dendritic cell (DC) populations are currently being investigated in various disease states, and the need to understand their functional roles is well recognized *(13–22)*. The field of DC biology has developed slowly over the past three decades. This is the result, in part, of their low frequency, the lack of specific markers, and their fast turnover rate upon isolation *(17,19,22,23)*. In spite of these limitations, the importance of DCs as the professional antigen-presenting cells of the immune system is well understood *(17,23–26)*. DCs present antigen to naive and memory T cells and are believed to control T-cell polarization depending on the DC subset, the type of DC maturation signal, and the duration of DC activation *(4,17–19,22,26–28)*. DCs affect T-cell responses not only by direct contact during antigen presentation but also through cytokine production *(21,29–31)*. The DC immune function assay is derived with modifications from methods that have been described earlier to measure cytokine secretion in T cells and in monocytes *(32–37)*. It presents a well-optimized tool to detect the immune

From: *Methods in Molecular Biology, vol. 215: Cytokines and Colony Stimulating Factors: Methods and Protocols*
Edited by: D. Körholz and W. Kiess © Humana Press Inc., Totowa, NJ

function of fresh peripheral blood DCs, eliminating extensive leukocyte processing and enrichment procedures *(1,2,4)*. The described protocols in **Subheading 3.** give examples of tumor necrosis factor (TNF)-α, interleukin (IL)-1-β, and IL-6 cytokines and their kinetics, as well as the detection of CD80 and CD83 well-established DC surface maturation and activation markers *(17,23,38)*.

The DC method is composed of four basic steps: the activation of whole peripheral blood (**Subheading 3.1.**), sample processing (**Subheading 3.2.**), flow-cytometric sample acquisition (**Subheading 3.3.**), and data analysis (**Subheading 3.4.**). During cell activation, blood is exposed to an activating agent, in this case bacterial lipopolysaccharide (LPS). To detect intracellular cytokines, the simultaneous addition of a secretion inhibitor is crucial. BrefeldinA (BFA) is used to disrupt the Golgi transport, which leads to intracellular accumulation of otherwise secreted cytokines. When surface antigens are chosen as a functional readout, Golgi-disruptive agents cannot be used because they also block the transport of newly synthesized integral proteins to the cell surface. Therefore, two different protocols apply for the detection of surface versus intracellular targets as functional readouts (**Subheadings 3.1.1.** and **3.1.3.**). The sample processing requires 2 h and is the most labor-intensive step of the procedure. The aliquoted blood volume can vary from 100 to 300 μL using the described 5-mL tube format. The author recommends the use of 300 μL blood, not older than 24 h. The sample processing step includes, in chronological order, cell surface staining, red blood cell lysis, cell permeabilization, and intracellular cytokine staining (**Subheading 3.2.1.**) or cell surface staining and red blood cell lysis for surface functional readouts (**Subheading 3.2.2.**). Washing steps are conducted after surface staining, permeabilization, and intracellular staining. Cell fixation takes place during the red cell lysis and cell permeabilization. The processed sample should be acquired within 1 h on a flow cytometer. Standard flow-cytometric procedures use two scatter parameters, representing cell size and cell granularity, and up to four fluorescence parameters to identify the cells of interest and to determine their cellular function, by changes in immunophenotype and protein secretion pattern, respectively. Here, the data analysis requires the combination of three fluorescence parameters to allow the identification of CD11c+ and CD11c– peripheral blood DCs, leaving one fluorescence parameter to determine the functional readout per sample. The time-to-answer of the methodology varies depending on the duration of sample activation. For TNF-α, the time is 4 h: 2 h for cell activation and 2 h for sample processing. The basics of the procedure can be applied to other intracellular targets, activation agents, and different subsets of DC populations. To address higher-throughput needs, it is possible to scale down the assay to a 96-microtiter-well plate format.

2. Materials

The immune function methodology has been optimized in great detail with the following materials. It is unknown if similar products provide the same results with the protocols described in **Subheading 3.**

2.1. Blood Activation

1. BD VACUTAINER® tubes with sodium heparin for blood collection (sterile 12 × 75-mm style) from BD Discovery Labware (Bedford, MA, cat. no. 367671).
2. Falcon® 5-mL polypropylene round-bottom tubes with caps (12 × 75-mm style) from BD Discovery Labware (Bedford, MA, cat. no. 352063).
3. Dimethyl sulfoxide (DMSO) from Sigma (St. Louis, MO, cat. no. D-8779).
4. LPS from Sigma (St. Louis, MO, cat. no. L-2654). Make an LPS stock solution of 0.5 mg/mL in DMSO and store 20-µL aliquots at –70°C. LPS dissolves easily in DMSO at room temperature. To make an LPS working solution, thaw a 20-µL aliquot of the 0.5 mg/mL LPS stock solution and dilute it 1 : 10 in 1X phosphate-buffered saline (PBS) to achieve a final working concentration of 0.05 mg/mL. Use this solution within 1 h.
5. Brefeldin A (BFA) from Sigma (St. Louis, MO, cat. no. B-7651). Make a BFA stock solution of 50 mg/mL in DMSO and store at –70°C. To make a BFA working solution, thaw the 50 mg/mL BFA stock solution and dilute a 5-µL aliquot 1 : 100 in 1X PBS to achieve a final working concentration of 0.5 mg/mL. Use this solution within 1 h.
6. Phosphate buffer saline (PBS) 1X.
7. Vortex.

2.2. Sample Processing

1. Lineage cocktail 1 (lin 1) FITC from BD Biosciences (San Jose, CA, cat. no. 340546). The lineage cocktail contains a mixture of antibodies that stain lymphocytes, monocytes, eosinophils, and neutrophils. Lin 1 includes antibodies to CD3, CD14, CD16, CD19, CD20, and CD56. The use of CD19 and CD20 in parallel provides a brighter staining of B-cells with lin 1. Peripheral blood dendritic cells and basophils can be distinguished from other leukocytes by their lack of staining with lin 1 FITC. The lin 1 antibody cocktail is provided at working concentration. Store the monoclonal antibody reagent at 2–8°C in the dark.
2. Anti-HLA-DR peridinin chlorophyll protein (PerCP), 0.0125 mg/mL (cat. no. 347364); CD11c allophycocyanin (APC), 0.025 mg/mL (cat. no. 340544); CD80 phycoerythrin (PE), 0.0032 mg/mL (cat. no. 340294); anti-TNF-α PE, 0.0032 mg/mL (cat. no. 340512); anti-IL-1-β PE, 0.003 mg/mL (cat. no. 340516); anti-IL-6 PE, 0.0015 mg/mL (cat. no. 340527); all from BD Biosciences, San Jose, CA. All monoclonal antibody solutions are provided at working concentration. Store reagents at 2–8°C in the dark.
3. CD83 PE custom product, working solution 0.05 mg/mL, from BD Biosciences (San Jose, CA). Store monoclonal antibody reagent at 2–8°C in the dark.

4. FACS™ Lysing Solution (10X) from BD Biosciences (San Jose, CA, cat. no. 349202). Make a 1X working dilution with deionized water. Store at room temperature (RT).
5. FACS Permeabilizing Solution 2 (10X) from BD Biosciences (San Jose, CA) cat. no. 340973). Make a 1X working dilution with deionized water. Store at RT.
6. Wash buffer (1X PBS with 0.5% bovine serum albumin [BSA]).
7. Falcon 5-mL polypropylene round-bottom tubes with caps (12 × 75-mm style) from BD Discovery Labware (Bedford, MA, cat. no. 352063).
8. Falcon 5-mL polystyrene round-bottom tubes (12 × 75-mm style) from BD Discovery Labware (Bedford, MA, cat. no. 352052).
9. Sorvall benchtop centrifuge.
10. Vortex.

2.3. Flow-Cytometric Sample Acquisition

The following components are a complementary system that is specifically designed for the multicolor analysis of fluoresceine isothiocyanate (FITC), phycoerythrin (PE), peridininchlorophyll protein (PerCP), and allophycocyanin (APC). Other fluorophore combinations cannot be used because they have alternate system requirements.

1. FACSCalibur™ dual-laser flow cytometer.
2. FACSComp™ software for instrument setup.
3. CellQuest™ or CellQuest Pro software.
4. CaliBRITE™ 3 beads, unlabeled, FITC-, PE-, PerCP-labeled for instrument setup (cat. no. 340486).
5. CaliBRITE APC beads for instrument setup (cat. no. 340487).

All products are distributed by BD Biosciences (San Jose, CA).

2.4. Flow-Cytometric Data Analysis

CellQuest, CellQuest Pro, or PAINT-A-GATE Pro™ software from BD Biosciences (San Jose, CA).

3. Method

3.1. Blood Activation Protocols

Collect venous blood of normal donors in sodium heparin VACUTAINER tubes. Other anticoagulants severely compromise the functional capacity of lymphocytes and are not recommended (*see* **Note 1**).

3.1.1. Blood Activation: Intracellular Cytokine Detection (TNF-α)

Peripheral blood is activated with LPS at 1 µg/mL final concentration in the presence of secretion inhibitor Brefeldin A (BFA) at 10 µg/mL final concentration. The baseline of TNF-α expression is determined with a resting control that contains BFA but lacks the stimulus LPS. Instead, stimulus solvent is added to balance the concentration of DMSO in activated and resting samples. Activated and resting samples are incubated in parallel.

Prepare activated control:

- Add 1 mL of whole human blood to a 5-mL polypropylene round-bottom tube.
- Add 20 µL of LPS working solution (0.05 mg/mL) to the same blood and vortex.
- Add 20 µL of BFA working solution (0.5 mg/mL) to the same blood and vortex.
- Incubate for 2 h.

Prepare resting control:

- Add 1 mL of whole human blood to a 5-mL polypropylene round-bottom tube.
- Add 20 µL of DMSO diluted 1:10 in PBS 1X to the same blood and vortex.
- Add 20 µL of BFA working solution (0.5 mg/mL) to the same blood and vortex.
- Incubate for 2 h.

The cell activation is performed at 37°C humidified atmosphere and 5% CO_2. During incubation, cap tubes loosely to prevent evaporation of the sample, but to allow gas exchange. The use of polypropylene labware for DC activation is important because it increases the DC recovery. DCs can stick to polystyrene surfaces and may be lost for further analysis.

3.1.2. Blood Activation: Cytokine Kinetic Assay (TNF-α, IL-1-β, IL-6)

For the determination of cytokine kinetics, LPS-activated whole blood is processed every hour in a time window from 0 to 8 h of incubation. Follow the basic procedure described in **Subheading 3.1.1.** for the positive activation control *(1)* with one variation: BFA is added to whole blood during only the last hour of sample incubation. This determines the cytokine production in 1-h time increments. Each time-point requires the preparation of one activated whole-blood sample.

3.1.3. Blood Activation: Functional Surface Antigen Detection (CD80, CD83)

Peripheral blood is activated with LPS at 1 µg/mL final concentration in the absence of secretion inhibitor BFA. The baseline of CD80 and CD83 expression is determined with a resting control that lacks LPS but contains stimulus solvent to balance the concentration of DMSO in activated and resting samples. Activated and resting samples are incubated in parallel.

Prepare activated control:

- Add 1 mL of whole human blood to a 5-mL polypropylene round-bottom tube.
- Add 20 µL of LPS working solution (0.05 mg/mL) to the same blood and vortex.
- Incubate for 4 h.

Prepare resting control:

- Add 1 mL of whole human blood to a 5-mL polypropylene round-bottom tube.
- Add 20 μL of DMSO diluted 1 ; 10 in PBS 1X to the same blood and vortex.
- Incubate for 4 h.

The cell activation is performed at 37°C humidified atmosphere and 5% CO_2. During incubation, cap tubes loosely to prevent evaporation of the sample, but to allow gas exchange. The use of polypropylene labware for DC activation is important because it increases the DC recovery. DCs can stick to polystyrene surfaces and may be lost for further analysis.

3.2. Sample Processing Protocols

The activated and resting control samples are processed in parallel for acquisition by flow cytometry. All steps of the procedure are performed at RT and in polypropylene tubes.

3.2.1. Sample Processing: Intracellular Cytokine Detection

Sample processing for intracellular antigens includes surface staining with fluorophore-conjugated antibodies, followed by red blood cell lysis, cell permeabilization, and intracellular cytokine staining using fluorophore-conjugated antibody. This protocol uses activated and resting samples from the procedure in **Subheadings 3.1.1.** and **3.1.2.**

1. Add 120 μL lin 1 FITC, 60 μL anti-HLA-DR PerCP, and 30 μL CD11c APC to an empty 5-mL polypropylene round-bottom tube.
2. Add 300 μL whole blood (WB) to the antibody mixture and vortex.
3. Incubate for 15 min at RT in the dark.
4. Add 3 mL FACS Lysing Solution 1X working concentration, cap tubes, and vortex.
5. Incubate for 10 min at RT in the dark.
6. Centrifuge 7 min at 500*g*.
7. Aspirate supernatant carefully and vortex gently to break off the cell pellet.
8. Resuspend surface-stained cells in 1 mL FACS Permeabilizing Solution 1X working concentration.
9. Incubate for 10 min at RT in the dark.
10. Add 2.5 mL wash buffer, cap tubes, and vortex gently.
11. Centrifuge 10 min at 500*g*.
12. Decant the supernatant instead of aspirating.
13. Resuspend permeabilized cells in remaining supernatant of the 5-mL tube gently.
14. Add 20 μL anticytokine PE reagent for intracellular staining.
15. Incubate for 30 min at RT in the dark.
16. Add 2 mL wash buffer and vortex gently.

17. Centrifuge 10 min at 500*g*.
18. Decant the supernatant.
19. Resuspend the now intracellular-stained and permeabilized cells in approx 200 µL wash buffer.
20. Acquire the samples in ≤1 h on the flow cytometer.
21. Store samples at 4°C if not acquired immediately (*see* **Note 2**).

3.2.2. Sample Processing: Functional Surface Antigen Detection

Sample processing for functional surface antigen detection includes cell surface staining with fluorophore-conjugated monoclonal antibodies followed by red blood cell lysis. The protocol uses activated and resting samples from the procedure in **Subheading 3.1.3.**

1. Add 60 µL lin 1 FITC, 30 µL anti-HLA-DR PerCP, 15 µL CD11c APC, and 20 µL CD83 PE to an empty 5-mL polypropylene round-bottom tube.
2. Add 60 µL lin 1 FITC, 30 µL anti-HLA-DR PerCP, 15 µL CD11c APC, and 60 µL CD80 PE to a second empty 5-mL polypropylene round-bottom tube.
3. The following processing steps apply to both sample tubes:
   a. Add 300 µL WB to the antibody mixture and vortex.
   b. Incubate for 15 min at RT in the dark.
   c. Add 3 mL of FACS Lysing Solution 1X working concentration; cap tubes and vortex.
   d. Incubate for 10 min at RT in the dark.
   e. Centrifuge 7 min at 500*g*.
   f. Aspirate the supernatant and vortex to break off the cell pellet.
   g. Add 2 mL wash buffer and vortex.
   h. Centrifuge 7 min at 500*g*.
   i. Aspirate the supernatant.
   j. Resuspend stained cells in approx 200 µL of wash buffer.
   k. Acquire the samples in ≤1 h on a flow cytometer.
   l. Store samples at 4°C if not acquired immediately (*see* **Note 2**).

3.3. Flow Cytometry Protocols

3.3.1. Instrument Calibration

The flow cytometer is set up for acquisition of fluorescently stained blood samples, derived from **Subheading 3.2.**, with CaliBRITE beads and FACS-Comp software. The four-color bead system is composed of unlabeled beads and FITC-, PE-, PerCP-, and APC-labeled beads that mimic autofluorescence of leukocytes and fluorescence emission of stained cells. Beads and software provide a system that automatically adjusts the instrument sensitivity and corrects for spectral overlap of the fluorescent dye emissions, known as fluorescent compensation, respectively.

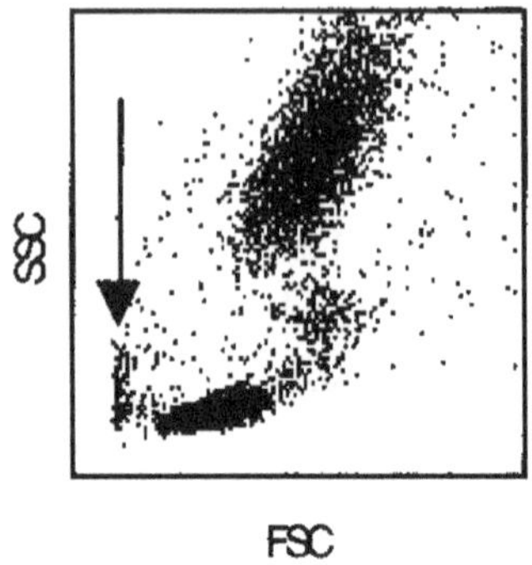

Fig. 1. Instrument setup using processed peripheral blood. The optimal FSC threshold is defined in a SSC vs FSC plot prior to sample acquisition (*see* arrow). This provides maximal exclusion of cellular debris in the data files. All leukocytes are shown as they appear during instrument setup. The SSC and FSC parameters are displayed in a 0 to 1024 linear scale.

1. Freshly prepare one tube with 1 mL of sheath fluid or wash buffer and a second tube with 3 mL of sheath fluid or wash buffer.
2. Then, add one drop of unlabeled beads and APC-labeled beads to the first tube and all five beads, unlabeled—FITC, PE, PerCP, APC—to the second tube.
3. Vortex and follow the FACSComp software instructions on the computer that is connected to the cytometer.
4. Manual instrument setup procedures are feasible but have not been validated for the DC immune function protocols.

3.3.2. Sample Acquisition

A flow cytometer counts particles, including cellular debris, as well as leukocytes of interest. It is the goal to include as little cellular debris as possible and to maximize the cell count in every data file. The forward scatter (FSC) parameter is a relative measure for particle size and, optimally adjusted, it reduces the number of small debris events in the data files. For this reason, we recommend fine-tuning the FSC threshold prior to sample acquisition (**Fig. 1**). The threshold setting might need to be readjusted when a different set of samples is run (i.e., LPS vs LPS and BFA-treated samples).

Dendritic cells are found at low frequency in peripheral blood. Therefore, many cells (events) have to pass through a flow cytometer to collect a statistically significant number of DCs (*see* **Note 3**). The more events that are recorded, the more space a data file takes for storage. Therefore, many flow-cytometric investigators choose live gating as an option in rare-event analysis during sample acquisition. The use of an acquisition (live) gate reduces the number of irrelevant leukocytes over the number of desired DCs in a data file.

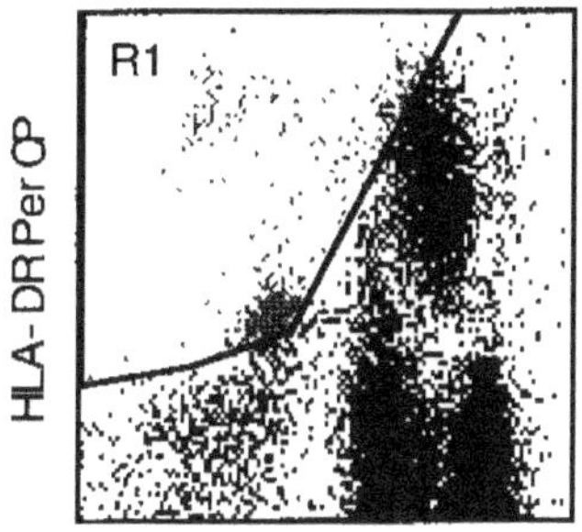

Fig. 2. Instrument setup using processed peripheral blood. The region R1 defines the HLA-DR vs lin 1 live gate, which needs to be activated during sample acquisition on the flow cytometer. Events positioned outside of the region are not stored in the data file. All leukocytes are shown as they appear during instrument setup. The fluorescent parameters are displayed in a four-decade logarithmic scale.

For DCs, the acquisition gate is defined in an HLA-DR PerCP vs lin 1 FITC dot plot by brightly stained HLA-DR and negative-to-dim lin-1-stained events (*see* **Fig. 2**). The boundaries of the acquisition gate need to be defined carefully because any event outside of it will not be recorded in the data file. For rare-cell populations, the data analysis can be inefficient without the use of live gating during sample acquisition because large data files (>50,000 events) significantly slow down the speed of data processing. Use the entire cell suspension of each sample for acquisition, which yields around 5000 to 8000 events. It might be helpful to prepare an additional sample to perform the FSC threshold adjustment and to define the acquisition gate.

3.4. Data Analysis

Dendritic cells can be characterized by low FSC and side scatter (SSC), high expression levels of HLA-DR, and lack of or minimal staining for lineage markers. Two different gating strategies have been developed for CD11c+ and CD11c– DCs (*see* **Figs. 3** and **4**). CD11c+ DCs respond to LPS stimulation, as shown by the detection of increased levels of TNF-α, CD80, and CD83, whereas CD11c– DCs remain unresponsive for TNF-α (*see* **Figs. 5** and **6**) (*see* **Note 4**).

3.4.1. CD11c+ DC Gating Strategy (Fig. 3)

1. Create an SSC vs CD11c dot plot and display all events of the data file.
2. Draw a region R1 to include CD11c bright and SSC low events as demonstrated in **Fig. 3**. The drawing of region R1 is not based on a discrete cluster resolution.

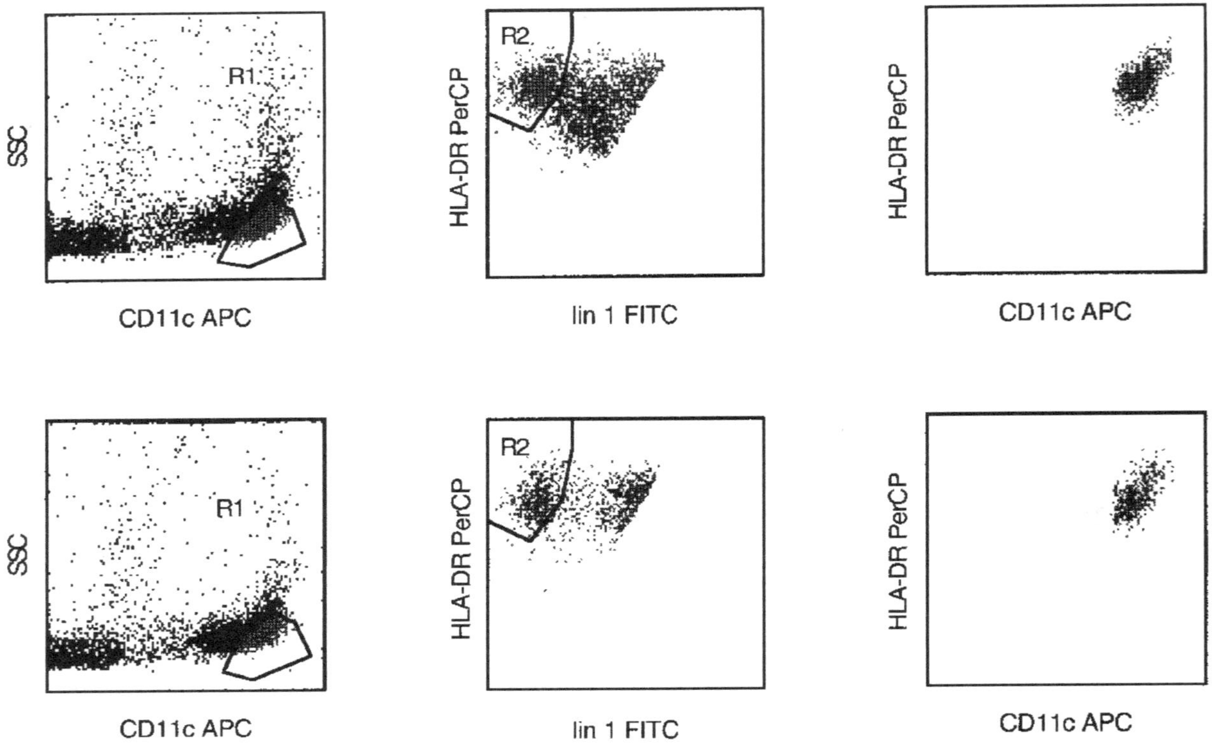

Fig. 3. Data analysis. Identification of CD11c+ DCs. The first row shows LPS-activated samples and the second row displays resting peripheral blood. From the left to the right, the plots are ungated, R1 gated, and "R3 = R1 and R2" gated. The drawing of region R1 is not based on a discrete cluster separation (*see* **Subheading 3.4.1.** for details). CD11c+ DCs are defined by low-scatter characteristics, high CD11c and high HLA-DR expression levels, and lineage cocktail 1 dim staining. The lineage cocktail includes CD3, CD14, CD16, CD19, CD20, and CD56. The HLA-DR versus CD11c dot plot displays CD11c+ DCs. The SSC is displayed in a 0 to 1024 linear scale, and the fluorescent parameters are displayed in a four-decade logarithmic scale.

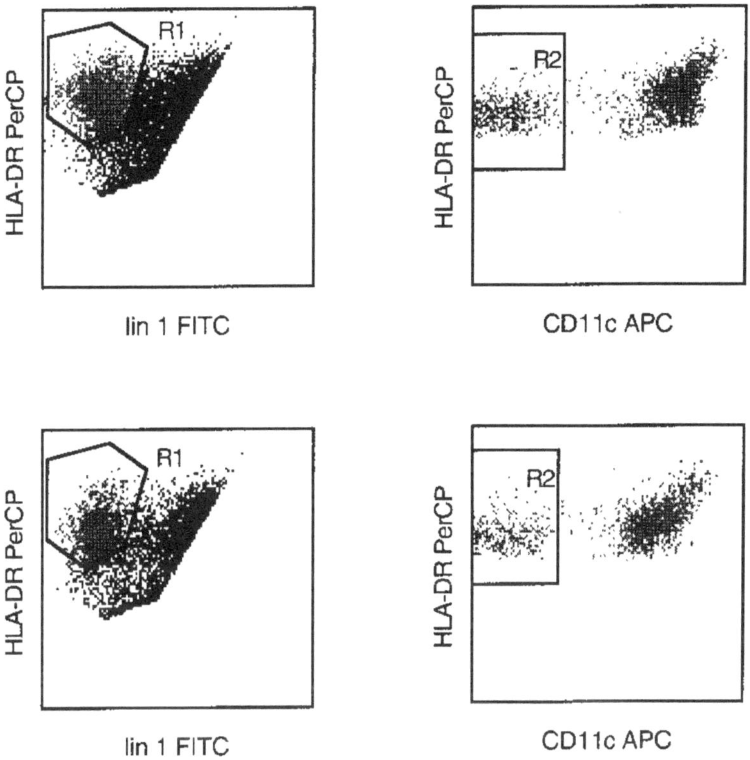

Fig. 4. Data analysis. Identification of CD11c– in LPS-activated blood. The first row shows LPS-activated samples and the second row displays resting peripheral blood. CD11c– DCs are defined by their high HLA-DR expression and their lack of lineage markers and CD11c antigen. The left plot is ungated. The right plot is gated on R1. The fluorescent parameters are displayed in a four-decade logarithmic scale.

3. Create an HLA-DR vs lin 1 dot plot that is gated on R1.
4. Go back to the SSC vs CD11c plot. Now, adjust the position and size of region R1 while observing the changing cluster separation of the HLA-DR bright lin 1 dim events in the R1 gated HLA-DR vs lin 1 dot plot. Choose the final definition for region R1 that reflects the best cluster resolution for the brightest HLA-DR and least lin-1-stained event.
5. Draw a region R2 around the cell cluster with minimal lin 1 and highest HLA-DR staining.
6. Create a logical gate "G3 = R1 and R2" in the gate list.
7. Create a TNF-α vs CD11c dot plot, a CD80 vs CD11c dot plot, and a CD83 vs CD11c dot plot that are each gated on "G3 = R1 and R2." *See* **Fig. 5**. Additionally, you can create a TNF-α histogram that is gated on "G3 = R1 and R2."

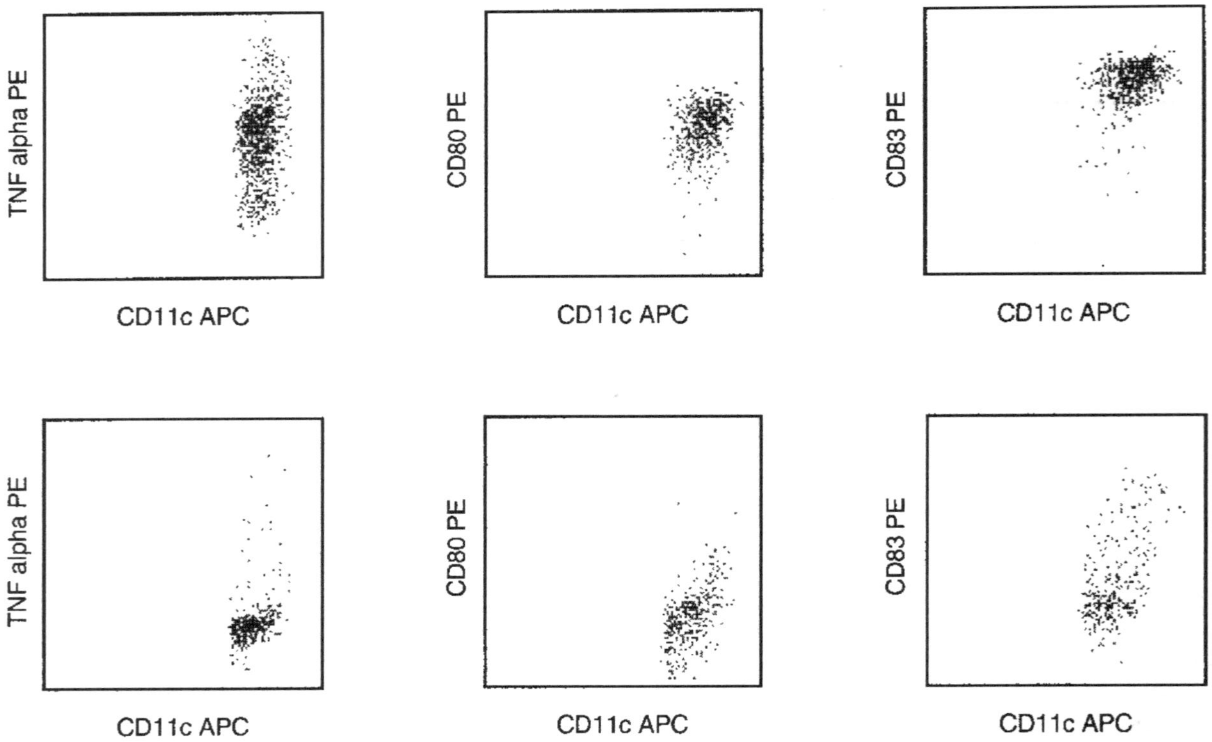

Fig. 5. Flow cytometric results of TNF-α, CD80, and CD83 in CD11c+ DCs. All plots display approx 500 cells. The upper row illustrates the upregulated expression of the functional markers in response to LPS stimulation, whereas the lower row displays the baseline expression of resting CD11c+ DCs. *See* **Fig. 3** for gating strategies. The fluorescent parameters are displayed in a four-decade logarithmic scale.

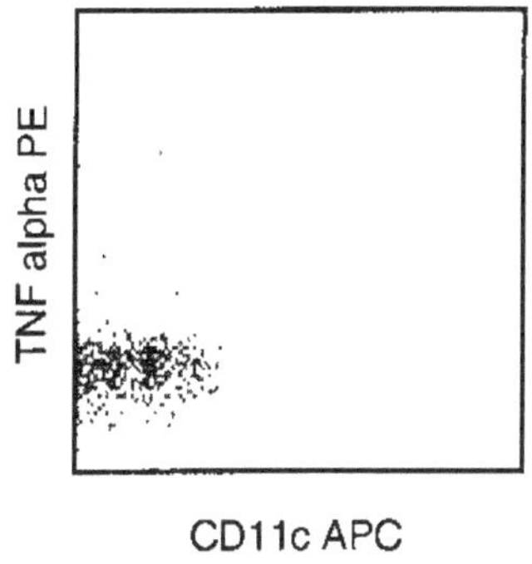

Fig. 6. Flow-cytometric results of TFN-α in LPS-activated CD11c– DCs. CD11c– DCs appear nonresponsive to LPS stimulation. Compare to **Fig. 5**. *See* **Fig. 4** for gating strategies. The fluorescent parameters are displayed in a four-decade logarithmic scale.

CD11c+ DCs have low-scatter characteristics and highest expression levels of CD11c among all leukocytes. This gating strategy does exclude CD11c– events from the R1-gated HLA-DR vs lin 1 dot plot, resulting in improved cluster separation of CD11c+ DCs and non-DC leukocytes *(1)*. Compare the resolution of the HLA-DR vs lin 1 dot plots in **Fig. 3** (R1 gated) and **Fig. 4** (ungated). These plots are generated from the same data file (*see* **Note 4**).

*3.4.2. CD11c– DC Gating Strategy (**Fig. 4**)*

1. Create an HLA-DR vs lin 1 dot plot and display all events.
2. Draw a region R1 to include events with minimal lin 1 and highest HLA-DR staining.
3. Create an HLA-DR vs CD11c dot plot that is gated on R1.
4. Draw a region R2 around CD11c– and HLA-DR-positive events.
5. Create a logical gate "G3 = R1 and R2" in the gate list.
6. Create a TNF-α vs CD11c dot plot that is gated on "G3 = R1 and R2." *See* **Fig. 6**.

For functional surface markers, this two-step gating strategy cannot sufficiently resolve CD11c– DCs all the time (**Subheading 3.1.3.**). In these samples, the resolution of HLA-DR bright and lin 1 dim DCs is decreased from non-DC leukocytes. For this reason, we do not recommend this gating strategy for the inexperienced user. Instead, the reader can use CD123 as a positive identification marker in combination with HLA-DR and lin 1. This requires the processing of an additional sample set (**Subheading 3.2.2.**), because CD11c+ DCs cannot be reliably identified by the use of CD123, HLA-DR, and lin 1 cocktail (*see* **Note 4**).

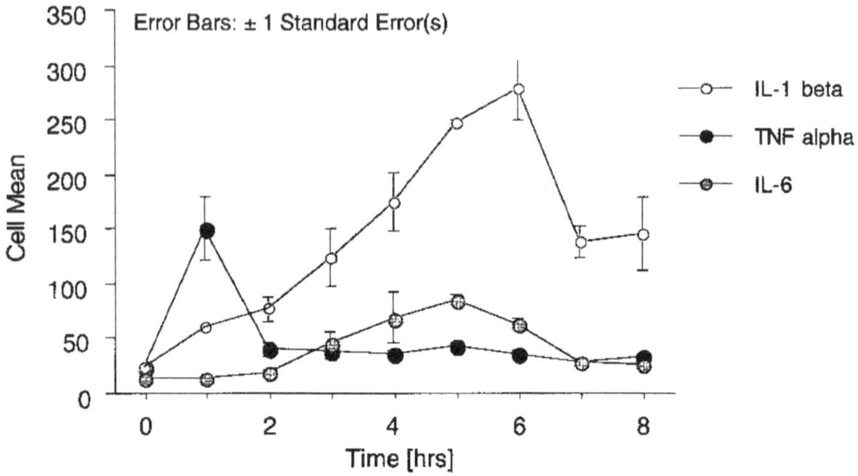

Fig. 7. Kinetics of TNF-α, IL-1β, and IL-6 in LPS-activated CD11c+ DCs. The time-course of LPS activation was 0 to 8 h. Intracellular cytokines are measured as PE median fluorescence intensity. TNF-α was produced first, then IL-1β and IL-6. *See* **Fig. 3** for gating strategies.

4. Notes

The low frequency of peripheral blood DCs and indirect staining procedures are creating technical challenges in the detection of DCs in unseparated samples. As demonstrated with these protocols and associated results shown in **Figs. 5–7**, it is possible to perform these tasks in a step-by-step procedure. Beyond the information already provided, the author would like to share a few additional comments for a successful performance of the assay:

1. The inability to detect a cytokine signal may be the result of insufficient permeabilization or inactive BFA or LPS solutions. Incorrect storage conditions may inactivate LPS or BFA solutions. It is imperative that correct dilutions and concentrations for solutions like FACS Permeabilizing, BFA, or LPS are used. Blood drawn in other anticoagulants than sodium heparin should not be used. The assay is incompatible with EDTA or ACD anticoagulants.
2. It might be necessary to transfer processed samples to polystyrene tubes for the flow-cytometric acquisition. Polypropylene tubes do not always fit tightly onto the sample injection system of the flow cytometer. Perform the transfer right before acquisition.
3. Dendritic cells have a slightly decreased density over other leukocytes, which can lead to a loss of DCs by insufficient pelleting during cell processing steps. Additionally, cell permeabilization procedures decrease the density of leukocytes. Therefore, if the yield of DCs appears too low, consider prolonging the centrifugation time or increasing the *g*-force of the centrifugation in **Subhead-**

ing **3.2.** It is also important to follow the requirement to use polypropylene tubes. DCs can stick to polystyrene material and might be lost from the cell suspension that is analyzed.

4. The staining intensity of the lineage cocktail in activated samples decreases over the length of the sample incubation time. The longer the blood is activated, the less sufficient is the separation of lin 1 dim versus lin 1 lineage-positive leukocytes—in particular, monocytes *(1)*. Monocytes have a very similar phenotype to CD11c+ DCs and are activated also by LPS, which affects the separation of lin-1-positive events from lin 1 dim DCs *(1)*. Best results are obtained for shortest activation times, ideally 2 h and maximally 4 h. Compare the resolution of HLA-DR vs lin 1 plots for LPS-activated and resting blood in **Figs. 3** and **4**. TNF-α is the intracellular readout of choice because of its early kinetics over other cytokines (*see* **Fig. 7**).

References

1. Willmann, K. and Dunne, J. F. (2000) A flow cytometric immune function assay for human peripheral blood dendritic cells. *J. Leuk. Biol.* **67,** 536–544.
2. Olweus, J., BitMansour, A., Warnke, R., Thompson, P. A., Carballido, J., Picker, L. J., et al. (1997) Dendritic cell ontogeny: a human dendritic cell lineage of myeloid origin. *Proc. Natl. Acad. Sci. USA* **94,** 12,551–12,556.
3. O'Doherty, U., Peng, M., Gezelter, S., Swiggard, W. J., Betjes, M., Bhardwaj, N., et al. (1994) Human blood contains two subsets of dendritic cells, one immunologically mature and the other immature. *Immunology* **82,** 487–493.
4. Liu, Y-J. and Blom, B. (2000) Focus on hematology. Introduction: TH2-inducing DC2 for immunotherapy. *Blood* **95,** 2482–2483.
5. Shapiro, H. M., ed. (1995) *Practical Flow Cytometry.* Wiley-Liss, New York.
6. Givan, A. L., ed. (1992) *Flow Cytometry First Principles.* Wiley-Liss, New York.
7. Young, N. S. and Beris, P. (2001) Flow cytometry in hematology and oncology. *Semin. Hematol.* **38,** 93–200.
8. Stein, D. S., Korvick, J. A., and Vermund S. H. (1992) CD4+ lymphocyte cell enumeration for prediction of clinical course of human immunodeficiency virus disease: a review. *J. Infect. Dis.* **166,** 1198–1199.
9. Maecker, H. T., Maino, V. C., and Picker, L. J. (2000) Immunofluorescence analysis of T-cell responses in health and disease. *J. Clin. Immunol.* **20,** 391–399.
10. Karanikas, V., Lodding, J., Maino, V. C., and McKenzie, I. F. C. (2000) Flow cytometric measurement of intracellular cytokines detects immune responses in MUC1 immunotherapy. *Clin. Cancer Res.* **6,** 829–837.
11. Bleesing, J. J. H. and Fleisher, T. A. (2001) Cell function-based flow cytometry. *Semin. Hematol.* **38,** 169–178.
12. Picker, L. J. and Maino, V. C. (2000) The CD4+ T cell resonse to HIV-1. *Curr. Opin. Immunol.* **12,** 381–386.
13. Auffermann-Gretzinger, S., Keeffe, E. B., and Levy, S. (2001) Impaired dendritic cell maturation in patients with chronic, but not resolved, hepatitis C virus infection. *Blood* **97,** 3171–3176.

14. Steinman, R. M. (1996) The dendritic cell in clinical immunology: the AIDS example. *J. Lab. Clin. Med.* **128,** 531–535.
15. Nouri-Shirazi, M., Banchereau, J., Fay, J., and Palucka, K. (2000) Dendritic cell based tumor vaccines. *Immunol. Lett.* **74,** 5–10.
16. Vuckovic, S., Fearnley, D. B., Gunningham, S., Spearing, R. L., Patton, W. N., and Hart, D. N. J. (1999) Dendritic cells in chronic myelomonocytic leukaemia. *Br. J. Hematol.* **105,** 974–985.
17. Hart, D. N. J. (1997) Dendritic cells: unique leukocyte populations which control the primary immune response. *Blood* **90,** 3245–3287.
18. Kapsenberg, M. L. Hilkens C. M. U., Wierenga, E. A., and Kalinski, P. (1999) The paradigm of type 1 and type 2 antigen-presenting cells. Implications for atopic allergy. *Clin. Exp. Allergy* **29,** 33–36.
19. Pulendran, B. Maraskovsky, E., Banchereau, J., and Maliszewski, C. (2001) Review. Modulating the immune response with dendritic cells and their growth factors. *Trends Immunol.* **22,** 41–47.
20. Sato, M., Iwakaba, K., Kimura, S., and Nishimura, T. (1998) Functional skewing of bone marrow-derived dendritic cells by Th1- or Th2-inducing cytokines. *Immunol. Lett.* **67,** 63–68.
21. Ghanekar, S., Zheng, L., Logar, A., Navratil, J., Borowski, L., Gupta, P., et al. (1996) Cytokine expression by human peripheral blood dendritic cells stimulated in vitro with HIV-1 and herpes simplex virus. *J. Immunol.* **157,** 4028–4036.
22. Kohrgruber, N., Halanek, N., Gröger, M., Winter, D., Rappersberger, K., Schmitt-Egenolf, M., et al. (1999) Survival, maturation, and function of CD11c– and CD11c+ peripheral blood dendritic cells are differentially regulated by cytokines. *J. Immunol.* **163,** 3250–3259.
23. Banchereau, J. and Steinman, R. M. (1998) Dendritic cells and the control of immunity. *Nature* **392,** 245–252.
24. Lanzavecchia, A. and Sallusto, F. (2000) Dynamics of T lymphocyte responses: intermediates, effectors, and memory cells. *Science* **290,** 92–97.
25. Hart, D. N. J., Clark, G. J., Dekker, J. W., Fearnley, D. B., Kato, M., Hock, B. D., et al. (1997) Dendritic cell surface molecules. *Dendritic Cells Fundam. Clin. Immunol.* 439–442.
26. Banchereau, J., Pulendran, B., Steinman, R., and Palucka, K. (2000) Commentary. Will the making of plasmacytoid dendritic cells in vitro help unravel their mysteries? *J. Exp. Med.* **192,** F39–F44.
27. Rissoan, M.-C., Soumelis, V., Kadowaki, N., Grouard, G., Briere, F., de Waal Malefyt, R., et al. (1999) Report. Reciprocal control of T helper cell and dendritic cell differentiation. *Science* **283,** 1183–1186.
28. Arpinati, M., Green, C. L., Heimfeld, S., Heuser, J. E., and Anasetti, C. (2000) Focus on hematology. Granulocyte-colony stimulating factor mobilizes T helper 2-inducing dendritic cells. *Blood* **95,** 2484–2490.
29. Heufler, C., Koch, F., Stanzl, U., Topar, G., Wysocka, M., Trinchieri, G., et al. (1996) Interleukin-12 is produced by dendritic cells and mediates T helper 1

development as well as interferon-γ production by T helper 1 cells. *Eur. J. Immunol.* **26,** 659–668.

30. Lapointe, R., Toso, J. F., Butts, H. A., Young, H. A., and Hwu, P. (2000) Human dendritic cells require multiple activation signals for the efficient generation of tumor antigen-specific T lymphocytes. *Eur. J. Immunol.* **30,** 3291–3298.
31. Zhou, L.-J. and Tedder, T. F. (1995) A distinct pattern of cytokine gene expression by human CD83+ blood dendritic cells. *Blood* **86,** 3295–3301.
32. Nomura, L. E., Walker, J. M., and Maecker, H. T. (2000) Optimization of whole blood antigen-specific cytokine assays for CD4+ T cells. *Cytometry* **40,** 60–68.
33. Suni, M. A., Picker, L. J., and Maino, V. C. (1998) Detection of antigen-specific T cell cytokine expression in whole blood by flow cytometry. *J. Immunol. Methods* **212,** 89–98.
34. Waldrop, S. L., Pitcher, C. J., Peterson, D. M., Maino, V. C., and Picker, L. J. (1997) Determination of antigen-specific memory/effector CD4+ T cell frequencies by flow cytometry. *J. Clin. Invest.* **99,** 1739–1750.
35. Jung, T., Schauer, U., Heusser, C., Neumann, C., and Rieger, C. (1993) Detection of intracellular cytokines by flow cytometry. *J. Immunol. Methods* **159,** 197–207.
36. Picker, L. J., Singh, M. K., Zdraveski, Z., Treer, J. R., Waldrop, S. L., Bergstresser, P. R., et al. (1995) Direct demonstration of cytokine synthesis heterogeneity among human memory/effector T cells by flow cytometry. *Blood* **86,** 1408–1419.
37. De Caestecker, M. P., Telfer, B. A., Hutchinson, I. V., and Ballardie, F. W. (1992) The detection of intracytoplasmic interleukin-1α, interleukin-β, and tumour necrosis factor a expression in human monocytes using two colour immunofluorescence flow cytometry. *J. Immunol Methods* **154,** 11–20.
38. Zhou, L.-J. and Tedder, T. F. (1995) Human blood dendritic cells selectively express CD83, a member of the immunoglobulin superfamily. *J. Immunol.* **154,** 3821–3835.

6

Evaluation of the Frequency of Virus-Specific CD8+ T Cells by Cytokine Flow Cytometry

Bodo Hoffmeister, Felix Kiecker, Ingolf Surel, Elham Khatamzas, Volker Schuster, Hans-Dieter Volk, and Florian Kern

1. Introduction

Since its introduction over 25 years ago, flow cytometry has evolved to be one of the most important methods in cellular diagnostics and research. Within recent years, new technology has dramatically expanded the range of parameters that can be analyzed by this technology. These include, for example, the production or secretion of soluble mediators ***(1–4)***, cell proliferation ***(5–8)***, and the detection of molecules expressed at very low density ***(9,10)***. One of the most prominent recent developments, however, is the analysis of antigen-specific T cells. The problems encountered when analyzing antigen-specific T cells are manyfold. One of the major problems is the fact that their frequencies tend to be rather low. Recent data show that, in certain situations (e.g., acute immune responses against viruses or bacteria), the actual frequency of specific CD8 cells can be much higher than previously estimated by limiting dilution assays; however, in most other responses, the frequencies of specific T cells are clearly below these values. Frequencies as low as 10^{-4}–10^{-5} have been reported ***(11,12)***.

Although previously the presence of antigen-specific T cells was noticeable through proliferation (e.g., ^{3}H-thymidin incorporation) or their effects on target cells (cytotoxicity testing), they may now be "visualized" directly as a result of detectable functional changes. One of these changes is the production of cyokines. Cytokines can be trapped intracellularly following administration of inhibitors of vesicular transport such as Brefeldin A or monensin. Subsequent fixation and permeabilization of the cell membrane with a detergent-containing permeabilization solution permits intracellular staining of cytokines with

From: *Methods in Molecular Biology, vol. 215:*
Cytokines and Colony Stimulating Factors: Methods and Protocols
Edited by: D. Körholz and W. Kiess © Humana Press Inc., Totowa, NJ

monoclonal antibodies. This development of "cytokine flow cytometry" (i.e., the detection of intracellularly stored cytokines in T cells following polyclonal stimulation) was one of the large milestones in flow cytometry in the last decade *(1–4)*. Although no information about antigen-specific yet nonreactive cells is obtained, this method has the major advantage that cell function can be studied along with phenotypes and frequencies. Moreover, cells may be studied within a reasonably natural composition of subsets (i.e., whole blood or peripheral blood mononuclear cells [PBMC]). In 1997, the use of pathogen-derived complex protein antigens for stimulating and visualizing antigen-specific peripheral blood CD4 T cells by multiparameter flow cytometry was reported *(13)*. This was the first report ever of a direct "observation" of antigen-specific CD4 T cells in peripheral blood. This new technology completely opened new avenues of research and clinical diagnosis. Because complex antigens are phagocytosed and processed and thus presented almost exclusively with class II major histocompatibility complex (MHC) molecules, it was, however, limited to CD4 T cells. Shortly afterward, though, the use of peptides was reported in a very similar assay *(14)*. Peptides may be loaded externally onto class I MHC (and class II MHC) molecules and thus stimulate antigen-specific CD8 (and CD4) T cells. Because many peptides can be combined in one single stimulation assay, even the response to complete proteins can now be analyzed efficiently and quickly using large peptide pools *(15)*. This may turn out to be very useful for the monitoring of immune responses (e.g., following vaccinations or in immunosuppressed patients).

Another recent milestone with respect to detecting antigen-specific T cells was the advent of (tetrameric) class I MHC/peptide complexes for the direct staining of T cells carrying the matching T-cell receptor (TCR) *(16–19)*. These tetramers stain antigen-specific T cells irrespective of function. Ideally, antigen-specific stimulation with peptides and tetramer staining are combined in one assay. Unfortunately, as a result of often pronounced TCR downregulation after TCR-dependent stimulation, MHC tetramers tend to not bind very well to recently activated T cells (unpublished results). Therefore, staining with tetramers following T-cell stimulation is not straightforward. Alternatively, tetramer is added prior to stimulation. In this way, the tetramer staining remains sufficiently bright for flow-cytometric identification following activation with peptide. In our hands, this approach gives the best results. The combined use of tetramer staining and peptide stimulation was recently used to analyze human immunodeficiency virus (HIV) and cytomegalovirus (CMV)-specific CD8 T cells in HIV+ patients *(20)*.

2. Materials

1. Ficoll–Paque (Pharmacia, Uppsala, Sweden).
2. RPMI 1640 medium (Biochrom KG, Berlin, Germany).

3. Fetal calf serum (Biochrom).
4. Cellstar™ polystyrene tissue culture tubes (Greiner, Frickenhausen, Germany).
5. Dulbecco's phosphate-buffered saline (PBS) (Gibco BRL, UK).
6. Brefeldin A (Sigma).
7. Bovine serum albumin (BSA) (Serva, Germany).
8. Na-Azide (Serva, Heidelberg, Germany).
9. Wash buffer/staining buffer: PBS containing 0.5% bovine serum albumin and 0.1% sodium azide.
10. Na–EDTA (Na–EDTA, Sigma).
11. Paraformaldehyde (Merck, Germany).
12. Permeabilizing solution (BD, Heidelberg, Germany).
13. Lysing solution (BD).

3. Methods

3.1. Stimulation of T Cells in PBMC or Whole Blood with CMV Antigens

Using whole blood is preferred by many because it is believed to reflect the natural (in vivo) environment more closely than PBMC suspensions. This view, although attractive at first glance, ignores that immune reactions do not take place in the large blood vessels. In fact, they take place in lymphatic and other tissues with many proteins and other factors not present that we find in whole blood. With regard to T-cell stimulation, we experienced that it often takes five times higher concentrations (or even more) of reagents for stimulation when whole blood is used compared with PBMC suspensions to obtain a comparable response (in percentage of interferon-γ-positive cells), regardless whether peptides or viral lysates are used for stimulation.

3.1.1. PBMC: Cells and Ex Vivo Stimulation with Peptides

White blood cells may be obtained from buffy coats of anticoagulated blood. Blood donors should be selected on the basis of HCMV seropositivity (seronegativity for controls). We generally use citrated or heparinized blood (Li–heparin or Na–heparin).

PBMC can be prepared by standard Ficoll–Paque density centrifugation and should be washed twice in sterile Dulbecco's PBS. A third, slow wash step should get rid of contaminating platelets that are supposed to be a source of inhibitory transforming growth factor-β (TGF-β).

1. Resuspend cells in culture media containing 10% fetal calf serum (FCS) (or autologous human serum), 2 mmol/L glutamine, and antibiotics, if desired (e.g., penicillin/streptomycin), and adjust to a cell concentration of 5×10^6/mL.

2. Place 100 µL of peptide solution in 16 × 125-mm Cellstar polystyrene tissue culture tubes. Alternatively, Falcon 2052 tubes (BD) may be used (*see* **Notes 1–3**).*
3. Add 400 µL of the cell suspension to the peptide solution and mix gently by hand.
4. Incubate tubes at 5° slant (from horizontal) in a standard incubator at 37°C in a humidified 5% CO_2 atmosphere.
5. After 2 h add 500 µL of RPMI 1640 containing the same additives as above plus 10 µg Brefeldin A (final concentration 10 µg/mL).
6. Reincubate at 5° slant in *identical* position as before.
7. After 6 h add 3 mL of ice-cold PBS and centifuge at 430*g* for 8 min at 4°C (for incubation time, *see* **Note 4**).
8. Remove supernatant by careful decanting or aspiration.
9. Add 2–3 mL of cold 2 m*M* EDTA/PBS and incubate for 10 min in a water-bath at 37°C.
10. Spin down (430 g for 8 min at 4°C) and remove supernatant by careful decanting.
11. Add 1 mL of cold PBS and vortex vigorously for approx 20 s to detach adhering cells from tube walls.
12. Pool tubes as desired, add 2–3 mL of wash buffer to each tube.
13. Spin down (430*g* for 8 min at 4°C).
14. Remove supernatant by carefully decanting and blotting the tubes dry on a paper towel (turn upside down and hold in this position until supernatant has been removed; *see* **Note 5**).
15. Add staining antibodies for surface staining (if desired) and an appropriate volume of wash buffer (we use 100 µL total staining volume; *see* **Note 6**) and incubate according to your standard surface-staining procedure. We stain for 30 min on melting ice in the dark, using phycoerythrin-conjugated anti-CD4 (CD4-PE), Peridinin chlorophyl protein-conjugated anti-CD4 (CD3-PerCP), and allophycocyanin-conjugated anti-CD8 (CD8-APC). *Some antigens require staining prior to (paraformaldehyde) fixation.*
16. Add 2–3 mL of 4% paraformaldehyde (PFA) (in PBS) and incubate for 5 min at 37°C in a water bath for fixation (*see* **Note 7**).
17. Spin down (430*g* for 8 min at 4°C) and decant (*see* **steps 13** and **14**).
18. Add 1 mL of permeabilizing solution and incubate for 10 min in the dark at room temperature (*see* **Note 7**).
19. Add 3 mL of PBS containing 0.1% Na–azide and 0.5% BSA.
20. Spin down and decant (430*g* for 8 min at 4°C).
21. Proceed to intracellular staining (same as **step 5**; *see* **Note 8**) (e.g., fluorescein isothiocyanate-conjugated anti-INF-γ [IFN-γ–FITC]).
22. Add 2–3 mL of wash buffer.
23. Spin down (430*g* for 8 min at 4°C) and decant (*see* **steps 13** and **14**; *see* **Note 5**).

*You may also use 96-well plates (I use 10^6 cells per well in 100 µL). Peptide loading is generally more efficient in a small volume. We have also successfully used culture bottles (25 mL) when stimulating a complete buffy coat. Efficiency, however, was not compared. If you use a culture bottle, use a scraper to recover your cells from the bottle wall.

24. Refix antibodies added following permeabilization (still unfixed) with 1% PFA/PBS if required (if samples are not analyzed immediately).

3.1.2. Whole-Blood Procedure

1. Place 1 mL of whole blood in 15-mL round-bottom culture tissue tube.*
2. Add highly concentrated peptide solution (e.g., 1 mg/mL of peptide in dimethyl sulfoxide [DMSO], add 5 µL; *see* **Notes 2** and **3**).
3. Incubate tubes at 5° slant (from horizontal) in a standard incubator at 37°C in a humidified 5% CO_2 atmosphere.
4. After 2 h, add 10 µg of Brefeldin A (e.g., 10 µL from a 1-mg/mL stock solution).
5. After a total of 6 h, add 100 µL of ice-cold 20 mmol EDTA solution (Na–EDTA in PBS) and incubate for 10 min at room temperature, vortex vigorously from time to time.
6. Add 10 mL of wash buffer.
7. Spin down (430*g* for 8 min at 4°C) and decant carefully.
8. Resuspend pellet by vortexing.
9. Add staining antibodies in PBS containing 0.1% Na–azide and 0.5% BSA for surface staining if desired (total staining volume = 100 µL) and incubate according to your standard surface staining procedure (we stain 30 min on melting ice in the dark; *see* **Notes 6** and **7**). You may also stain just an aliquot of 100 µL or so in a smaller tube (we use Falcon 2052 tubes) and then proceed with the respective smaller volumes.
10. Add a 20-fold volume of BD lysing solution to your tube and incubate for 10 min at room temperature.
11. Spin down and remove supernatant by careful decanting or aspiration.
12. Add 1 mL of permeabilizing solution and incubate for 10 min in the dark at room temperature.
13. Complete like PBMC procedure.

3.2. Reagent Preparation and Storage (Selected Reagents)

Brefeldin A should be dissolved in DMSO and stored in small aliquots at high concentration at –80°C. The working solution should be used up rapidly. Peptides should be dissolved in DMSO under nitrogen atmosphere if possible (as DMSO is a powerful oxidant) and stored in small aliquots at –80°C.

3.3. Flow Cytometry

3.3.1. Data Acquisition

The use of a four-color instrument permits the use of two T-cell parameters (e.g., CD3 *and* CD4 or CD3 *and* CD8) along with two activation markers (IFN-γ plus CD69 or IFN-γ plus interleukin [IL]-2). Fewer colors obviously

*Fifty-milliliter conical tubes can also be used if desired; other tubes/dishes were not tested in our lab.

reduce data resolution. We generally use IFN-γ to identify antigen-specific T cells; however, these protocols can, of course, be used with other cytokines.

3.3.2. Data Analysis

In this type of experiment, the main objective is the distinction and quantification of positive and negative events in a percentage of a reference population or in absolute numbers. Essentially, it is a question of correctly identifying and sizing *populations* rather than measuring varying degrees of expression of a particular parameter. Accurate quantification of intracelluar cytokines using quantification beads is not straightforward at this time, as it depends too much on the exact reproducibility of membrane permeabilization.

1. Following data acquisition, a region is set around the lymphocytes in an sideward scatter (SSC) vs forward scatter (FSC) scatter plot according to standard procedure (*see* **Fig. 1A**).
2. Next, the gated lymphocytes are analyzed in a CD3 vs CD4 (or CD8) dot plot, and a region is drawn around CD3+ lymphocytes (*see* **Fig. 1B**).
3. CD4 or CD8+ T cells may be gated in a CD4 (or CD8) vs IFN-γ dot plot. Importantly, activated T cells downregulate the T-cell receptor including CD4 or CD8 to variable degrees. This must be considered when gating the CD4 or CD8– population. The gate drawn in **Fig. 1C** (solid line) takes this into account; however, it does not include potentially activated (and downregulated) CD4– T-cells that do not produce IFN-γ. In this way, if the frequency of IFN-γ producers is estimated, the reference population (CD4– T cells) is made a little smaller than it really is, however, the overall mistake is generally very small. In contrast to this, drawing a gate as indicated by the dotted line in **Fig. 1C** would massively increase the CD4 reference population by including many non-CD4 events. Working out the frequency this way would lead to a very great mistake.
4. CD4 (or CD8) T cells can then be viewed in a CD69 versus IFN-γ (or isotype-control staining) dot plot (**Fig. 1D**). The limits between IFN-γ (isotype)-positive and -negative events can be set using the CD69-negative population, which should be entirely IFN-γ negative. In most cases, clearly distinct populations can be discriminated (i.e., CD69 positive/IFN-γ negative, CD69 positive/IFN-γ positive, CD69 negative/IFN-γ negative). A CD69-negative/IFN-γ positive population usually results from unspecific staining rather than IFN-γ production in CD69-negative T-cells. The CD69-positive/IFN-γ-positive or isotype-control-positive events are counted and expressed in percentage of the total respective CD4+ (or CD8+) T-cell population.
5. The percentages of positive events should be determined as follows:
   a. In the stimulated sample, the percent isotype-control-positive events are subtracted from the percent IFN-γ-positive events.
   b. In the unstimulated sample the percent isotype-control-positive events are subtracted from percent IFN-γ-positive events.

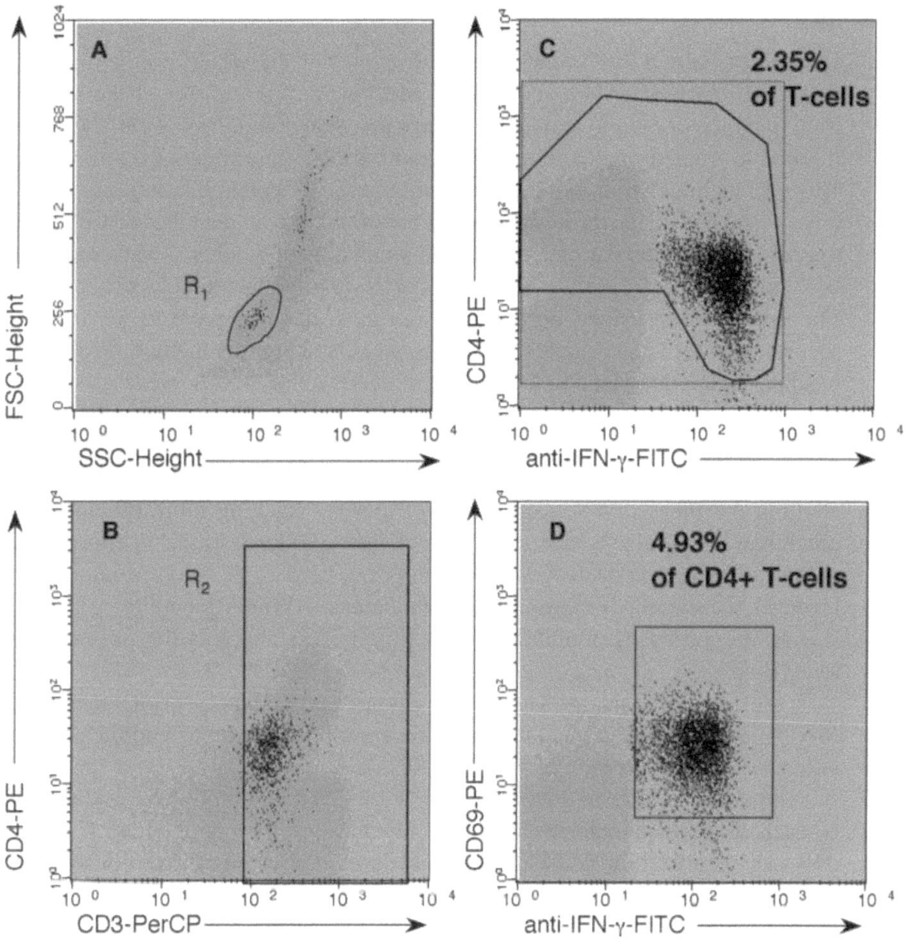

Fig. 1. Flow-cytometric analysis of intracellular cytokine staining. Freshly prepared PBMC from a CMV seropositive (IgG) healthy adult were stimulated ex vivo with a CMV-derived 15-amino-acid peptide. The diagrams show the gating procedure including a forward/side scatter gate (**A**), a CD3 vs CD4 gate (**B**), and a CD4 vs IFN-γ gate (**C**); 2.35% of CD3 T lymphocytes and 4.93% of the CD4 T lymphocytes (**D**) were identified to be INF-γ positive.

c. The IFN-γ true positive events in the control sample are subtracted from the respective true positive events in the stimulated sample.
d. The difference is the true frequency of T cells that respond to antigen specific stimulation with IFN-γ production.

4. Notes

1. The use of Falcon 2052 tubes allows to carry out the whole protocol using a single tube including data acquisition. Meanwhile, the transfer of the samples from one tube or dish to another is always associated with loss of cells.
2. If possible prepare small aliquots of reagents, such as one aliquot per stimulation. This will facilitate your work and reduces the risk of contamination.
3. It is advisable to place the antigens used for stimulation in the tubes or wells prior to adding cells. This is especially useful when a large number of samples has to be stimulated, because it allows you to begin stimulation simultaneously.
4. The protocols described herein are examples that will have to be modified according to the requirements of the researchers using this method. Prolonging incubation time may lead to higher background "noise" levels, whereas shortening it will reduce responses. In our experience, maximum responses with peptides and lysates are obtained with incubation times clearly less than 20 h. Some very useful additional information is provided by Waldrop et al. ***(21)***.
5. Blotting the tubes dry on a paper towel makes decanting more efficient. The combination of cell surface staining and intracellular staining is appropriate, unless antigens are destroyed by the fixative used (e.g., paraformaldehyde); however, the staining background is often increased. This is generally no problem if only the gating of populations is required. This shortens the protocol and saves time.
6. It is useful to prepare an "antibody cocktail" immediately prior to staining. Mix the antidodies you wish to use in the appropriate amounts and the required volume of the mix to each of your samples.
7. It is important to refer to the permeabilization and fixation procedures described by each antibody manufacturer. Most companies test their antibodies using their own permeabilization and fixation procedures, these may differ considerably between companies.
8. Isotype controls are of limited use and do not always have to be performed. Additional donor material is required for such controls. If there is no doubt that the antibody used for staining (e.g., IFN-γ really does stain IFN-γ [and not something else]), it is reasonable to use unstimulated samples stained for IFN-γ as the only control. The background when using PBMC is typically very small (**Fig. 2**). When stimulating with whole antigen (viral lysate) the responding cells are predominantly CD4+. When stimulating with peptides, the responding cells may be CD4+ or CD8+. This should be considered when choosing monoclonal antibodies for staining the samples.

Acknowledgments

The above protocols describe procedures as currently used in our lab; however, the following individuals have directly or indirectly contributed work, experience, and knowledge to these protocols: Louis Picker, Sharon Waldrop, and Christine Pitcher (OHSU, Portland, Oregon), and Skip Maino and Maria

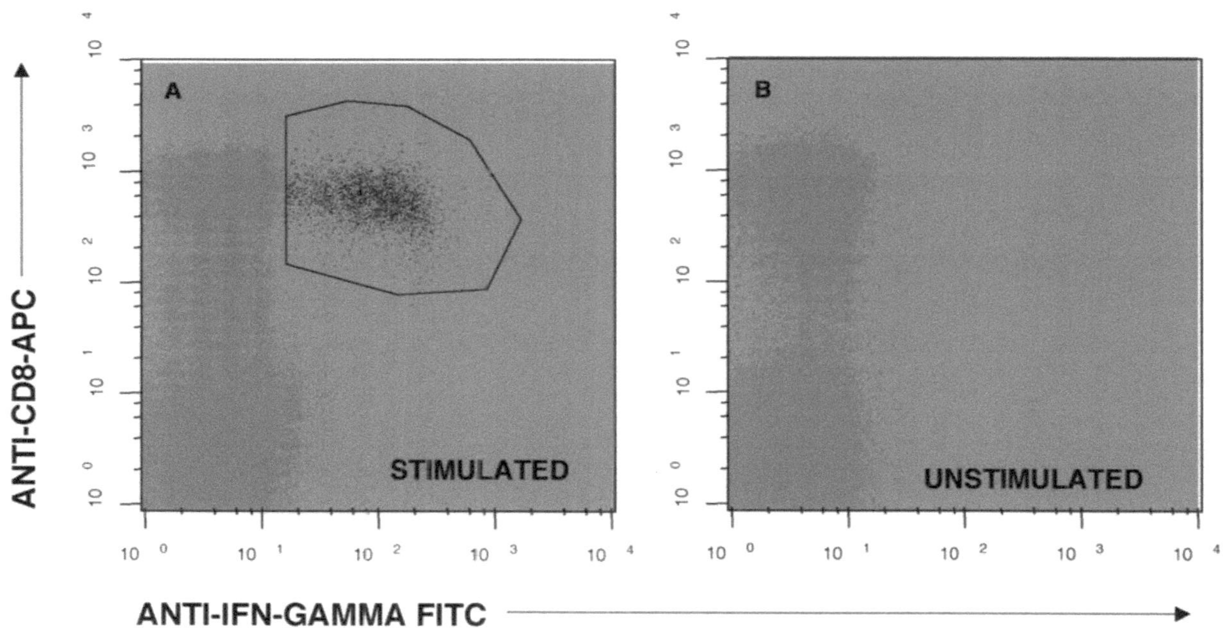

Fig. 2. Unstimulated samples serve as control for peptide stimulation. In **(A)**, freshly prepared PBMC from a CMV-seropositive (IgG) healthy adult were stimulated ex vivo with a single CMV-IE-1-derived 15-amino-acid peptide LSEFCRVLCCYVL for 6 h. Cytokine production was then visualized using anti-IFN-γ-FITC following permeabilization of the cell membrane; 1.32% of gated T lymphocytes were identified to be IFN-γ positive. **(B)** displays the unstimulated control.

Suni (BD Biosciences, San Jose, California). Louis Picker and Skip Maino in particular have perfected the technology of cytokine flow cytometry (CFC) and established the first CFC-based detection assay for antigen-specific T cells in peripheral blood.

References

1. Jung, T., Schauer, U., Heusser, C., Neumann, C., and Rieger, C. (1993) Detection of intracellular cytokines by flow cytometry. *J. Immunol. Methods* **159,** 197–207.
2. Kreft, B., Singer, G. G., Diaz-Gallo, C., and Kelley, V, R. (1992) Detection of intracellular interleukin-10 by flow cytometry. *J. Immunol. Methods* **156,** 125–128.
3. Maino, V. C. and Picker, L. J. (1998) Identification of functional subsets by flow cytometry: intracellular detection of cytokine expression. *Cytometry* **34,** 207–215.
4. Picker, L. J., Singh, M. K., Zdraveski, Z., Treer, J. R., Waldrop, S. L., Bergstresser, P. R., et al. (1995) Direct demonstration of cytokine synthesis heterogeneity among human memory/effector T cells by flow cytometry. *Blood* **86,** 1408–1419.
5. Latt, S. A., George, Y. S., and Gray, J. W. (1977) Flow cytometric analysis of bromodeoxyuridine-substituted cells stained with 33258 Hoechst. *J. Histochem. Cytochem.* **25,** 927–934
6. Lyons, A. B. and Parish, C. R. (1994) Determination of lymphocyte division by flow cytometry. *J. Immunol. Methods* **171,** 131–137.
7. Lyons, A. B. (1999) Divided we stand: tracking cell proliferation with carboxyfluorescein diacetate succinimidyl ester. *Immunol. Cell. Biol.* **77,** 509–515.
8. Lyons, A. B. (2000) Analyzing cell division in vivo and in vitro using flow cytometric measurement of CFSE dye dilution. *J. Immunol. Methods* **243,** 147–154.
9. Scheffold, A., Miltenyi, S., and Radbruch, A. (1995) Magnetofluorescent liposomes for increased sensitivity of immunofluorescence. *Immunotechnology* **1,** 127–137.
10. Scheffold, A., Assenmacher, M., Reiners-Schramm, L., Lauster, R., and Radbruch, A. (2000) High-sensitivity immunofluorescence for detection of the pro- and anti-inflammatory cytokines gamma interferon and interleukin-10 on the surface of cytokine-secreting cells. *Nat. Med.* **6,** 107–110.
11. Brosterhus, H., Brings, S., Leyendeckers, H., Manz, R. A., Miltenyi, S., Radbruch, A., et al. (1999) Enrichment and detection of live antigen-specific CD4(+) and CD8(+) T cells based on cytokine secretion. *Eur. J. Immunol.* **29,** 4053–4059.
12. Novak, E. J., Liu, A. W., Nepom, G. T., and Kwok, W. W. (1999) MHC class II tetramers identify peptide-specific human CD4(+) T cells proliferating in response to influenza A antigen. *J. Clin. Invest.* **104,** R63–R67.
13. Waldrop, S. L., Pitcher, C. J., Peterson, D. M., Maino, V. C., and Picker, L. J. (1997) Determination of antigen-specific memory/effector CD4+ T cell frequencies by flow cytometry: evidence for a novel, antigen-specific homeostatic mechanism in HIV-associated immunodeficiency. *J. Clin. Invest.* **99,** 1739–1750.
14. Kern, F., Surel, I. P., Brock, C., Freistedt, B., Radtke, H., Scheffold, A., et al. (1998) T-Cell epitope mapping by flow cytometry. *Nat. Med.* **4,** 975–978.

15. Kern, F., Faulhaber, N., Frömmel, C., Khatamzas, E., Prösch, S., Schönemann, C., et al. (2000) Analysis of CD8 T cell reactivity to cytomegalovirus using protein-spanning pools of overlapping pentadecapeptides. *Eur. J. Immunol.* **30,** 1676–1682.
16. Altman, J. D., Moss, P. A. H., Goulder, P. J. R., Barouch, D. H., McHeyzer-Williams, M. G., Bell, J. I., et al. (1996) Phenotypic analysis of antigen-specific T lymphocytes [published erratum appears in *Science* 1998, **280(5371)**:1821]. *Science* **274,** 94–96.
17. Murali-Krishna, K., Altman, J. D., Suresh, M., Sourdive, D., Zajac, A., and Ahmed, R. (1998) In vivo dynamics of anti-viral CD8 T cell responses to different epitopes. An evaluation of bystander activation in primary and secondary responses to viral infection. *Adv. Exp. Med. Biol.* **452,** 123–142.
18. Murali-Krishna, K., Altman, J. D., Suresh, M., Sourdive, D. J., Zajac, A. J., Miller, J. D., et al. (1998) Counting antigen-specific CD8 T cells: a reevaluation of bystander activation during viral infection. *Immunity* **8,** 177–187.
19. Tan, L. C., Gudgeon, N., Annels, N. E., Hansasuta, P., O'Callaghan, C. A., Rowland-Jones, S., et al. (1999) A re-evaluation of the frequency of CD8+ T cells specific for EBV in healthy virus carriers. *J. Immunol.* **162,** 1827–1835.
20. Appay, V., Nixon, D. F., Donahoe, S. M., Gillespie, G. M., Dong, T., King, A., et al. (2000) HIV-specific CD8(+) T cells produce antiviral cytokines but are impaired in cytolytic function. *J. Exp. Med.* **192,** 63–75.
21. Waldrop, S. L., Davis, K. A., Maino, V. C., and Picker, L. J. Normal human CD4+ memory T cells display broad heterogeneity in their activation threshold for cytokine synthesis. *J. Immunol.* **161,** 5284–5295.

7

Detection of Common Cytokine and Colony Stimulating Factor Gene Polymorphisms

Thomas Lehrnbecher and Stephen J. Chanock

1. Introduction and General Considerations

With the completion of the first map of the entire human genome, it is estimated that there are approx 35,000 genes, which encode for translated RNA and protein products *(1,2)*. The difference between any two human genomes is estimated to be less than 0.1% *(3–5)*. However, in light of the magnitude of the genome (approx 3.2 billion bases), the number of variations is still very large *(6)*. The most common variation is the single-nucleotide polymorphism (SNP), which, by definition, has a frequency of greater than 1%. Therefore, it is possible that the number of SNPs in a given individual could number in the millions *(6)*. Variable nucleotide repeats (useful for microsatellite studies), deletions, and substitutions are rarer in frequency, but still useful tools for genetic analysis (e.g., linkage studies and whole-genome scans). SNPs occur at an interval of approximately once every 1.3 kb of DNA *(3,7–9)*. In the investigation of polymorphisms, particular emphasis is directed not only at the coding region, but also at 5′ and 3′ regulatory regions of the candidate genes, such as the promoter which plays an important part in controlling gene transcription. The 3′-untranslated region (3′UTR) can determine RNA half-life or ribosomal translation of RNA species. Nonsynonymous SNPs (viz. those that change the coding amino acid) constitute less than 5% of SNPs in coding regions. Variations in introns or exons are probably less frequent than variations in intergenic regions *(8,9)*.

By definition, polymorphisms do not significantly alter expression or function of a gene, unlike highly penetrant mutations [i.e., the sickle cell mutation

From: *Methods in Molecular Biology, vol. 215: Cytokines and Colony Stimulating Factors: Methods and Protocols*
Edited by: D. Körholz and W. Kiess © Humana Press Inc., Totowa, NJ

in the β-globin gene ***(10)***]. The phenotype of a common polymorphism barely affects the gene, or, in many cases, not at all. However, under stress conditions, such as a primary immunodeficiency, it has been shown that polymorphisms can act as modifying genes for specific outcomes in monogenic disorders ***(11,12)***. In a similar manner, common polymorphisms contribute to the susceptibility to develop a wide range of complex disorders ***(13)***.

Genetic diversity provides a rich resource for studying evolutionary and biological questions. Genetic polymorphisms, such as SNPs, can be useful tools for candidate gene association studies, which seek to determine the contribution of one or more variant SNPs to either disease susceptibility or outcome. These studies may provide a foundation for identifying and applying genetic risk factors in clinical medicine. However, a number of challenges must be met before implementation in clinical medicine can be confidently achieved. These include addressing complex ethical questions, refinement of biostatistical methods, and development of cost-effective, high-throughput platforms.

The explosion of knowledge, specifically in the form of possible SNPs, has created a daunting challenge, the curation of polymorphisms in a pathway or field (i.e., cell or immunology cycle). Public efforts have been promulgated to identify and validate SNPs by searching public databases for predicted SNPs ***(14,15)*** as well as sequences generated as part of the public genome project, known as the SNP Consortium (http://snp.cshl.org/). The goal of the Genetic Annotation Initiative of the Cancer Genome Anatomy Project of the National Cancer Institute (GAI, http://lpg.nci.nih.gov/GAI) is to systematically identify and catalog common polymorphisms in several thousand genes of interest to cancer biology and immunology by resequence technology ***(16)***. The GAI has also developed Web-based tools to search public databases for SNPs. Of possible interest to the immunology community is a website, known as I-SNP (http://www-dcs.nci.nih.gov/pedonc/ISNP/), which has collected PubMed references and polymorphisms reported in genes of significance in immune function. **Table 1** displays a set of public sites, useful for identifying polymorphisms in known genes.

Genetic diversity in immune genes is considered a significant determinant of host response to environmental and pathogen challenges. Many of the already identified polymorphisms in cytokines and other genes of innate immunity most likely reflect responses to significant challenges ***(17)***. There is a growing body of evidence to suggest that variations in key cytokines can influence host response, which, in turn, affect disease outcome and susceptibility to a range of disorders.

Table 1
Websites for Selecting SNPs in Candidate Genes

Name of website	Type of resource
PubMed (www. ncbi.nlm.nih.gov/entrez/query.fcgi)	Published literature (search by gene) Resource for searching published data
dB SNP (www.ncbi.nlm.nih.gov/SNP/)	NCBI database of deposited SNPs Central repository of SNPs
The SNP Consortium (snp.cshl.org/)	Public database of predicted SNPs Deposited in db-SNP
LocusLink (www.ncbi.nlm.nih.gov/LocusLink)	NCBI database linked to RefSeq Known variants linked to reference sequence
Cancer Genome Anatomy Project-GAI (lpg.nci.nih.gov/GAI/)	NCI-based SNP Discovery Project Includes gene lists, tools for SNP analysis
SNP pipeline (lpgws.nci.nih.gov:82/perl/snp/snp_cgi.p)	CGAP–GAI search of EST/Unigene Tool for analyzing EST sequences for SNPs
Leelab SNP Database (www.bioinformatics.ucla.edu/snp/)	UCLA search of EST/Unigene Tool for analyzing EST sequences for SNPs
HG Base (Hgbase.cgr.ki.se/)	International database Repository for SNP
Imunology–SNP Database (www-dcs.nci.nih.gov/pedonc/ISNP/)	Curated collection of immunologically significant SNPs SNP database of known genes with SNPs
University of Utah Genome Center GeneSNPs (http://www.genome.utah.edu/genesnps_old/)	Curated collection of SNPs derived from public databases

1.1. Genetic Association Studies

The genetic association study is designed to evaluate the contribution of one or more polymorphisms to well-defined clinical end points, such as disease susceptibility or outcome *(18)*. In practice, these studies compare the distribution of genotypes in two groups, with one serving as a "control." The candidate gene approach selects variants on the basis of biological effects, either demonstrated or predicted, and the findings usually correspond with a plausible hypothesis or mechanism. Pilot studies are often prone to false positives, but this is acceptable because genetic association studies must be confirmed in several populations. Furthermore, findings are often population-specific, reflecting differences between geographically defined groups. Size is also an important determinant of statistical significance and confidence limits *(19)*. Because many chromosomal regions contain dense clusters of genes in close proximity to one another, an observed association might not be the result of the studied candidate gene, but, instead, to a separate loci in linkage disequilibrium. This has been the central focus of whole-genome scans, which seek to identify a region of a chromosome by linkage analysis, an approach that has been highly successful in the investigation of highly penetrant, rare disorders, but, for association studies, it is problematic. The field is quickly moving toward the utilization of haplotypes and possibly "intelligent" scans, which utilize collections of SNPs restricted to known genes in a pathway or that reside on a particular chromosome. To this end, large collections of polymorphisms will provide a foundation for investigating the role of human diversity in mapping complex diseases, the role of modifying genes in monogenetic disorders (such as cystic fibrosis), and evolutionary questions.

1.2. Statistical Considerations

In order to keep pace with the increasing complexity in the analysis of association studies, informatic tools and statistical support are evolving to handle immense data sets. Many have argued that classical correction factors in association studies are not practical, especially in pilot studies when a prior hypothesis has not been well established *(20)*. The statistical analysis and criteria of statistical significance become increasingly difficult when combinations of genotypes are examined. In the common disease–common variant model, multiple genetic loci probably contribute to pathogenesis *(21,22)*. Thus, the possibility of searching with collections of genes will require modifications of classic correction factors *(20)*.

1.3. Technical Considerations

The classic gel-based technologies for polymorphism detection (e.g., allele-specific amplification, restriction fragment length polymorphism [RFLP], artificial introduction of restriction sites, allele-specific oligonucleotide hybridization [ASO], and single-strand conformational polymorphism [SSCP]) are predicated on amplification of DNA in a polymerase chain reaction (PCR). Many laboratories have applied automated sequence technology for detection of variation in sequences facilitated by the use of robotics for increasing throughput. Newer technologies such as real-time PCR (Taqman®), chip-based technologies, and mass spectrometry have already begun to revolutionize the field and, hopefully, will permit the simultaneous analysis of multiple genes within large patient populations *(7)*. These high-throughput platforms are efficient, but, currently, very expensive and labor intensive; however, they have the advantage of multiplexing SNPs, thus allowing large collections of SNPs to be interrogated.

1.4. Key Genes in Cytokine and Hematopoietic Growth Factor Pathways

Cytokines and growth factors are regulatory proteins of the immune response. Disruptions in their expression pattern and/or function have been implicated in a wide spectrum of diseases, including autoimmune disorders, cancer, infectious diseases, and neurodegenerative processes. **Table 2** highlights selected polymorphisms in cytokines and growth factors. Many of them have been associated with clinical endpoints. Unlike expression array studies, which can evaluate mRNA profiles for any given cell type, analysis of genotypes or alleles requires specific amplification of the region immediately adjacent to the nucleotide changes of interest. Therefore, the investigator must design specific assays that amplify the region of interest. It is beyond the scope of this chapter to present the method for choosing oligonucleotides for amplification. Instead, we will present two representative assays—one that interrogates two common variants, a promoter SNP in *IL6*, and the other is a variable number of tandem repeats (VNTR) in the interleukin1 receptor antagonist (*IL1RN*) gene.

2. Materials

2.1. Isolation of Genomic DNA

1. PUREGENE® DNA Isolation Kit (Gentra Systems, Inc, Minneapolis, MN)
2. TE buffer (10 m*M* Tris-HCl [pH 8.0], 1 m*M* EDTA [pH 8.0]).
3. Isopropanol (2-propanol).
4. 70% Ethanol.
5. Spectrophotometer.

2.2. PCR Amplification

1. Enzyme and dNTPs originate from Amersham Pharmacia Biotech (includes *Taq* DNA polymerase, reaction buffer, and nucleotides [dNTP, 10 m*M* each dATP, dCTP, dGTP, dTTP]).

Table 2
Selected Polymorphisms of Cytokines and Growth Factors

Cytokine/ growth factor (chromosome)	Polymorphism	Primers (5′ → 3′; F = forward; R = reverse)	Annealing temp. (°C)	Detection method	Ref.
TNF (6p21.3)	Promoter (G-376A)	F: CAA ACA CAG GCC TCA GGA CT R: CCT ATT GCC TCC ATT TCT TTT G	61	Sequencing	***23***
	Promoter (G-308A)	F: CAA AAG AAA TGG AGG CAA TAG GTT TTG AGG GCC AT R: AGG GCG GGG AAA GAA TCA TTC AAC CAG CGG AAA AC	63	*Nco*I digest, 3% agarose	***24***
		F: TCT CGG TTT CTT CTC CAT CG R: ATA GGT TTT GAG GGG CAT GG/A	62	Sequence-specific PCR, 2% agarose	***25***
		F: TCC TGC ATC CTG TCT GGA AG R: GTC TTC TGG GCC ACT GAC TG	59	Heteroduplex analysis	***26***
		F: ACT CAA CAC AGC TTT TCC CTC CA R: TCC TCC CTG CTC CGA TTC CG	66	ASO	***27***
	Promoter (G-238A)	F: AAG GAA ACA GAC CAC AGA CCT G R: GGA TAC CCC TCA CAC TCC CC	63	Sequencing	***23***
		F: GTT CAG CCT CCA GGG TCC TAC ACA R: GGG ATT TGG AAA GTT GGG GAC ACA	58	Heteroduplex analysis	***26***
LTA (lymphotoxin) (6p21.3)	Intron 1 (A252G)	F: CCG TGC TTC GTG CTT TGG ACT A R: AGA GCT GGT GGG GAC ATG TCT G	64	*Nco*I digest, 2% agarose	***24***
IL1A (2q14)	Promoter (G-889T)	F: GGG GGC TTC ACT ATG TTG CCC ACA CTG GAC TAA R: GAA GGC ATG GAT TTT TAC ATA TGA CCT TCC ATG	58	*Nco*I digest, 2% agarose	***24***

	Intron 5 [Dinucleotide repeat $(AC)_{19}$]	F: CCT GCC TAG TGA GTG TGG AAG R: GTG TTG ATG TAG ATT GTG TGT GC	57	Sequencing	***28***
IL1B (2q14)	Promoter (C-511T)	F: TGG CAT TGA TCT GGT TCA TC R: GTT TAG GAA TCT TCC CAC TT	55	*Ava*I digest, 9% PAGE	***29***
	Exon 5 (T3953C) Tyr ↔ Lysin	F: CTC AGG TGT CCT CGA AGA AAT CAA A R: GCT TTT TTG CTG TGA GTC CCG	58	*Taq*1 digest, 3% agarose	***24***
		F: GAG GCC TGC CCT TCT GAT T R: CGG AGC GTG CAG TTC AGT	60	Heteroduplex analysis	***26***
IL2 (4q26)	3′ Flanking region (dinucleotide repeat)	F: AAA GAG ACC TGC TAA CAC A R: CCT ATG TTG GAG ATG TTT AT	55	Size fractionation, 6% polyacrylamide gel	***30***
IL3 (5q31.1)	Promoter (T-68C)	F: CAT GGA TGA ATA ATT ACG TCT GTG G R: GGA GCA CGA CGG GCA GGC GGC TCA T	58	Size fractionation, polyacrylamide gel	***31***
IL1RN (2q14.2)	Intron 2 (VNTR)	F: CTC AGC AAC ACT CCT AT R: TCC TGG TCT GCA GGT AA	58	Size fractionation, 2% agarose	***24***
IL4 (5q31.1)	Promoter (C-498T)	F: GCC TCT ATG CAG AGA AGG AGC CCC R: CCT GTG AAA TCA GAC CAA TAG G	58	Size fractionation, polyacrylamide gel	***31***
	Promoter (C-285T)	F: ACA AAT TCG GAC ACC TGC R: GTG AGG CAA TTA GTT TAT CAG	58	*Mae*I digest, 3% agarose	***32***
	Promoter (A-81G)	F: ACA AAT TCG GAC ACC TGC R: GTG AGG CAA TTA GTT TAT CAG	58	*Tsp45*I digest, 3% agarose	***32***
IL5 (5q31.1)	Promoter (T-746C)	F: GCT CAT GAA CAG AAT ACG TA R: GAA GGT ATT GGC TCA TAG TAC	52	*Rsa*I digest, 3% agarose	***33***
IL6 (7p21)	Promoter (G-597A)	F: AAG TAA CTG CAC GAA ATT TGA GGA/G R: TGC AAT GTG ACG TCC TTT AGC ATC/G	65	Sequence-specific PCR, 1% agarose	***34***
IL6 (7p21)	Promoter (G-572C)	F: GGC CAG GCA GTT CTA CAA CAG CCG/C R: TGC AAT GTG ACG TCC TTT AGC ATC/G	65	Sequence-specific PCR,1% agarose	***34***
	Promoter (G-174C)	F: TTG TCA AGA CAT GCC AAG TGC R: CAG AAT GAG CCT CAG AGA CAT CTC C	67	*Nla*III digest, 3% agarose gel	***23***
		F: GCT TCT TAG CGC TAG CCT CAA TG R: TGG GGC TGA TTG GAA ACC TTA TTA	60	Heteroduplex analysis	***26***

(continued)

Table 2 *(Continued)*

Cytokine/ growth factor (chromosome)	Polymorphism	Primers (5′ → 3′; F = forward; R = reverse)	Annealing temp. (°C)	Detection method	Ref.
	3′ Flanking region (VNTR)	F: GCA ACT TTG AGT GTG TCA CG R: TGA CGT GAT GGA TGC AAC AC	60	Size fractionation, 7% polyacrylamide	***35***
IL8 (4q13–q21)	Promoter (A-352T)	F: GTG GAA CTG ATT TCT ATG TGA A R: CCA CAA TTT GGT GAA TTA TCA AT/A	65	Sequence-specific PCR, 1.5% agarose	***36***
	Intron 1 (G293T)	F: CAT TCT CAC TGT GTG TAA ACA T R: ACG TTA AAT ATA TGC ATG CTA CC/A	65	Sequence-specific PCR,1.5% agarose	***36***
	Intron 1 (T678C)	F: AGT TGA GCA AAA GGT AAC TCA GA R: GTC ATA ACT GAC AAC ATT GAA CA/G	65	Sequence-specific PCR,1.5% agarose	***36***
IL9 (5q31.1)	Promoter (A-351C)	F: GCA ACC TCA GTC TTA CTA TGC R: GTT GAG TAC TGA AAT GCT GAA GG	58	Size fractionation, polyacrylamide gel	***31***
IL10 (1q31–q32)	Promoter (T-3575A)	F: CTG TAG AAT GCA CCC TCC AAA ATC T R: TTT GAG ACA GAG TCT CGC TCT G	Not specified	Sequencing	***37***
	Promoter (G-2849A)	F: CTG TAG AAT GCA CCC TCC AAA ATC T R: TTT GAG ACA GAG TCT CGC TCT G	Not specified	Sequencing	***37***
	Promoter (A-2776G)	F: CTG TAG AAT GCA CCC TCC AAA ATC T R: TTT GAG ACA GAG TCT CGC TCT G	Not specified	Seuencing	***37***
	Promoter (C-2763A)	F: CTG TAG AAT GCA CCC TCC AAA ATC T R: TTT GAG ACA GAG TCT CGC TCT G	Not specified	Sequencing	***37***
	Promoter (G-1082A)	F: ATC CAA GAC AAC ACT ACT AA R: TAA ATA TCC TCA AAG TTC C	55.5	ASO	***27***
		F: AAT CCA AGA CAA CAC TAC TAA GGC R: CTG GAT AGG AGG TCC CTT AC	57	Heteroduplex analysis	***26***
		F: CAC CAC AAA TCC AAG ACA ACA CTA CTA AG R: AAA TAA CAA GGA AAA GAA GTC AGG ATT C	57	5′ Nuclease-based PCR	***38***

	Promoter (C-819T)	F: ATC CAA GAC AAC ACT ACT AA R: TAA ATA TCC TCA AAG TTC C	55.5	ASO	***27***
		F: TAC AGT AGG GTG AGG AAA CC R: GGT AGT GCT CAC CAT GAC CC	57	Heteroduplex analysis	***26***
	Promoter (C-592A)	F: GAA ATC CGG GTA AAG GAG CC R: AGT TCC CAA GCA GCC CTT CC	62	Heteroduplex analysis	***26***
IL11 (19q13.3–q13.4)	5′ Flanking region (dinucleotide repeat)	F: GTC TCT GTC TCA GTC TGT CC R: GAC GGA AAG GCA GAG AAA GG	60	Size fractionation, 6% polyacrylamide	***39***
IL13 (5q31)	Promoter (A-1512C)	F: CAA CCG CCG CGC CAG CGC CTT CTC R: CCG CTA CTT GGC CGT GTG ACC GC	54	*Bst*UI digest, 3% agarose	***33***
	Promoter (C-1112T)	F: GGA ATC CAG CAT GCC TTG TGA GG R: GTC GCC TTT TCC TGC TCT TCC CGC	54	*Bst*UI digest, 3% agarose	***33***
	Exon 4 (G2044A) Arg ↔ Gln	F: CTT CCG TGA GGA CTG AA TGA GAC GGT C R: GCA AAT AAT GAT GCT TTC GAA GTT TCA GTG GA	55	*Nla*IV digest, 3% agarose	***33***
G-CSFR (1 p35–34.3)	Intron 14 (G-A)	F: AGA GGC AGG TGA GCA ACA GC R: GGG ACT GAC TTT GAA TCC CCT GGC C	65	*Bal*1 digest, 4% agarose	***40***

Abbreviations: TNF, tumor necrosis factor; IL, interleukin; G-CSFR, granulocyte colony stimulating factor receptor; VNTR, variable number of tandem repeats; ASO, allele-specific oligohybridization.

2. Specific oligonucleotides (primers) for the PCR reaction must be generated based on the adjacent sequence. Many commercial groups provide the service of synthesizing high-grade oligonucleotides (*see* **Note 1**).
3. Thermocycler. The PCR thermal cycler is required for both PCR amplifications and direct sequencing. To facilitate the high throughput of samples, a thermal cycler capable of running 96 or 384 samples is desirable. We use the MultiCycler PTC 220 Dyad (Biozym, Hessisch Oldendorf, Germany).

2.3. Electrophoresis

1. Horizontal electrophoresis system (small unit for 10 samples, large unit for 96 samples + 8 DNA size standards).
2. High-voltage power supply.
3. Agarose (*see* **Note 2**).
4. Tris-borate buffer (TBE; 5X: 54 g Tris base, 27.5 mL boric acid, 20 mL of 0.5 *M* EDTA [pH 8.0]).
5. Ethidium bromide (10 mg/mL, Amersham Pharmacia Biotech; stock solutions should be stored in light-tight containers at room temperature).
6. DNA size standard (100-bp ladder, Amersham Pharmacia Biotech).
7. 5X Loading buffer: 0.05% orange G, 20 m*M* EDTA (pH 8.0), 20% Ficoll (Type 400) in water (store at room temperature).
8. Ultraviolet (UV) transilluminator with documentation system.

2.4. Sequencing

1. Shrimp alkaline phosphatase and exonuclease I (Amersham Pharmacia Biotech).
2. Thermo Sequenase Cycle Sequencing Kit (Amersham Pharmacia Biotech; includes Thermal Sequenase DNA polymerase 4 U/μL; reaction buffer; dGTP termination master [7.5 μ*M* dATP, dCTP, dGTP, dTTP], stop solution).
3. [α-^{33}P]ddNTPs (Amersham Pharmacia Biotech). (**Caution:** To work with radioactive materials, laboratory personnel must be properly trained in the effective use of special precaution, which includes shields, special storage boxes, and disposal.)
4. Sequencing electrophoresis system (includes gel apparatus, glass plates, spacer set, comb).
5. High-voltage power supply.
6. Glycerol-tolerant buffer (20X glycerol-tolerant buffer: 216 g Tris-HCl base, 72 g taurine, 4 g EDTA, double-distilled water [ddH_2O] ad 1000 mL [Amersham Pharmacia Biotech]).
7. Acrylamide/bis-acrylamide 19:1 solution 40% (w/v) (Amersham Pharmacia Biotech).
8. Urea (ultrapure) (Amersham Pharmacia Biotech).
9. 10% Ammonium persulfate (APS; ammonium persulfate 1 g [Amersham Pharmacia Biotech] and H_2O to 10 mL; the solution may be stored at 4°C for several weeks).

10. TEMED (*N,N,N′,N′*-tetramethylethylenediamine) (Amersham Pharmacia Biotech).
11. Acrylease™ Nonstick Plate Coating (Stratagene, La Jolla, CA).
12. Tape.
13. Gel-loading duck bill tips (Anachem Ltd, Luton, UK).
14. 3MM Whatman paper.
15. Saran-Wrap.
16. Autoradiography films (Kodak).
17. Autoradiography cassettes.
18. Vacuum gel dryer system.
19. Film processor.

3. Methods

3.1. Isolation of Genomic DNA

Screening for polymorphisms is performed on genomic DNA (gDNA), which can be purified from fresh, frozen, or dried whole blood or bone marrow. Unlike RNA, gDNA is relatively stable and, thus, can be extracted from archived material as well as fresh samples. Genomic DNA can be isolated from whole blood or bone marrow by standard methodologies described in the literature. The salt precipitation extraction method used by the PUREGENE® DNA Isolation Kit yields large quantities of moderate grade of DNA, suitable for PCR amplification (*see* **Notes 3–6**). Phenol–chloroform extraction and ethanol precipitation are critical for high-quality gDNA ***(41)***.

3.1.1. Isolation of Genomic DNA by PUREGENE® DNA Isolation Kit

1. To isolate DNA from human whole-blood samples using the PUREGENE® DNA Isolation Kit, add 300 μL blood to a 1.5-mL-tube containing 900 μL RBL lysis solution provided by the manufacturer.
2. Mix gently and incubate 10 min at room temperature. After centrifugation for 20 s at 13,000*g* in a microfuge, remove supernatant with a micropipet, leaving behind the visible white pellet and 10–20 μL of residual liquid.
3. Vortex the tube vigorously to resuspend the white blood cells in the residual supernatant, add 300 μL of cell lysis solution and pipet up and down to resuspend the cells.
4. After RNase treatment (for this step, add 1.5 μL RNase A solution, mix the sample by inverting 25 times, and incubate at 37°C for 15 min), cool the samples down to room temperature and add 100 μL protein precipitation solution.
5. Vortex vigorously for 20 s and centrifuge at 13,000*g* for 3 min, so that the precipitated proteins will form a tight, dark brown pellet.
6. Pour the supernatant containing the DNA into a clean 1.5-mL tube containing 300 μL isopropanol.
7. Mix the sample gently by inverting the tube 50 times and centrifuge at 13,000*g* for 1 min. The DNA will be visible as a small white pellet.

8. Pour off the supernatant, add 300 µL of 70% ethanol and wash the DNA pellet by inverting the tube several times.
9. Centrifuge at 13,000*g* for 1 min and carefully pour off the ethanol. Air-dry 15–30 min before adding 50 µL of TE buffer. The gDNA can be stored at 4°C.

3.1.2. Isolation of Genomic DNA by Phenol–Chloroform Extraction

1. For phenol–chloroform extraction, add 0.8 mL of 1X SSC buffer (20X SSC: 3 *M* NaCl, 300 m*M* sodium citrate, pH 7.0) to 1 mL blood.
2. Centrifuge for 1 min at 13,000*g* in a microfuge and remove 1 mL of the supernatant.
3. Add 1 mL of 1X SSC buffer, vortex, centrifuge as in **step 2**, and remove all of the supernatant.
4. Add 375 µL of 0.2 M sodium acetate, vortex briefly, mix the sample with 25 µL of 10% SDS (sodium dodecyl sulfate, Amersham Pharmacia Biotech) and 5 µL of proteinase K (20 mg/mL H_2O), and incubate for 1 h at 55°C. Add 120 µL H_2O-saturated phenol–chloroform–isoamyl alcohol and vortex for 30 s. **Caution:** Because phenol and chloroform are hazardous chemicals, wear protective apparel and work with these reagents in a chemical fume hood.
5. Centrifuge the sample for 2 min at 13,000*g*.
6. Carefully remove the aqueous layer to a new 1.5-mL tube, add 1 mL of cold 100% ethanol, mix, and incubate for 15 min at –20°C.
7. Centrifuge for 2 min at 13,000*g*, pour off the supernatant, and resuspend pellet in 180 µL TE buffer.
8. Incubate for 10 min at 55°C. Mix the sample with 20 µL of 2 *M* sodium acetate.
9. Add 500 µL of cold 100% ethanol and centrifuge for 1 min at 13,000*g*.
10. Decant the supernatant and wash the DNA pellet with 300 µL of 70% ethanol by inverting the tube several times.
11. Centrifuge at 13,000*g* for 1 min and carefully pour off the ethanol. Air-dry 15–30 min before adding 50 µL of TE buffer.
12. Measure the absorbance of the DNA at 260 nm and 280 nm. An OD_{260} of 1 corresponds to approximately a concentration of double-stranded DNA equivalent to 50 µg/mL. A final concentration of gDNA between 20 and 100 ng/µL is useful. The quotient $OD_{260}:OD_{280}$ should be greater than 1.75. A lower ratio is an indication that significant protein contamination is present and could undermine amplification. Phenol–chloroform extraction should be performed if the 260/280 ratio is especially low.

3.2. IL-6 Promoter Polymorphism (G-174C)

3.2.1. PCR Amplification

A common polymorphic site has been identified at position –174 of the human interleukin-6 *(IL6)* promoter, a transition G → C ***(42)***. This variant is interesting because of the functional implications, an alteration in gene

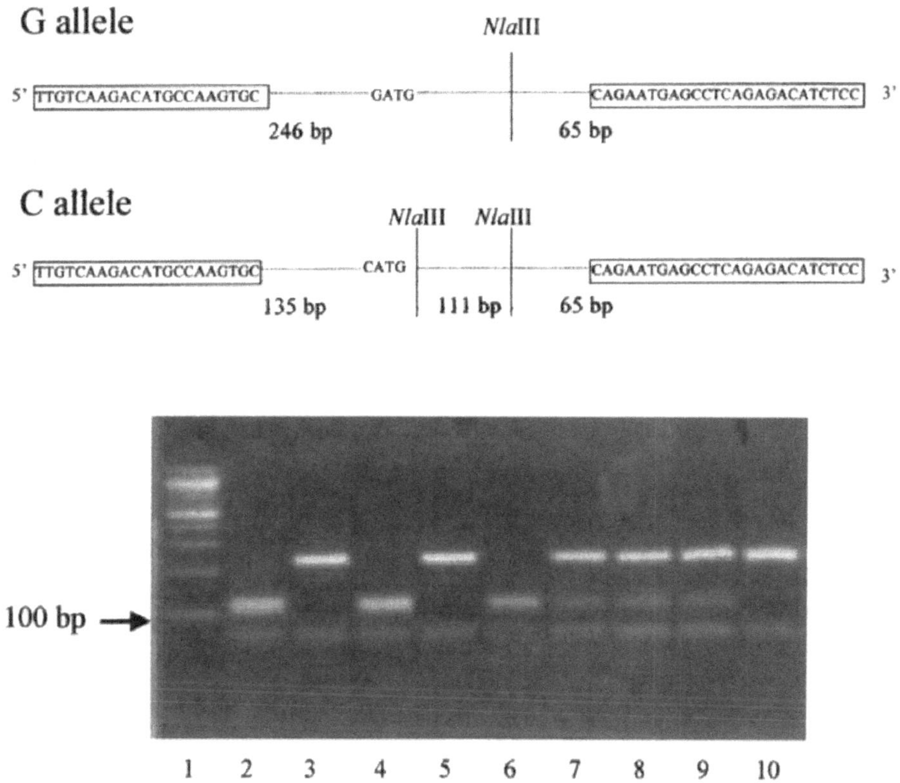

Fig. 1. The G→C *IL6 promoter* polymorphism at position –174. After digestion of the 311-bp PCR product with *Nla*III, allele G gives two fragments (65 bp and 246 bp). On the 3% agarose gel, homozygotes for allele G are seen on lanes 3, 5, and 10. The allele C creates a second recognition site for *Nla*III and the fragments are 65 bp, 111 bp, and 135 bp in length. Homozygotes for allele C are seen on lanes 2, 4, and 6 (the fragments 111 bp and 135 bp in length are not separated). In heterozygotes, *Nla*III digest gives fragments of 65 bp, 111 bp, 135 bp, and 246 bp. Heterozygotes are shown on lanes 7–9. The 100-bp ladder is loaded on lane 1.

expression. The *IL6* promoter SNP has been studied in a wide spectrum of genetic association studies and been shown to be significant in autoimmune disorders (inflammatory bowel disease), neurodegenerative diseases (multiple sclerosis), bone density/mineralization, and inflammatory/malignant disorders (i.e., Kaposi sarcoma in human immunodeficiency virus [HIV]-infected men). Moreover, the distribution of this SNP varies greatly among populations studied to date. The variant C allele can easily be detected by RFLP because the C creates a second recognition site for the restriction enzyme *Nla*III (*see* **Fig. 1**).

After PCR amplification and restriction digest, the polymorphism can be screened by gel electrophoresis (*see* **Fig. 1**).

To facilitate rapid throughput, it is recommended that multiple PCR reactions (e.g., in a 96-well microtiter plate) be performed simultaneously. Therefore, it is preferred to generate a PCR master mix containing all common components, except template DNA (*see* **Notes 7** and **8**). Aliquot 25 μL of the master mix into the 96-well microtiter plate.

	One reaction	Master mix for a 96-well plate
10X reaction buffer	2 μL	200 μL
dNTP (100 m*M*)	0.5 μL	50 μL
Primer IL6-F*	0.5 μL	50 μL
Primer IL6-R†	0.5 μL	50 μL
H_2O	21 μL	2100 μL
Taq DNA polymerase	0.25 μL	25 μL

*IL6-F: 5′ TTG TCA AGA CAT GCC AAG TGC 3′ (stock: 100 ng/μL).
†IL6-R: 5′ CAG AAT GAG CCT CAG AGA CAT CTC C 3′ (stock: 100 ng/μL)

1. Between 20 and 50 ng of gDNA should be used as the template.
2. Seal the 96-well microtiter plate and place it into the thermal cycler. After an initial denaturation step for 10 min at 95°C, run 25 cycles of 95°C for 1 min, 63°C for 1 min, and 72°C for two min (*see* **Note 9**). In this assay, it is critical to perform a 10-min extension at 72°C.

3.2.2. Digestion of PCR Products

1. For the restriction digest of the PCR product, the following is suitable (*see* **Note 10**):

PCR product	8 μL
H_2O	10 μL
10X reaction buffer	2 μL
*Nla*III	1 μL

   *Nla*III is purchased from New England Biolabs, MA.
2. Incubate at 37°C for at least 2 h.

3.2.3. Electrophoresis

After PCR amplification and restriction digest, the genotype of the *IL6 promoter* polymorphism is directly determined by size fractionation.

1. Add 20 g of agarose (SFR, Amresco, Solon, OH) to 100 mL of 0.5X TBE in a 250-mL Erlenmeyer flask. Heat the solution in a microwave until the solution just starts to boil. Allow solution to cool. It is important to stir agarose while cooling to prevent “lumps” from forming. Once agarose has cooled to 65°C, add ethidium bromide to a final concentration of 0.5 μg/mL and mix gently (wear gloves because ethidium bromide is highly mutagenic).

2. Pour the molten agarose solution into the casting tray. Immediately push any air bubbles to edges of the template using a pipet tip. Insert comb and allow gel to polymerize for approx 1 h to room temperature.
3. Add 1000 mL of 0.5X TBE (running buffer) to the electrophoresis tank. Place gel in tank and carefully remove combs from gel. Ensure that the gel is covered by buffer to at least a depth of 3–5 mm (*see* **Note 11**).
4. Transfer 8 µL of each digested PCR product into a 96-well microtiter plate. Ensure that product is in each well. Then, add 2 µL of 5X loading buffer. Using a multichannel pipet, load carefully 10 µL of sample/loading buffer into each well of the gel, starting with the fifth well of each row. The first four wells are reserved for the size marker, followed by a negative control (without DNA) and known genotypes of homozygous and heterezygous control samples (positive controls).
5. Place the lid of the electrophoresis system on top of the electrophoresis tank, connect the electrodes to the power supply, and electrophorese the samples at a voltage of 1–5 V/cm (measured as the distance between the electrodes). Run the gel until the bromphenol blue has migrated the appropriate distance through the gel (bromphenol blue migrates at approximately the same rate as linear double-stranded 300 bp in length).
6. Once electrophoresis is complete, remove gel from the tank and place it under UV light. Photograph if separation of bands is clear; otherwise, run the gel longer (*see* **Fig. 1**).

3.2.4. Sequencing

For quality control, it is often useful to confirm the validity of the results by direct sequence analysis. Nonradioactive, fluorescence-based automated sequencing systems, such as the MegaBACE DNA Sequencing System (Amersham Pharmacia Biotech) or ABI Prism (Applied Biosystems), are quick and accurate, but expensive. We present manual radioactive sequencing, which is still used in many laboratories. It is possible to use the same oligonucleotides to amplify a fragment for sequence analysis. However, it is preferable to design at least one oligonucleotide to include a 5′ universal tag sequence, which will serve as the site for priming the sequence reaction. Often, a bacteriophage M13 universal sequence is used for this purpose.

3.2.4.1. Sequencing Reaction

1. For sequencing PCR products, *pretreat* DNA with exonuclease I and shrimp alkaline phosphatase. This removes residual single-stranded primers and remaining dNTPs from the PCR mixture and prevents any interference with the PCR reaction.

PCR product	5 µL
Exonuclease I (10 U/µL)	1 µL
Shrimp alkaline phosphatase (2 U/µL)	1 µL

Mix and incubate at 37°C for 15 min. Inactivate enzymes by heating to 80°C for 15 min. It is convenient to perform these steps in the thermal cycler.

2. For the preparation of the *reaction mixture*, prepare an appropriate master mix containing all the components except template DNA.

	One reaction	Master mix for a 96-well plate
Primer (0.5 pmol/μL) (*see* **Note 12**)	2 μL	200 μL
dH_2O	13 μL	1300 μL
Reaction buffer	2 μL	200 μL
Thermal sequenase	2 μL	200 μL

3. Transfer 19 μL of the reaction mixture (prepared in **step 2**) and 1 μL of the pretreated DNA (prepared in **step 1**) into a labeled tube (*see* **Note 13**). Mix and place the samples on ice.
4. Mix 2 μL of termination master mix (7.5 μ*M* dATP, dCTP, dGTP, dTTP) and 0.5 μL of [α-^{33}P]-labeled terminators ddNTP to produce a *termination mix* for each ddNTP (*see* **Note 14**). Label, fill, and cap four tubes (G, A, T, C). It is more accurate and convenient to prepare batches of termination mixes sufficient for all sequences to be performed.
5. Dispense 2.5 μL from each termination mix to a 96-well microtiter plate. Start with ddGTP to the first well and continue with ddATP, ddTTP, and ddCTP. Keep this order for all samples.
6. Transfer 4.5 μL of the first pretreated DNA sample/reaction mixture (prepared in **step 3**) to the wells already containing 2.5 μL of the termination mix (G, A, T, C). Continue with the other samples in the same way. Keep a record of the order of templates to facilitate reading the data later.
7. Seal the plate and place it in the thermal cycler. Run 35 cycles at the following conditions: 95°C for 30 s, 55°C for 30 s, and 72°C for 1 min. Add 4 μL of stop solution.
8. Cover the samples with aluminum foil and freeze them until electrophoresis is performed (*see* **Note 15**).

3.2.4.2. Sequencing Gel and Electrophoresis

1. Sequencing gels should be prepared at least 2 h before use and can be made up to 24 h before needed. To prepare the gel, mix the following:

20X glycerol tolerant buffer	5 mL
Acrylamide/bis-acrylamide 19:1 solution 40%	14.4 mL
dH_2O	42.0 mL
Urea (ultrapure)	48 g

Wear gloves because acrylamide, which can be absorbed through the skin, is a potent neurotoxin. Mix until urea is dissolved. This takes approx 30–45 min. Place solution on ice.

2. Wash glass plates and spacers thoroughly with water and ethanol. Treat the surface of one plate with Acrylease™ Nonstick Plate Coating; this prevents the

gel from sticking to both plates and reduces the possibility that the gel will tear when it is removed from the plates.

3. Arrange the glass plates with the spacers and make certain that both sides and the bottom are water-tight (e.g., use a sealing tape).
4. Add 400 µL 10% APS and 90 µL TEMED right before pouring the gel.
5. Slowly pour the solution with a 50-mL pipet down one side of the gel mold while holding the frame at an angle of approx 45° to the horizontal. Pour the solution in a continuous stream to avoid air bubbles (*see* **Note 16**).
6. Insert the comb and allow the gel to polymerize for 1–2 h.
7. After polymerization, insert the gel into the electrophoresis apparatus. Fill the reservoirs with running buffer (1X glycerol-tolerant buffer).
8. Preheat gel for approximately half an hour (*see* **step 11**).
9. Turn off the power. Remove the comb (*see* **Note 17**).
10. Heat sequencing reactions at 80°C for 10 min for denaturing. Place samples on ice and load approx 2 µL in one well with gel-loading duck bill tips. Carefully avoid letting samples bleed into the adjoining lane. Again, keep a record of the order of the templates and the load of the reactions.
11. Run the gel at a voltage gradient between 1 and 8 V/cm until the marker dyes have migrated the desired distance (e.g., xylene cyanol FF comigrates in a 8% denaturing polyacrylamide gel with DNA of 76 bp in length, bromphenol blue with DNA of 19 bp length).
12. At the end of electrophoresis, remove the frame from the electrophoresis apparatus. Using the end of a metal spatula, separate both glass plates. That gel will remain attached to the lower plate. Cut off a small piece of the upper right corner of the gel to distinguish left and right side after exposure.
13. Remove gel from glass by placing a 3MM Whatman paper on top of the gel. Slowly lift the Whatman paper with the attached gel. Place Saran-Wrap over the gel and place the gel into the dryer. Dry the gel under vacuum at 80°C for 1–2 h.
14. Remove the Saran-Wrap and place the gel with autoradiography film in a cassette.
15. Develop autoradiograph after overnight exposure and read the sequence.

3.3. *IL1RN* Polymorphism

A common polymorphism in the *IL1RN* gene lies in intron 2, where there is a variable number of identical tandem repeats. These can be screened by a PCR amplification technique and the variants determined by size fractionation via gel electrophoresis. The most common allele contains four repeats (allele 1, frequency 0.74) ***(43)***. Allele 2 contains two repeats and its frequency is 0.21 ***(43)***. Variants of the *IL1RN* have been associated with a series of disorders, including auto-immune diseases, alcoholism, cancer, and infectious diseases.

3.3.1. PCR Amplification

1. The PCR amplification for screening the polymorphism in the *IL1RN* gene is performed as outlined in **Subheading 3.2.1.** For the PCR reaction, mix the following:

	One reaction	Master mix for a 96-well plate
10X reaction buffer	2 µL	200 µL
dNTP (100 m*M*)	0.5 µL	50 µL
Primer IL1RN-F*	0.5 µL	50 µL
Primer IL1RN-R†	0.5 µL	50 µL
H_2O	21 µL	2100 µL
Taq DNA polymerase	0.25 µL	25 µL

*IL1RN-F: 5′ CTC AGC AAC ACT CCT AT 3′ (stock: 100 ng/µL)
†IL1RN-R: 5′ TCC TGG TCT GCA GGT AA 3′ (stock: 100 ng/µL)

2. Add between 20 ng and 50 ng of the appropriate genomic DNA.
3. Thermal cycling parameters are as follows: After an initial denaturation step for 2 min at 96°C, run 35 cycles of 94°C for 30 s, 58°C for 60 s, and 72°C for 2 min (*see* **Note 9**).

3.3.2. Electrophoresis

After PCR amplification, the genotype of *IL1RN* is directly determined by size fractionation. As outlined in **Subheading 3.2.3.**, mix 8 µL of DNA with 2 µL of loading buffer and run the samples on a 3% agarose gel (TreviGel 500 [Trevigen, Gaithersburg, MD]). Load next to the molecular-weight marker and the negative control digested PCR samples of known genotypes as positive controls.

3.3.3. Sequencing

Perform quality control in a double-blinded manner for the first 50 samples as outlined in **Subheading 3.2.4.**

4. Notes

1. For a PCR-RFLP assay, a gene fragment containing the polymorphic site is amplified using a unique pair of primers. It is useful to test primer sequences for homologies to publicly available genomic sequence using the Basic Local Alignment Search Tool (http://www.ncbi.nlm.nih.gov/BLAST/). This avoids coamplifying highly homologous regions.
2. An effective separation of DNA is achieved on agarose gels, which can vary in concentration and density, both of which influence the migration patterns

of fragments. For optimal separation of small PCR products (e.g., <0.5 kb), Agarose SFR (Amresco, Solon, OH) or TreviGel 500 (Trevigen, Gaithersburg, MD) should be used at a concentration of at least 2%.

3. Genomic DNA purified from blood is generally useful, even if the patient has received a transfusion within the last several weeks.
4. The yield of gDNA depends greatly on the number of white blood cells. This has to be considered especially for samples from (neutropenic) cancer patients.
5. For purification of gDNA of clotted blood, add 50 μL clotted blood to a sterile 1.5 mL microtube containing 550 μL Cell Lysis Solution (PUREGENE® DNA Isolation Kit). Pipet up and down several times to mix. Add 3 μL Proteinase K Solution (20 mg/mL) and mix by inverting 25 times. Incubate at 55°C for 3 h to overnight until clots have dissolved. Cool the samples to room temperature and continue with the RNase treatment, protein precipitation, DNA precipitation, and DNA hydration. Because the PUREGENE® DNA Isolation Kit is a reagent-based system, the reagent volumes may be scaled up proportionately according to the sample size.
6. If necessary, red blood cell lysis may be repeated.
7. All PCR reactions are set up on ice. To avoid contamination, PCR reactions should be set up in a special area designed only for PCR. In addition, we suggest the use of filter tips.
8. Most experts agree that before calling a genotype, at least or more separate amplification assays should be performed and matched.
9. The purity and fidelity of PCR amplicons are dependent on optimization of the assay. However, there is room for variation in conditions; for example, reported PCR conditions for *IL1RN* may differ in cycle number, cycle length, and annealing temperature ***(43–45)***. For each individual laboratory, the optimal PCR conditions might differ slightly from conditions reported here.
10. The enzyme content of the restriction probe should not be more than 1 : 10 of the entire volume, because the enzymes are diluted in a very highly concentrated glycerol solution that inhibits the enzyme activity.
11. If gels are covered with running buffer, they can be prepared in advance and left overnight.
12. For diluting the sequencing primer to a concentration of 0.5 pmol/μL, determine the concentration of primer by reading the optical density at 260 nm (OD_{260}). The approximate concentration (pmol/μL) is given by the following formula: Concentration (pmol/μL)=$OD_{260}/0.01N$), where N is the number of bases.
13. If the PCR reaction results in good quantity and quality, we typically use (0.5 – 1 μL of pretreated PCR products (length < 500 bp) for the sequencing reaction.
14. If secondary structures produce gel artifacts resulting in a loss of information, dITP master (7.5 μ*M* dATP, dCTP, dGTP, and 37.5 μ*M* dITP) can be used instead of dGTP master (7.5 μ*M* dATP, dCTP, dGTP, dTTP). Then, the termination temperature should be reduced from 72°C to 60°C and the extension time should be increased to approx 5 min.

15. Samples can be kept frozen for several days before running. Consider that the half-life of P^{33} is approx 25 d.
16. In the presence of air bubbles, avoid loading these lanes for sequencing electrophoresis. If there are too many bubbles in the gel, prepare a new one.
17. It is important to wash out the wells thoroughly as soon as the comb is removed. We flush out the wells with running buffer using a fine pipet or a Pasteur pipet.

Acknowledgment

T.L. was supported by a Dr. Mildred Scheel Stipendium, Deutsche Krebshilfe e.V.

References

1. Lander, E. S., Linton, L. M., Birren, B., Nusbaum, C., Zody, M. C., Baldwin, J., et al. (2001) Initial sequencing and analysis of the human genome. *Nature* **409,** 860–921.
2. Venter, J. C., Adams, M. D., Myers, E. W., Li, P. W., Mural, R. J., Sutton, G. G., et al. (2001) The sequence of the human genome. *Science* **291,** 1304–1351.
3. Masood, E. (1999) As consortium plans free SNP map of human genome. *Nature* **398,** 545–546.
4. Marth, G. T., Korf, I., Yandell, M. D., Yeh, R. T., Gu, Z., Zakeri, H., et al. (1999) A general approach to single-nucleotide polymorphism discovery. *Nat. Genet.* **23,** 452–456.
5. Smigielski, E. M., Sirotkin, K., Ward, M., and Sherry, S. T. (2000) dbSNP: a database of single nucleotide polymorphisms. *Nucleic Acids Res.* **28,** 352–355.
6. Kruglyak, L. and Nickerson, D. A. (2001) Variation is the spice of life. *Nat. Genet.* **27,** 234–236.
7. Wang, D. G., Fan, J. B., Siao, C. J., Berno, A., Young, P., Sapolsky, R., et al. (1998) Large-scale identification, mapping, and genotyping of single-nucleotide polymorphisms in the human genome. *Science* **280,** 1077–1082.
8. Cargill, M., Altshuler, D., Ireland, J., Sklar, P., Ardlie, K., Patil, N., et al. (1999) Characterization of single-nucleotide polymorphisms in coding regions of human genes. *Nat. Genet.* **22,** 231–238.
9. Halushka, M. K., Fan, J. B., Bentley, K., Hsie, L., Shen, N., Weder, A., et al. (1999) Patterns of single-nucleotide polymorphisms in candidate genes for blood-pressure homeostasis. *Nat. Genet.* **22,** 239–247.
10. Saiki, R. K., Scharf, S., Faloona, F., Mullis, K. B., Horn, G. T., Erlich, H. A., et al. (1985) Enzymatic amplification of beta-globin genomic sequences and restriction site analysis for diagnosis of sickle cell anemia. *Science* **230,** 1350–1354.
11. Foster, C. B., Lehrnbecher, T., Mol, F., Steinberg, S. M., Venzon, D. J., Walsh, T. J., et al. (1998) Host defense molecule polymorphisms influence the risk for immune-mediated complications in chronic granulomatous disease. *J. Clin. Invest.* **102,** 2146–2155.

12. Garred, P., Pressler, T., Madsen, H. O., Frederiksen, B., Svejgaard, A., Hoiby, N., et al. (1999) Association of mannose-binding lectin gene heterogeneity with severity of lung disease and survival in cystic fibrosis. *J. Clin. Invest.* **104,** 431–437.
13. Chanock, S. J. and Foster, C. B. (1999) SNPing away at innate immunity. *J. Clin. Invest.* **104,** 369–370.
14. Buetow, K. H., Edmonson, M. N., and Cassidy, A. B. (1999) Reliable identification of large numbers of candidate SNPs from public EST data. *Nat. Genet.* **21,** 323–325.
15. Irizarry, K., Kustanovich, V., Li, C., Brown, N., Nelson, S., Wong, W., et al. (2000) Genome-wide analysis of single-nucleotide polymorphisms in human expressed sequences. *Nat. Genet.* **26,** 233–236.
16. Strausberg, R. L., Buetow, K. H., Emmert-Buck, M. R., and Klausner, R. D. (2000) The cancer genome anatomy project: building an annotated gene index. *Trends Genet.* **16,** 103–106.
17. Hill, A. V. (1999) Genetics and genomics of infectious disease susceptibility. *Br. Med. Bull.* **55,** 401–413.
18. Collins, F. S. and McKusick, V. A. (2001) Implications of the Human Genome Project for medical science. *JAMA* **285,** 540–544.
19. Lehrnbecher, T., Foster, C. B., Zhu, S., Leitman, S. F., Goldin, L. R., Huppi, K., et al. (1999) Variant genotypes of the low affinity Fc-gamma receptors in two control populations and a review of low affinity Fc-gamma receptor polymorphisms in control and disease populations. *Blood* **94,** 4220–4232.
20. Kruglyak, L. (1997) The use of a genetic map of biallelic markers in linkage studies. *Nat. Genet.* **17,** 21–24.
21. Lander, E. S. (1996) The new genomics: global views of biology. *Science* **274,** 536–539.
22. Risch, N. and Merikangas, K. (1996) The future of genetic studies of complex human diseases. *Science* **273,** 1516–1517.
23. Mira, J. P., Cariou, A., Grall, F., Delclaux, C., Losser, M. R., Heshmati, F., et al. (1999) Association of TNF2, a TNF-alpha promoter polymorphism, with septic shock susceptibility and mortality: a multicenter study. *JAMA* **282,** 561–568.
24. Foster, C. B., Lehrnbecher, T., Samuels, S., Stein, S., Mol, F., Metcalf, J. A., et al. (2000) An IL6 promoter polymorphism is associated with a lifetime risk of development of kaposi sarcoma in men infected with human immunodeficiency virus. *Blood* **96,** 2562–2567.
25. Warzocha, K., Ribeiro, P., Bienvenu, J., Roy, P., Charlot, C., Rigal, D., et al. (1998) Genetic polymorphisms in the tumor necrosis factor locus influence non-Hodgkin's lymphoma outcome. *Blood* **91,** 3574–3581.
26. Morse, H. R., Olomolaiye, O. O., Wood, N. A., Keen, L. J., and Bidwell, J. L. (1999) Induced heteroduplex genotyping of TNF-alpha, IL-1beta, IL-6 and IL-10 polymorphisms associated with transcriptional regulation. *Cytokine* **11,** 789–795.
27. Perrey, C., Pravica, V., Sinnott, P. J., and Hutchinson, I. V. (1998) Genotyping for polymorphisms in interferon-gamma, interleukin-10, transforming growth factor-

beta 1 and tumour necrosis factor-alpha genes: a technical report. *Transplant. Immunol.* **6,** 193–197.
28. Epplen, C., Frank, G., Gomolka, M., Albert, E., Nurnberg, P., and Epplen, J. T. (1994) Dinucleotide repeat polymorphism in the human IL1A gene. *Hum. Mol. Genet.* **3,** 1710.
29. Hurme, M. and Santtila, S. (1998) IL-1 receptor antagonist (IL-1Ra) plasma levels are co-ordinately regulated by both IL-1Ra and IL-1beta genes. *Eur. J. Immunol.* **28,** 2598–2602.
30. Parkes, M., Satsangi, J., and Jewell, D. (1998) Contribution of the IL-2 and IL-10 genes to inflammatory bowel disease (IBD) susceptibility. *Clin. Exp. Immunol.* **113,** 28–32.
31. Borish, L., Mascali, J. J., Klinnert, M., Leppert, M., and Rosenwasser, L. J. (1994) SSC polymorphisms in interleukin genes. *Hum. Mol. Genet.* **3,** 1710.
32. Hook, S., Cheng, P., Holloway, J., Riley, G., Sawyer, G., Le Gros, G., et al. (1999) Analysis of two IL-4 promoter polymorphisms in a cohort of atopic and asthmatic subjects. *Exp. Clin. Immunogenet.* **16,** 33–35.
33. Graves, P. E., Kabesch, M., Halonen, M., Holberg, C. J., Baldini, M., Fritzsch, C., et al. (2000) A cluster of seven tightly linked polymorphisms in the IL-13 gene is associated with total serum IgE levels in three populations of white children. *J. Allergy Clin. Immunol.* **105,** 506–513.
34. Terry, C. F., Loukaci, V., and Green, F. R. (2000) Cooperative influence of genetic polymorphisms on interleukin 6 transcriptional regulation. *J. Biol. Chem.* **275,** 18,138–18,144.
35. Schmidt, S., Papassotiropoulos, A., Bagli, M., Harzheim, M., Heun, R., and Klockgether, T. (2000) No association of serum levels of interleukin-6 and its soluble receptor components with a genetic variation in the 3′ flanking region of the interleukin-6 gene in patients with multiple sclerosis. *Neurosci. Lett.* **294,** 139–142.
36. Renzoni, E., Lympany, P., Sestini, P., Pantelidis, P., Wells, A., Black, C., et al. (2000) Distribution of novel polymorphisms of the interleukin-8 and CXC receptor 1 and 2 genes in systemic sclerosis and cryptogenic fibrosing alveolitis. *Arthritis Rheum.* **43,** 1633–1640.
37. Gibson, A. W., Edberg, J. C., Wu, J., Westendorp, R. G., Huizinga, T. W., and Kimberly, R. P. (2001) Novel single nucleotide polymorphisms in the distal il-10 promoter affect il-10 production and enhance the risk of systemic lupus erythematosus. *J. Immunol.* **166,** 3915–3922.
38. Maurer, M., Kruse, N., Giess, R., Toyka, K. V., and Rieckmann, P. (2000) Genetic variation at position –1082 of the interleukin 10 (IL10) promotor and the outcome of multiple sclerosis. *J. Neuroimmunol.* **104,** 98–100.
39. Bellingham, J., Gregory-Evans, K., and Gregory-Evans, C. Y. (1998) A polymorphic dinucleotide repeat in the 5′ flanking region of the human interleukin 11 (IL11) gene. *Immunogenetics* **47,** 499–500.
40. Shibata, S., Asano, Y., Yokoyama, T., Shimoda, K., Nakashima, H., Okamura, S., et al. (1998) Analysis of the granulocyte colony-stimulating factor receptor gene

structure using PCR-SSCP in myeloid leukemia and myelodysplastic syndrome. *Eur. J. Haematol.* **60,** 197–201.

41. Wyman, A. R. and White, R. (1980) A highly polymorphic locus in human DNA. *Proc. Natl. Acad. Sci. USA* **77,** 6754–6758.
42. Olomolaiye, O., Wood, N. A., and Bidwell, J. L. (1998) A novel *Nla*III polymorphism in the human IL-6 promoter. *Eur. J. Immunogenet.* **25,** 267.
43. Tarlow, J. K., Blakemore, A. I., Lennard, A., Solari, R., Hughes, H. N., Steinkasserer, A., et al. (1993) Polymorphism in human IL-1 receptor antagonist gene intron 2 is caused by variable numbers of an 86-bp tandem repeat. *Hum. Genet.* **91,** 403–404.
44. Danis, V. A., Millington, M., Hyland, V. J., and Grennan, D. (1995) Cytokine production by normal human monocytes: inter-subject variation and relationship to an IL-1 receptor antagonist (IL-1Ra) gene polymorphism. *Clin. Exp. Immunol.* **99,** 303–310.
45. Langabeer, S. E. and Linch, D. C. (1998) IL-1 receptor antagonist gene polymorphism in patients with secondary acute myeloid leukaemia. *Cytokines Cell. Mol. Ther.* **4,** 7–9.

8

Apoptosis Induction by TRAIL

Angelika Eggert, Hauke Sieverts, and Henning Walczak

1. Introduction

1.1. Background

In 1995, the tumor necrosis factor (TNF)-related apoptosis-inducing ligand (TRAIL) was identified on the basis of sequence homology to the other members of the TNF family *(1)*. Like TNF and CD95 ligand (CD95L/FasL/APO-1L), TRAIL (APO-2L) has been been shown to be a potent inducer of apoptosis in various cancer cell lines *(2–4)*. However, in contrast to injection of TNF or CD95L, which are both lethal to mice *(5)*, TRAIL exerts potent antitumor activity in vivo without exhibiting systemic toxicity *(6–10)*. Thus, treatment with TRAIL might be a promising therapeutic approach for many solid tumors.

Two agonistic receptors for TRAIL have been identified: TRAIL-R1 (DR4) *(11)* and TRAIL-R2 (KILLER, DR5, TRICK2, APO-2) *(12–16)*. Both contain a cytoplasmic death domain that is essential for the transmission of the death signal. Although TRAIL and its apoptosis-inducing receptors TRAIL-R1 and TRAIL-R2 are widely expressed in human tissues, most normal cells are not sensitive to TRAIL-mediated killing *(1)*, whereas other normal cells exhibit reduced sensitivity *(17)*. Two potentially antagonistic surface receptors, the membrane-anchored TRAIL-R3 (DcR1, TRID) *(11,14,18)*, which lacks a cytoplasmic domain, and TRAIL-R4 (DcR2, TRUNDD) *(19,20)*, which contains a truncated, nonfunctional death domain are unable to transduce a death signal). The initially observed preferential expression of these antagonistic receptors in normal tissues led to the hypothesis that both proteins may protect normal tissues from TRAIL-induced apoptosis by competing with the agonistic

From: *Methods in Molecular Biology, vol. 215:*
Cytokines and Colony Stimulating Factors: Methods and Protocols
Edited by: D. Körholz and W. Kiess © Humana Press Inc., Totowa, NJ

receptors for limited amounts of ligand ***(11,14,19)***. However, recent studies suggest that the mechanisms regulating TRAIL-mediated apoptosis are more complex than this model ***(21–32)***.

TRAIL induces apoptosis via a caspase-dependent signaling cascade ***(33)***. FADD/Mort1 and caspase-8 are recruited to the apoptosis-inducing TRAIL receptors ***(17,34–37)***. Activation of caspase-8 and caspase-3 is followed by late dissipation of the mitochondrial membrane potential and cytochrome-*c* release and induction of DNA fragmentation ***(38)***. TRAIL can bypass the antiapoptotic effect of Bcl-2 or BCL-xL overexpression in some cell types ***(39,40)*** but not in others ***(41)***. Tumor cells that lack caspase-8 because of deletion or hypermethylation of the gene promoter are resistant to TRAIL-induced apoptosis ***(42–44)***.

As induction of tumor cell death by direct activation of TRAIL receptors might be an attractive new therapeutic strategy for many human tumors, the expression and function of the TRAIL system deserves further investigation in a wide range of human malignancies.

1.2. Methods

Methods for analyzing the role of the TRAIL system in a certain cell type can be classified as quantitative or functional. Each approach has advantages and limitations. The most appropriate method depends on the cell population in question. In our opinion, reverse transcription–polymerase chain reaction (RT-PCR), flow cytometry, and Western blot analysis are the most useful assays for the quantitative analysis of the TRAIL system in cell lines. For the analysis of the TRAIL system in the tissue of mixed cell populations (i.e., tumor tissue), methods like flow cytometry and, if possible, immunohistochemistry are adequate because of the ability to differentially determine surface and intracellular expression on subpopulations of cells. Thus far, functional assays using recombinant TRAIL for the induction of apoptosis are limited to cell cultures (short-term culture of primary cells or established cell lines).

1.2.1. Quantitative Analysis

1.2.1.1. RT-PCR

One quantitative approach is the analysis of TRAIL and TRAIL-receptor mRNA expression by semiquantitative RT-PCR. RT-PCR is a sensitive analytical method and is a suitable approach for the analysis of mRNA expression in cell lines. Because of the extreme sensitivity of PCR, there is a risk in PCR-based measurements of tumor tissue that the mRNA detected may derive from a minor contaminating cell population like lymphocytes or fibroblasts.

For quantitative RT-PCR protocols, a sensitive detection of PCR products is essential. PCR has to be performed at a reduced number of cycles to ensure that the amplification is still in the exponential phase. Under these conditions, the amount of PCR products generated is too small to be detectable by commonly used ethidium bromide staining. About 500-fold more sensitive detection of PCR products is achieved by the blotting-based technique using biotinylated PCR primers described here. The housekeeping gene glyceraldehyde-3-phosphate dehydrogenase (*GAPD*), which is transcribed constitutively in most cell types and tissues, is used as an invariant internal control for variations in RT, PCR, and product detection steps. A modification of *GAPD* primers (application of biotinylated vs nonbiotinylated primers in a ratio of 1:49) reduced the signal intensity of *GAPD* that was initially too strong and, thereby, allowed the analysis of both *GAPD* and target gene signals within the linear range of X-ray film detection ***(45)***.

1.2.1.2. Western Blot and Immunohistochemistry

Immunohistochemistry is a useful technique for the analysis of protein expression in primary tissues. Originally, most antibodies required frozen tissue. More recently, an increasing number of antibodies can be applied to formalin-fixed, paraffin-embedded tissue, which is the universally applied method of tissue preservation. This has opened the door to the testing of archival material. To ensure the quality of the analysis, immunohistochemistry should always be performed in cooperation with a pathologist. However, immunohistochemical detection of TRAIL and its receptors will not be described here, as it is still unclear whether currently available antibodies that can be used for the specific detection of TRAIL and its receptors by flow cytometry analysis and/or Western blot are indeed suitable for the specific detection of these antigens in tissue sections.

Western blot analysis is a useful method for detecting the different constituents of the TRAIL system in cell lines. Although an antibody suitable for the detection of TRAIL in cell lysates is available, the TRAIL receptor-specific antibodies that are currently available work best after affinity precipitation of the ligand. However, upon very high expression of the specific receptors, these antibodies may, in fact, also work directly on cell lysates. This has to be thoroughly controlled.

1.2.1.3. Flow Cytometry

Flow cytometry (FCM) allows cell-by-cell analysis and is attractive for the analysis of cell lines as well as primary tissue. Two major advantages in using FCM to analyze TRAIL-receptor expression are (1) the ability to determine actual *surface* expression of the receptors and (2) the ability to determine differential surface expression on subpopulation of cells, using multiparameter FCM. In the context of tumor tissues, this usually requires the inclusion of a tumor marker for the malignant population. Recently, monoclonal antibodies for sensitive analysis of cell surface expression of TRAIL receptors have become available. For this analysis, adherent cells have to be detached nonenzymatically, whereas nonadherent cells can be used as such.

1.2.2. Functional Analysis: Assessment of Cell Viability After TRAIL Treatment Using MTT Assay

As mentioned earlier, functional assays using recombinant TRAIL for the induction of apoptosis are limited to cell cultures (short-term culture of primary cells or established cell lines). Functional tests are most useful when the cells have been shown to express the relevant receptors (preferably by FCM analysis; RT-PCR and Western blot should be performed in addition), to determine whether they are capable of responding. This requires a functional TRAIL signaling pathway in the cells to be analyzed. The functional approach is the most relevant to the analysis of malignant cells because the mere expression of the potentially apoptosis-inducing TRAIL receptors can be without functional consequence, as it is necessary but not sufficient for the apoptotic response of tumor cells to treatment with TRAIL.

2. Materials

2.1. Semiquantitative RT-PCR

2.1.1. Tissues and RNA Extraction (see **Note 1**)

1. Pelleted tumor cells (approx 1×10^7 cells) or frozen tumor tissue (at least 20 mg).
2. RNeasy Mini or Midi Kit (Qiagen, Valencia, CA).
3. Ethanol 100% and 70% (*see* **Note 2**).
4. β-Mercaptoethanol.
5. 1 mL Syringes and 21G needles.

2.1.2. RT-PCR

1. Primer sequences (*see* **Table 1**) (*see* **Note 3**). All primers and probes should be aliquoted and stored at –80°C (*see* **Note 4**).
2. SuperScript™ Preamplification System (Gibco-BRL, Gaithersburg, MD) containing random hexamers (50 ng/µL), diethyl-pyrocarbonate (DEPC)-treated H_2O, 10 m*M* dNTP mix, 0.1 *M* dithiothreitol (DTT) (*see* **Note 5**), 10X PCR buffer (200 m*M* Tris-HCl [pH 8.4], 500 m*M* KCl), 25 m*M* $MgCl_2$, SuperScript II RT enzyme (200 U/µL), RNAse H (2 U/µL), and control RNA.

Table 1
Primer Sequences

Primer	Sequence	PCR product size (bp)
TRAIL sense	GACCCCAATGACGAAGAGAG	299
TRAIL antisense	CTCAGGAATGAATGCCCACT	
TRAIL-R1 sense	CTGAGCAACGCAGACTCGCTGTCCAC	506
TRAIL-R1 antisense	TCCAAGGACACGGCAGAGCCTGTGCCAT	
TRAIL-R2 sense	GCCTCATGGACAATGAGATAAAGGTGGCT	502
TRAIL-R2 antisense	CCAAATCTCAAAGTACGCACAAACGG	
TRAIL-R3 sense	GAAGAATTTGGTGCCAATGCCACTG	612
TRAIL-R3 antisense	CTCTTGGACTTGGCTGGGAGATGTG	
TRAIL-R4 sense	CTTTTCCGGCGGCGTTCATGTCCTTC	453
TRAIL-R4 antisense	GTTTCTTCCAGGCTGCTTCCCTTTGTAG	
Caspase-8 sense	AGAGAGAAGCAGCAGCCTTG	302
Caspase-8 antisense	GGGGCTTGATCTCAAAATGA	
cFLIP sense	TGATGGCAGAGATTGGTGAG	255
cFLIP antisense	CTTGTCCCTGCTCCTTGAAC	
GAPDH sense	CATCAAGAAGGTGGTGAAGC	160
GAPDH antisense	GAGCTTGACAAAGTGGTCGT	

3. AmpliTaq Gold™ 5 U/µL with GeneAmp 10X PCR-Buffer II (100 m*M* Tris-HCl [pH 8.3], 500 m*M* KCl) and 25 m*M* $MgCl_2$ (Perkin-Elmer, Applied Biosystems, Foster City, CA).
4. Mineral oil (Sigma, St. Louis, MO).
5. Eppendorf microcentrifuge.
6. 37°C, 42°C, and 70°C water baths or heat blocks.
7. Eppendorf tubes, 0.5 or 1 mL, and PCR tubes, 0.2 mL; autoclaved.
8. Aerosol-resistant barrier pipet tips.
9. Pipets.
10. Programmable thermal cycler (i.e., PTC-100 Programmable Thermal Controller [MJ Research Inc,Waltham, MA]).

2.1.3. Gel Electrophoresis and Southern Analysis

1. 5X Tris-borate electrophoresis buffer (TBE): 54 g Tris base, 27.5 g boric acid, and 20 mL of 0.5 *M* ethylenediaminetetraacetic acid (EDTA). Made up to 1 L and autoclave. Store at room temperature.
2. 6X Loading dye: 0.25% xylene cyanol, 0.25% bromophenol blue, and 30% glycerol.
3. Chloroform.

4. 40% Polyacrylamide/bis-acrylamide 19:1 (Bio-Rad, Hercules, CA) (*see* **Note 6**).
5. Tetramethylethylenediamine (TEMED, Bio-Rad).
6. Ammonium persulfate (25%) (Sigma, St. Louis, MO).
7. Biotinylated molecular-weight marker ϕX174/Hind I fragments (Gibco-BRL, Gaithersburg, MD).
8. Mini Protean II or III electrophoresis cell with 0.75-mm spacers and 15-well combs (Bio-Rad).
9. Mini Trans-Blot Transfer Cell (Bio-Rad).
10. Nylon membrane N+ (Amersham Corp., Arlington Heights, IL).
11. Long-wavelength ultraviolet (UV) light source (360 nm).
12. Power supply.
13. Southern Light Detection Kit (Tropix, Bedford, MA).
14. 10% Sodium dodecyl sulfate (SDS).
15. 10X MBS (20.9 g MOPS, 88 g NaCl, 5 mL of 10 *M* NaOH, deionized H_2O to 1000 mL).
16. Blocking buffer (0.6 g I-Block Reagent from Southern Light Detection Kit, 30 mL of 10X MBS, 15 mL of 10% SDS, deionized H_2O to 300 mL) (*see* **Note 7**).
17. Wash buffer (50 mL of 10X MBS, 25 mL of 10% SDS, deionized H_2O to 500 mL).
18. X-omat AR Film (Kodak) and suitable film cassette.

2.2. Western Blot

1. Nonidet P-40 lysis buffer: 1% NP40, 20 m*M* Tris-HCl (pH 8.0), 137 m*M* NaCl, 0.5 m*M* EDTA, 10% glycerol, 1 m*M* phenylmethylsulfonyl fluoride. Add fresh before use: 0.15 U/mL aprotinin, 20 μ*M* leupeptin, and 1 m*M* sodium vanadate.
2. Bradford reagent (Bio-Rad, Hercules, CA).
3. Tissue culture dishes, 10 cm.
4. Mini Protean II or III electrophoresis cell with 1.0-mm spacers and 10-well combs (Bio-Rad).
5. Mini Trans-Blot Transfer Cell (Bio-Rad) and Power supply.
6. Nitrocellulose membrane Hybond ECL (Amersham Corp., Arlington Heights, IL).
7. ECL Western blotting analysis system (Amersham).
8. Primary antibodies according to **Table 6**.
9. 5% Bovine serum albumin (BSA) in phosphate-buffered saline (PBS).
10. Wash buffer: PBS containing 0.05% Tween-20.

2.3. Flow-Cytometric Analysis

1. 2 m*M* EDTA in PBS.
2. FCM buffer (PBS containing 5% fetal calf serum [FCS]).
3. PBS.
4. MAbs according to **Table 7**.

2.4. Functional Analysis: Assessment of Cell Viability After TRAIL Treatment Using MTT Assay

1. TRAIL apoptosis kit (Upstate Biotechnology, Lake Placid, NY), containing 10 µg TRAIL and 150 µg mouse monoclonal IgG-potentiator). Alternatively, IgG potentiator, rhsTRAIL Set (Alexis Corp., cat. no. 850-018) or His-TRAIL (Alexis Corp., cat. no. 201-073).
2. 96-Well flat-bottom tissue culture plates.
3. Appropriate cell culture media.
4. Reagent reservoirs.
5. Multichannel pipet.
6. MTT (Sigma, St. Louis, MO).
7. 1 *N* HCl (*see* **Note 8**) and isopropanol.
8. Microplate reader with 570-nm filter.

3. Methods

3.1. RNA Extraction

Successful cDNA synthesis and subsequent PCR begins with the isolation of high-quality intact RNA (*see* **Note 9**). The quality of the RNA dictates the maximum amount of sequence information that can be converted into cDNA. Because RNA is easily degraded, extreme care should be taken when handling samples (*see* **Note 1**). Historically, we have used standard methods of phenol extraction and ethanol precipitation for larger samples that give the purest RNA *(46)*. However, these methods are time-consuming, use toxic agents, and may not give good recovery for small samples. The method of choice in our lab is the use of the Qiagen RNeasy Kit (Qiagen, Valencia, CA) to extract total RNA from tumor tissue or cell lines. As this method is exactly as outlined in the instruction pamphlet that accompanies the kit, it will not be described in detail here.

3.2. RT-PCR

3.2.1. Controls

Each RT-PCR includes RNA extractions from several tumors/cell lines. Appropriate controls are as follows:

1. A negative RNA control. This is a tube that had no cells added prior to the extraction. It is used to ensure that all of the RNA extraction solutions are free of contaminating nucleic acids.
2. A no RT control. It is used to ensure that no contaminating genomic DNA is present in the RNA sample used.

Table 2
Composition of RT Reaction Mixes

Reagents	Amount per reaction	Final concentration
RNA	1 µg	
Random hexamers	3 µL	150 ng
10X PCR buffer	2 µL	1X
$MgCl_2$ (25 m*M* stock)	2 µL	2.5 m*M*
dNTPs (10 m*M* stock)	1 µL	0.5 m*M*
DTT (0.1 *M* stock)	2 µL	10 m*M*
Superscript reverse transcriptase	1 µL	200 U
DEPC-treated water	up to 20 µL	

3. A H_2O control. It is used to ensure that there is no cross-contamination between tubes.
4. A positive RNA control. RT-PCR of these tissue RNAs should be positive for the indicated genes.
   TRAIL: spleen, lung, prostate, thymus, placenta;
   TRAIL-R1: liver, thymus, spleen, small intestine, peripheral blood leukocytes (PBL);
   TRAIL-R2: ovary, spleen, small intestine, PBL;
   TRAIL-R3: PBL, spleen, placenta, pancreas, skeletal muscle;
   TRAIL-R4: testis, colon, PBL;
   Caspase8: PBL, thymus, spleen;
   cFLIP: T-lymphocytes, heart;
   If no PCR product is present in these controls, it indicates that the reagent used, often an oligo, was not good and renders all other results uninterpretable.

3.2.2. RT Reaction

Reverse transcription (RT) of 1 µg total RNA is done in a 20 µL total volume according to the manufacturer's instructions (*see* composition of RT mixes in **Table 2**):

1. Mix and briefly centrifuge each component before use, keep RNA on ice, and wear gloves all the time.
2. Prepare RNA/random hexamers mixtures in sterile 0.5-mL tubes: 1 µg RNA + 3 µL random hexamers + DEPC–H_2O to 12 µL (*see* **Note 10**).
3. Denature each sample at 70°C for 10 min and incubate on ice for at least 1 min.
4. Prepare a master mix of the following reaction mixture, adding each component in the indicated order:
   10X PCR buffer: 2 µL per reaction, 25 mM MgC12: 2 µL per reaction,
   10 mM dNTP mix: 1 µL per reaction, and 0.1 M DTT: 2 µL per reaction
   You need to prepare the master mix for your number of samples + three additional reactions.

Table 3
Composition of PCR Reaction Mixes

Reagent	Amount per reaction	Final concentration
DEPC–H_2O	2.3 µL	
10X PCR buffer	1.0 µL	1X
$MgCl_2$ (25 m*M* stock)	0.8 µL	2.0 m*M*
dNTPs (2.5 m*M* stock)	0.8 µL	200 µ*M*
GAPD primer sense	1.0 µL	0.4 µ*M*
GAPD primer antisense	1.0 µL	0.4 µ*M*
Target gene primer sense	1.0 µL	0.4 µ*M*
Target gene primer antisense	1.0 µL	0.4 µ*M*
AmpliTaq Gold Polymerase	0.1 µL	0.5 U

5. Add 7 µL of the prepared reaction mixture to each RNA/random hexamers mixture, mix gently, collect by brief centrifugation, and incubate at room temperature for 5 min.
6. Add 1 µL (200 U) of SuperScript II RT to each tube, mix, and incubate at room temperature for 10 min.
7. Transfer the tubes to 42°C and incubate for 50 min.
8. Terminate the reactions at 70°C for 15 min. Chill on ice.
9. Collect the samples by brief centrifugation. Add 1 µL of RNase H to each tube and incubate for 20 min at 37°C.

3.2.3. PCR Reaction

While the RNA is being reverse transcribed, the PCR oligo mixes can be set up. PCR is carried out in a final volume of 10 µL containing 0.5 U Taq Gold Polymerase, 200 µ*M* dNTPs, 0.4 µ*M* of each primer, in a buffer consisting of 50 m*M* KCl, 10 m*M* Tris-HCl (pH 8.3), 2.0 m*M* $MgCl_2$ and 1 µL of the RT product (reverse transcribed total RNA).

1. Keep all reagents and samples on ice.
2. Make up a master mix for all reactions (*n* samples + 3) according to **Table 3**.
3. Add 1 µL of the cDNA (RT product) to be analyzed to each sample.
4. Overlay every sample with 10 µL mineral oil (*see* **Note 11**).
5. Close cap and place in PCR cycler.
6. Perform amplification on a PTC-100 Programmable Thermal Controller (MJ Research Inc.) (*see* **Note 12**) and use the PCR program in **Table 4**.

3.2.4. Gel Electrophoresis and Southern Analysis

1. Add 2 µL of 6X loading dye to each PCR sample.
2. Add 50 µL chloroform to each PCR sample (*see* **Note 13**).

Table 4
PCR Program

Denaturation	Annealing	Extension	No. of cycles
1. 95°C / 12 min			1
2. 95°C / 30 s	55°C / 30 s	72°C / 1.5 min	20
3.		72°C / 5 min	1

Table 5
Nondenaturing 6% Polyacrylamide Gel (Sufficient for Three Gels)

Reagent	Amount (3 gels)
ddH_2O	13 mL
5X TBE buffer	4 mL
40% Polyacrylamide/ bis-acrylamide 19:1	3 mL
TEMED	15 µL
Ammonium persulfate 25%	200 µL

3. A blue bubble containing the PCR product has formed. Pipet the blue bubble completely into a new tube.
4. Run up to 14 samples in parallel to a biotinylated molecular-weight marker (Gibco-BRL) on a nondenaturing 6% polyacrylamide gel (*see* **Table 5**) in 1X TBE buffer at 75 V for approx 1.5 h (*see* **Note 14**).
5. Southern-transfer DNA from the gel to a nylon membrane (Hybond N+, Amersham) in 0.25X TBE buffer at 30 V overnight (*see* **Note 15**).
6. Immobilize DNA on the membrane by UV crosslinking (1.5 min).
7. Detect biotin-labeled DNA according to the "Southern-Light-Protocol" (Tropix, Bedford, MA). Perform all steps at room temperature:
   - Wash blot 2X 5 min in blocking buffer (0.5 mL/cm^2).
   - Incubate for 10 min in blocking buffer (1 mL/cm^2).
   - Dilute Avidx-AP conjugate (from kit) 1:5000 in blocking buffer (3 µL/15 mL per membrane) and incubate blot for 20 min in this conjugate solution.
   - Wash 1X 5 min in blocking buffer (0.5 mL/cm^2), wash 3X 5 min in wash buffer (1 mL/cm^2) and wash 1X 5 min in 1X assay buffer (0.5 mL/cm^2, diluted from 10X assay buffer in the kit).
   - Drain blots by touching a corner on a paper towel and place in small tray (flat surface).
   - Pipet a thin layer of CSPD Ready-to-Use solution (3 mL per membrane) onto the blot and incubate for 5 min (*see* **Note 16**).

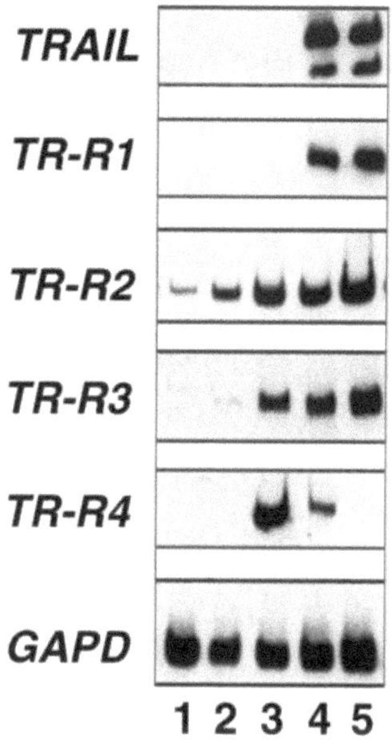

Fig. 1. Representative example of semiquantitative RT-PCR showing expression levels of TRAIL, TRAIL-R1, TRAIL-R2, TRAIL-R3, and TRAIL-R4 in neuroblastoma cell lines.

- Drain excess CSPD solution and place blot in Southern Light Development Folders (alternatively: Saran-Wrap). Smooth out any bubbles.

8. Develop after 30 min exposure to X-ray film (*see* **Note 17**).
9. Quantification of RNA transcript expression can be performed by densitometric analysis on X-ray films using NIH 1.55 Image software (*see* **Fig. 1**). A modification of the GAPD primers (biotinylated:nonbiotinylated at a ratio of 1:49) allows accurate quantification within the linear range of X-ray detection of both the target transcript and GAPD *(45)*.

3.3. Western Blot Analysis

There is an enormous diversity of TRAIL and TRAIL-receptor antibodies available from different companies. However, many of these antibodies are unspecific. The antibodies indicated here have been thoroughly controlled. Specificity of the respective anti-TRAIL and anti-TRAIL-R MAbs was determined by Western blot analysis of TRAIL and TRAIL-R1 to TRAIL-R4 in lysates prepared from cells transfected with expression vectors coding for the individual TRAIL receptors (data not shown). We use the following general protocol with the primary antibodies listed in **Table 6**.

Table 6
Suitable Antibodies for Western Blot Analysis

Antigen	Name of antibody	Supplier	Conc. (μg/mL)	Size of protein
Anti-TRAIL	Clone HS501	Alexis Corp., San Diego, CA	1	28 kD
Anti-TRAIL-R1 (DR4)	PSC-M39	Alexis Corp., San Diego, CA	1	55 kD
Anti-TRAIL-R2 (DR5/ TRICK2/APO-2/KILLER)	PSC-2019	Alexis Corp., San Diego, CA	1	55 kD
Anti-TRAIL-R3 (DcR1)	210-774-R100	Alexis Corp., San Diego, CA	1	36 kD
Anti-TRAIL-R4 (DcR2)	PSC-2021			
Anti-Caspase-8 (FLICE, MACH)	Clone C-15	UBI, Lake Placid, NY	1	55/53 kD 18 kD (cleaved)
Anti-FLIP (long and short)	Clone NF6	Alexis Corp., San Diego, CA	1	55 kD 25 kD (cleaved)

1. Grow cells almost confluent in a 10-cm tissue culture dish.
2. Lyse cells rapidly in 800 μL/dish of Nonidet P-40 lysis buffer.
3. Measure protein content of each sample using Bradford reagent and normalize samples for total protein content.
4. Separate samples (20 μg protein or proteins eluted from beads after ligand-affinity immunoprecipitation) by electrophoresis on denaturing 4–12% NuPage Bis-Tris gradient gels (Novex, San Diego, CA) in MOPS buffer according to the manufacturer's instructions and electroblot proteins from the gel to a nitrocellulose membrane.
5. Block nonspecific sites by incubation of the membrane in 5% BSA/PBS for at least 2 h. Five percent blocking reagent (from ECL kit) in 0.1% Tween-20/PBS or 5% nonfat dry milk (NFDM) in PBS/Tween (PBS containing 0.05% Tween-20) is also suitable blocking buffers.
6. Immunostain for 1–2 h at room temperature (or overnight at 4°C) with the primary antibody at the appropriate dilution according to **Table 6** (*see* **Note 18**).
7. Wash the membrane with wash buffer 6X 5 min.
8. Incubate the membrane in the appropriate horseradish peroxidase (HRPO)-conjugated isotype-specific secondary antibody diluted 1:20,000 in PBS/Tween for 1 h at room temperature.
9. Wash the membrane with wash buffer 6X 5 min.
10. Detect immunocomplexes using the ECL chemiluminescence system (Amersham Corp., Arlington Heights, IL) according to the manufacturer's instructions.

Table 7
Suitable Antibodies for FCM Analysis

Antigen	Name of antibody	Supplier	Conc. (μg/mL)
TRAIL	TRAIL Clone 2E5	Alexis Corp. San Diego, CA	10
TRAIL-R1 (DR4)	HS101	Alexis Corp. San Diego, CA	10
TRAIL-R2 (DR5/TRICK2, APO-2/KILLER)	HS201	Alexis Corp. San Diego, CA	10
TRAIL-R3 (DcR1)	HS301	Alexis Corp. San Diego, CA	10
TRAIL-R4 (DcR2)	HS402	Alexis Corp. San Diego, CA	10

3.4. Flow Cytometric Analysis

1. Detach cells nonenzymatically with EDTA.
2. After washing with PBS, incubate the cells with the MAbs against the four surface-expressed TRAIL receptors or a control mIgG1 for 20 min in PBS containing 5% FCS and 10 μg/mL of the specific antibody (according to **Table 7**).
3. Wash cells twice with 200 μL PBS.
4. Incubate cells for 20 min with biotinylated secondary goat anti-mouse IgG according to the manufacturer's instructions.
5. Wash as before and incubate in the presence of streptavidin-PE for 20 min (Pharmingen, Hamburg, Germany) according to the manufacturer's instructions.
6. Wash again as in **step 3**, resuspend cells in 200 μL of FACs buffer containing 1 μg/mL F-AAD (Amino-Actinomycin D) and analyze cells on a flow cytometer (*see* **Note 19**).

3.5. Functional Analysis: Assessment of Cell Viability After TRAIL Treatment Using MTT Assay

1. Prepare stock solution of TRAIL: 10 μg/mL in sterile H_2O + 0.1% BSA. Store aliquots of 100 μL at –20°C.
2. Seed 100 μL tumor cells into 96-well plates at a density of 5×10^4 cells per well and culture them for 24 h at 5% CO_2 in their appropriate growth medium (i.e., RPMI 1640 medium supplemented with 10% fetal bovine serum, L-glutamine, penicillin, and streptomycin).
3. After 24 h, carefully suck off medium and add 50 μL of recombinant-soluble human TRAIL in medium to each well for a final concentration of 200 ng/mL medium (final volume is 100 μL).

4. For study of dose dependency, treat cells with increasing concentrations of TRAIL (i.e., 0, 3, 6, 12.5, 25, 50, 100, and 200 ng/mL).
5. Add 50 μL potentiator to each well for a final concentration of 1.5 μg/mL.
6. Perform each condition in triplicate and use the following controls:
   Positive control 1: seeded tumor cells, no TRAIL, no potentiator;
   Positive control 2: seeded tumor cells, potentiator only, no TRAIL;
   Blanks: four wells with medium only, no cells.
7. Incubate cell cultures at 37°C, 5% CO_2, for the following treatment times: 4 h, 8 h, 12 h, 24 h, 48 h.
8. Perform a colorimetric 3-(4,5-dimethylthiazol-2-yl)-2,5-diphenyl tetrazolium bromide (MTT) assay as described ***(47)***:
   - Add 10 μL of MTT (5 mg/mL in PBS) to each well using a multichannel pipet.
   - Incubate for 4 h at 37°C, 5% CO2, in the dark.
   - Add 150 μL isopropanol + 2% HCl (0.1N) to each well (*see* **Note 18**).
   - Mix with pipet until MTT crystals dissolve (approx 5 min).
   - Read in microplate reader at 570 nm.
   - Use a multiwell scanner to measure the absorbance at 570 nm.
9. Calculate percentage of survival compared to untreated control cells. Cell lines can be considered highly sensitive to TRAIL-induced cell death if there is <50% viability after 24 h of TRAIL treatment at a concentration of 200 ng/mL.

3.5.1. Modifications

3.5.1.1. Cycloheximide or Actinomycin D

Inhibition of protein synthesis by cycloheximide (CHX) or actinomycin (Act D) has previously been shown to sensitize cancer cell lines to death-receptor-induced apoptosis ***(44,48–50)***. In this respect, the apoptosis-inhibitory effect of certain Bcl-2 family members and proteins like cFLIP may be neutralized by prior treatment of the cells with protein synthesis inhibitors like CHX or Act D. This can, for example, be achieved by comparing the effects on survival upon treatment of cells with TRAIL (e.g., 200 ng/mL) alone as compared to treatment of cells with TRAIL at the same conentrations in the presence of CHX (titrate around 1 μg/mL; Sigma) or Act D (titrate around 10 ng/mL; Sigma). As a control, the cells also need to be treated with CHX and Act D at the above-mentioned subtoxic concentration in the absence of TRAIL.

3.5.1.2. Caspase Inhibitors

The requirement of certain caspases for TRAIL-induced apoptosis signaling can be analyzed using specific caspase inhibitors. Analyze the effects of caspase inhibitors by measuring percentages of survival of TRAIL (200 ng/mL)-treated cells in the presence of certain caspase inhibitors. The pancaspase inhibitor carbobenzyloxy-Val-Ala-Asp (OMe) fluoromethyl ketone (zVAD-fmk)

(Enzyme Systems Products, Livermore, CA) or the specific caspase-8 inhibitor Z-IETD-FMK and the caspase-9 inhibitor Z-LEHD-FMK (Calbiochem, San Diego, CA) are dissolved in dimethyl sulfoxide (DMSO) and used at a final concentration of 50 μ*M* in culture medium. However, specificity can only be shown when these inhibitors are also used at concentration of 10 μ*M* and 1 μ*M*.

3.5.1.3. Demethylation with 5′ Aza-deoxycytidine

Lack of caspase-8 expression because of hypermethylation of the gene promoter has been shown to correlate with resistance to TRAIL-induced apoptosis in certain tumor cells ***(42–44)***. Treatment of these cells with the demethylating agent 5-aza-2′-deoxycytidine could restore TRAIL sensitivity. To analyze the role of gene hypermethylation for TRAIL resistance, cells should be treated prior to functional TRAIL assays with 5-aza-2′-deoxycytidine at a concentration of 0.1–10 μ*M* for 4–7 d with the medium and drug renewed twice weekly.

3.6. Analysis of Specific Apoptosis

In addition to the MTT assay, which provides only information about the number of living cells, apoptotic cell death has to be confirmed by at least two different apoptosis-specific assays. We suggest the use of one of the following methods. All the methodologies listed can be applied using commercially available kits from several companies. As all technical details are provided there, we just describe the basic principles of each method.

3.6.1. Subdiploid DNA Content Analysis

Flow cytometry on the basis of cell DNA content provides quantitative data on the percentage of apoptotic cells in a cell population. Because of their content of degraded DNA, apoptotic cells appear as a hypodiploid population that is eluted before the normal 2N cells in the G1 phase of the cell cycle and, therefore, has been named the "sub-G1" region of the elution pattern ***(51)***.

3.6.2. Annexin-V Binding

The anticoagulant annexin-V binds to various phospholipid species with the highest affinity for phosphatidyl-serine (PS), which, in normal cells, is located in the inner leaflet of the plasma membrane. PS is translocated to the outer layer of the cell mebrane in the early phases of apoptosis, during which the cell membrane itsself remains intact. At this point, PS can be revealed by ligation with fluoresceine isothiocyanate (FITC)-labeled annexin-V and used to quantify apoptotic cells by flow cytometry or fluorescence microscopy.

3.6.3. Cell Death ELISA

To assess specific apoptosis, the extent of DNA fragmentation in TRAIL-treated cells can be quantitated (OD_{405} nm) using the Cell Death Detection ELISA (Roche Molecular Biochemicals, Indianapolis, IN). The assay is based on a quantitative sandwich enzyme immunoassay principle, which detects histone-associated DNA fragments in the cytoplasmatic fraction of apoptotic cell lysates.

3.6.4. TUNEL Assay

The method of TdT-mediated dUTP nick-end labeling (TUNEL) is based on the specific binding of deoxynucleotidyl transferase (TdT) to the 3′OH ends of DNA strand breaks. The enzyme catalyzes a template-independent addition of deoxyribonucleoside triphosphates to DNA. The exposure of nuclear DNA in histological sections to TdT after proteolytic treatment is used to incorporate biotinylated deoxyuridine at sites of DNA breaks. The signal is then amplified by avidin-peroxidase. We have used the Apoptosis Detection System (Promega, Madison, WI).

3.6.5. Activation of Caspases

A number of assay kits for fluorometric detection of a specific caspase activity are available. The assays are designed to detect the shift in fluorescence emission of the molecule 7-amino-4-trifluoromethyl coumarin (AFC) conjugated to a specific caspase substrate. Upon proteolytic cleavage of the substrate by the caspase to be analyzed, the free AFC emits a yellow-green fluorescence at 505 nm. Colorimetric assays are also available, but are less sensitive.

4. Notes

1. Ideally, RNA should be extracted in an area dedicated solely for that purpose and away from areas used for cloning and DNA work. Gloves should be worn at all times, and work areas should be decontaminated with bleach. Sterile disposable plasticware should be used whenever possible. Because of the extreme sensitivity of the RT-PCR used here, it is necessary to take every precaution to reduce the risk of cross-contamination of samples. PCR product should not be brought back into the area where RT-PCR is set up. The lab should have dedicated areas for RNA exctraction, RT-PCR amplification, and postamplification analysis.
2. All solutions, with the exception of Tris, should be treated with the RNase inhibitor diethylpyrocarbonate (DEPC). For DEPC-treated water, add 200 µL of DEPC (Sigma) to 1 L of ddH_2O. Shake well, then loosen the cap and incubate for 2–8 h at 37°C. Autoclave; then shake again to remove DEPC breakdown products.

3. Frequently, RNA preparations contain small amounts of genomic DNA that may be amplified along with the target cDNA. If possible, PCR primers should be designed to bracket cDNA sequences that cross an intron–exon boundary in genomic DNA. For our protocol, all PCR primers must be biotinylated at their 5′ ends. *GAPD* primers are used as internal control to detect *GAPD* transcripts, which are ubiquitously expressed at high levels, in order to check the quality of the RNA and to control for variations in RT, PCR, and product detection steps. Primers for TRAIL-R1, TRAIL-R2, TRAIL-R3, and TRAIL-R4 have been described previously and their specificity has been confirmed ***(49)***.
4. Commercially synthesized primers are redissolved in sterile H_2O as stock solutions of 100 μ*M*. Primers used for PCR have to be diluted to 4 μ*M*. With the use of the primers described here, it should not be necessary to optimize PCR conditions. However, with the use of different primer sequences, PCR conditions might have to be changed. Key factors for amplification by PCR appear to be pH of the buffer, MgCl, dNTP, primer, and *Taq* polymerase concentrations and also the cycling parameters (especially the annealing temperature).
5. Dithiothreitol is required for stabilization of some enzymes.
6. Do not use polyacrylamide/bis-acrylamide 37.5:1, as the gel will stick to the glass plate. **Caution:** Unpolymerized polyacrylamide is neurotoxic!
7. Fresh blocking buffer should be prepared for each PCR detection. Otherwise, the buffer appears viscous after 3 d, and if used, the blots show a gray background.
8. The HCl used has to be of analytical grade to avoid precipitates in the wells.
9. One key to successful RT-PCR is the quality of the starting RNA. Because RNA is easily degraded once it is in solution, RT-PCR is best performed on samples recently extracted. PCR products can be analyzed at once, kept on 4°C overnight or on –20°C for a week.
10. A first strand cDNA synthesis reaction may be primed using different methods. The most nonspecific of the primers, random hexamers, are typically used when a particular mRNA is difficult to copy in its entirety. With this method, all RNAs in a cell population are templates for cDNA synthesis, and the PCR primers confer the needed specificity during the PCR amplification reaction. A more specific priming method is to use oligo(dT) to hybridize to 3′ poly(A) tails. The amount and complexity of cDNA is considerably less than when random hexamers are used.
11. Samples have to be overlaid with sterile mineral oil (even when using new PCR cyclers) because of the small volume of 10 μL. Otherwise, a large part of the sample will be lost by Verdunstung.
12. A hot start is often used to improve specificity by preventing priming and extension at low temperatures. Amplitaq Gold gives good results.
13. The chloroform is necessary to separate the PCR product from the mineral oil.
14. When pipetting three gels in a row, you have to be fast, otherwise the gel polymerizes before you can place the comb. When loading the gel, you have to be careful not to contaminate neighboring wells with the sample.

15. The gel is very thin and you have to be careful not to break it. We place the glass plate with the gel into the buffer, place Whatman paper under the swimming gel, carefully take it out of the buffer, and place it on the pads.
16. The CSPD solution must stay on the membrane for 5 min. To ensure an even and continuous spread on the membrane, the tray should be moved back and forth by hand once in a while.
17. Signals should be of sufficient intensity to give a good autoradiograph following 20–30 min of exposure, although longer exposure is sometimes required. When exposing the blot twice, the second exposure time is much shorter, as the maximum light emission occurs around 20 min after use of CSPD.
18. The appropriate dilution might vary in different experiments. The optimal dilution should be determined for each application.
19. Specificity in FACS staining of the respective anti-TRAIL and anti-TRAIL-R MAbs was determined by staining of TRAIL-R1 to TRAIL-R4 and TRAIL on CV1/EBNA cells transfected with pCDNA3.1-based expression vectors coding for the individual surface-bound TRAIL receptors or TRAIL itself.

References

1. Wiley, S. R., Schooley, K., Smolak, P. J., Din, W. S., Huang, C.-P., Nicholl, J. K., et al. (1995) Identification and characterisation of a new member of the TNF family that induces apoptosis. *Immunity* **3,** 673–682.
2. Pitti, R. M., Masters, S. A., Ruppert, S., Donahue, C. J., Moore, A., and Ashkenazi, A. (1996) Induction of apoptosis by Apo-2 ligand,a new member of tumor necrosis factor cytokine family. *J. Biol. Chem.* **271,** 12,687–12,690.
3. Jeremias, I., Herr, I., Boehler, T., and Debatin, K.-M. (1998) TRAIL/Apo-2-ligand-induced apoptosis in human T cells. *Eur. J. Immunol.* **28,** 143–152.
4. Cuello, M., Ettenberg, S. A., Nau, M. M., and Lipkowitz, S. (2001) Synergistic induction of apoptosis by the combination of trail and chemotherapy in chemoresistant ovarian cancer cells. *Gynecol. Oncol.* **81,** 380–390.
5. Ogasawara, J., Watanabe-Fukunaga, R., Adachi, M., Matsuzawa, A., Kasugai, T., Kitamura, Y., et al. (1993) Lethal effect of the anti-Fas antibody in mice. *Nature* **364,** 806–809.
6. Ashkenazi, A. and Dixit, V. M. (1999) Apoptosis control by death and decoy receptors. *Curr. Opin. Cell Biol.* **11,** 255–260.
7. Walczak, H., Miller, R. E., Ariail, K., Gliniak, B., Griffith, T. S., Kubin, M., et al. (1999) Tumoricidal activity of tumor necrosis factor-related apoptosis-inducing ligand in vivo. *Nat. Med.* **5,** 157–163.
8. Ashkenazi, A., Pai, R. C., Fong, S., Leung, S., Lawrence, D. A., Marsters, S. A., et al. (1999) Safety and antitumor activity of recombinant soluble Apo2 ligand. *J. Clin. Invest.* **104,** 155–162.
9. Lawrence, D., Shahrokh, Z., Marsters, S., Achilles, K., Shih, D., Mounho, B., et al. (2001) Differential hepatocyte toxicity of recombinant Apo2L/TRAIL versions. *Nat. Med.* **7,** 383–385.

10. Roth, W., Isenmann, S., Naumann, U., Kugler, S., Bahr, M., Dichgans, J., et al. (1999) Locoregional Apo2L/TRAIL eradicates intracranial human malignant glioma xenografts in athymic mice in the absence of neurotoxicity. *Biochem. Biophys. Res. Commun.* **265,** 479–483.
11. Pan, G., Ni, J., Wei, Y. F., Yu, G., Gentz, R., and Dixit, V. M. (1997) An antagonist decoy receptor and a death domain-containing receptor for TRAIL. *Science* **277,** 815–818.
12. Walczak, H., Degli-Esposti, M. A., Johnson, R. S., Smolak, P. J., Waugh, J. Y., Boiani, N., et al. (1997) TRAIL-R2: a novel apoptosis-mediating receptor for TRAIL. *EMBO J.* **16,** 5386–5397.
13. Pan, G., O'Rourke, K., Chinnaiyan, A. M., Gentz, R., Ebner, R., Ni, J., et al. (1997) The receptor for the cytotoxic ligand TRAIL. *Science* **276,** 111–113.
14. Sheridan, J. P., Masters, S. A., Pitti, R. M., Gurney, A., Skubatch, M., Baldwin, D., et al. (1997) Control of TRAIL-induced apoptosis by a family of signaling and decoy receptors. *Science* **277,** 818–821.
15. Schneider, P., Bodmer, J. L., Thome, M., Hofmann, K., Holler, N., and Tschopp, J. (1997) Characterization of two receptors for TRAIL. *FEBS Lett.* **416,** 329–334.
16. Screaton, G. R., Mongkolsapaya, J., Xu, X. N., Cowper, A. E., McMichael, A. J., and Bell, J. I. (1997) TRICK2, a new alternatively spliced receptor that transduces the cytotoxic signal from TRAIL. *Curr. Biol.* **7,** 693–696.
17. Leverkus, M., Neumann, M., Mengling, T., Rauch, C. T., Broecker, E.-B., Krammer, P. H., et al. (2000) Regulation of tumor necrosis factor-related apoptosis-inducing ligand sensitivity in primary and transformed human keratinocytes. *Cancer Res.* **60,** 553–559.
18. Degli-Esposti, M. A., Smolak, P. J., Walczak, H., Waugh, J., Huang, C.-P., DuBose, R. F., et al. (1997) Cloning and characterisation of TRAIL-R3, a novel member of the emerging TRAIL receptor family. *J. Exp. Med.* **1997,** 1165–1170.
19. Pan, G., Ni, J., Yu, G.-L., Wei, Y.-F., and Dixit, V. M. (1998) TRUNDD, a new member of the TRAIL receptor familythat antagonizes TRAIL signalling. *FEBS Lett.* **424,** 41–45.
20. Marsters, S. A., Sheridan, J. P., Pitti, R. M., Huang, A., Skubatch, M., Baldwin, D., et al. (1997) A novel receptor for APO2L/TRAILcontains a truncated death domain. *Curr. Biol.* **7,** 1003–1006.
21. Bernard, D., Quatannens, B., Vandenbunder, B., and Abbadie, C. (2001) Rel/NF-kappaB transcription factors protect against tumor necrosis factor (TNF)-related apoptosis-inducing ligand (TRAIL)-induced apoptosis by up-regulating the TRAIL decoy receptor DcR1. *J. Biol. Chem.* **276,** 27,322–27,328.
22. Guan, B., Yue, P., Clayman, G. L., and Sun, S. Y. (2001) Evidence that the death receptor DR4 is a DNA damage-inducible, p53-regulated gene. *J. Cell Physiol.* **188,** 98–105.
23. Hao, C., Beguinot, F., Condorelli, G., Trencia, A., Van Meir, E. G., Yong, V. W., et al. (2001) Induction and intracellular regulation of tumor necrosis factor-related apoptosis-inducing ligand (TRAIL) mediated apotosis in human malignant glioma cells. *Cancer Res.* **61,** 1162–1170.

24. Lacour, S., Hammann, A., Wotawa, A., Corcos, L., Solary, E., and Dimanche-Boitrel, M. T. (2001) Anticancer agents sensitize tumor cells to tumor necrosis factor-related apoptosis-inducing ligand-mediated caspase-8 activation and apoptosis. *Cancer Res.* **61,** 1645–1651.
25. Meng, R. D., McDonald, E. R., 3rd, Sheikh, M. S., Fornace, A. J., Jr., and El-Deiry, W. S. (2000) The TRAIL decoy receptor TRUNDD (DcR2, TRAIL-R4) is induced by adenovirus-p53 overexpression and can delay TRAIL-, p53-, and KILLER/DR5-dependent colon cancer apoptosis. *Mol. Ther.* **1,** 130–144.
26. Mongkolsapaya, J., Cowper, A. E., Xu, X. N., Morris, G., McMichael, A. J., Bell, J. I., et al. (1998) Lymphocyte inhibitor of TRAIL (TNF-related apoptosis-inducing ligand): a new receptor protecting lymphocytes from the death ligand TRAIL. *J. Immunol.* **160,** 3–6.
27. Nagane, M., Pan, G., Weddle, J. J., Dixit, V. M., Cavenee, W. K., and Huang, H. J. (2000) Increased death receptor 5 expression by chemotherapeutic agents in human gliomas causes synergistic cytotoxicity with tumor necrosis factor-related apoptosis-inducing ligand in vitro and in vivo. *Cancer Res.* **60,** 847–853.
28. Sheikh, M. S., Burns, T. F., Huang, Y., Wu, G. S., Amundson, S., Brooks, K. S., et al. (1998) p53-Dependent and -independent regulation of the death receptor KILLER/DR5 gene expression in response to genotoxic stress and tumor necrosis factor alpha. *Cancer Res.* **58,** 1593–1598.
29. Takeda, K., Hayakawa, Y., Smyth, M. J., Kayagaki, N., Yamaguchi, N., Kakuta, S., et al. (2001) Involvement of tumor necrosis factor-related apoptosis-inducing ligand in surveillance of tumor metastasis by liver natural killer cells. *Nat. Med.* **7,** 94–100.
30. Wen, J., Ramadevi, N., Nguyen, D., Perkins, C., Worthington, E., and Bhalla, K. (2000) Antileukemic drugs increase death receptor 5 levels and enhance Apo-2L-induced apoptosis of human acute leukemia cells. *Blood* **96,** 3900–3906.
31. Wu, G. S., Burns, T. F., McDonald, E. R., 3rd, Jiang, W., Meng, R., Krantz, I. D., et al. (1997) KILLER/DR5 is a DNA damage-inducible p53-regulated death receptor gene. *Nat. Genet.* **17,** 141–143.
32. Zhang, X. D., Nguyen, T., Thomas, W. D., Sanders, J. E., and Hersey, P. (2000) Mechanisms of resistance of normal cells to TRAIL induced apoptosis vary between different cell types. *FEBS Lett.* **482,** 193–199.
33. Mariani, S. M., Matiba, B., Armandola, E. A., and Krammer, P. H. (1997) Interleukin 1 beta-converting enzyme related proteases/caspases are involved in TRAIL-induced apoptosis of myeloma and leukemia cells. *J. Cell Biol.* **137,** 221–229.
34. Kischkel, F. C., Lawrence, D. A., Chuntharapai, A., Schow, P., Kim, K. J., and Ashkenazi, A. (2000) Apo2L/TRAIL-dependent recruitment of endogenous FADD and caspase-8 to death receptors 4 and 5. *Immunity* **12,** 611–620.
35. Sprick, M. R., Weigand, M. A., Rieser, E., Rauch, C. T., Juo, P., Blenis, J., et al. (2000) FADD/MORT1 and caspase-8 are recruited to TRAIL receptors 1 and 2 and are essential for apoptosis mediated by TRAIL receptor 2. *Immunity* **12,** 599–609.

36. Bodmer, J. L., Holler, N., Reynard, S., Vinciguerra, P., Schneider, P., Juo, P., et al. (2000) TRAIL receptor-2 signals apoptosis through FADD and caspase-8. *Nat. Cell Biol.* **2,** 241–243.
37. Schneider, P., Thome, M., Burns, K., Bodmer, J. L., Hofmann, K., Holler, N., et al. (1997) TRAIL receptors 1 (DR4) and 2 (DR5) signal FADD-dependent apoptosis and activate NF-KB. *Immunity* **7,** 831.
38. Thomas, W. D., Zhang, X. D., Franco, A. V., Nguyen, T., and Hersey, P. (2000) TNF-Related apoptosis-inducing ligand-induced apoptosis of melanoma is associated with changes in mitochondrial membrane potential and perinuclear clustering of mitochondria. *J. Immunol.* **165,** 5612–5620.
39. Walczak, H., Bouchon, A., Stahl, H., and Krammer, P. H. (2000) Tumor necrosis factor-related apoptosis-inducing ligand retains its apoptosis-inducing capacity on Bcl-2- or Bcl-xL-overexpressing chemotherapy-resistant tumor cells. *Cancer Res.* **60,** 3051–3057.
40. Keogh, S. A., Walczak, H., Bouchier-Hayes, L., and Martin, S. J. (2000) Failure of Bcl-2 to block cytochrome c redistribution during TRAIL- induced apoptosis. *FEBS Lett.* **471,** 93–98.
41. Hinz, S., Trauzold, A., Boenicke, L., Sandberg, C., Beckmann, S., Bayer, E., et al. (2000) Bcl-XL protects pancreatic adenocarcinoma cells against CD95- and TRAIL-receptor-mediated apoptosis. *Oncogene* **19,** 5477–5486.
42. Teitz, T., Wei, T., Valentine, M. B., Vanin, E. F., Grenet, J., Valentine, V. A., et al. (2000) Caspase 8 is deleted or silenced preferentially in childhood neuroblastomas with amplification of MYCN. *Nat. Med.* **6,** 529–535.
43. Hopkins-Donaldson, S., Bodmer, J. L., Bourloud, K. B., Brognara, C. B., Tschopp, J., and Gross, N. (2000) Loss of caspase-8 expression in highly malignant human neuroblastoma cells correlates with resistance to tumor necrosis factor-related apoptosis-inducing ligand-induced apoptosis. *Cancer Res.* **60,** 4315–4319.
44. Eggert, A., Grotzer, M. A., Zuzak, T. J., Wiewrodt, B. R., Ho, R., Ikegaki, N., et al. (2001) Resistance to tumor necrosis factor-related apoptosis-inducing ligand (TRAIL)-induced apoptosis in neuroblastoma cells correlates with a loss of caspase-8 expression. *Cancer Res.* **61,** 1314–1319.
45. Eggert, A., Brodeur, G. M., and Ikegaki, N. (2000) A relative quantitative RT-PCR protocol for *TrkB* expression in neuroblastoma using *GAPD* as an internal control. *Biotechniques* **28,** 681–691.
46. Chomczynski, P. and Sacchi, N. (1987) Single-step method of RNA isolation by acid guanidinium thiocyanate–phenol–chloroform extraction. *Anal. Biochem.* **62,** 156–159.
47. Mosmann, T. (1983) Rapid colorimetric assay for cellular growth and survival: application to proliferation and cytotoxicity assays. *J. Immunol. Methods* **65,** 55–63.
48. Weller, M., Malipiero, U., Rensing-Ehl, A., Barr, P. J., and Fontana, A. (1995) Fas/Apo-1 gene transfer for human malignant glioma. *Cancer Res.* **55,** 2936–2944.
49. Griffith, T. S., Chin, W. A., Jackson, G. C., Lynch, D. H., and Kubin, M. Z. (1998) Intracellular regulation of TRAIL-induced apoptosis in human melanoma cells. *J. Immunol.* **161,** 2833–2840.

50. Mori, S., Murakami-Mori, K., Nakamura, S., Ashkenazi, A., and Bonavida, B. (1999) Sensitization of AIDS–Kaposi's sarcoma cells to Apo-2 ligand-induced apoptosis by actinomycin D. *J. Immunol.* **162,** 5616–5623.
51. Nicoletti, I., Migliorati, G., Pagliacci, M. C., Grignani, F., and Riccardi, C. (1991) A rapid and simple method for measuring thymocyte apoptosis by propidium iodide staining and flow cytometry. *J. Immunol. Methods* **139,** 271–279.

9

Angiogenic Cytokines

Quantitative and Functional Analysis

Angelika Eggert and John H. Maris

1. Introduction

1.1. Background

Following the recognition that angiogenesis is essential for tumor growth and metastasis formation ***(1,2)***, evidence has accumulated that angiogenesis is a component of many pathologies. Numerous angiogenic factors that regulate this complex process alone or in synergy have been identified. Vascular endothelial growth factor (VEGF) ***(3)***, basic fibroblast growth factor (bFGF) ***(4)***, transforming growth factor-α (TGF-α), platelet-derived growth factor (PDGF), and angiopoietin-1 and -2 (Ang-1 and Ang-2) have been shown to induce angiogenesis in a variety of experimental models ***(5–7)***. VEGF is an important angiogenic agent and endothelial specific mitogen, which has been implicated in the neovascularization of a wide variety of tumors ***(9)***. VEGF acts via a paracrine mechanism mainly through two specific receptors on the surface of endothelial cells: Flt 1 and KDR ***(10,11)***. Although encoded by a single gene, VEGF has several isoforms generated by alternative splicing ***(12,13)***. Of these, the main isoforms $VEGF_{121}$ and $VEGF_{165}$ are secreted soluble glycoproteins, whereas $VEGF_{189}$ and $VEGF_{206}$ remain bound to heparan sulfate proteoglycans at the cell surface ***(14)***. The importance of VEGF as a potential target for antineoplastic therapy has been demonstrated in several studies in which neutralizing antibodies to VEGF or VEGF receptors inhibited tumor growth and vascularization in vivo ***(15,16)***.

From: *Methods in Molecular Biology, vol. 215:*
Cytokines and Colony Stimulating Factors: Methods and Protocols
Edited by: D. Körholz and W. Keiss © Humana Press Inc., Totowa, NJ

VEGF-B and VEGF-C are two recently discovered members of the VEGF family *(17–19)*, that are expressed in many tissues and have mitogenic and/or chemotactic actions on endothelial cells, indicating that they may also contribute to the induction or maintenance of angiogenesis. VEGF-C was discovered as the ligand for the third member of the VEGF receptor family (Flt-4 or VEGFR-3), which is expressed mainly on lymphatic endothelium of adult tissues *(19,20)*. Expression of VEGF-B and VEGF-C has been detected in a variety of human tumors *(21)* and VEGF-C has been shown to be lymphangiogenic in vivo *(22)*.

Ang-1, the ligand for TIE2, a receptorlike tyrosine kinase expressed almost exclusively in endothelial cells, seems to be important to maintain vessel integrity by mediating interactions between the endothelium and the surrounding matrix *(23)*. It stabilizes the structure of newly formed vessels and has a later role in the process of neovascularization than VEGF. Its naturally occurring antagonist Ang-2 binds with similar affinity to TIE2, but it does not activate the receptor *(24)*. Inhibition of Ang-1 by Ang-2 has been suggested to drive angiogenesis in the presence of angiogenic inducers like VEGF by loosening contacts between endothelial and periendothelial cells, thus rendering endothelial cells accessible to angiogenic inducers *(25)*. Basic FGF is a mitogenic, angiogenic, and neurotrophic factor expressed by many tumor cells *(8,26)*. PDGF consists of two related polypeptides (A- and B-chain) *(27,28)*. It originally was discovered to be involved in the regulation of cell migration and proliferation, but it has more recently been found to possess an angiogenic capability both in vitro and in vivo *(6)*. TGF-α has been shown to induce VEGF expression *(29)* and also has an angiogenic role in vivo *(5,30)*.

As inhibition of angiogenesis might be an attractive therapeutic strategy for many human tumors, the expression and function of cytokines stimulating angiogenesis deserves further investigation in a wide range of human malignancies. Establishing "angiogenic profiles" might increase our understanding of angiogenesis-associated diseases and permit the design of individual antiangiogenic therapies, but it may also provide parameters for diagnosis, prognosis and monitoring response to therapies.

1.2. Methods

A wide variety of angiogenesis assays have been established in the past 5–10 yr. Many sophisticated in vivo models are relatively expensive and demand high levels of surgical skills. In this chapter, relatively straightforward, inexpensive protocols to assess the "angiogenic profile" of a disease will be introduced and briefly discussed. For more advanced in vitro and in vivo protocols, the readers are referred to *Angiogenesis Protocols*, edited by J. C. Murray.

Methods for analyzing angiogenic cytokines and angiogenesis in a certain cell type can be classified as quantitative or functional. Each approach has advantages and limitations. The most appropriate method depends on the cell population in question. In our experience, reverse transcription–polymerase chain reaction (RT-PCR) and enzyme-linked immunosorbent assay (ELISA) analyses are the most useful assays for the quantitative analysis of angiogenic stimulators in cell lines. For the analysis of angiogenic cytokines in tissue of mixed-cell populations (i.e., tumor tissue), immunohistochemistry may be a more adequate method, because differential expression by subpopulations of cells can be determined. Simple functional assays include endothelial cell proliferation and migration assays to assess the angiogenic potential in vitro as well as a murine Matrigel plug assay to determine the amount of angiogenesis in vivo.

1.2.1. Quantitative Analysis

1.2.1.1. Semiquantitative RT-PCR

Measurement of angiogenic ligand and receptor mRNA expression levels is a semiquantitative approach to analyze neovascularization. RT-PCR is a sensitive analytical method and is a suitable approach for the analysis of mRNA expression in cell lines. Because of the extreme sensitivity of PCR, there is a risk in PCR-based measurements of tumor tissue that the mRNA detected may derive from a minor contaminating cell population like lymphocytes or fibroblasts or surrounding vessels.

1.2.1.2. Determination of VEGF and bFGF Protein Levels in Conditioned Medium: Western Blot Analysis and ELISA

Protein levels of angiogenic cytokines in the conditioned medium (CM) of cell lines can be determined by Western blot analysis or a specific ELISA assay. In our experience, Western blot analysis can be difficult because of a high salt concentration in concentrated CM. Dialysis may be necessary before sodium dodecyl sulfate–polyacrylamide gel electrophoresis (SDS–PAGE). A better alternative are Quantikine Immunoassays (R&D Systems), solid-phase ELISAs designed to measure levels of angiogenic cytokines in cell culture, serum, and plasma. Specific ELISAs for accurate quantification are available for VEGF, bFGF, PDGF-A and -B, and TGF-β.

1.2.1.3. Microvessel Count and Angiogenic Cytokine Expression in Tumor Tissue

For the measurement of angiogenesis in tissue sections most studies have employed a method based on the one developed by Weidner et al. ***(31)***, in which blood vessels are immunohistochemically highlighted and the number of microvessels is quantified in the most vascular areas (so-called "hot spots") of the tumor. Because endothelium is highly heterogeneous, the choice of antibody profoundly influences the number of microvessels available for assessment. Many antibodies suffer from low specificity and their antigens may be present on many nonendothelial elements. Antibodies to factor-VIII-related antigen identify only a proportion of capillaries and also detect lymphatic endothelium. The most specific and sensitive human endothelial marker currently available is CD31, which is present on most capillaries and is a reliable epitope for immunostaining in routinely handled formalin-fixed paraffin-embedded tissues ***(32)***. A good alternative antigen in human tissues is CD34, although this antigen is also expressed by some stromal cells ***(33)***. In animal models, however, the newly formed tumor blood vessels derive from the host animal even when the engrafted tumor cells are human. In this setting, antibodies raised against human endothelial epitopes are difficult to use. Rat and mouse endothelial cells express on their surface α-D-galactosyl residues that can be detected by lectin histochemistry with the lectin Bandeiraea Simplicifolia agglutinin (BSI) ***(34)***. In addition to immunostaining of microvessels with CD31 or BSI lectin, a protocol for immunohistochemical analysis of VEGF and bFGF protein expression in paraffin-embedded tissue will be described.

1.2.2. Functional Analysis

To assess the potential of expressed angiogenic cytokines to stimulate angiogenesis in vitro and in vivo, a functional analysis is required.

1.2.2.1. In Vitro: Endothelial Cell Proliferation and Migration

Endothelial cell (EC) proliferation and migration are important steps in the process of angiogenesis. Activated ECs migrate through their basement membrane into the surrounding tissue stroma, where they form solid sprouts. The newly formed sprouts move toward the source of angiogenic stimulus by a combination of cell migration and cell proliferation. The effect of soluble angiogenic stimulators (or conditioned media obtained from cell lines producing angiogenic factors) on EC proliferation and migration is measured in dose-response curves, which serve as an initial screening method to define the angiogenic potential of the material to be analyzed. Considering the well-documented heterogeneity of ECs, it is clear that the effects of different angiogenic stimulators or inhibitors can only be compared using the same EC target population (i.e., human umbilical vein endothelial cells [HUVEC]).

1.2.2.2. In Vivo: Matrigel Assay

A relatively simple and rapid in vivo method to determine the angiogenic potential of compounds or conditioned media is desirable to augment in vitro findings. One such quantitative method is the murine Matrigel plug assay *(35)*. Matrigel, an extract of the Englebreth–Holm–Swarm tumor composed of basement membrane components, is liquid at 4°C and forms a gel at 37°C. Matrigel is injected subcutaneously into the ventral region of mice, either alone or mixed with potential angiogenic compounds (or angiogenic-factor-producing tumor cells). It solidifies forming a "Matrigel plug," and when angiogenic factors are contained, endothelial cells migrate into the plug and form vessels. The level of angiogenesis can be viewed by embedding and sectioning the plugs in paraffin and staining using Masson's Trichrome, which stains the Matrigel blue and the endothelial cells/vessels red (not described here). A simpler and faster alternative is to measure hemoglobin levels in the plugs with the Drabkin assay (Sigma). Hemoglobin reacts with Drabkin's reagent which contains potassium ferricyanide, potassium cyanide, and sodium bicarbonate. Most forms of hemoglobin are converted into methemoglobin by the action of ferricyanide. The methemoglobin then reacts with cyanide to form cyanmethemoglobin. The absorbance of this derivative at 530–550 nm is proportional to the hemoglobin content in the sample.

2. Materials

2.1. Semiquantitative RT-PCR

Reagents and protocols for semiquantitative RT-PCR have been described in detail in Chapter 8. The same RT-PCR protocol can be used for detecting mRNA expression of angiogenic stimulators. The specific primer sequences are listed in **Table 1**. *See* **Note 1**.

2.2. ELISA

1. Quantikine® Immunoassay (i.e., VEGF or bFGF; R&D Systems, Minneapolis, MN).
2. Microplate reader capable of measuring absorbance at 450 nm, with the correction wavelength set at 540 nm or 570 nm.
3. Pipets, 50 µL–200 µL for the assay; 1 mL, 5 mL, 10 mL, and 25 mL for reagent preparation.
4. Multichannel pipet or automated microplate washer.
5. 12 mm × 75-mm Polypropylene test tubes.

2.3. Endothelial Proliferation Assay

1. Human umbilical vein endothelial cells (HUVEC, Clonetics Corporation, Walkersfield, MD). (*See* **Note 2**.)
2. HUVEC growth medium: EBM and EGM (Clonetics Corporation) (*see* **Note 3**).
3. 24-Well flat-bottom tissue culture plates.
4. 1% Fetal bovine serum (FBS).

Table 1
Primer Sequences (*see* Note 1)

Primer	Sequence	PCR product size (bp)
VEGF sense	AGGCCAGCACATAGGAGAGA	104, 236, 308, 359
VEGF antisense	ACCGCCTCGGCTTGTCACAT	
VEGF-B sense	GACAGTGCTGTGAAGCCA	225, 326
VEGF-B antisense	TAGCCTCTGAGGCAAG	
VEGF-C sense	GAAAGGAGGCTGGCAACATA	228
VEGF-C antisense	ACGGACACACATGGAGGTTT	
Angiopoietin-1 sense	AGGAGCAAGTTTTGCGAGAG	263
Angiopoietin-1 antisense	CTGCAGAGCGTTTGTGTTGT	
Angiopoietin-2 sense	CCACAAATGGCATCTACACG	202
Angiopoietin-2 antisense	CCCAGCCAATATTCTCCTGA	
bFGF sense	GTGTGTGCTAACCGTTACCT	238
bFGF antisense	GCTCTTAGCACAGACATTGGAAG	
TGF-α sense	CGCCCTGTTCGCTCTGGGTAT	241
TGF-α antisense	AGGAGGTCCGCATGCTCACAG	
PDGF-A sense	CCCCTGCCCATTCGGAGGAAGAGA	228
PDGF-A antisense	TTGGCCACCTTGACGCTGCGGTG	
GAPDH sense	CATCAAGAAGGTGGTGAAGC	160
GAPDH antisense	GAGCTTGACAAAGTGGTCGT	

5. Conditioned medium (CM) of the cell lines to be tested (50-fold concentrated supernatant, diluted 1:3 in EBM + 1% FBS) (*see* **Note 4**).
6. Phosphate-buffered saline (PBS).
7. 3% Formaldehyde/2% sucrose solution.
8. Cresyl violet in 20% methanol.
9. Phase-contrast microscope or Coulter counter.
10. Negative control: EBM + 1% FBS.
11. Positive control: recombinant VEGF (100 ng/mL in EBM + 1% FBS) or bFGF (10 ng/mL in EBM + 1% FBS).

2.4. Endothelial Migration Assay

1. Transwell inserts (Costar) containing a polycarbonate membrane with 8-μm pores.
2. 600 μL CM to be tested (50-fold concentrated, diluted 1:3 in EBM + 1% FBS).
3. 100 μL of HUVEC (2×10^4 cells).
4. 4% Paraformaldehyde in PBS.
5. Cotton swab.

6. Quick-Dip (Mercedes Medical Inc., Sarasota, FL).
7. Glass slides and mounting medium.
8. Phase-contrast microscope.

2.5. Immunohistochemistry

1. Silane-coated microscope slides.
2. Dry incubator/oven at 37°C.
3. Xylene.
4. Graded alcohols (100%, 90%, and 70% ethanol).
5. 5% H_2O_2 in methanol.
6. Phosphate-buffered saline (PBS).
7. 0.1 *M* Tris-HCl, pH 7.6.
8. 2% Donor horse serum (DHS) in 0.1 *M* Tris-HCl, pH 7.6.
9. 0.01 *M* Citrate buffer, pH 6.0 (1.92 g citric acid, H_2O to 1 L).
10. Anti-VEGF rabbit polyclonal antibody A-20 (Santa Cruz Biotechnology), 1:400.
11. Anti-bFGF rabbit polyclonal antibody F5537 (Sigma), 1:200.
12. Monoclonal antibody against human PECAM/CD31 (JC70a, Dako, UK, 10 µg/µL) or against mouse CD31 (Pharmingen, 1:30). Alternatively biotinylated BSI–Lectin (Sigma) 1 mg/mL, diluted 1:20.
13. Anti-rabbit secondary antibody.
14. StrAviGen Super Sensitive kit (BioGenex Laboratories, San Ramon, CA) or StreptABC kit (Dako, UK).
15. Hematoxillin.
16. Normal mouse IgG (Vector Laboratories, Burlington, CA).
17. Aqueous mountant.
18. Phase-contrast microscope.

2.6. Matrigel Assay

1. Five or six female mice for each test group (i.e., athymic NCR [nu/nu] mice) with suitable cages and supplies.
2. Matrigel (Collaborative Biochemical, Bedford, MA); store at –20°C. 0.5 mL per mouse.
3. 1-mL Syringes and 25-gage needles.
4. Tumor cells or angiogenic compound to be analyzed.
5. MatriSperse (Collaborative Biochemical).
6. Drabkin's reagent kit (Sigma Diagnostics, St. Louis, MO).
7. Spectrophotometer capable of measuring absorbance at 530–550 nm.

3. Methods

3.1. Semiquantitative RT-PCR

Reagents and protocols for semiquantitative RT-PCR have been described in detail in Chapter 8. The same RT-PCR protocol can be used for detecting mRNA expression of angiogenic stimulators.

3.2. Collection of Conditioned Media for ELISA, EC Proliferation, and Migration Assays

1. Grow the cell culture to be analyzed to 80% confluency
2. Starve the cell culture for 24–48 h in serum-free RPMI medium + 1% heat-inactivated FBS (*see* **Note 5**).
3. If the effect of hypoxia has to be tested, control cells can be grown in a hypoxic chamber or hypoxia can be mimicked by addition of 100 μ*M* desferoxamine mesylate ***(36)*** (*see* **Note 6**).
4. Collect cell supernatants.
5. Remove particulates by centrifugation at 1400*g* for 10 min and store at –80°C.
6. Concomitantly, harvest cell pellets and determine cell number by counting the cells with a Coulter counter or in a hemocytometer (*see* **Note 7**).

3.3. ELISA

1. Bring all reagents of the Quantikine Immunoassay to room temperature and prepare all reagents and standards as directed in the instruction pamphlet of the kit.
2. Add 50 μL of assay diluent to each well.
3. Add 200 μL of standard or sample (cell culture supernatant) per well.
4. Cover with the adhesive strip provided and incubate for 2 h at room temperature.
5. Aspirate each well and wash, repeating the process twice for a total of three washes. Wash by filling each well with wash buffer (400 μL) using a multichannel pipet or autowasher (*see* **Note 8**). After the last wash, remove any remaining wash buffer by aspirating.
6. Add 200 μL of conjugate (i.e., VEGF) to each well. Cover with a new adhesive strip and incubate for 2 h at room temperature.
7. Repeat the aspiration/wash as in **step 5**.
8. Add 200 μL of substrate solution to each well and incubate for 20 min at room temperature.
9. Add 50 μL of stop solution to each well (*see* **Note 9**).
10. Determine the optical density of each well within 30 min, using a microplate reader set to 450 nm with wavelength correction set to 540 nm or 570 nm (*see* **Note 10**).
11. Generate a standard curve and calculate the results as directed in the instruction pamphlet of the kit. The amount of angiogenic cytokine protein in the conditioned medium analyzed should be expressed as pg protein/10^6 cells/24 h.

3.4. Immunohistochemistry: Microvessels, VEGF, bFGF

3.4.1. Staining Procedure

1. Cut 4-μm formalin-fixed paraffin-embedded sections of the representative tumor block onto silane-coated slides (*see* **Note 11**).
2. Dry at 37°C overnight in an incubator.

3. Dewax using xylene for 2X 5 min before passing through graded alcohols into water.
4. Place in PBS for 5 min and block endogenous peroxidase by incubation in 0.5% H_2O_2 in methanol for 30 min.
5. Wash sections in running destilled water for 10 min.
6. Pretreat sections with microwave: 15 min, 1000 W in freshly prepared 0.01 *M* citrate buffer.
7. Cool down rapidly to room temperature with running distilled water.
8. Immerse sections for 5 min in 0.1 *M* Tris-HCl, pH 7.6.
9. Immerse sections for 10 min in 2% DHS in 0.1 *M* Tris-HCl, pH 7.6.
10. Rinse and apply the primary antibody diluted in 2% DHS in 0.1 *M* Tris-HCl, pH 7.6 (appropriate dilution according to **Subheading 2.5.**; *see* **Note 12**).
11. Incubate in a humidified chamber at 4°C overnight.
12. Detect antibody reaction with StrAviGen Super Sensitive kit or standard Strept-ABC immunohistochemistry according to the manufacturer's instructions (*see* **Note 13**).
13. Counterstain sections with hematoxillin.
14. Negative control: Omission of primary antibody and substitution with normal mouse IgG.
15. Positive control: Suitable tissue with known expression of the angiogenic cytokine to be analyzed or tissue with normal vasculature (*see* **Note 14**).

3.4.2. Quantification of Angiogenic Cytokine and Microvessel Staining

Evaluation of immunostained sections is usually carried out by two observers over a conference microscope. The intensity of VEGF and bFGF cytoplasmatic staining in representative tissue sections can be graded on a scale of 0–3+: 0 = no staining, 1+ = weak staining, 2+ = moderate staining, and 3+ = strong staining (*see* **Fig. 1**).

Microvessel counts are determined by a method based on that developed by Weidner et al. *(31)*. The three most vascular areas ("hot spots") of the tumor sections containing the maximum number of discrete microvessels are identified at low magnification (40×–100×). The number of vessels is then counted at high magnification (200×–400×) in these regions. Any EC cluster separate from adjacent microvessels is considered a countable vessel. Larger vessels containing more than eight erythrocytes or a thick lamina muscularis are excluded from the count.

3.5. Endothelial Cell Proliferation

1. Plate HUVEC in gelatinized 24-well tissue culture plates at a concentration of 5.0×10^4/well.
2. After growing for 24 h in EGM medium, starve HUVEC overnight in EBM medium supplemented with 1% FBS (*see* **Note 15**).

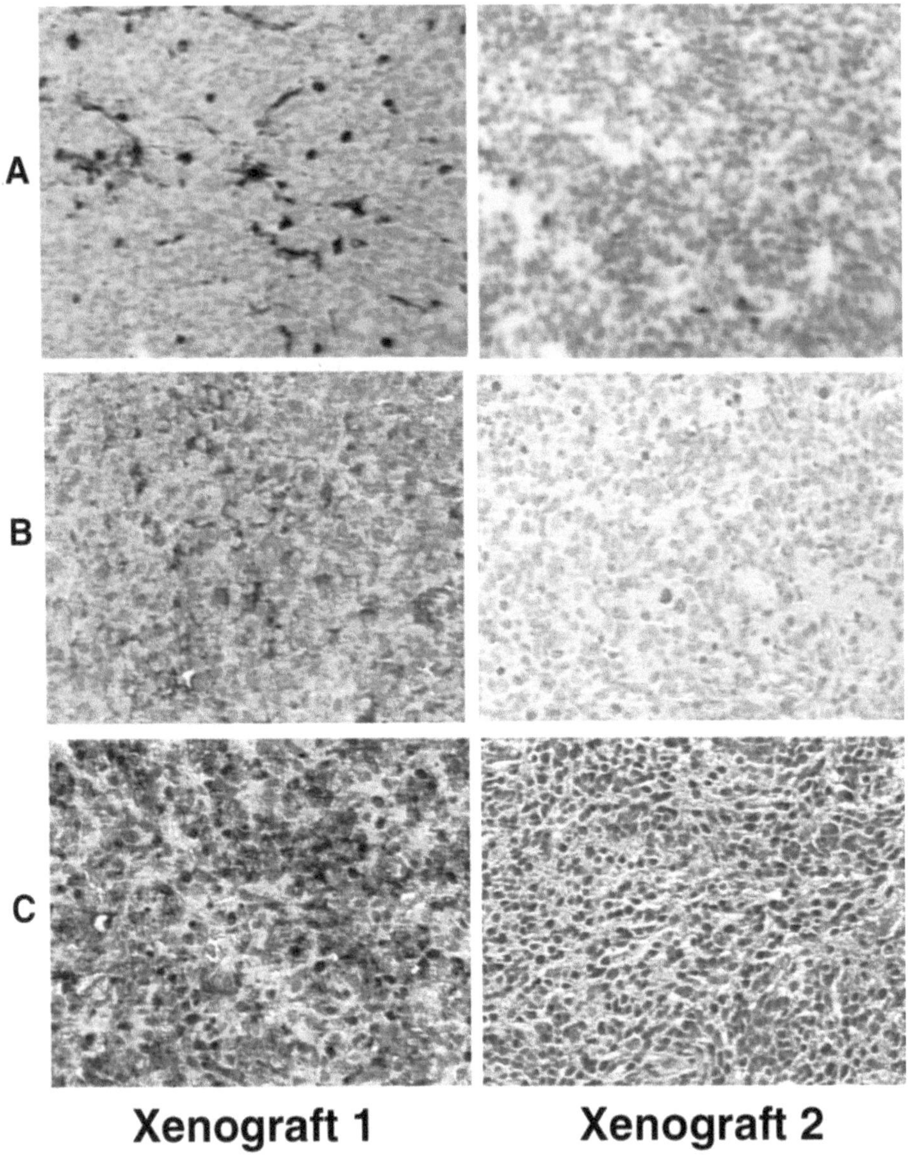

Fig. 1. Immunohistochemistry of neuroblastoma xenografts. Representative example of immunohistochemistry of microvessels with BSI–Lectin (**A**), VEGF expression (**B**), and bFGF expression (**C**) in two different neuroblastoma xenografts in nude mice. Left: Highly vascular with strong VEGF and bFGF staining; right: poor vascularization with very weak VEGF and bFGF staining. Magnification: 100×.

3. Concentrate the collected CM of the cell line to be tested 50-fold and dilute the concentrate 1:3 in EBM + 1% FBS (*see* **Note 4**).
4. Incubate HUVEC in duplicate cultures of this prepared CM at 37°C for 96 h.
5. Include positive controls of recombinant VEGF (100 ng/mL) or bFGF (10 ng/mL).
6. Use EBM + 1% FBS as a negative control.
7. Following incubation wash HUVEC with PBS.
8. Fix HUVEC with a 3% formaldehyde/2% sucrose solution.
9. Stain with a solution of cresyl violet in 20% methanol.
10. Count HUVEC in 10 different high-power fields at 320× magnification (*see* **Fig. 2**).
11. Alternatively, the number of HUVEC (unstained) can be counted in a Coulter counter.

3.6. Endothelial Cell Migration

We use a modified Boyden chamber assay for the quantification of EC migration *(37,38)*:

1. Place Transwell inserts (Costar) in 24-well culture dishes.
2. Add 600 µL of the CM to be tested (50-fold concentrated, diluted 1:3 in EBM + 1% FBS) to the bottom well.
3. Plate 100 µL of HUVEC (2×10^4 cells) on the top of the Transwell membrane in the upper compartment.
4. Incubate HUVEC at 37°C and 5% CO_2 for 4 h.
5. Fix the membranes in 4% paraformaldehyde in PBS.
6. Wipe the cells that had not migrated through the pores from the upper surface of the membrane with a cotton swab (*see* **Note 16**).
7. Cut membranes out of the Transwell insert.
8. Stain membranes with Quick-Dip (Mercedes Medical Inc., Sarasota, FL) and mount them on glass slides.
9. Quantitate migration by counting the cells that migrated to the bottom of the membrane in 10 high-power fields.

*3.7. In Vivo Angiogenesis: Matrigel Assay (*see Note 16*)*

1. Day 0: Inject athymic NCR (nu/nu) mice sc in the abdominal midline with 0.5 mL Matrigel alone or mixed on ice with the tumor cells to be analyzed as the angiogenic stimulus *(39)*.
2. Days 10–14: harvest Matrigel plugs for hemoglobin assay. Take photos of plugs to document macroscopic angiogenesis. (*See* **Note 17**.)
3. Perform quantitation of hemoglobin in the plugs by the Drabkin method *(40)*:
   - Dissolve harvested Matrigel pellets in MatriSperse at 4°C overnight (*see* **Note 18**).
   - Assay supernatant for hemoglobin content using the Drabkin's reagent kit according to the manufacturer's instructions.

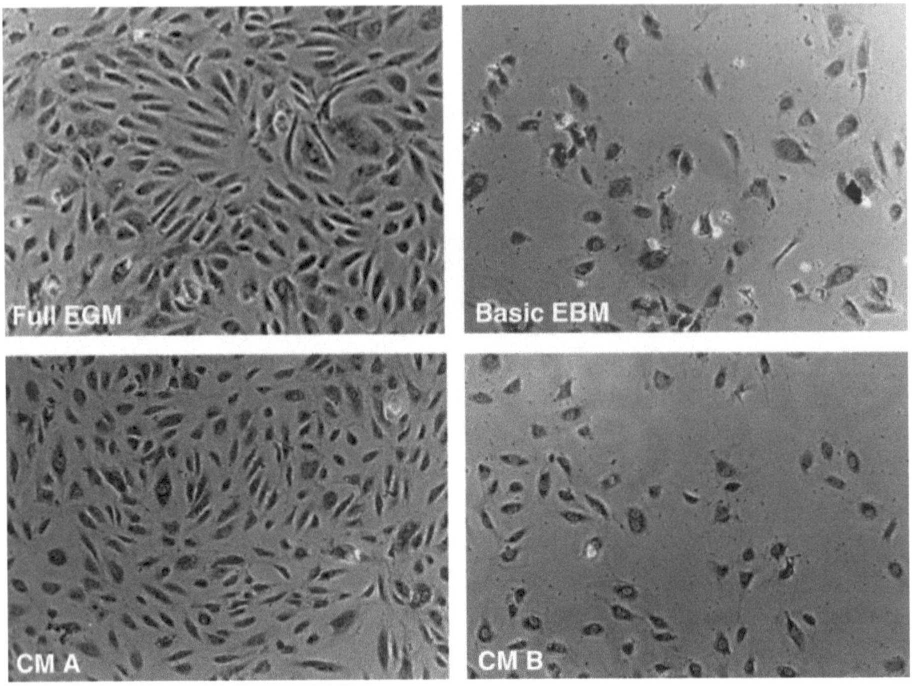

Fig. 2. Effect of NB cell supernatant on the proliferation of human vascular endothelial cells (HUVEC). Representative example of HUVEC cultured with conditioned medium A (lower left, highly stimulatory) and conditioned medium B (lower right, no stimulatory effect) in comparison to full EGM medium (upper left, containing EC growth factors) and basic EBM medium (upper right, containing no growth factors). Phase-contrast microscopy; 320×.

4. Notes

1. Frequently, RNA preparations contain small amounts of genomic DNA that may be amplified along with the target cDNA. If possible, PCR primers should be designed to bracket cDNA sequences that cross an intron–exon boundary in genomic DNA. For our protocol, all PCR primers must be biotinylated at their 5′ ends. *GAPD* primers are used as internal control to detect *GAPD* transcripts, which are ubiquitously expressed at high levels, in order to check the quality of the RNA and to control for variations in RT, PCR, and product detection steps. Primer sequences for VEGF are able to detect all four different molecular species produced by alternative splicing: $VEGF_{121}$, $VEGF_{165}$, $VEGF_{189}$, and $VEGF_{201}$. Primers for VEGF-B are able to detect the two different isoforms $VEGF\text{-}B_{167}$ and $VEGF\text{-}B_{186}$.
2. A great deal of heterogeneity exists in the morphology of the endothelium and in the phenotype displayed by individual ECs. Other ECs than HUVEC can be

used (like bovine aortic ECs or bovine brain ECs), but the effects of a particular angiogenic factor compared to another should always be defined in the context of the same target cell population.

3. EBM: Endothelial basal medium, containing no growth factors, not stimulatory for EC proliferation or migration. EGM: complete EC growth medium, containing all growth factors and supplements necessary for the propagation of EC.
4. The optimal concentration and dilution has to be determined for each cell type and CM. If the concentration is too low, the amount of angiogenic factors might not be sufficient to see differences in the effect on EC proliferation or migration. If the concentration is too high, dialysis might be necessary before incubating ECs in it, as a high salt concentration is cytotoxic for ECs.
5. The optimal time for the collection of CM might vary. Some cell types cannot stand more than 24 h in medium with 1% FBS; for others, it might be necessary to collect CM for several days to obtain sufficient amounts of angiogenic factors.
6. Hypoxia is known to upregulate production of angiogenic factors like VEGF. Thus, it might be of interest to compare the amounts of angiogenic factors in normoxic and hypoxic cell cultures. If a hypoxic chamber is not available, hypoxia can be mimicked by addition of cobalt chloride or desferoxamine ***(36)***. Both activate the hypoxia inducible factor 1 (HIF-1), like hypoxia itself.
7. The amount of angiogenic factor production should always be calculated for the same cell count, as different cell density can influence the results. A comparison of the same protein amounts might not give accurate results.
8. Complete removal of liquid at each step is essential to good performance.
9. If color change does not appear uniform, gently tap the plate to ensure thorough mixing.
10. If wavelength correction is not available, subtract readings at 540 nm or 570 nm from the readings at 450 nm. This subtraction will correct for optical imperfections in the plate. Readings made directly at 450 nm without corrections may be higher and less accurate.
11. Eight-micrometer cryostat sections can also be used, but the area of tumor assessed is less representative.
12. The optimal dilution might vary in different tissues and has to be determined for each experiment.
13. This method uses the high affinity of streptavidin for biotin. After application of the primary antibody, a biotinylated secondary antibody is overlayed. The tertiary antibody complex of streptavidin–biotin–horseradish peroxidase (HRP) is then applied. The brown end product is formed as 3′3-diamino benzidine HCl (DAB) is oxidized when it donates electrons to acticate the HRP/H_2O_2 reaction.
14. A control slide of the same tissue section should be analyzed together with each batch of samples to ensure comparable staining results.
15. One percent FBS has to be added for the stability of VEGF.
16. The upper surface of the membrane has to be wiped with a cotton swap to prevent counting of ECs, that had not migrated through the membrane, but just attach to its upper surface.

17. The optimal time for the harvest of the plugs has to be determined for each individual set of experiments. If it is too short, the vascularization is not sufficient to be measured. If it is too long, the plugs have solidified and cannot be dissolved in Matrisperse any more.
18. Different results in experiments can be the result of differences in Matrigel preparations (use the same lot for the whole series of experiments), age and gender of the mice (vessel formation in 6-mo-old mice is reduced as compared to mice 12–24 mo old ***(40)*** as well as different injection sites (one of the best areas in terms of angiogenic response is the ventral side of the mouse in the groin area close to the dorsal midline). Use the same conditions for all experiments that you intend to compare.

References

1. Folkman, J. (1986) How is blood vessel growth regulated in normal and neoplastic tissue? *Cancer Res.* **46,** 467–473.
2. Folkman, J. (1989) What is the evidence that tumors are angiogenesis dependent? *J. Natl. Cancer Inst.* **82,** 4–6.
3. Leung, D. W., Cachianes, G., Kuang, W. J., Goeddel, D. V., and Ferrara, N. (1989) Vascular endothelial growth factor is a secreted angiogenic mitogen. *Science* **246,** 1306–1309.
4. Folkman, J. and Klagsbrun, M. (1987) Angiogenic factors. *Science* **235,** 442–447.
5. Rappolee, D. A., Mark, D., Banda, M. J., and Werb, Z. (1991) Wound macrophages express TGF-α and other growth factors in vivo: analysis by mRNA phenotyping. *Science* **241,** 708–712.
6. Risau, W., Drexler, H., Mironov, V., Smits, A., Siegbahn, A., Funa, K., et al. (1992) Platelet-derived growth factor is angiogenic in vivo. *Growth Factors* **7,** 261–266.
7. Suri, C., Jones, P. F., Patan, S., Bartunkova, S., Maisonpierre, P. C., Davis, S., et al. (1996) Requisite role of angiopoietin-1, a ligand for the tie-2 receptor, during embryonic angiogenesis. *Cell* **87,** 1171–1180.
8. Smith, K., Fox, S. B., Whitehouse, R., Taylor, M., Greenall, M., Clarke, J., et al. (1999) Upregulation of basic fibroblast growth factor in breast carcinoma and its relationship to vascular density, oestrogen receptor, epidermal growth factor receptor and survival. *Ann. Oncol.* **10,** 707–713.
9. Plate, K. H., Breier, G., Weich, H. A., and Risau, W. (1992) Vascular endothelial growth factor is a potential tumor angiogenesis factor in human gliomas in vivo. *Nature* **359,** 845–848.
10. de Vries, C., Escobido, J. A., Ueno, H., Houck, K., Ferrara, N., and Williams, L. T. (1992) The fms-like tyrosine kinase, a receptor for vascular endothelial growth factor. *Science* **255,** 989–991.
11. Terman, B. I., Dougher-Vermazen, M., Carrion, M. E., Dimitrov, D., Armellino, D. C., Gospodarowicz, D., et al. (1992) Identification of the KDR tyrosine kinase as a receptor for vascular endothelial growth factor. *Biochem. Biophys. Res. Commun.* **187,** 1579–1586.

12. Tischer, E., Mitchell, R., Hartman, T., Silva, M., Gospodarowicz, D., Fiddes, J. C., et al. (1991) The human gene for vascular endothelial growth factor. Multiple protein forms are encoded through alternative exon splicing. *J. Biol. Chem.* **266,** 11,947–11,954.
13. Neufeld, G., Tessler, S., Gitay-Goren, H., Cohen, T., and Levi, B. Z. (1994) Vascular endothelial growth factor and its receptors. *Prog. Growth Factor Res.* **5,** 89–97.
14. Houck, K. A., Leung, D. W., Rowland, A. M., Winer, J., and Ferrara, N. (1992) Dual regulation of vascular endothelial growth factor bioavailability by genetic and proteolytic mechanisms. *J. Biol. Chem.* **267,** 26,031–26,037.
15. Asano, M., Yukita, A., Matsumoto, T., Kondo, S., and Suzuki, H. (1995) Inhibition of tumor growth and metastasis by an immunoneutralizing monoclonal antibody to human vascular endothelial growth factor/vascular permeability factor 121. *Cancer Res.* **55,** 5296–5301.
16. Kim, K. J., Li, B., Winer, J., Armanini, M., Gillett, N., Philipps, H. S., et al. (1993) Inhibition of vascular endothelial growth factor-induced angiogenesis suppresses tumour growth in vivo. *Nature* **362,** 841–844.
17. Olofsson, B., Pajusola, K., Kaipainen, A., von Euler, G., Joukov, V., Saksela, O., et al. (1996) Vascular endothelial growth factor B, a novel growth factor for endothelial cells. *PNAS* **93,** 2576–2581.
18. Olofsson, B., Pajusola, K., von Euler, G., Chilov, D., Alitalo, K., and Eriksson, U. (1996) Genomic organisation of the mouse and human genes for vascular endothelial growth factor B (VEGF B) and characterization of a second splice isoform. *J. Biol. Chem.* **271,** 19,310–19,317.
19. Joukov, V., Pajusola, K., Kaipainen, A., Chilov, D., Lahtinen, I., Kukk, E., et al. (1996) A novel vascular endothelial growth factor, VEGF-C, is a ligand for the Flt4 (VEGFR-3) and KDR(VEGFR-2) receptor tyrosine kinases. *EMBO J.* **15,** 290–298.
20. Kaipainen, A., Korhonen, J., Mustonen, T., van Hinsbergh, V. W., Fang, G. H., Dumont, D., et al. (1995) Expression of the fms-like tyrosine kinase 4 gene becomes restricted to lymphatic endothelium during development. *Proc. Natl. Acad. Sci.* **92,** 3566–3570.
21. Salven, P., Lymboussaki, A., Heikkila, P., Jaaskela-Saari, H., Enholm, B., Aase, K., et al. (1998) Vascular endothelial growth factors VEGF-B and VEGF-C are expressed in human tumors. *Am. J. Pathol.* **153,** 103–108.
22. Skobe, M., Hawighorst, T., Jackson, D. G., Prevo, R., Janes, L., Velasco, P., et al. (2001) Induction of tumor lymphangiogenesis by VEGF-C promotes breast cancer metastasis. *Nat. Med.* **7,** 192–198.
23. Davis, S., Aldrich, T. H., Jones, P. F., Acheson, A., Compton, D. L., Jain, V., et al. (1996) Isolation of angiopoietin-1, a ligand for the tie-2 receptor, by secretion-trap expression cloning. *Cell* **87,** 1161–1169.
24. Maisonpierre, P. C., Suri, C., Jones, P. F., Bartunkova, S., Wiegand, S. J., Radziejewski, C., et al. (1997) Angiopoietin-2, a natural antagonist for tie-2, that disrupts in vivo angiogenesis. *Science* **277,** 55–60.

25. Hanahan, D. (1997) Signaling vascular morphogenesis and maintenance. *Science* **277,** 48–50.
26. Schweigerer, L., Neufeld, G., Friedman, J., Abraham, J. A., Fiddes, J. C., and Gospodarowicz, D. (1987) Capillary endothelial cells express basic fibroblast growth factor, a mitogen that promotes their own growth. *Nature* **325,** 257–259.
27. Heldin, C.-H. (1992) Structural and functional studies on platelet-derived growth factor. *EMBO J.* **11,** 4251–4259.
28. Ross, R., Raines, E. W., and Bowen-Pope, D. F. (1986) The biology of platelet-derived growth factor. *Cell* **46,** 155–169.
29. Gille, J., Swerlick, R. A., and Caughman, S. W. (1997) Transforming growth factor-α-induced transcriptional activation of the vascular permeability factor (VPF/VEGF) gene requires AP-2-dependent DNA binding and transactivation. *EMBO J.* **18,** 50–59.
30. Ciardiello, F., Kim, N., McGeady, M. L., Liscia, D. S., Saeki, T., Bianco, C., et al. (1991) Expression of transforming growth factor α (TGFα) in breast cancer. *Ann. Oncol.* **2,** 169–182.
31. Weidner, N., Semple, J. P., Welch, W. R., and Folkman, J. (1991) Tumor angiogenesis and metastasis—correlation in invasive breast carcinoma. *N. Engl. J. Med.* **324,** 1–8.
32. Parums, D. V., Cordell, J. L., Micklem, K., Heryet, A. R., Gatter, K. C., and Mason, D. Y. (1990) JC70: a new monoclonal antibody that detects vascular endothelium associated antigen on routinely processed tissue sections. *J. Clin. Pathol.* **43,** 752–757.
33. Li, V. W., Folkerth, R. D., Watanabe, H., Yu, C., Rupnick, M., Barnes, P., et al. (1994) Microvessel count and cerebrospinal fluid basic fibroblast growth factor in children with brain tumours. *Lancet* **344,** 82–86.
34. Hayes, C. E. and Goldstein, I. J. (1974) An alpha-D-galactosyl-binding lectin from Bandeiraea simplicifolia seeds. Isolation by affinity chromatography and characterization. *J. Biol. Chem.* **249,** 1904–1914.
35. Malinda, K. M. (2000) In vivo Matrigel migration and angiogenesis assays, in *Angiogenesis Protocols* (Murray, J. C., ed.), Humana, Totowa, NJ, pp. 47–53.
36. Wang, G. L. and Semenza, G. L. (1993) Desferrioxamine induces erythropoietin gene expression and hypoxia-inducible factor 1 DNA-binding activity: implications for models of hypoxia signal transduction. *Blood* **82,** 3610–3615.
37. Kwiatkowski, J. L., Rutkowski, J. L., Yamashiro, D. J., Tennekoon, G. I., and Brodeur, G. M. (1998) Schwann cell-conditioned medium promotes neuroblastoma survival and differentiation. *Cancer Res.* **58,** 4602–4606.
38. Eggert, A., Grotzer, M. A., Ikegaki, N., Liu, X. G., Evans, A. E., and Brodeur, G. M. (2000) Expression of neurotrophin receptor TrkA inhibits angiogenesis in neuroblastoma. *Med. Pediatr. Oncol.* **35,** 569–572.
39. Coughlin, C. M., Salhany, K. E., Wysocka, M., Aruga, E., Kurzawa, H., Chang, A. E., et al. (1998) Interleukin-12 and interleukin-18 synergistically induce murine

tumor regression which involves inhibition of angiogenesis. *J. Clin. Invest.* **101,** 1441–1452.

40. Passaniti, A., Taylor, R. M., Pili, R., Guo, R., Long, P. V., Haney, J. A., et al. (1992) A simple quantitative method for assessing angiogenesis and antiangiogenic agents using reconstituted basement membrane, heparin and fibroblastic growth factor. *Lab. Invest.* **67,** 519–528.

III

Protocols on Gene Therapy

10

Experimental Strategies for Combined Suicide and Immune Cancer Gene Therapy

An Overview

Christof M. Kramm, Tim Niehues, and Nikolai G. Rainov

1. Definition of Suicide Gene Therapy

Suicide gene therapy systems are characterized by the transfer of therapeutic transgenes, which encode for enzymes of various origins. These enzymes are able to convert nontoxic prodrugs into highly cytotoxic metabolites. Thus, all cells that are transduced by suicide genes will be destroyed when the corresponding prodrug is applied. The cytotoxicity is mostly limited to transduced cells, but nontransduced cells may also be affected by the so-called bystander effect. The bystander effect *(1)* describes the phenomenon that in cell culture only a relatively small amount of cells has to be transduced with the suicide gene to gain complete cell killing upon treatment with the corresponding prodrug.

2. Suicide Gene Therapy Systems

To date, there are at least 13 different enzyme gene/prodrug systems that have been tested for cancer gene therapy (for review, *see* **ref.** *2*). The most widely used suicide gene therapy systems are the herpes simplex virus type 1 thymidine kinase gene (*HSV-tk*)/ganciclovir (GCV) and the *Escherichia coli* cytosine deaminase (cd)/5-fluorocytosine (5FC) systems.

The *HSV-tk* enzyme phosphorylates nucleoside analogs like acyclovir and ganciclovir to their monophosphorylated metabolites *(3)*. Only these monophosphorylated forms can then be further phosphorylated by cellular

From: *Methods in Molecular Biology, vol. 215:*
Cytokines and Colony Stimulating Factors: Methods and Protocols
Edited by: D. Körholz and W. Kiess © Humana Press Inc., Totowa, NJ

mammalian enzymes. The resulting triphosphates are the main cytotoxic substrates that are inserted into the growing DNA chain, thereby leading to a premature chain determination. Cell death then occurs as a result of apoptosis induction. Because of its close relation to DNA synthesis, the HSV-tk system appears to induce cytotoxicity only in proliferating cells.

In contrast, the cd system is supposed to kill proliferating and nonproliferating cells because 5-fluorouracil, a common chemotherapeutic agent, represents the main cytotoxic metabolite of 5FC and interacts with DNA and RNA synthesis *(4)*. Subsequent cell death also occurs for the cd/5FC system via induction of apoptosis.

3. Immune Response Induction by Suicide Gene Therapy

Initially, suicide gene therapy has been exclusively regarded as a purely cytotoxic anticancer treatment strategy, but in recent years, suicide gene therapy has been more and more linked to the induction of an antitumor-specific immune response. First hints for an essential involvement of the immune system arose from suicide gene therapy studies in immunocompromised animals where treatment results were less impressive than in immunocompetent animals *(5)*. The induction of an immunological antitumor memory was also reported for the HSV-tk and the cd systems *(6,7)*. As immunological correlates, CD4+ and CD8+ cells, macrophages, and natural killer cells were shown to infiltrate tumors after suicide gene therapy *(8–10)*. Thus, it was hypothesized that suicide gene therapy can change the immunoinhibitory microenvironment of tumors into an immunostimulatory one *(11)*. This immunostimulatory microenvironment was linked to the induction of a Th1-like cytokine profile (interleukin [IL]-1, IL-2, IL-6, IL-12, interferon [IFN]-γ, tumor necrosis factor [TNF]-α, granulocyte–macrophage colony-stimulating factor [GM-CSF]) upon HSV-tk/GCV treatment *(12–15)*.

4. Combined Suicide and Immune Gene Therapy

4.1. Choice of Transgene Systems

Regarding the immune responses induced by suicide gene therapy, it seems consequent that several attempts have been made to treat tumors by combinations of suicide and immune gene therapy (*see* **Table 1**). Whereas HSV-tk or cd have been exclusively employed on side of the suicide genes, the variety of immune modifications includes different cytokine, chemokine, and antigen-presenting cell (APC) strategies. The transfer of cytokine genes was, by far, the most frequently used immune gene therapy approach. Here, the Th1 cytokine genes *IL-2*, *GM-CSF*, and *IL-12* were clearly favored to enhance the immune response induced by suicide gene therapy. However, not all attempts to combine immune and suicide gene therapy were successful. Some studies ***(14,16,41)*** could not demonstrate an improved therapeutic outcome upon combined treatment with *IL-2* and *HSV-tk* gene therapy. However, these are single reports, in general, a clearly enhanced therapeutic efficiency was found by combined suicide/immune gene therapy (*see* **Table 1**).

4.2. Choice of Vector Systems

As vector systems for combined suicide and immune gene therapy (*see* **Table 1**), adenovirus (ADV) vectors have been clearly preferred to retroviral vectors. Herpes simplex virus (HSV) vectors also appear to gain increasing importance ***(43,44,46)***, nonviral vector systems like liposomes seem to play only a minor role.

4.2.1. Retroviral Vectors

Retroviral vectors represent the most widely used vector system in suicide gene therapy, but the results of clinical phase I–III trials for brain tumor patients ***(47,48)*** have shown no significant therapeutic success. This is most likely the result of the relatively low transduction efficiency of retroviral vectors. Retroviral vectors represent instable particles with a short half-life and a high tendency to rapid degradation (*see* **ref. *49***). Furthermore, serum factors (e.g., complement) also have a high potential for inactivation of retroviral vectors ***(50)***.

Usually so-called retroviral vector producer cells (VPC) are used for in vivo application. Retroviral VPC are genetically modified cells (e.g., fibroblasts) of human or murine origin that express structural and functional retroviral genes (*gag*, *pol*, *env*) that are necessary for retroviral replication (*see* **Fig. 1**). However, retroviral VPC lack the packaging signal and the long terminal repeats (LTR), which are also essential for the generation and packaging of infectious retroviral particles. Packaging signal and LTR are components of retroviral vectors in which *gag*, *pol*, and *env* genes have been replaced by transgenes. Thus, neither VPC nor retroviral vectors alone can lead to the generation of retroviral particles. However, when VPC are transduced with retroviral vectors, all elements for vector production are reunited, and, as a consequence, retroviral vectors are generated, packaged, and released into the extracellular space (*see* **Fig. 1**).

When transduced VPC are implanted into a tumor, retroviral vectors are continuously generated and released *in situ*. Thus, a higher transduction efficiency is expected than by direct use of ex vivo-generated retroviral particles. However, transduction efficiency via VPC is also limited because VPC represent xenologous cells that are attacked and destroyed by the host's immune system. In brain tumors of rats, murine VPC showed only an average survival time of

Table 1
Previous Animal Studies with Combined Suicide and Immune Cancer Gene Therapy

Pharmacogene therapy system	Immunogene therapy system	Rodent tumor model	Vector system	Therapeutic efficiency	Ref.
HSVtk/GCV	IL-2	Gliosarcoma	Retrovirus	Not enhanced	*16*
HSV-tk/GCV	IL-2	Colon carcinoma	ADV	Enhanced	*17*
HSV-tk/GCV	IL-2	Colon carcinoma	Liposomes	Enhanced	*18*
HSV-tk/GCV	IL-2	Mouth squamous cell carcinoma	ADV	Enhanced	*19, 20*
HSV-tk/GCV	IL-2	Lung cancer	ADV	Enhanced	*21*
HSV-tk/GCV	IL-2	Salivary gland cancer	ADV	Enhanced	*22*
HSV-tk/GCV	1. IL-2 2. IL-6 3. B7-1	Various tumors	ADV	Not enhanced	*14*
HSV-tk/GCV	1. IL-2 2. IL-2/GM-CSF	Metastatic colon carcinoma	1. ADV 2. ADV + ex vivo vaccine	Enhanced	*23*
HSV-tk/GCV	IL-4	Glioma	Retrovirus	Enhanced	*24*
HSV-tk/GCV	1. IL-7 2. IL-7 + DC	Nonsmall-cell lung cancer	Ex vivo vaccine	1. No effect 2. Enhanced	*25*
HSV-tk/GCV	GM-CSF	Colon carcinoma	Retrovirus	Enhanced	*26*
HSV-tk/GCV	IFN-α	Erythroleukemia Mammary adenocarcinoma	Ex vivo vaccine	Enhanced	*27*
cd /5FC	IL-2	Erythroleukemia	ADV	Enhanced	*28, 29*
cd/5FC	IL-2	Melanoma	ADV	Enhanced	*30*
cd/5FC	IL-4	Melanoma	Liposomes + ex vivo vaccines	No effect	*31*

HSV-tk/GCV	1. IL-2 2. IL-2/GM-CSF	Melanoma	ADV	Enhanced	***32***
cd/5FC	GM-CSF	Melanoma	ADV	Enhanced	***33***
5FC	IFN-γ	Mammary carcinoma	Ex vivo vaccine	Enhanced	***34***
HSV-tk/GCV	TNF-α	Fibrosarcoma Glioma	HSV	Enhanced only in fibrosarcoma	***35***
HSV-tk	IL-2	Medullary thyroid carcinoma	Ex vivo vaccines	Enhanced	***36***
cd/5FC	Lymphotactin	Colon carcinoma	ADV	Enhanced	***37***
HSV-tk	IL-12	Hepatocellular carcinoma	ADV	Enhanced	***38***
cd/5FC	SCF + GM-CSF	Colon carcinoma	ADV	Enhanced	***39***
HSV-tk/GCV	1. IL-2 2. GM-CSF 3. IL-2 + GM-CSF	Metastatic breast carcinoma	ADV	Enhanced	***40***
HSV-tk/GCV	IL-2	Bladder carcinoma	ADV	Not enhanced	***41***
HSV-tk/GCV	1. IL-12 2. GM-CSF	Peritoneal carcinomatosis of colon carcinoma	Retrovirus	Enhanced	***42***
HSV-tk/GCV + radiosurgery	TNF-α	Glioblastoma multiforme	HSV	Enhanced	***43***
HSV-tk/GCV	GM-CSF	Colon carcinoma	HSV	Enhanced	***44***
cd/5FC	IL-18	Melanoma	ADV	Enhanced	***45***
HSV-tk/GCV	IL-12	Colon carcinoma	HSV	Enhanced	***46***

Abbreviations: ADV, adenovirus vector; DC, dendritic cells; IFN-α, interferon-α; IFN-γ, interferon-γ; IL-2, interleukin-2; IL-4, interleukin-4; IL-6, interleukin-6; IL-7, interleukin-7; IL-12, interleukin-12; IL-18, interleukin-18; GM-CSF, granulocyte–macrophage colony-stimulating factor; HSV, herpes simplex virus vector; SCF, stem cell factor; TNF-α, tumor necrosis factor-α.

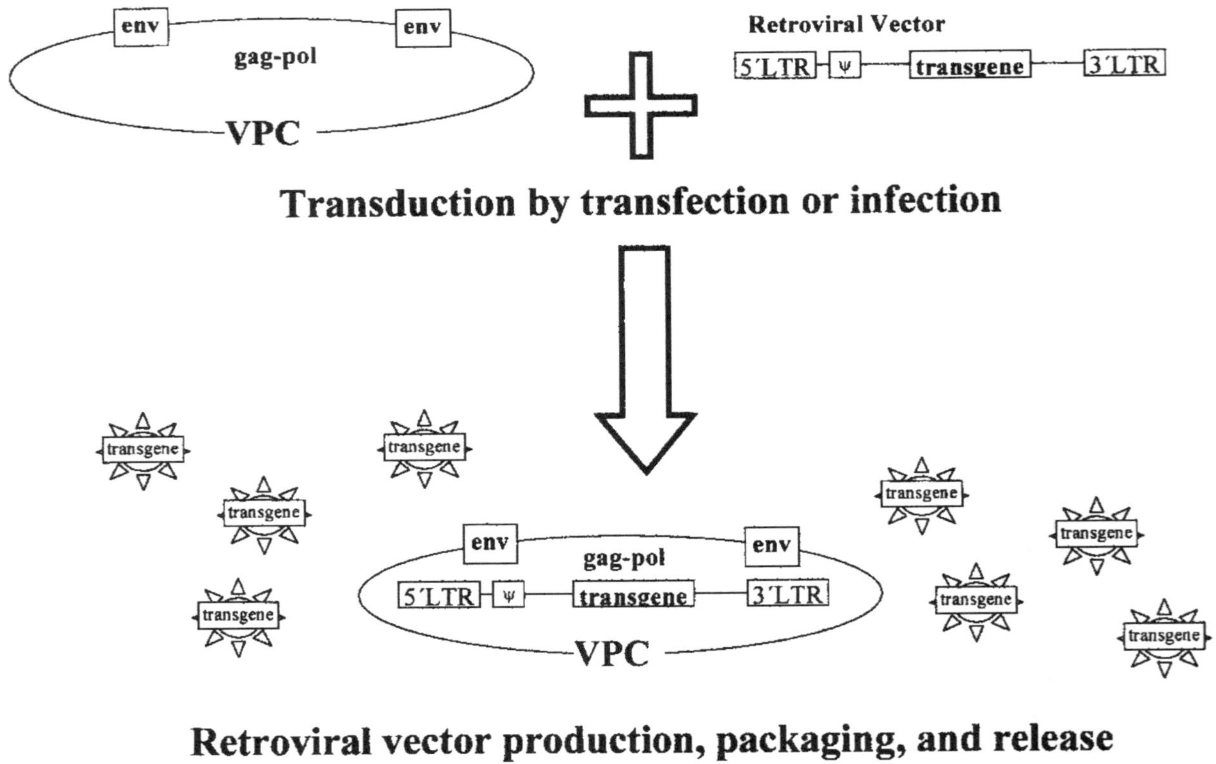

Fig. 1. Generation of replication-incompetent retroviral vectors by use of vector producer cells. Retroviral vector producer cells (VPC) are genetically modified mammalian cells that express the retroviral structural and functional genes *gag*, *pol*, and *env*. When VPC are transduced with a retroviral vector carrying the transgene (suicide or cytokine gene or both together), all elements for specific vector production (*env*, *gag*, *pol*, packaging signal ψ, 5′ and 3′ LTRs) are present within the VPC. Thus, retroviral vector generation, packaging, and release occur.

5–14 d *(51)*. Another disadvantage of a sufficient transduction efficiency might also be the absent migration potential of the VPC. Common retroviral VPC always stay at the site of implantation. Released retroviral vectors can only diffuse a few hundred micrometers between the densely packed tumor cells. Thus, despite a continuous vector production, the transduction efficiency is spatially limited. All of these disadvantages may explain why retroviral vectors seemed to have lost some of their initial appeal for cancer gene therapy.

However, retroviral vectors have been proved in many trials to be safe when used in patients. To date, the only successful clinical applications of gene therapy were all performed with retroviral vectors [i.e., suicide gene therapy as biological safety mechanisms for adoptive immune therapy with donor lymphocyte infusions after allogeneic stem cell transplantation *(52)* and the successful genetic correction of an X-linked severe combined immune deficiency *53)*]. In these cases, sufficient transduction efficiency in target cells was achieved by retroviral gene transfer ex vivo, not in vivo as in all previous clinical suicide cancer gene therapy trials.

4.2.2. Adenoviral Vectors

In contrast to retroviral vectors, adenovirus vectors offer a significantly higher transduction efficiency. They are more stable than retroviral vectors and they can be generated in higher titers (10^{12} ADV vs 10^7–10^8 retroviral vector particles per milliliter). Adenovirus vectors can be injected intravenously/intra-arterially or directly into tumors. In animal models, resulting gene transfer efficiency into tumor cells usually exceeds by far the outcome achieved after retroviral transduction. The duration of adenovirus-mediated transgene expression usually only lasts a few weeks, which, *per se*, does not automatically appear as a crucial disadvantage for cancer gene therapy where short transgene expression is usually sufficient for the therapeutic effect.

Although several clinical trials have demonstrated the safety of adenovirus vectors, one patient with an urea cycle deficiency died from hepatic coma after injection of genetically modified adenovirus particles into his hepatic artery *(54)*. The liver was probably destroyed by a severe immune response to adenovirus particles. This immune response may have occurred because the genetic backbone of clinically used adenovirus vectors is derived from human pathogenic adenovirus serotypes, and potentially harmful immune responses may be elicited in some people. To avoid the risk of inducing an overwhelming immune reaction, a new generation of adenovirus vectors, so-called high-capacity or gutless adenovirus vectors *(55)*, have been developed in which all adenoviral genes, including those with a high immunogenic potential, have been replaced. This new vector generation may offer a solution to avoid fatal immune reactions after adenovirus application.

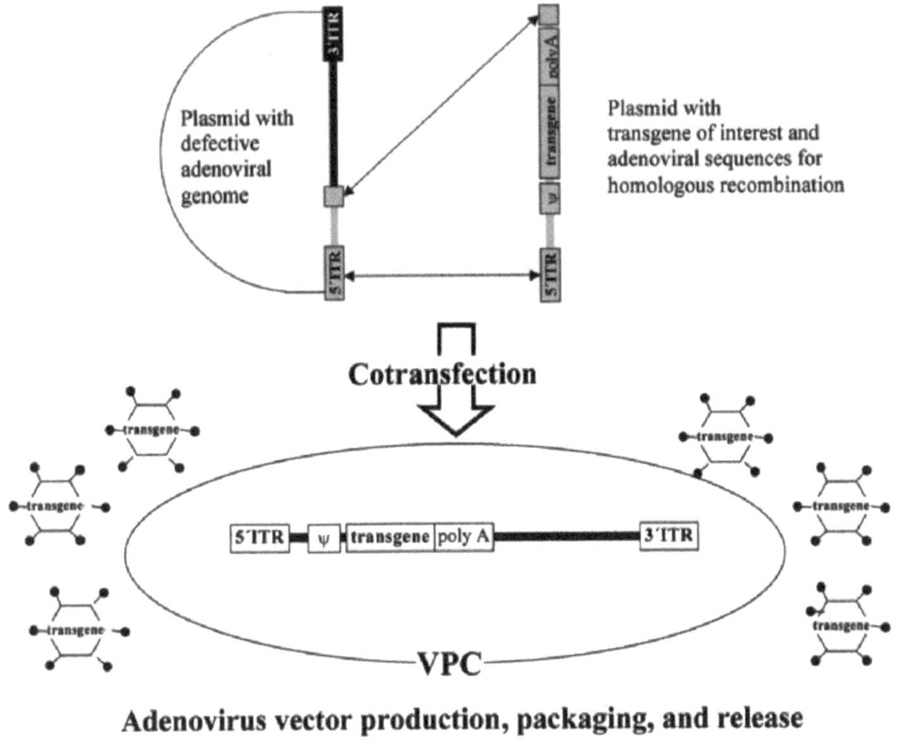

Fig. 2. Generation of replication incompetent adenovirus vectors. Adenovirus vectors can be generated by homologous recombination between two plasmids carrying either the deleted adenoviral genome or the transgene (suicide or cytokine gene) flanked by adenoviral sequences. Homologous recombination with production, packaging, and release of adenovirus vectors occur after cotransfection of permissive cells, which then serve as vector packaging cells (VPC). Convenient kits for adenovirus vector generation are commercially available.

Regarding the combinational use of suicide and cytokine gene therapy, adenovirus vectors provide a high transduction efficiency, which is crucial for a sufficient therapeutic antitumor response. Furthermore, adenoviral vectors with different transgenes are easy to generate as shown in **Fig. 2**.

4.2.3. Recombinant Herpes Vectors

Recombinant herpes simplex virus (HSV) vectors were used in recent studies to transfer cytokine and suicide genes into tumor cells. HSV vectors bear similar characteristics for gene therapy as adenovirus vectors. They can be also generated in high titers ex vivo (10^{10}–10^{11} infectious particles per milliliter) and transduction efficiencies are comparable ***(56)***. Recombinant HSV vectors

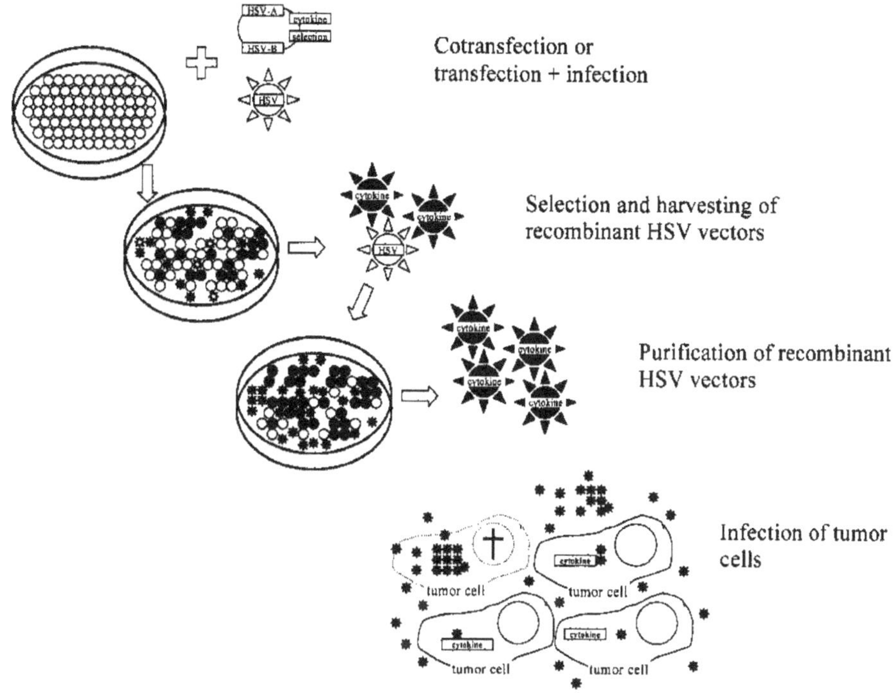

Fig. 3. Generation of recombinant HSV vectors. Recombinant HSV vectors are generated by homologous recombination after cotransduction of permissive cells with HSV viral genomic DNA containing the *HSV-tk* gene and plasmid DNA containing the cytokine gene and a selection marker flanked by HSV sequences. Resulting HSV vectors with suicide and cytokine genes are selected and purified after repeated infections of permissive cells. Purified HSV vector stocks are harvested from supernatants after cell lysis and concentrated by centrifugation.

are more difficult to generate than retroviral or adenoviral vectors (*see* **Fig. 3**). Contamination of virus stocks with wild-type HSV is a non-neglectible event and may be fatal for gene therapy purposes. Thus, applied HSV vector stocks have to be regularly checked for wild-type contamination.

Recombinant HSV vectors are often replication conditional, which means that they can replicate in proliferating cells, but not in nonproliferating cells. The cell proliferation-dependent vector generation is achieved by inactivation of viral genes whose replicational function can be compensated by cellular genes during cell proliferation. Replication-conditional HSV vectors can generate new vector particles in proliferating tumor cells, thereby increasing the amount of vectors *in situ*. Because the replication cycle of HSV is obligatorily accompanied by cell lysis, replication-conditional HSV vectors can also destroy tumor cells. Thus, the use of replication-conditional HSV vectors for combined suicide and immune gene therapy bears the advantage that not only does the suicide gene therapy approach lead to a selective killing of transduced tumor cells, but also the infection by HSV vectors.

Herpes simplex virus vectors also induce a local inflammatory response with the influx of immune cells and release of immunostimulatory cytokines and chemokines. Thus, the basis for an immunomodulatory treatment strategy is well prepared. However, like adenovirus vectors HSV vectors are derived from a human pathogen. Thus, an immunity to HSV often exists in patients. Overwhelming immune responses to local applications of HSV vectors may induce severe side effects. In our own animal experiments, intrathecal injections of replication-conditional HSV vectors were not only highly efficient in destroying disseminated tumor cells within the cerebrospinal fluid, but also elicited a strong inflammatory response with a considerable morbidity and, in some cases, also the death of the animals ***(57)***. Only few clinical trials have been performed to date with replication-conditional HSV vectors to treat brain tumor patients. The observed toxicity was negligible but the achieved therapeutic efficiency was also only moderate ***(58)***.

4.3. Choice of Vector Design

For combined suicide and immune gene therapy, the mode by which the different therapeutic genes are expressed within the different vectors may also be of importance. Most studies used mixtures of different retroviral VPC or adenovirus vectors to transduce tumor cells with suicide and cytokine genes. The intratumoral application of mixed vectors (or mixed VPC) expressing either the suicide or the cytokine gene represents an easy way to achieve a sufficient expression of both therapeutic genes within the tumor. For the enhancement of an antitumor immune response, it is not obligatory that the suicide and the cytokine genes are expressed within the same cells. It may even be disadvantageous because cells that are killed by suicide gene therapy will not longer secrete the cytokine. Thus, if all cells simultaneously carry the suicide and the cytokine gene, the cytokine effect may be abrogated by the onset of prodrug treatment. However, if many tumor cells carry either the suicide or the cytokine gene, tumor antigens are released after suicide gene therapy mediated tumor cell killing while cytokine release continues from nonaffected cells.

However, few studies used vectors where suicide and cytokine genes are expressed within the same HSV vector ***(43,44,46)*** or within the same retroviral expression cassette ***(59)***. In HSV vectors, the *HSV-tk* gene is already part of the viral genome, thus, the cytokine gene has only to be inserted into a pre-existing HSV backbone. In bicistronic retroviral vectors, the suicide and cytokine genes are both expressed under the control of one promoter via an internal ribosome entry site (IRES). This assures that both genes are expressed within the same cell after retroviral transduction. However, the gene downstream of the IRES usually shows a significantly lower expression than the gene upstream of the IRES. Thus, by using bicistronic vectors for a combined suicide and cytokine gene therapy approach, sufficient expression of both therapeutic genes has to be confirmed in all cases. Interestingly, this construct was chosen for a first clinical trial *(59)*.

5. Conclusions

The combination of suicide and immune gene therapy offers an interesting new concept for the treatment of tumors. Selective tumor cell killing with specific antigen release prepares the ground for an effective *in situ* antitumor vaccination achieved by local cytokine secretion. Adenovirus vectors with their known high transduction efficiency appear to be the most promising vector system to date for this purpose. The use of third-generation high-capacity adenovirus vectors will reduce the risk of harmful immune responses observed in previous trials. The application of different vectors carrying either the suicide or the cytokine gene is supposed to be advantageous to vector constructs with both genes together.

The concept of combined suicide and immune gene therapy may help to circumvent some of the limitations that have led to the previous failure of suicide gene therapy as a novel treatment approach for tumor patients. The induction and enhancement of an immune response may allow for a good therapeutic outcome despite a relatively low transduction efficiency of the vector systems available to date.

References

1. Moolten, F. L. and Wells, J. M. (1990) Curability of tumors bearing herpes thymidine kinase genes transferred by retroviral vectors. *J. Natl. Cancer Inst.* **82,** 297–300.
2. Aghi, M., Hochberg, F., and Breakefield, X. O. (2000) Prodrug activating enzymes in cancer gene therapy. *J. Gene Med.* **2,** 148–164.
3. Elion, G. B. (1983) The biochemistry and mechanism of action of acyclovir. *J. Antimicrob. Chemother.* **12 (Suppl. B),** 9–17.
4. Polak, A. (1990) Mode of action studies. B. Target nucleic acid synthesis and cell division. I. 5-Fluorocytosine, in *Chemotherapy of Fungal Diseases* (Ryley, J. F., ed.), Springer-Verlag, Berlin, pp. 155–159.
5. Gagandeep, S., Brew, R., Green, B., Christmas, S. E., Klatzmann, D., Poston, G. J., et al. (1996) Prodrug-activated gene therapy: involvement of an immunological component in the "bystander effect." *Cancer Gene Ther.* **3,** 83–88.

6. Barba, D., Hardin, J., Sadelain, M., and Gage, F. H. (1994) Development of anti-tumor immunity following thymidine kinase mediated killing of experimental brain tumors. *PNAS* **91,** 4348–4352.
7. Mullen, C. A., Coale, M. M., Lowe, R., and Blaese, R. M. (1994) Tumors expressing the cytosine deaminase suicide gene can be eliminated in vivo with 5-fluorocytosine and induce protective immunity to wild type tumor. *Cancer Res.* **54,** 1503–1506.
8. Caruso, M., Panis, Y., Gagandeep, S., Houssin, D., Salzmann, J. L., and Klatzmann, D. (1993) Regression of established macroscopic liver metastases after in situ transduction of a suicide gene. *PNAS* **90,** 7024–7028.
9. Consalvo, M., Mullen, C. A., Modesti, A., Musiani, P., Allione, A., Cavallo, F., et al. (1995) 5-Fluorocytosine-induced eradication of murine adenocarcinomas engineered to express the cytosine deaminase suicide gene requires host immune competence and leaves an efficient memory. *J. Immunol.* **154,** 5302–5312.
10. Hall, S. J., Sanford, M. A., Atkinson, G., and Chen, S. H. (1998) Induction of a potent antitumor natural killer cell activity by herpes simplex virus-thymidine kinase and ganciclovir therapy in an orthotopic mouse model of prostate cancer. *Cancer Res.* **58,** 3221–3225.
11. Freeman, S. M., Ramesh, R., and Marrogi, A. J. (1997) Immune system in suicide-gene therapy. *Lancet* **349,** 2–3.
12. Ramesh, R., Marrogi, A. J., Munshi, A., Abboud, C. A., and Freeman, S. M. (1996) In vivo analysis of the "bystander effect": a cytokine cascade. *Exp. Hematol.* **24,** 829–838.
13. Vile, R. G., Castleden, S., Marshall, J., Camplejohn, R., Upto, C., and Chong, H. (1997) Generation of an anti-tumour immune response in a non-immunogenic tumour: HSVtk killing in vivo stimulates a mononuclear cell infiltrate and a TH1-like profile of intratumoural cytokine expression. *Int. J. Cancer* **71,** 267–274.
14. Felzmann, T., Ramsey, W. J., and Blaese, R. M. (1997) Characterization of the antitumor immune response generated by treatment of murine tumors with recombinant adenoviruses expressing HSVtk, IL-2, IL-6 or B7-1. *Gene Ther.* **4,** 1322–1329.
15. Mullen, C. A., Anderson, L., Woods, K., Nishino, M., and Petropoulos, D. (1998) Ganciclovir chemoablation of herpes thymidine kinase suicide gene-modified tumors produces tumor necrosis and induces systemic immune responses. *Hum. Gene Ther.* **9,** 2019–2030.
16. Ram, Z., Walbridge, S., Heiss, J. D., Culver, K. W., Blaese, R. M., and Oldfield, E. H. (1994) In vivo transfer of the human interleukin-2: negative tumoricidal results in experimental brain tumors. *J. Neurosurg.* **80,** 535–540.
17. Chen, S. H., Chen, X. H., Wang, Y., Kosai, K., Finegold, M. J., Rich, S. S., et al. (1995) Combination gene therapy for liver metastasis of colon carcinoma in vivo. *PNAS* **92,** 2577–2581.
18. Coll, J. L., Mesnil, M., Lefebvre, M. F., Lancon, A., and Favrot, M. C. (1997) Long-term survival of immunocompetent rats with intraperitoneal colon carcinoma

tumors using herpes simplex thymidine kinase/ganciclovir and IL-2 treatments. *Gene Ther.* **4,** 1160–1166.
19. O'Malley, B. W., Cope, K. A., Chen, S. H., Li, D., Schwarta, M. R., and Woo, S. L. (1996) Combination gene therapy for oral cancer in a murine model. *Cancer Res.* **56,** 1737–1741.
20. O'Malley, B. W., Jr., Sewell, D. A., Li, D., Kosai, K., Chen, S. H., Woo, S. L., et al. (1997) The role of interleukin-2 in combination adenovirus gene therapy for head and neck cancer. *Mol. Endocrinol.* **11,** 667–673.
21. Kwong, Y. L., Chen, S. H., Kosai, K., Finegold, M., and Woo, S. L. (1997) Combination therapy with suicide and cytokine genes for hepatic metastases of lung cancer. *Chest* **112,** 1332–1337.
22. O'Malley, B. W., Jr. and Li, D. (1998) Combination gene therapy for salivary gland cancer. *Ann. NY Acad. Sci.* **842,** 163–170.
23. Chen, S. H., Kosai, K. K., Xu, B., Pham-Nguyen, K., Constant, C., Finegold, M. J., et al. (1996) Combination suicide and cytokine gene therapy for hepatic metastases of colon carcinoma: sustained antitumor immunity prolongs animal survival. *Cancer Res.* **56,** 3758–3762.
24. Benedetti, S., Dimeco, F., Pollo, B., Cirenei, N., Colombo, B. M., Bruzzone, M. G., et al. (1997) Limited efficacy of the HSV-Tk/GCV system for gene therapy of malignant gliomas and perspectives for the combined transduction of the interleukin-4 gene. *Hum. Gene Ther.* **8,** 1345–1353.
25. Sharma, S., Miller, P. W., Stolina, M., Zhu, L., Huang, M., Paul, R. W., et al. (1997) Multicomponent gene therapy vaccines for lung cancer: effective eradication of established murine tumors in vivo with interleukin-7/herpes simplex thymidine kinase-transduced autologous tumor and ex vivo activated dentritic cells. *Gene Ther.* **4,** 1361–1370.
26. Hayashi, S., Emi, N., Yokoyama, I., Namii, Y., Uchida, K., and Takagi, H. (1997) Inhibition of establishment of hepatic metastasis in mice by combination gene therapy using both herpes simplex virus-thymidine kinase and granulocyte macrophage-colony stimulating factor genes in murine colon cancer. *Cancer Gene Ther.* **4,** 339–344.
27. Santodonato, L., D'Agostino, G., Santini, S. M., Carlei, D., Musiani, P., Modesti, A., et al. (1997) Local and systemic antitumor response after combined therapy of mouse metastatic tumor cells expressing IFN-alpha and HSVtk: perspectives for the generation of cancer vaccines. *Gene Ther.* **4,** 1246–1255.
28. Ju, D. W., Wang, B. M., and Cao, X. (1998) Adenovirus-mediated combined suicide gene and interleukin-2 gene therapy for the treatment of established tumor and induction of antitumor immunity. *J. Cancer Res. Clin. Oncol.* **124,** 683–689.
29. Ju, D., Cao, X., and Wang, B. (1998) Therapeutic effects of combined suicide gene and cytokine gene therapy on erythroleukemia-bearing mice. *Zhonghua Xue Ye Xue Za Zhi* **19,** 294–298.
30. Ju, D., Cao, X., and Wang, B. (1998) Antitumor effect of combined therapy with adenovirus-mediated CD suicide gene and interleukin 2 gene transfer and its immunological mechanism. *Zhonghua Zhong Liu Za Zhi* **20,** 108–111.

31. Missol-Kolka, E., Sochanik, A., and Szala, S. (1998) Combined therapy of B16(F10) murine melanoma using E. coli cytosine deaminase gene and murine interleukin-4 gene. *Neoplasma* **45,** 305–311.
32. Bonnekoh, B., Greenhalgh, D. A., Chen, S. H., Block, A., Rich, S. S., Krieg, T., et al. (1998) Ex vivo and in vivo adenovirus-mediated gene therapy strategies induce a systemic anti-tumor immune defence in the B16 melanoma model. *J. Invest. Dermatol.* **110,** 867–871.
33. Cao, X., Ju, D. W., Tao, Q., Wang, J., Wan, T., Wang, B. M., et al. (1998) Adenovirus-mediated GM-CSF gene and cytosine deaminase gene transfer followed by 5-fluorocytosine administration elicit more potent antitumor response in tumor-bearing mice. *Gene Ther.* **5,** 1130–1136.
34. Nanni, P., De Giovanni, C., Nicoletti, G., Landuzzi, L., Rossi, I., Frabetti, F., et al. (1998) The immune response elicited by mammary adenocarcinoma cells transduced with interferon-gamma and cytosine deaminase genes cures lung metastases by parental cells. *Hum. Gene Ther.* **9,** 217–224.
35. Moriuchi, S., Oligino, T., Krisky, D., Marconi, P., Fink, D., Cohen, J., et al. (1998) Enhanced tumor cell killing in the presence of ganciclovir by herpes simplex virus type 1 vector-directed coexpression of human tumor necrosis factor-alpha and herpes simplex virus thymidine kinase. *Cancer Res.* **58,** 5731–5737.
36. Soler, M. N., Milhaud, G., Lekmine, F., Treilhou-Lahille, F., Klatzmann, D., and Lausson, S. (1999) Treatment of medullary thyroid carcinoma by combined expression of suicide and interleukin-2 genes. *Cancer Immunol. Immunother.* **48,** 91–99.
37. Ju, D. W., Tao, Q., Cheng, D. S., Zhang, W., Zhang, M., Hamada, H., et al. (2000) Adenovirus-mediated lymphotactin gene transfer improves therapeutic efficacy of cytosine deaminase suicide gene therapy in established murine colon carcinoma. *Gene Ther.* **7,** 329–338.
38. Drozdzik, M., Qian, C., Xie, X., Peng, D., Bilbao, R., Mazzolini, G., et al. (2000) Combined gene therapy with suicide gene and interleukin-12 is more efficient than therapy with one gene alone in a murine model of hepatocellular carcinoma. *J. Hepatol.* **32,** 279–286.
39. Cao, X., Huang, X., Ju, D. W., Zhang, W., Hamada, H., and Wang, J. (2000) Enhanced antitumoral effect of adenovirus-mediated cytosine deaminase gene therapy by induction of antigen-presenting cells through stem cell factor/granulocyte-macrophage colony-stimulating factor gene transfer. *Cancer Gene Ther.* **7,** 177–186.
40. Majumdar, A. S., Zolotorev, A., Samuel, S., Tran, K., Vertin, B., Hall-Meier, M., et al. (2000) Efficacy of herpes simplex virus thymidine kinase in combination with cytokine gene therapy in an experimental metastatic breast cancer model. *Cancer Gene Ther.* **7,** 1086–1099.
41. Freund, C. T., Sutton, M. A., Dang, T., Contant, C. F., Rowley, D., and Lerner, S. P. (2000) Adenovirus-mediated combination suicide and cytokine gene therapy for bladder cancer. *Anticancer Res.* **20,** 1359–1365.

42. Lechanteur, C., Delvenne, P., Princen, F., Lopez, M., Fillet, G., Gielen, J., et al. (2000) Combined suicide and cytokine gene therapy for peritoneal carcinomatosis. *Gut* **47,** 343–348.
43. Niranjan, A., Moriuchi, S., Lunsford, L. D., Kondziolka, D., Flickinger, J. C., Fellows, W., et al. (2000) Effective treatment of experimental glioblastoma by HSV vector-mediated TNF alpha and HSV-tk gene transfer in combination with radiosurgery and ganciclovir administration. *Mol. Ther.* **2,** 114–120.
44. Jones, R. K., Pope, I. M., Kinsella, A. R., Watson, A. J., and Christmas, S. E. (2000) Combined suicide and granulocyte–macrophage colony-stimulating factor gene therapy induces complete tumor regression and generates antitumor immunity. *Cancer Gene Ther.* **7(12),** 1519–1528.
45. Ju, D. W., Yang, Y., Tao, Q., Song, W. G., He, L., Chen, G., et al. (2000) Interleukin-18 gene transfer increases antitumor effects of suicide gene therapy through efficient induction of antitumor immunity. *Gene Ther.* **7,** 1672–1679.
46. Toda, M., Martuza, R. L., and Rabkin, S. D. (2001) Combination suicide/cytokine gene therapy as adjuvants to a defective herpes simplex virus-based cancer vaccine. *Gene Ther.* **8,** 332–339.
47. Ram, Z., Culver, K. W., Oshiro, E. M., Viola, J. J., DeVroom, H. L., Otto, E., et al. (1997) Therapy of malignant brain tumors by intratumoral implantation of retroviral vector-producing cells. *Nat. Med.* **3,** 1354–1361.
48. Rainov, N. G. (2000) A phase III clinical evaluation of herpes simplex virus type 1 thymidine kinase and ganciclovir gene therapy as an adjuvant to surgical resection and radiation in adults with previously untreated glioblastoma multiforme. *Hum. Gene Ther.* **11,** 2389–2401.
49. Kramm, C. M., Sena-Esteves, M., Barnett, F. H., Rainov, N. G., Schuback, D. E., Yu, J. S., et al. (1995) Gene therapy for brain tumors. *Brain Pathol.* **5,** 345–381.
50. Agrawal, R. S., Karhu, K., Laukkanen, J., Kirkinen, P., Yla-Herttuala, S., and Agrawal, Y. P. (1999) Complement and anti-alpha-galactosyl natural antibody-mediated inactivation of murine retrovirus occurs in adult serum but not in umbilical cord serum. *Gene Ther.* **6,** 146–148.
51. Short, M. P., Choi, B. C., Lcc, J. K., Malick, A., Breakefield, X. O., and Martuza, R. L. (1990) Gene delivery to glioma cells in rat brain by grafting of a retrovirus packaging cell line. *J. Neurosci. Res.* **27,** 427–439.
52. Bonini, C., Ferrari, G., Verzeletti, S., Servida, P., Zappone, E., Ruggieri, L., et al. (1997) HSV-TK gene transfer into donor lymphocytes for control of allogeneic graft-versus-leukemia. *Science* **276,** 1719–1724.
53. Cavazzana-Calvo, M., Hacein-Bey, S., de Saint Basile, G., Gross, F., Yvon, E., Nusbaum, P., et al. (2000) Gene therapy of human severe combined immunodeficiency (SCID)-X1 disease. *Science* **288,** 669–672.
54. Marshall, E. (1999) Gene therapy death prompts review of adenovirus vector. *Science* **286,** 2244–2245.
55. Kochanek, S. (1999) High-capacity adenoviral vectors for gene transfer and somatic gene therapy. *Hum. Gene Ther.* **10,** 2451–2459.

56. Boviatsis, E. J., Chase, M., Wei, M. X., Tamiya, T., Hurford, R. K., Jr., Kowall, N. W., et al. (1994) Gene transfer into experimental brain tumors mediated by adenovirus, herpes simplex virus, and retrovirus vectors. *Hum. Gene Ther.* **5,** 183–191.
57. Kramm, C. M., Rainov, N. G., Sena-Esteves, M., Barnett, F. H., Chase, M., Herrlinger, U., et al. (1996) Long-term survival in a rodent model of disseminated brain tumors by combined intrathecal delivery of herpes vectors and ganciclovir treatment. *Hum. Gene Ther.* **7,** 1989–1994.
58. Markert, J. M., Medlock, M. D., Rabkin, S. D., Gillespie, G. Y., Todo, T., Hunter, W. D., et al. (2000) Conditionally replicating herpes simplex virus mutant, G207 for the treatment of malignant glioma: results of a phase I trial. *Gene Ther.* **7,** 867–874.
59. Palu, G., Cavaggioni, A., Calvi, P., Franchin, E., Pizzato, M., Boschetto, R., et al. (1999) Gene therapy of glioblastoma multiforme via combined expression of suicide and cytokine genes: a pilot study in humans. *Gene Ther.* **6,** 330–337.

11

Gene Therapy with Plasmids Encoding Cytokine– or Cytokine Receptor–IgG Chimeric Proteins

Ciriaco A. Piccirillo and Gérald J. Prud'homme

1. Introduction

Cytokine therapy can influence the outcome of autoimmune diseases by altering either T-helper 1 (Th1) vs T-helper 2 (Th2) balance or antigen-presenting cell (APC) function, or by shifting the balance between inflammatory and regulatory cytokines *(1)*. However, cytokine and soluble cytokine-receptor therapy have been limited by the short half-life ($T_{1/2}$) of these proteins and the necessity to administer relatively large boluses of recombinant proteins *(2)*. This results in transient high systemic levels and, in the case of cytokines, toxicity and poor therapeutic efficiency. The in vivo blockade of cytokine function by monoclonal antibody therapy, although feasible, has also faced therapeutic limitations *(3)*. Moreover, the isolation and production of highly purified and stable therapeutic proteins is laborious and expensive. Gene therapy has significant advantages, allowing long-term and relatively constant delivery of cytokines or their receptors at therapeutic levels. This can be accomplished with viral gene therapy vectors, as well as plasmid DNA expression vectors (nonviral approach). Our laboratory has been particularly interested in the delivery of vectors encoding cytokines and cytokine receptors for the prevention or treatment of autoimmune diseases.

1.1. Naked Plasmid DNA

Simple needle injection of naked plasmid DNA in saline into skeletal muscle results in the nonintegrating transfection of 3–5% of myocytes. The process

From: *Methods in Molecular Biology, vol. 215: Cytokines and Colony Stimulating Factors: Methods and Protocols*
Edited by: D. Körholz and W. Kiess © Humana Press Inc., Totowa, NJ

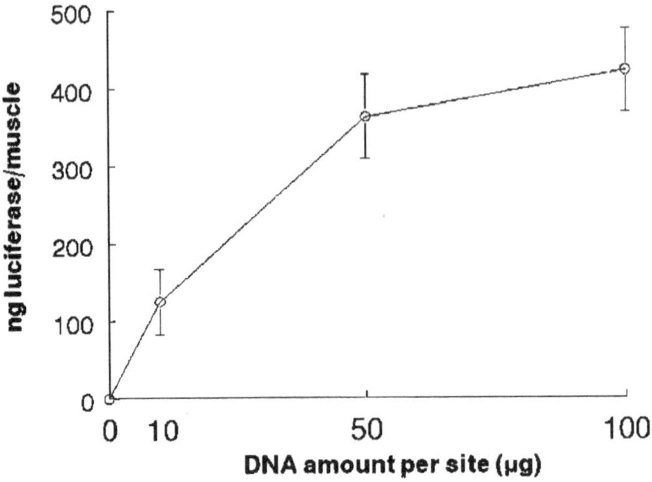

Fig. 1. Luciferase expression in skeletal muscle is a DNA dose-dependent process. Female BALB/c mice (4–6 wk, n = 8) were injected with increasing amounts of VR1255. Luciferase was measured 7 d following plasmid expression.

of plasmid DNA entry into skeletal muscle fibers is not well understood. Several uptake mechanisms have been postulated, including plasma membrane disruption, receptor-mediated endocytosis, and potocytosis ***(4–8)***. Direct injection of naked plasmids DNA has also been performed in other tissues, including skin, liver, tongue, thyroid, tumors, and lung (reviewed in **ref. *9***). Expression levels in these tissues were considerably lower than in skeletal muscle. Possibly, myocytes differ by having DNA uptake mechanisms dependent on the T-tubule system or caveolae ***(5,9,10)***. The nuclei of skeletal muscle cells are normally located peripherally, apposed to the cell membrane, which may facilitate entry of plasmids into nuclei.

The injection by needle of a 50-μL dose (the maximum recommended volume) of fluorescence-labeled plasmid into the tibialis anterior (TA) muscle of mice results in the rapid dissemination of DNA throughout the muscle ***(11)***. DNA is internalized by myocytes within 5 min and, over several hours, by mononuclear cells (perhaps macrophages or dendritic cells) located along muscle fibers and in the draining lymph nodes. Of note, the transgene appears to be expressed primarily by myocytes despite uptake of DNA by other cells ***(11)***. The uptake of plasmid DNA in muscle appears to occur by a saturable process, because luciferase expression levels increased to about 100 μg of DNA (*see* **Fig. 1**), but at higher doses, expression declined (not shown). Dowty and co-workers ***(6)*** have suggested that this saturation is associated with a receptorlike uptake mechanism, although it remains uncharacterized.

Interestingly, APCs have toll-like or other receptors that bind bacterial DNA and increase its uptake *(12)*, but this DNA appears to be rapidly degraded *(11)*.

Under optimal conditions, 50 µg of naked plasmid DNA injected into a mouse muscle (without any enhancing maneuver) can lead to the synthesis of >300 ng of nonsecreted reporter protein (luciferase) *(10)*. In our hands, serum values of secreted proteins have ranged from few picograms per milliliter to over 300 ng/mL *(13)*. Evidently, several factors have an effect on these results, from the vector components (promoters, introns, and terminator sequences) to the rate of protein turnover. The human cytomegalovirus (CMV) immediate–early enhancer–promoter (IE-EP) is highly effective in most mammalian cell types and is the most widely used in gene therapy studies.

Peak protein levels are usually observed 1–2 wk after im DNA administration. As discussed below, expression is increased by pretreating muscles with anesthetic agents such as bupivacaine, which induce a phase of regeneration from local satellite cells. Typical results are shown in **Figs. 1** and **2**. The persistence of expression varies greatly (2 wk to over 1 yr) depending on the antigenicity of the product and probably other factors. For example, genes encoding xenogeneic proteins are often expressed for only 2–3 wk in normal mice (*see* **Fig. 2**), but for several months in severe combined immune deficiency (SCID) mice (not shown). Moreover, production of nonimmunogenic (self) proteins in immunocompetent mice can persist at significant levels for >6 mo following DNA administration (*see* **Fig. 3**).

The technique of naked DNA injection has been most widely applied for immunization purposes. Plasmids encoding an antigen (i.e., DNA vaccines) have been used to elicit specific humoral and cellular immune responses against a wide variety of antigens in several species, including humans (reviewed in **ref.** *9*). Responses involve all arms of the immune system (humoral, Th cells, and cytotoxic T lymphocytes [CTLs]) and are both potent and long-lived, at least in certain animal models. Several investigators have codelivered cytokine genes, on the same or separate plasmids, to either attenuate or potentiate immune responses to the encoded antigen *(14)*.

1.2. Systemic Delivery of Therapeutic Proteins by Direct im DNA Injection

Although the initial focus was on vaccination, recently im gene transfer for the production of therapeutic proteins has gained more attention. Skeletal muscles are easily accessible for injection, and the large mass and nondividing (postmitotic) nature of these cells provide advantageous features *(6,7)*. Skeletal muscles are highly vascularized, so that secreted proteins can readily enter the circulation. Moreover, the persistence of gene expression has generally been much greater in muscle than other tissues. Therefore, skeletal muscles can

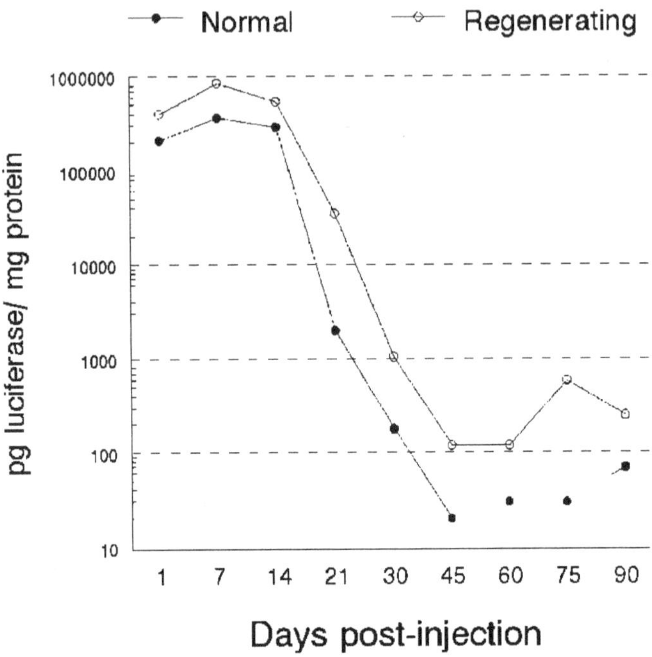

Fig. 2. Time-course of reporter gene expression in normal and regenerating muscle. Female BALB/c mice (4–6 wk, n = 6–8) were injected with 50 μg of VR1255 7 d after a 0.4% bupivacaine muscle pretreatment (bupivacaine induces regeneration from satellite cells). Control mice did not receive bupivacaine and were injected with equal amounts of VR1255. Muscles were excised at various time-points and tissue extracts were assayed for luciferase activity. Normal, untreated muscle: closed circles; bupivacaine pretreated muscle: open circles.

be used as in vivo biological factories for stable and long-term synthesis of therapeutic proteins ***(15,16)***.

1.3. Autoimmune Diabetes in NOD Mice

The nonobese diabetic (NOD) mouse develops diabetes spontaneously, through an autoimmune process that shares many features with human insulin-dependent diabetes mellitus (IDDM or type 1 diabetes) ***(17)***. Strong evidence suggests that Th1 cells and macrophages, which infiltrate islets of Langerhans (insulitis) and secrete inflammatory cytokines, play an important role in beta islet-cell destruction ***(18–20)***. We focused on the potential immunoregulatory roles of transforming growth factor-β1 (TGF-β1) and interleukin (IL)-4 in this disease. TGF-β1 is a cytokine with multiple suppressive effects on T-cells, B-cells, natural killer (NK) cells and macrophages ***(21–23)***. IL-4 promotes Th2

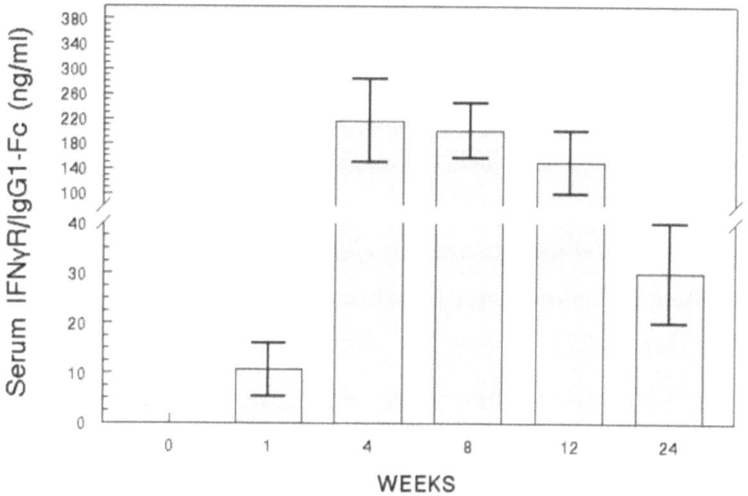

Fig. 3. IFN-γR/IgG1-Fc serum levels after im plasmid delivery. CD-1 mice received two inoculations of 200 μg of plasmid encoding IFNγR/IgG1-Fc (100 μg/muscle) spaced 1 wk apart (wk 0 and 1). The results represent the serum levels of IFN-γR/IgG1-Fc fusion protein (± SEM) in these mice, as determined with an enzyme-linked immunosorbent assay. The fusion protein was not detectable in the serum of blank vector-treated mice.

cell differentiation and, like TGF-β1, suppresses Th1 cytokine production and inhibits macrophage activity *(24–26)*.

Intramuscular injection of plasmid DNA-encoding latent TGF-β1 (pCMV–TGF-β1) resulted in uptake and expression of this vector by muscle cells *(27)*. TGF-β1 serum levels were increased by twofold to fourfold for >2 wk. The production of TGF-β1 protected NOD mice from autoimmune insulitis and diabetes, with or without cyclophosphamide (CYP)-mediated acceleration. pCMV–TGF-β1-treated NOD mice had low insulitis scores and depressed IL-12 and IFN-γ mRNA expression. A plasmid encoding an IL-4/IgG1-Fc fusion protein was also protective *(28)*, presumably through its ability to favor Th2 responses and/or suppress macrophage activity.

1.4. Experimental Allergic Encephalomyelitis

Experimental allergic encephalomyelitis (EAE) is characterized by perivascular leukocytic infiltration (CD4+ T-cells and macrophages) and demyelination of the central nervous system (CNS). There is compelling evidence for a role of inflammatory cytokines produced by Th1 cells and macrophages in the pathogenesis *(29–32)*. We found that administration of plasmid DNA-encoding either TGF-β1 or IL-4/IgG1-Fc delayed the onset and considerably reduced the severity of EAE *(33)*. Gene transfer of either cytokine had suppressive effects on myelin basic protein (MBP)-stimulated T-cell proliferation and cytokine production. Protection from EAE correlated with a significant decrease in CNS histopathology and inflammatory cytokine production (IL-12 and IFN-γ), with upregulation of endogenous TGF-β1 and IL-4.

1.5. Protection Against Lupus or Autoimmune Diabetes with IFN-γR/IgG1-Fc-Expressing Plasmids

Many cytokine abnormalities have been described in lupus, with increased levels of IFN-γ in serum, lymphoid organs, and afflicted tissues being the most salient feature *(13)*. IFN-γ or IFN-γ-receptor (IFN-γR) gene ablation or, alternatively, IFN-γ neutralization with monoclonal antibodies (MAbs) or soluble receptors, prevents or delays onset of murine lupus. In contrast, administration of IFN-γ accelerates development of lupus.

For our immunogene therapy studies toward neutralizing IFN-γ, we constructed an expression plasmid (VR–IFNγR/IgG1) encoding an IFN-γR/IgG1-Fc (all murine) fusion protein *(28,34)*. As in a previous study *(33)*, the cDNA segments were inserted into VR1255 (Vical Inc., San Diego, CA), a plasmid selected for its high expression in muscle *(16)*. In transfected cells, IFN-γR/IgG1-Fc was secreted as a disulfide-linked homodimer that strongly inhibited IFN-γ activity. Fusion proteins with Fc segments offer advantages over native proteins, such as prolonged half-life, higher avidity for the ligand, and ease of purification. Murine IgG1-Fc was chosen because IgG1 does not activate the complement and has a long half-life in serum. However, other IgG isotypes can be mutated to remove their complement-activating ability.

Intramuscular injection of a total of 100–200 μg of VR–IFN-γR/IgG1 plasmid into the TA and rectus femoris (RF) muscles of CD-1 or NOD mice *(28,34)* resulted in serum IFN-γR/IgG1-Fc levels ranging from 100 to 300 ng/mL for >4 mo and detectable levels for >6 mo after the last DNA injection (*see* **Fig. 3**). These protein levels are superior to those we have obtained with vectors encoding other proteins. This may be partly explained by the nonimmunogenic nature of the murine IFN-γR/IgG1-Fc fusion protein. Furthermore, improved expression may be related to the neutralization of IFN-γ, because IFN-γ can suppress transcription promoted by the CMV enhancer–promoter, which is contained in the VR1255 plasmid.

Although results were comparable in CD-1 and NOD mice, there were much lower levels of soluble receptor in the Fas-deficient lupus-prone MRL-Fas^{lpr} strain, which has severe early onset lupus, and massive lymphoaccumulation (primarily of double-negative T cells), and produces strikingly large amounts of IFN-γ. In the latter mice, the soluble receptor is possibly sequestered into tissues or vector expression is suppressed by residual IFN-γ.

To enhance systemic fusion protein expression in MRL-Fas^{lpr} mice, we coupled im injections of naked DNA with local electroporation *(35)*. Shortly after the injection of DNA, application of low-field-strength, square-wave electric pulses through external or invasive electrodes augments gene transfer efficiency by 1 to 3 orders of magnitude *(36)*. When electroporation was applied to MRL-Fas^{lpr} mice, the serum IFN-γR/IgG1-Fc levels exceeded 100 ng/mL and IFN-γ serum levels were consequently reduced *(35)*. Thus, electroporation enhanced gene transfer in these mice and it is likely that this technique will be even more relevant to other species. Indeed, in larger mammals, including primates, gene transfer following the im administration of naked DNA is not as efficient as in mice, but it is greatly improved by electroporation *(36)*.

Intramuscular IFN-γR/IgG1-Fc plasmid treatments, especially when enhanced by electroporation, protected prediseased MRL-Fas^{lpr} mice from early lethality and reduced autoantibody titers, renal disease, and histologic markers of lupus *(35)*. Most importantly, when therapy was initiated in 4-mo-old diseased mice, survival was extended beyond expectations, with 100% of mice alive at 14 mo of age compared to none in the control group. Remarkably, disease was arrested or even ameliorated in the experimental group. The effectiveness of this treatment in late disease is, to our knowledge, unprecedented.

Interferon-γ appears to play an important pathogenic role in some models of diabetes, including the NOD mouse *(18–20)*. We observed that im gene transfer of our IFN-γR/IgG1-Fc-encoding plasmid prevented insulitis and diabetes in either the multiple-low-dose streptozotocin (STZ)-induced diabetes (MDSD) model in CD-1 and C57BL/6 mice or in NOD mice (CYP-accelerated or spontaneous) *(28,34)*. This is consistent with the finding that administration of either STZ (CD-1, C57BL/6, or NOD mice) or CYP (NOD mice) induces a burst of IFN-γ release and rapid onset of diabetes.

1.6. Other Studies

A number of laboratories have successfully used im injection of cytokine plasmid DNA for in vivo modulation of immune responses in models of lupus, arthritis, autoimmune diabetes, colitis, and other immunologic disorders *(13,37–40)*. TGF-β1, IL-4, and IL-10 have been most frequently protective in these diseases.

1.7. Plasmid DNA Expression Vector Design and Construction

The essential components of plasmid DNA vectors designed for vaccination or therapeutic protein delivery are the same as for conventional eukaryotic expression vectors *(16)*, namely a coding DNA sequence (cDNA), an upstream enhancer–promoter and a downstream polyadenylation signal (poly A)/mRNA termination sequence (which may be isolated from a range of mammalian and viral sources), and a bacterially derived plasmid backbone containing both an origin of replication (ORI) for *Escherichia coli* and an antibiotic resistance selectable marker. Many plasmid expression vectors also contain intronic sequences, which have been shown to greatly increase gene expression within transfected cells. The enhancer–promoters that have been most widely used for DNA vaccination and systemic protein delivery are the CMV IE-EP and the Rous sarcoma virus long terminal repeat (RSV-LTR). Compared to most other promoters, these are excellent for mediating the highest levels of protein expression by in vitro transfection of cell lines. In vivo and in vitro expression may differ, and plasmids should be tested in vivo to confirm their efficacy.

2. Materials

2.1. Plasmids

1. pCI-neo obtained from Promega (Promega, Madison, WI).
2. pCDNA-1, pCDNA-3 plasmids purchased from Invitrogen Corp. (Carlsbad, CA).
3. VR1255, VR1012 kindly provided by VICAL Inc. (San Diego, CA).

2.2. Enzymes

Restriction enzymes, ligases, PfU polymerases and reverse transcriptases (Superscript preamplification system) can be obtained from a number of reliable sources: Pharmacia (Baie d'Urfe, Montreal, Canada), Gibco-BRL (Gaithersburg, MD), Promega (Madison, WI), New England Biolabs (Beverly, MA), and Stratagene (La Jolla, CA).

2.3. Luciferase Assays

A highly sensitive luciferase assay kit is available at Promega (Madison, WI). Purified luciferase may be obtained at Analytical Luminescence Labs (San Diego, CA).

2.4. Nucleic Acid Purification

Purification columns may be purchased from Qiagen (Valencia, CA). The Limulus amoebocyte endotoxin detection kit can be purchased at BioWhittaker Inc (East Rutherford, NJ). Total RNA may be isolated from stimulated cells using RNAzol B reagent.

2.5. Electroporation

An Electro Square Porator model ECM830 can be purchased from BTX Corp. (San Diego, CA). Conductive gel can be purchased from Perrier Medical (PER100/1; Montreal, Quebec, Canada) and calliper electrodes can be obtained from BTX Corp. (San Diego, CA).

2.6. Antibodies

Antibodies can be obtained from the following sources: purified anti-IFN-γR MAb (clone GR-20; obtained from ATCC, Rockville, MD), rat biotinylated anti-mouse IgG1 (PharMingen Canada, Mississauga, Ontario), and alkaline phosphatase–streptavidin conjugate (Bio-Rad Laboratories, Hercules, CA).

3. Methods

3.1. Plasmid DNA Vector Construction and Expression

1. A number of plasmid DNA expression vectors have been described for im injection protocols, with varying degrees of success. In our hands, pCI-neo and pCDNA-1/3 have generated satisfactory results. For optimal intramuscular expression, we resorted to VR1255 ***(16)***. In the latter plasmid, a luciferase (Luc+) gene is under the transcriptional control of the human CMV IE-EP and downstream of the human CMV intron A. Transcription is terminated by the minimal rabbit β-globin terminator. Luc+ encodes a modified luciferase, which preferentially localizes in the cytoplasm. VR1255 has been shown to be far superior (150-fold) to several other plasmid vectors for direct im injection of naked plasmid DNA ***(16)***. As control plasmids, we used either VR1012 or a blank VR1255 (Luc+ gene deleted) which do not encode any protein in mammalian cells. Expression capability of a given plasmid DNA expression vector for im injection may be assessed by performing luciferase assays from injected skeletal muscle tissue.
2. Cytokine and cytokine receptor coding sequences may be produced by reverse transcription–polymer chain reaction (RT-PCR) from ConA-stimulated splenocytes using gene-specific primers containing proper restriction sites at their 5′ and 3′ ends to allow cloning into the appropriate vector.
3. Total RNA may be isolated from stimulated cells using RNAzol B reagent. PCR amplification of cytokine/cytokine receptor cDNA should be done with Pfu polymerase, which possesses proofreading capabilities. In the case of murine TGF-β1, we cloned its coding sequence into compatible enzyme restriction sites of pCI-neo to generate pCMV–TGF-β1. The murine TGFβ1 cDNA is under the transcriptional control of a CMV IE-EP and downstream of a chimeric intron. For optimal im expression, we subcloned into compatible enzyme restriction sites of VR1255, in place of the Luc+ gene, to generate pVR–TGF-β1, which encodes the latent form of TGF-β1.
4. Before using a new construct for im injection, it is suggested to test it for expression and bioactivity. Plasmid expression is confirmed by transiently transfecting COS-7 cells and assaying the supernatants (48–72 h, enzyme-linked immunosorbent assay [ELISA] or Western blot) for secreted plasmid-derived products. RT-PCR analysis is also performed to detect vector-derived mRNA. The latent TGF-β1 was activated by acidification and detected by ELISA, and its bioactivity was confirmed by the CCL64 mink cell line proliferation assay. Both the ELISA and bioassay detect only active TGF-β1. The sequence integrity of the subcloned cDNA should be confirmed by DNA sequencing.

5. Plasmid DNA expression vectors encoding IFN-γ have also been constructed in a similar fashion. In addition, one can also simultaneously express two different coding sequences in the same expression cassette. This is achieved by subcloning the respective coding elements on either side of an internal ribosomal entry site (IRES), which allows for translation of two coding regions from the same mRNA. In our laboratory, this approach was applied for the cloning and expression of IL-12 p35, and p40 subunits.

3.2. Construction of IFN-γR/IgG1 Expression Vector

1. In some instances, cytokine/cytokine-receptor IgG1-Fc fusion products are required because they offer numerous benefits such as increased serum half-life and higher avidity to its ligand. Our laboratory has constructed murine IL-4/IgG1-Fc and IFN-γR/IgG1-Fc fusion plasmid expression vectors.
2. For the production of such plasmid constructs, the coding sequences are fused to a murine IgG1 heavy-chain segment (part of CH1, hinge, CH2 and CH3) by overlap PCR gene assembly ***(28,34)***.
3. Murine IgG1-Fc is used because it does not activate the complement system. The resulting fusion product was then subcloned into VR1255. COS-7 cells transfected with these plasmids secrete a disulfide-linked homodimer.
4. The construction of an interferon-receptor/IgG1 construct is described as an example. Mouse IFN-γR α-chain cDNA was prepared by RT-PCR, from the RNA of lipopolysaccharide (LPS)-stimulated mouse splenic cells, and amplified with the following pair of primers: 5′-TAG TAG GAT ATC CTG TCA GAG GTG TCC CTC GCG CAG GAA-3′ (sense) and 5′-AAC GTT GCA GGT GAC GGT CTC GCT AGG AGG GAT ACA GAC GTC TTT CGA T-3′ (antisense).
5. Mouse IgG1 constant heavy-chain cDNA was produced by RT-PCR, from RNA extracted from an IgG1-secreting hybridoma, and amplified with the following pair of primers: 5′-ATC GAA AGA CGT CTG TAT CCC TCC TAG CGA GAC CGT CAC CTG CAA CGT T-3′ (sense) and 5′-TAT TAT GAA TTC GAG GTA GGT GTC AGA GTC CTG TAG G-3′ (antisense).
6. In all cases, PCR amplification was performed with Pfu DNA polymerase. These PCR-generated cDNA fragments were designed for overlap, as we have described ***(28,34)***. The IFN-γR/IgG1-encoding sequence was generated by PCR amplification of the overlapping DNA segments.
7. This segment was then inserted into *Eco*RV and *Eco*RI restriction sites of the VR1255 vector, after deletion of the original luciferase cDNA segment.

3.3. Transfection of COS-7 Cells

1. COS-7 cells were transfected with VR1255–IFNγR/IgG1 plasmid DNA (example) using a standard DNA–calcium phosphate coprecipitation technique.
2. Cells were maintained in Dulbecco's modified Eagle's medium (DMEM) with 5% fetal bovine serum (FBS) at 37°C in 5% CO_2.

3. Supernatants were collected at 72 h and stored at <–80°C. The production of the secreted IFN-γR/IgG1-Fc fusion protein was confirmed by detection in supernatants with a specific ELISA (*see* **Subheading 3.4.**).
4. Biological activity was demonstrated by neutralizing IFN-γ-stimulated NO production in RAW 264.7 cell lines.

3.4. IFN-γR/IgG1-Fc Fusion Protein ELISA Assays

Enzyme-linked immunosorbent assay detecting IFNγR/IgG1-Fc fusion protein, but not native IFN-γR, which lacks IgG1 determinants, was performed as follows:

1. 96-Well plates were coated overnight with purified anti-IFN-γR MAb (clone GR-20) in carbonate buffer, pH 9.6, at 4°C.
2. After washing with phosphate-buffered saline tween (PBST), the wells of the plate were blocked with 3% bovine serum albumin (BSA) for 1 h at 37°C.
3. As a second layer, culture supernatants or sera were applied to the plate for 1.5 h at 37°C.
4. After additional washings, a third layer consisting of rat biotinylated anti-mouse IgG1 was applied to the plate.
5. This was followed by washing and incubation with alkaline phosphatase–streptavidin conjugate. The reaction was developed and read by a standard method.

3.5. Plasmid DNA Preparation

1. Large-scale plasmid DNA preparations are produced by the alkaline lysis method using a Qiagen Endofree Plasmid Mega kit.
2. All plasmid preparations for im injections are resuspended in sterile 0.85% saline. Spectrophotometric analysis should reveal 260/280-nm ratios of 1.80 or higher.
3. Plasmid DNA preparations should be free of bacterial RNA or genomic DNA, as visualized on a 1% agarose gel. Gel electrophoresis is used to confirm the percentage of supercoiled plasmid (greater than 95%).
4. Plasmid identity is confirmed by restriction endonuclease analysis or PCR using gene-specific primers.
5. A limulus endotoxin assay should be performed to ensure minimal LPS contamination in the final plasmid DNA preparation (*see* **Note 1**).

3.6. Intramuscular Injection of Plasmid DNA

1. Mice are anesthetized by ip injection with xylazine (10 mg/kg) and ketamine (200 mg/kg).
2. The RF and/or TA muscle(s) of each mouse are injected with a 0.5-cm^3 sterile 29G1/2 insulin syringe, fitted with a plastic collar to limit needle penetration to 2 mm. Mice usually receive 25–100 μg of plasmid DNA per muscle in a

maximum volume of 50 μL of sterile saline. Other muscle groups may also be targeted, but RF and TA muscle groups produce more consistent results.
3. Control mice receive equal amounts of an appropriate control (usually blank) vector. When two plasmids are mixed, all mice must receive the same total amount of plasmid DNA, as the plasmid backbone may have immunostimulatory effects (*see* **Note 2** below).
4. To ensure that the im injection technique is properly performed, we recommend preliminary injections with luciferase-encoding plasmids followed by luciferase assays on treated muscle groups. (*See* **Notes 2** and **3**.)
5. To increase gene expression, one may also induce muscle regeneration with local anesthetic agents such as bupivacaine (Marcaine) ***(41)***. (*See* **Note 4**.)
6. Briefly, mice are anesthetized and TA muscles were injected with 50 μL of a 0.4% bupivacaine solution.
7. Five to seven days later, 50 μg of plasmid DNA is injected in each treated TA muscle. (*See* **Note 5**.)

3.7. Electroporation

Electroporation is currently the most effective method to increase gene transfer. We perform this technique as we have previously described ***(35)*** in anesthetized mice except that instead of invasive electrodes, we now apply caliper-type electrodes to the skin of the legs coated with conductive gel.

As in other cases, DNA injection in the TA and/or RF muscles is done with the mice on their back. Gently stretch the hind leg and hold it down with the foot. To inject the TA, locate the tibia, the injection site is approx 1 mm away from this bone in the middle of the length of the TA. To inject the RF, locate the knee, the injection site is to the side, approx 5 mm above the knee. The RF muscle can be injected at more than one site depending on the age and size of the mice used. After injection, the muscle is covered with conductive gel and a caliper electrode is positioned on each side of the leg to be electroporated. Electroporation is done using an ECM 830 Electro Square Porator set at the low-voltage mode (LV) at 200 V/cm with a pulse length of 20 ms for eight pulses at 1-s intervals as recommended ***(2)***. Generally, the expression of a gene under the control of the CMV IE-EP starts to appear 3–7 d postinjection.

3.8. Extraction of Luciferase from Skeletal Muscles

1. Mice were killed at various time-points postinjection in a CO_2 chamber.
2. The entire tibialis anterior muscle was excised from each leg and immediately put in a ice-cooled 1.5-mL microcentrifuge tube.
3. Tissue samples were then quickly stored at –80°C until needed.
4. Frozen muscles were placed in frozen lysis reagent cut into small pieces with sterile scissors and completely homogenized with a hand-held tissue grinder.
5. The entire homogenization process was conducted on ice. Sterile Milli-Q water was added to the mixture, vortexed briefly, and centrifuged for 30 min at 4°C.

6. The upper phase was transferred to a clean, sterile, 1.5-mL microcentrifuge tube and stored at –80°C until assayed.

3.9. Luciferase Assays

Total luciferase activity is determined using a commercial kit.

1. Briefly, 100 µL of luciferase substrate are added to 50 µL of muscle extract, properly mixed, and quickly placed in a luminometer.
2. Light units are recorded many times within 10 s after addition of the substrate.
3. The total luciferase content of the muscle sample is determined from relative light units using a standard curve of purified luciferase, which is diluted in muscle extract from uninjected muscles.

3.10. Lymphocyte Proliferation and Effector Function

1. To assess immunologic effects of cytokine/cytokine-receptor gene transfer, lymph node cells (LNC) or splenocytes (SPL) are prepared from draining lymph nodes of treated mice.
2. Proliferative responses were assessed by incubating LNC or purified T-cell populations ([5–8 × 10^5 cells/well) with specific antigen, anti-CD3 (2C11, 0.5 µg/mL) or Concanavalin A (ConA, 5 mg/mL).
3. The cultures are maintained in 96-well flat-bottom microtiter plates for 72–96 h at 37°C in humidified 5% CO_2 air.
4. The wells are pulsed with 1 mCi/well of [^{3}H]thymidine (NEN) for the final 8–12 h of culture. Results are reported as the mean counts per minute (cpm) of triplicate cultures ± SEM. Prior to harvesting, supernatants are obtained for cytokine production determination.

3.11. Determination of Plasma TGF-β1 Levels

In order to minimize the activation of platelets and subsequent release of endogenous TGF-β1, platelet-poor plasma was obtained as follows:

1. Whole blood was mixed with a 1.5% EDTA solution, mixed thoroughly, and put on ice immediately after blood collection.
2. The blood–EDTA mixture was then layered gently on 20% sucrose and centrifuged for 30 min at 12,000*g*.
3. The upper two-thirds of the upper phase (platelet-poor plasma fraction) was collected without disturbing the interface.
4. Once separated, the platelet-poor plasma was frozen at –80°C until assayed for TGF-β1 by ELISA.

3.12. PCR and RT-PCR Analysis

1. To assess physical presence of plasmid DNA in injected muscles, PCR analysis is performed with plasmid-specific primers.

2. Mice are killed and their TA muscles are excised, immediately frozen in liquid nitrogen, and stored at –80°C.
3. Total genomic DNA is isolated from thawed muscle specimens using conventional nucleic acid isolation protocols.
4. The PCR reactions is performed in a 50-μL reaction volume containing 2.5 μL of genomic DNA, 10 m*M* Tris-HC1 (pH 8), 50 m*M* KC1, 2 m*M* dNTP, 5 m*M* $MgC1_2$ 2μ*M* of each primer, and 1.5 U of *Taq* DNA polymerase.
5. General PCR cycling conditions may include the following: one cycle at 94°C; 30–40 cycles at 94°C for 1 min, at 55°C for 2 min, and at 72°C for 2 min; one final extension cycle at 72°C for 10 min.
6. The PCR amplimers are analyzed on a 1.5% agarose gel containing 0.5 μg/mL ethidium bromide. Vector-derived mRNA in treated muscle is detected by RT-PCR.
7. Briefly, total RNA is extracted from injected muscles and reverse transcribed with the Superscript preamplification system. One to two microliters of the reverse-transcription reaction is then used for PCR amplification.
8. Semiquantitative, competitive, or quantitative PCR (Real-Time/TaqMan PCR) is then performed using cytokine-specific primers such as IL-12 p40, IL-4, IFN-γ, TGF-β1, and β-actin. Optimal nonsaturating amplification conditions are carried out (PCR reactions are terminated in the linear portion of the amplification reaction) and PCR products are analyzed on 1.5% agarose gel containing 0.5 μg/mL ethidium bromide.
9. In some instances, RT-PCR may be coupled with Southern blotting to achieve greater detection sensitivity. In this case, PCR products are transferred onto a Hybond-N+ nylon membrane after electrophoresis by vacuum blotting, and hybridized by incubation with ^{32}P-labeled cDNA probes.
10. Cytokine probes consisted of 25 nucleotide oligonucleotides that bound to internal portions of PCR amplimers. Probes were prepared by labeling 50 ng DNA with [α-^{32}P]dCTP using a T4 polynucleotide kinase kit.
11. Blots were either exposed to autoradiographic film or subjected to PhosphorImager analysis. Semiquantitative/quantitative mRNA analysis is conducted by calculating relative quantities of RT-PCR signals for each cytokine, normalized to the β-actin signal of each sample.

4. Notes

1. DNA made from bacterial cultures contains important quantities of endotoxin (LPS), which can act as an adjuvant in vivo. Different plasmid DNA preparations can vary significantly in LPS content. To ensure that the LPS content is low in the plasmid preparation, we recommend that each lot be tested with a limulus amoebocyte lysate assay. Routinely, we use commercial low-endotoxin DNA purification columns. Generally, cesium-chloride purification protocols also yield low levels of LPS contamination. However, cesium-chloride purification is

labor intensive and requires the use of ethidium bromide, which is a carcinogenic compound.

2. The CpG-containing immunostimulatory sequences (ISS) of plasmid vectors induce production of inflammatory cytokines. ISS are numerous in most plasmids, but some authors have produced plasmids that are partially depleted of these sequences. The ISS have a beneficial adjuvant effect in DNA vaccines, presumably by stimulating production of IL-12 and IFN-γ. However, viral promoters (such as CMV IE-EP) are cytokine sensitive and we believe ISS to have a negative impact on the duration of gene expression. We and others have often observed a substantial decrease in expression by 21–30 d after DNA injection. In particular, we have noted this effect with various cytokine genes. However, as shown in **Fig. 3**, long-term expression can be achieved with some proteins, particularly if they are non-immunogenic and/or anti-inflammatory, or if the recipient mice are immunodeficient. Notably, in DNA vaccination experiments, responding T-cells have been reported to kill antigen-producing myocytes. It seems likely that several mechanisms can terminate vector expression.
3. The TA and RF muscles are the most frequently targeted in naked plasmid DNA injection protocols.
4. Muscle regeneration induced by anesthetic agents, such as bupivacaine, can greatly improve plasmid DNA uptake and expression in mouse skeletal muscles, presumably by increasing the proliferation of satellite cells in the muscle *(41)*. Following bupivacaine injection, muscle necrosis occurs within 48 h and is followed by satellite cell proliferation and myotube formation. Our results show that bupivacaine-treated muscles are superior to untreated muscles for foreign gene expression, because reporter levels are significantly increased at every time-point postinjection. Despite its obvious effect on the magnitude of luciferase levels, bupivacaine-induced muscle regeneration did not significantly prolong foreign gene expression. Electroporation can also damage muscle and, in addition to facilitating the entry of plasmids into myocytes, it may have effects similar to bupivacaine.
5. Muscle type, age, and sex are all factors that may influence gene uptake or expression. For instance, different muscle groups express considerably different levels of reporter protein. Gene transfer is much more efficient in rodents that in larger mammals, although gene expression has been reported in humans in DNA vaccination and other studies. The factors determining gene expression in various species have not been elucidated, but they may include the type of muscle fiber transfected, muscle anatomy, or physiology *(8)*.

Acknowledgments

Our studies were funded by the Juvenile Diabetes Foundation International, the Canadian Diabetes Association, the National Cancer Institute of Canada, and the Fraser Fund of the Royal Victoria Hospital, Montreal, Canada. C.A.P.

is supported by a Fogarty Foreign Scientist Postdoctoral Fellowship (NIAID, NIH). We thank Vical Inc. (San Diego, CA) for providing the VR1255 vector used in our studies.

References

1. Liblau, R. S., Singer, S. M., and McDevitt, H. O. (1995) Thl and Th2 CD4+ T cells in the pathogenesis of organ-specific autoimmune diseases. *Immunol. Today* **1,** 34–38.
2. Lotze, M. (1985) In-vivo administration of purified human interleukin 2. I. Half-life and immunologic effects of the Jurkat cell line-derived interleukin 2. *J. Immunol.* **134,** 157–166.
3. Olsson, T. (1995) Critical influences of the cytokine orchestration on the outcome of myelin Ag-specific T cell autoimmunity in experimental allergic encephalomyelitis and multiple sclerosis. *Immunol. Rev.* **144,** 245–268.
4. Wolff, J. A., Malone, R. W., Williams, P., Chong, W., Ascadi, G., Jani, A., et al. (1990) Direct gene transfer into mouse muscle. *Science* **247,** 1465–1468.
5. Wolff, J. A., Dowty, M. E., Jiao, S., Repetto, G., Berg, R. K., Ludtke, J. J., et al. (1992) Expression of naked plasmids by cultured myotubes abd entry of plasmids into T tubules and caveolae of mammalian skeletal muscle. *J. Cell Sci.* **103,** 1249–1259.
6. Dowty, M. E., Williams, P., Zhang, G., Hagstrom, J. E., and Wolff, J. A. (1995) Plasmid DNA entry into postmitotic nuclei of primary rat myotubes. *Proc. Natl. Acad. Sci. USA* **92,** 4572–4576.
7. Wolff, J. A., Ludtke, J. J., Ascadi, G., Williams, P., and Jani, A. (1992) Long-term persistence of plasmid DNA and foreign gene expression in mouse muscle. *Hum. Mol. Genet.* **1,** 363–369.
8. Doh, S. G., Vahlsing, H. L., Hartikka J., Liang, X., and Manthorpe, M. (1997) Spatial-temporal patterns of gene expression after injection of lacZ plasmid DNA. *Gene Ther.* **4,** 648–663.
9. Donnelly, J. J., Ulmer, J. B., Shiver, J. W., and Liu, M. A. (1997) DNA vaccines. *Annu. Rev. Immunol.* **15,** 617–648.
10. Manthorpe, M. F., Cornefert-Jensen, J., Hartikka, J., Felgner, J. Rundell, A., Margalith, M., et al. (1993) Gene therapy by intramusclular injection of plasmid DNA; studies on firefly luciferase gene expression in mice. *Hum. Gene Ther.* **4,** 419–431.
11. Dupuis, M., Denis-Mize, K., Woo, C., Goldbeck, C., Selby, M. J., Chen, M. C., et al. (2000) Distribution of DNA vaccines determines their immunogenicity after intramuscular injection in mice *J. Immunol.* **165,** 2850–2858.
12. Hemmi, H., Takeuchi, O., Kawai, T., Kaisho, T., Sato, S., Sanjo, H., et al. (2000) A Toll-like receptor recognizes bacterial DNA. *Nature* **408,** 740–745.
13. Prud'homme, G. J., Lawson, B. R., and Theofilopoulos, A. N. (2001) Anticytokine gene therapy of autoimmune diseases. *Exp. Opin. Biol. Ther.* **1,** 359–373.

14. Prud'homme, G. J., Lawson, B. R., Chang, Y., and Theofilopolous, A. N. (2001) Immunotherapeutic gene transfer into muscle. *Trends Immunol.* **22,** 149–155.
15. Levy, M. Y., Barron, L. G., Meyer, K. B., and Szoka, F. C., Jr. (1996) Characterization of plasmid DNA transfer into mouse skeletal: evaluation of uptake mechanism, expression and secretion of gene products into blood. *Gene Ther.* **3,** 201–211.
16. Hartikka J., Sawdey, M., Cornefert-Jensen, F., Margalith, M., Barnhart, K., Nolasco, M., et al. (1996) An improved plasmid DNA expression vector for direct injection into skeletal muscle. *Hum. Gene Ther.* **7,** 1205–1217.
17. Kikutani, H. and Makino, S. (1992) The murine autoimmune diabetes model: NOD and related strains. *Adv. Immunol.* **51,** 285–322.
18. Rabinovitch, A. (1998) An update on cytokines in the pathogenesis of insulin-dependent diabetes mellitus. *Diabetes/Metab. Rev.* **14,** 129–151.
19. Rabinovitch, A. (1993) Immunology and diabetes mellitus: roles of cytokines in IDDM pathogenesis and islet β-cell destruction. *Diabetes Rev.* **1,** 215–240.
20. Delovitch, T. L. and Singh, B. (1997) The nonobese diabetic mouse as a model of autoimmune diabetes: immune dysregulation gets the NOD. *Immunity* **7,** 728–738.
21. Roberts, A. B. and Sporn, M. B. (1990) The transforming growth factor-betas, in *Peptide Growth Factors and Their Receptors* (Sporn, M. and Roberts, A., eds.), Springer-Verlag, New York, pp. 419–472.
22. Letterio, J. J. and Roberts, A. B. (1998) Regulation of immune responses by TGF-β. *Ann. Rev. Immunol.* **16,** 137–161.
23. Prud'homme, G. J. and Piccirillo, C. A. (2000) The inhibitory effects of transforming growth factor-beta-1 (TGF-beta-1) in autoimmune diseases. *J. Autoimmun.* **14,** 23–42.
24. Gautam, S., Tebo, J. M., and Hamilton, T. A. (1992) IL-4 suppresses cytokine gene expression induced by IFN-γ and/or IL-2 in murine peritoneal macrophages. *J. Immunol.* **148,** 1725–1730.
25. Inobe, J. I., Chen, Y., and Weiner, H. L. (1996) In vivo administration of IL-4 induces TGF-β-producing cells and protects animals from experimental autoimmune encephalomyelitis. *Ann. NY Acad. Sci. USA* **778,** 390–392.
26. Falcone, M., Rajan, A. J., Bloom, B. R., and Brosnan, C. F. (1998) A critical role for IL-4 in regulating disease severity in experimental allergic encephalomyelitis as demonstrated in IL-4-deficient C57BL/6 mice and BALB/c mice. *J. Immunol.* **161,** 4822–4830.
27. Piccirillo, C. A., Chang, Y., and Prud'homme, G. J. (1998) TGF-β1 somatic gene therapy prevents autoimmune disease in non-obese diabetic mice. *J. Immunol.* **161,** 3950–3956.
28. Chang, Y. and Prud'homme, G. J. (1999) Intramuscular administration of expression plasmids encoding interferon-gamma-receptor/IgG1 or IL-4/IgG1 chimeric proteins protects from autoimmunity. *J. Gene Med.* **1,** 415–423.
29. Martin, R. and McFarland, H. F. (1995) Immunological aspects of experimental allergic encephalomyelitis and multiple sclerosis. *Crit. Rev. Clin. Lab. Sci.* **2,** 121–182.

30. O'Garra, A., Steinman, L., and Gijbels, K. (1997) CD4+ T-cell subsets in autoimmunity. *Curr. Opin. Immunol.* **9,** 872–883.
31. Racke, M. K., Bonomo, A., Scott, D. E., Cannella, B., Levine, A., Raine, C. S., et al. (1994) Cytokine-induced immune deviation as a therapy for inflammatory autoimmune disease. *J. Exp. Med.* **180,** 1961–1966.
32. Renno, T., Krakowski, M. K., Piccirillo, C., Lin, J., and Owens, T. (1995) TNFα expression by resident microglia and infiltrating leukocytes in the central nervous system of mice with experimental allergic encephalomyelitis. *J. Immunol.* **154,** 944–953.
33. Piccirillo, C. A. and Prud'homme, G. J. (1999) Prevention of experimental allergic encephalomyelitis by intramuscular gene transfer with cytokine-encoding plasmid vectors. *Hum. Gene Ther.* **10,** 1915–1922.
34. Prud'homme, G. J. and Chang, Y. (1999) Prevention of autoimmune diabetes by intramuscular gene therapy with a nonviral vector encoding an interferon-γ receptor/IgG1 fusion protein. *Gene Ther.* **6,** 771–777.
35. Lawson, B. R., Prud'homme, G. J., Chang, Y., Gardner, H. A., Kuan, J., Kono, D. H., et al. (2000) Treatment of murine lupus with cDNA encoding IFN-γR/Fc. *J. Clin. Invest.* **106,** 207–215.
36. Mir, L. M., Bureau, M. F., Gehl, J., Rangara, R., Rouy, D., Caillaud, J. M., et al. (1999) High-efficiency gene transfer into skeletal muscle mediated by electric pulses. *Proc. Natl. Acad. Sci. USA* **96,** 4262–4267.
37. Giladi, N. E., Raz, E., Karmeli, F., Okon, E., and Rachmilewitz, D. (1995) Transforming growth factor-beta gene therapy ameliorates experimental colitis in rats. *Eur. J. Gastroent. Hepatol.* **7,** 341–347.
38. Croxford, J. L., Triantaphyllapoulos, K., Podhajcer, O. L., Feldmann, M., Baker, D., and Chernajovsky, Y. (1998) Cytokine gene therapy in experimental allergic encephalomyelitis by injection of plasmid DNA-cationic liposome complex into the central nervous system. *J. Immunol.* **160,** 5181–5187.
39. Raz., E., Lotz, M., Bairs, S. M., Berry, C. C., Eisenberg, R. A., and Carson, D. (1995) Modulation of disease activity in murine systemic lupus erythematosus by cytokine gene delivery. *Lupus* **4,** 286–292.
40. Tokui, M., Takei, I., Tashiro, F., Shimada, A., Kusaga, A., Ishii, M., et al. (1997) Intramuscular injection of expression plasmid DNA is an effective means of long-term systemic delivery of interleukin-5. *Biochem. Biophys. Res. Comm.* **233,** 527–531.
41. Danko, I., Fritz, J. D., Jiao, S, Hogan, K., Latendresse, J. S., and Wolff, J. A. (1994) Pharmacological enhancement of in vivo foreign gene expression in muscle. *Gene Ther.* **1,** 114–121.

12

Interferon-β Gene Therapy for the Treatment of Arthritis

Yuti Chernajovsky, Hanna Dreja, Kostas Triantaphyllopoulos, and David Gould

1. Introduction

Interferons were originally described as antiviral agents by Isaacs and Lindenmann *(1)*. From its subtypes, fibroblast interferon (IFN-β) *(2)* was also found to have important cytostatic *(3,4)* as well as immunomodulatory functions *(5)* that have been advantageous for its clinical use both in cancer *(6)* and in autoimmune disease conditions such as multiple sclerosis *(7,8)* and, recently, in rheumatoid arthritis *(9)*. As with many other cytokines, its local production is well controlled and environmental factors affect its expression. IFN-β has direct effects on the majority of cell types in the body, as its receptors are widely expressed in all tissues. Its use as a recombinant protein is limited because of its short half-life and production costs and its systemic delivery (subcutaneously) is inefficient, done at high doses and for long times using periodic administration. We and others have shown that the delivery of secreted biological compounds such as cytokines or their inhibitors (e.g., their soluble receptors) by gene therapy *(10–13)* can in a single administration achieve long-term therapeutic effects at doses that are various logs of magnitude lower than those used with protein therapy. Gene therapy could also be designed to deliver the gene product locally, or its transcriptional control could be engineered so that is regulated using exogenously added drugs *(14)*. IFN-β, as some other cytokines (e.g., interleukin [IL]-4, IFN-γ) is species-specific, and in order to obtain appropriate biological responses in mice, we had to clone the mouse gene by polymerase chain reaction (PCR). Although the mouse gene was originally cloned in the 1980s *(15)*, reagents against mouse IFN-β have become

From: *Methods in Molecular Biology, vol. 215: Cytokines and Colony Stimulating Factors: Methods and Protocols*
Edited by: D. Körholz and W. Kiess © Humana Press Inc., Totowa, NJ

commercially available only recently. Both recombinant mouse IFN-β and antibodies were not accessible, as there was not extensive biological research in mouse models and most data reported were directly from clinical trials or using human cell lines. As the studies on the immunoregulatory functions of IFN-β increased using animals models, the need for mouse reagents arose. Recent studies described differences is signaling pathways used in mice ***(16)*** and a gene IFN-β knockout mouse has been produced ***(17)***.

Here, we describe in detail the way we have used IFN-β in a mouse model of rheumatoid arthritis both as a plasmid vector and in retroviral vectors and how we have assessed its biological activity.

2. Materials

2.1. Tissue Culture

1. HAM-F12 medium (Life Technologies Ltd., Paisley, UK).
2. Fetal bovine serum (FBS) (Life Technologies Ltd.) supplemented with 2.5 U/mL penicillin, 2.5 μg/mL streptomycin and 2 m*M* glutamine (Life Technologies Ltd., Paisley, UK).
3. Trypsin-EDTA (Life Technologies Ltd., Paisley, UK).
4. Alpha-DMEM medium without nucleosides (PAA Laboratories, Linz, Austria).
5. Dulbecco's modified Eagle's medium (DMEM) supplemented with glutamax-1 (Life Technologies).
6. For metabolic labeling, ^{35}S–methionine–cysteine mix (Amersham-Pharmacia Biotech, Bucks, UK) and cysteine–methionine-free medium (Life Technologies Ltd.).
7. Thymidine (at a final concentration of 10 μg/mL, L-proline (150 μg/mL), and glycerol (10% w/v in medium without serum) were from Sigma, Poole, UK. Phleomycin (stock of 2.5 mg/mL) was from Sigma (G418, Geneticin; Life Technologies Ltd.). These were filtered sterilized on Sartorious membranes (0.2 μm) prior to use.
8. Di-hydrofolate reductase (DHFR)-deficient Chinese hamster ovary (CHO) cells were kindly provided by L. Chasin (Columbia University, NY); monkey COS-7 cells were developed by Y. Gluzman (Cold Spring Harbor, NY), and mouse LTK- (thymidine kinase deficient) were a gift from I. Kerr (Imperial Cancer Research Fund, London). α-BOSC23 was provided by W. S. Pear (Rockefeller University, NY).
9. Sterile plasticware (pipets and plates) were from Falcon (Becton Dickinson, Rutherford, NJ).

2.2. Plasmid-, DNA-, and RNA-Modifying Enzymes

1. Plasmid pcDNA3 was from In-Vitrogen (Groningen, The Netherlands).
2. T4 DNA ligase, restriction enzymes, and Rnase A were from either New England Biolabs (Hitchin, Herts, UK) or Boehringer-Mannheim (Lewes, East Sussex, UK). These are used according to the supplier's instructions.

2.3. Biochemistry

1. Serine–protease inhibitors (SPI) were from Calbiochem (Beeston, UK. Pepstatin-A at 10 μg/mL, aprotinin at 1 μg/mL, chymostatin at 10 μg/mL, leupeptin at 10 μg/mL, and AEBSF [4-(2-aminoethyl) benzene sulfonyl-fluoride, HCl] at 200 μ*M*. They were all prepared as a 1000-fold concentrated stock.
2. Most salts and buffers were from Sigma (Poole, UK). Protein-G–Sepharose (Amersham Pharmacia Biotech, Bucks, UK).
3. Rat–anti-mouse IFN-β (7F-D3; AMS, Abingdon, UK).
4. NP-40 (BDH, Poole, UK), trichloroacetic acid (TCA) (Sigma, Poole, UK).
5. Phenol (water saturated) was from Life Technologies Ltd. (Paisley, UK).
6. Acrylamide/bis-acrylamide (Protogel) for sodium dodecyl sulfate–polyacrylamide gel electrophoresis (SDS-PAGE) was from National Diagnostics (Atlanta GA).
7. Sodium dodecyl sulfate (SDS) was from BDH (Poole, UK).
8. Lysozyme was from Boehringher-Mannheim (Lewes, UK).
9. Polymyxin B–agarose columns were from Pierce (Rockford, IL).
10. The detergent *n*-Octyl-β-D-thioglucopyranoside (ULTROL®) was from Calbiochem (La Jolla, CA).

2.4. In Vivo Work

1. Freund's complete adjuvant (FCA) was prepared by grinding 100 mg *Mycobacterium tuberculosis* in 30 mL incomplete Freund's adjuvant (Difco Laboratories, East Molsey, UK).
2. Anesthetic was fentanyl-fluanisone (Hypnorm, Janssen Animal Health, High Wycombe Bucks, UK).

3. Methods

3.1. Permanent Transfection into DHFR-Deficient Chinese Hamster Ovary Cells, Selection and Ring Cloning of Resistant Clones

1. DHFR-deficient CHO cells are maintained in HAM-F12 medium with 10% FBS, penicillin/streptomycin, and glutamine.
2. pcDNA3 (20 μg)-expressing mouse IFN-β is linearized with *Pvu*I and ligated with *Pvu*I-cut pSV_2DHFR (1 μg) ***(18)***. After phenol extraction, the plasmids are ligated in 300 μL with T4 DNA ligase at 16°C. Covalent ligation in vitro of the pSV_2DHFR and the plasmid-expressing IFN-β guarantees that upon subsequent amplification of the DHFR minigene, by selection with methotrexate, the IFN-β gene also will get coamplified.
3. The DNA is precipitated in 0.4 *M* NH_4 acetate and resuspended in water and added as 1 mL calcium phosphate coprecipitate on 0.5×10^6 CHO cells on 10-cm plates seeded 24 h earlier. Four hours later, the cells are treated with 10% glycerol in HAM-F12 without FBS. Cells are washed in FBS-free media and let to recover for 48 h.
4. Transfected cells are then trypsinized and split into six 10-cm plates. Selection is carried out in Alpha-DMEM medium without nucleosides, 10% dialyzed

FBS *(18)*, and 1 mg/mL G418. Medium is changed twice weekly until 2–3 wk later clearly defined clones can be isolated by ring cloning or all the cells are trypsinized and maintained as a population.

5. For ring cloning, the position of clones is marked with waterproof marker underneath the plastic plate. Plates are washed with PBS (phosphate-buffered saline) twice and the liquid is well removed by suction from the plate. Plastic tips of 20–200 μL are cut with scissors and their broader base is sterilized by autoclaving. Similarly, vacuum silicon grease is sterilized in a glass Petri dish. Using sterile tweezers, the base of the tip is carefully coated with silicon grease and adhered to the marked clones forming a well. Fifty microliters of trypsin is added and the detachment of the cells is monitored under the microscope. Once detached, the cells are removed from the "well" using 50 μL of selection medium and the cells are plated on six well plates containing 1 mL of medium. Plating efficiency depends on the size of the clones. We recommend picking clones that have more than 120 cells.

3.2. Transient Transfection into Monkey COS-7 Cells

1. Twenty micrograms of plasmid DNA is transfected as described in **Subheading 3.1.3.** in duplicates to 0.5×10^6 COS-7 cells seeded in 10-cm plates. The DNA coprecipitate is left overnight. COS-7 cells were grown in DMEM with antibiotics and 10% FBS.
2. Forty-eight hours after osmotic glycerol shock, with 10% glycerol in serum-free DMEM, the supernatants were collected for IFN antiviral assay.

3.3. IFN-β Biological Assay

Interferon-β was assayed for its antiviral properties, inhibiting the cytopathic effect of encephalomyocarditis (EMC) virus infection of murine LTK^- fibroblasts.

1. Briefly, 1×10^4 LTK^- cells are added (50 μL/well) in a 96-well plate and incubated overnight. It is important to check that the cells have attached properly and that a homogeneous layer of cells covers the well. Sometimes we have had problems with cells not attaching properly and this affects the reading of the assay.
2. Supernatants from IFN-β-expressing cells were serially diluted (1:2) in a different plate and added to the cells (50 μL/well).
3. The following day, culture medium was replaced by fresh 100 μL of medium containing the EMC virus (kindly provided by Dr. I. Kerr, Imperial Cancer Research Fund, London), diluted $1:10^5$ (to 0.6 plaque-forming units/cell).
4. The next day, the cytopathic activity was assessed by light microscopy and the dilution of samples yielding 50% protection from lysis by IFN-β was determined 24 h postinfection *(19)*. For details about growing and isolation of the EMC virus, refer to **ref. *20***.

3.4. Metabolic Labeling of CHO Cells

1. Confluent plates of permanently transfected or nontransfected CHO cells were washed with cysteine–methionine free medium containing 10% dialyzed FBS and supplemented with 10 μg/mL thymidine, 2 m*M* glutamine, penicillin/streptomycin, and 150 μg/mL L-proline. Labeling was either overnight or for 48 h in the presence of ^{35}S–methionine–cysteine mix at 1 Ci/mmol using 250 mCi/plate in 5 mL media.
2. Supernatants were collected, cell debris spun down, and clear supernatants supplemented with SPI.

3.5. Immunoprecipitation

1. The ^{35}S-labeled supernatants are precleared with protein-G–Sepharose in PBS with 0.1% NP-40 (using 400 μL of 50% beads/vol).
2. The supernatant volume corresponding to 25×10^6 counts per minute (cpm) of TCA total precipitated protein is incubated with the monoclonal rat–anti-mouse IFN-β (7F-D3 at a dilution of 1:250) for 3–4 h at 4°C.
3. The antigen–antibody complexes are then bound to protein-G–Sepharose (700 μL of 50% solution) by mixing overnight at 4°C rolling end-over-end. Protein-G–Sepharose beads are washed three times with 5 mL of 0.1% NP-40 in PBS. Proteins bound to beads are split into fractions of 50-μL beads in small tubes and directly resuspended in Laemmli SDS-PAGE loading buffer ***(21)***.
4. Gels are fixed for 30 min in 7% acetic acid and 10% methanol, washed in water for 30 min, and treated with 1 *M* sodium salicylate for 1 h before drying and exposing to autoradiography with X-ray film.

3.6. Preparation of CII in Freund's Complete Adjuvant

1. Bovine articular cartilage was used as a source of CII and had been purified by pepsin digestion and salt fractionation, as described by Miller ***(22)***.
2. Freund's complete adjuvant is prepared as described in **Subheading 2.4.** (in vivo work). One volume of CII (4 mg/mL in 0.1 *M* acetic acid) is emulsified with 1 vol of FCA prior to injection.

3.7. Induction of Arthritis in DBA/1 Mice

1. DBA/1 mice were maintained under germ-free conditions. Eight- to twelve-week-old male DBA/1 mice are anaesthetized with 100 μL fentanyl-fluanisone, diluted 1:10 in sterile water and injected intraperitoneally.
2. The base of the tail is shaved prior to immunization. One hundred microliters of 2 mg/mL CII in FCA were injected subcutaneously at the base of the tail ***(23)***. Two weeks later, the mice were boosted with a similar injection using incomplete FA. Macroscopically visible signs of arthritis developed 2–3 wk after immunization and the incidence was >90%.

3.8. Isolation of LPS-Free Plasmid DNA from Spheroplasts

To reduce the levels of contaminating bacterial LPS during injections of naked DNA therapy, the method described by Sandri-Goldin was followed *(24,25)*:

1. Overnight, bacterial culture was centrifuged (5000*g* for 10 min) and the pellet was resuspended in 50 mL of cold 20% sucrose in 50 m*M* Tris-HCl (pH 8.0).
2. Ten milliliters of 5 mg/mL lysozyme was added to digest the outer cell wall, which is the source of LPS. The mixture was left on ice for 5 min.
3. Twenty milliliters of cold 0.25 *M* EDTA (pH 8.0) is added and the tube was left on ice for a further 5 min. Twenty milliliters of 0.25 *M* Tris-HCl (pH 8.0) is added and the solution is incubated for 5 min at 37°C.
4. The spheroplasts are pelleted (1000*g* for 10 min at 4°C), resuspended in 30 mL Sandri-Goldin resuspension buffer (50 m*M* glucose, 25 m*M* Tris-HCl (pH 8.0), 10 m*M* EDTA [pH 8.0] with 100 μg/mL RNase A (Sigma, Poole, UK) and incubated for 15 min at 37°C.
5. Proteins and contaminating chromosomal DNA are removed by centrifugation (5000*g* for 10 min). The supernatant is filtered through a nylon sieve and the plasmid solution precipitated with 0.6 vol of isopropanol.
6. After centrifugation (5000*g* for 10 min), the pellet is washed with ethanol, dried, resuspended in TE (Tris-HCl, pH 8, 10 m*M* and 1 m*M* EDTA, pH 8.0), extracted twice with phenol–chloroform (1:1) (v/v), followed by one extraction with chloroform, and, finally, precipitated in ethanol (2.5–3 v/v).
7. The pellet is washed and resuspended in TE, containing 1% of the detergent *n*-octyl-β-D-thioglucopyranoside (ULTROL) and incubated for 30 min at room temperature. The micelles are spun down; the supernatant is precipitated with ethanol and resuspended in 10 mL of 0.1 *M* NaCl.
8. Polymyxin B–agarose columns are rinsed with sterile water and three rounds of 0.1 *M* sodium acetate (pH 4.0), alternating with 0.1 *M* sodium borate (pH 8.0). After flushing with sterile water, the column was equilibrated with 0.1 *M* NaCl. Using a peristaltic pump, the DNA was circulated through the column overnight. After collecting the eluate, remaining DNA in the column is rinsed out with 20 mL of 0.1 *M* NaCl and the DNA is precipitated with ethanol.

3.9. Intramuscular DNA Injection

1. DNA (10–100 μg in up to 100 μL PBS) is injected intramuscularly using a Hamilton syringe and a 26G needle. The DNA is maintained at 1–2 mg/mL in sterile water (it is important to maintain the DNA at this concentration to avoid precipitates forming in solution) and mixed 1:1 with PBS two times prior to injection.
2. Reproducible injection into the quadriceps is performed by holding the mouse in the left hand, the scruff of the neck between the index and thumb. Its hind left

leg is immobilized against the base of the thumb with the ring finger, and its hind right foot is accommodated on top of the ring finger and held in place with the little finger. In this position, the right quadriceps is easily accessible. We inject in three sites in the same muscle. We found that injections in the tibialis muscle are not reproducible, as this muscle has a thick myofascial sheet and only very small volumes can be injected.

3.10. Retroviral Vector Packaging

The cloning and expression of mouse IFN-β in the retrovirus vector pBabe-Bleo *(26)* has been previously described *(19)*. As IFN is an antiviral known to affect retrovirus maturation and assembly, we decided to avoid packaging it as a retroviral particle in mouse packaging cell lines, which could be affected by its antiviral effect via interaction with its cellular receptors. Hence, we packaged the retroviral vector in a different species packaging cell line.

1. The human amphotropic packaging cell line α-BOSC23 *(27)* is grown and maintained in DMEM containing 10% (v/v) FBS, 2.5 U/mL penicillin, and 2.5 μg/mL streptomycin supplemented with glutamax-1.
2. Transient transfections of α-BOSC23 packaging cell line are performed as for COS-7 cells as described in **Subheading 3.2.** Supernatants (5 mL/plate) containing packaged retroviral particles are cleared from cellular debris by centrifugation for 10 min at 3k and then snap frozen at –70°C until use.

3.11. Retrovirus Infection

1. Cell supernatants containing packaged retrovirus were thawed at 37°C and supplemented with 8 μg/mL polybrene (Sigma) made as a stock in serum-free DMEM at 8 mg/mL and filter-sterilized. We established an immortalized DBA/1 mouse fibroblast cell line *(13)*, after infection with a retrovirus expressing a temperature-sensitive SV40 large T-antigen also containing a U19 mutation *(28)*. These fibroblasts are grown and maintained in DMEM containing 10% (v/v) FBS, with glutamine, penicillin/streptomycin, and 1 mg/mL G418.
2. To infect the DBA/1 immortalized fibroblasts with the pBabe-Bleo mouse IFN-β vector and select for infected cells, their sensitivity to the selective drug phleomycin has to be assessed. Twenty-five μg/mL of phleomycin is sufficient to cause cell death during a 5-d period; higher concentrations are toxic and do not allow for appropriate selection. Similarly, for any other selectable drug for cells in culture, drug concentrations that select cells and are not toxic overnight have to be carefully assessed by doubling dilutions on cells that are seeded at a concentration that allows growth in the log phase.
3. For viral infection, DBA/1 cells are plated at 5×10^5 per 10-cm plate 1 d prior to infection. Medium is removed and 5 mL viral supernatants supplemented with polybrene is added. Then, 4–16 h later, the supernatant is washed away and

cells were left to recover for 48 h. Then, they were trypsinized 1:4 and put in selection media with phleomycin.

4. Selection media was changed every 3 d and after 2–3 wk, growing clones can be isolated or are maintained as a population (**Subheading 3.1.**). The IFN-β expression level is assessed by the antiviral assay (**Subheading 3.3.**).
5. For use in the collagen-induced arthritis model, cells are trypsinized, counted in an hemocytometer, and washed in DMEM with FBS and in Hank's medium. Cell are resuspended in Hank's medium in a final volume not exceeding 200 μL for intraperitoneal injection containing 4–12 million cells that are injected either at different time-points postimmunization or at onset of disease *(13)*.

References

1. Isaacs, A. and Lindenmann, J. (1957) Virus interference. I. The interferon. *Proc. R. Soc. London Ser. B* **147,** 258–267.
2. Knight, E. (1976) Interferon: purification and initial characterization from human diploid cells. *Proc. Natl. Acad. Sci. USA* **73,** 520–523.
3. Knight, E. J. (1976) Antiviral and cell growth inhibitory activities reside in the same glycoprotein of human fibroblast interferon. *Nature* **262,** 302–303.
4. Einat, M., Resnitzky D., and Kimchi A. (1985) Close link between reduction of c-myc expression by interferon and, G0/G1 arrest. *Nature* **313,** 597–600.
5. Devajyothi, C., Kalvakolanu, I., Babcock, G. T., Vasavada, H. A., Howe, P. H., and Ransohoff, R. M. (1993) Inhibition of IFN γ-induced MHC class II gene transcription by IFN β and Type β-1 transforming growth factor in human astrocytoma cells. *J. Biol. Chem.* **268,** 18,794–18,800.
6. Liberati, A., De Angelis, V., Fizzotti, M., Schippa, M., Cecchini, M., Adiuto, D., et al. (1994) Natural-killer-stimulatory effect of combined low-dose interleukin-2 and IFN β in hairy-cell leukemia patients. *Cancer Immunol. Immunother.* **38,** 323–331.
7. Arnason, B. G. W., Dayal, A., Qu, Z. X., Jensen, M. A., GenÁ, K., and Reder, A. T., eds. (1996) *Mechanisms of Action of IFN β in Multiple Sclerosis.* Springer Seminars in Immunopathology, Vol. 18, Springer-Verlag, Berlin, pp. 125–148.
8. Panitch, H. S. and Bever, C. T. (1993) Clinical trials of interferons in multiple sclerosis: what have we learned? *J. Neuroimmunol.* **46,** 155–164.
9. Tak, P. P., 't Hart, B. A., Kraan, M. C., Jonker, M., Smeets, T. J., and Breedveld, F. C. (1999) The effects of IFNβ treatment on arthritis. *Rheumatology* **38,** 362–369.
10. Chernajovsky, Y., Adams, G., Podhajcer, O., Mueller, G., Robbins, P., and Feldmann, M. (1995) Inhibition of transfer of collagen-induced arthritis into SCID mice by ex-vivo infection of spleen cells with retroviruses expressing soluble human TNF receptor. *Gene Ther.* **2,** 731–735.
11. Croxford, J. L., Triantaphyllopoulos, K., Podhajcer, O. L., Feldmann, M., Baker, D., and Chernajovsky, Y. (1998) Cytokine gene therapy in experimental allergic encephalomyelitis by injection of plasmid DNA-cationic liposome complex into the central nervous system. *J. Immunol.* **160,** 5181–5187.

12. Chernajovsky, Y., Adams, G., Triantaphyllopoulos, K., Ledda, M. F., and Podhajcer, O. L. (1997) Pathogenic lymphoid cells engineered to express TGF β 1 ameliorate disease in a collagen-induced arthritis model. *Gene Ther.* **4,** 553–559.
13. Triantaphyllopoulos, K. A., Williams, R. O., Tailor, H. and Chernajovsky, Y. (1999) Amelioration of collagen-induced arthritis and suppression of IFN-γ, IL-12, and TNFα production by IFNb gene therapy. *Arthritis Rheum.* **42,** 90–99.
14. Gould, D. J., Berenstein, M., Dreja, H., Ledda, F., Podhajcer, O. L., and Chernajovsky, Y. (2000) A novel doxycycline inducible autoregulatory plasmid which displays "on/off" regulation suited to gene therapy applications. *Gene Ther.* **7,** 2061–2070.
15. Higashi, Y., Sokawa, Y., Watanabe, Y., Kawade, Y., Ohno, S., Takaoka, C., et al. (1983) Structure and expression of a cloned cDNA mouse IFN β. *J. Biol. Chem.* **258,** 9522–9529.
16. Farrar, J., Smith, J., Murphy, T., Leung, S., Stark, G. and Murphy, K. (2000) Selective loss of type I interferon-induced STAT4 activation caused by a minisatellite insertion in mouse Stat2. *Nat. Immunol.* **1,** 65–69.
17. Deonarain, R., Alcamí, A., Alexiou, M., Dallman, M., Gewert, D., and Porter, A. (2000) Impaired antiviral response and alpha/beta interferon induction in mice lacking β interferon. *J. Virol.* **74,** 3404–3409.
18. Chernajovsky, Y., Mory, Y., Chen, L., Marks, Z., Novick, D., Rubinstein, M., et al. (1984) Efficient constitutive production of human fibroblast interferon by hamster cells transformed with the IFN-β 1 gene fused to an SV40 early promoter. *DNA* **3,** 297–308.
19. Triantaphyllopoulos, K., Croxford, J. L., Baker, D., and Chernajovsky, Y. (1998) Cloning and expression of murine IFNb and a TNF antagonist for gene therapy of experimental allergic encephalomyelitis. *Gene Ther.* **5,** 253–263.
20. Minor, P. (1985) Growth, assay and purification of picornaviruses, in *Practical Approach* (Mahy, B., ed.), IRL, Oxford, pp. 25–42.
21. Laemmli, U. K. (1970) Cleavage of structural proteins during the assembly of the head of bacteriophage T4. *Nature* **227,** 680–685.
22. Miller, E. J. (1972) Structural studies on cartilage collagen employing limited cleavage and solubilization with pepsin. *Biochemistry* **11,** 4903–4909.
23. Williams, R. O., Feldmann, M., and Maini, R. N. (1992) Anti-TNF ameliorates joint disease in murine collagen-induced arthritis. *Proc. Natl. Acad. Sci. USA* **89,** 9784–9788.
24. Sandri-Goldin, R. M., Goldin, A. L., Levine, M., and Glorioso, J. C. (1981) High-frequency transfer of cloned herpes simplex virus type 1 sequences to mammalian cells by protoplast fusion. *Mol. Cell. Biol.* **1,** 743–752.
25. Wicks, I. P., Howell, M. L., Hancock, T., Kohsaka, H., Olee, T., and Carson, D. A. (1995) Bacterial lipopolysaccharide copurifies with plasmid DNA: implications for animal models and human gene therapy. *Hum. Gene Ther.* **6,** 317–323.
26. Morgenstern, J. P. and Land, H. (1990) Advanced mammalian gene transfer: high titre retroviral vectors with multipledrug selection markers and a complementary helper-free packaging cell line. *Nucleic Acids Res.* **18,** 3587–3596.

27. Pear, W. S., Nolan, G. P., Scott, M. L., and Baltimore, D. (1993) Production of high-titer helper-free retroviruses by transient transfection. *Proc. Natl. Acad. Sci. USA* **90,** 8392–8396.
28. Almazan, G. and McKay, R. (1992) An oligodendrocyte precursor cell line from rat optic nerve. *Brain Res.* **579,** 234–245.

13

Gene Gun-Based In Vivo Gene Transfer

Application to DNA Vaccination

Tohru Sakai and Kunisuke Himeno

1. Introduction

The introduction of genetic material into tissue of interest remains a rate-limiting step for molecular investigations in many fields. Many different methods have been developed for the delivery of a gene in vivo, such as virus-, cationic liposome-, and particle-mediated gene transfer or direct injection of DNA.

The particle-mediated method for gene delivery with a gene gun utilizes a shock wave to accelerate DNA-coated gold particles into target cells or tissues. The application of gene transfer by particle bombardment was first described by Sanford et al. and was shown to be an efficient method for transformation of a plant *(1)*. This gene delivery method is also effective in various somatic tissues in vitro and in vivo. High levels of transgene activity have been readily detected both in tissue extracts and at cellular levels from tissues of animals into which various reporter genes have been bombarded *(2,3)*. This in vivo gene delivery method has been used for DNA vaccination, because it is simple and only a small amount of DNA is sufficient to develop both cellular and humoral immune responses, which can lead to protection from various infectious pathogens *(4)*. Recently, the gene-gun system has been employed in gene therapy. In fact, treatment with a particle-mediated in vivo cytokine gene after implantation of tumor cells has been shown to inhibit tumor growth and to prolong the survival of tumor-bearing mice *(5)*.

This chapter deals with the preparation of microcarriers for particle-mediated gene transfer, optimization of the conditions for in vivo gene transfer, and the

From: *Methods in Molecular Biology, vol. 215: Cytokines and Colony Stimulating Factors: Methods and Protocols*
Edited by: D. Körholz and W. Kiess © Humana Press Inc., Totowa, NJ

conditions for DNA vaccination using a Helios Gene Gun System (Bio-Rad, Hercules, CA).

1.1. Preparation of Microcarriers

Prior to gene transfection, the plasmid DNA must be attached to the gold particle. This is accomplished by precipitation of the DNA from solution in the presence of gold microcarriers and polycation spermidine by the addition of $CaCl_2$. The DNA/microcarrier solution is coated onto the inner wall of gold-coat tubing and dried. The tubing is then cut into 0.5-in.-long cartridges. These cartridges, when inserted into the cartridge holder of a Helios Gene Gun, are the source of the DNA that enters the target cells of tissues by helium discharge (*see* **Fig. 1**).

2. Materials

2.1. Precipitation of DNA onto Microcarriers

1. Plasmid DNA: Plasmid DNA should be prepared by using a commercial plasmid purification kit or by carrying out cesium chloride gradient centrifugation twice. Plasmid DNA is dissolved in TE buffer (pH 8.0, 10 m*M* Tris-HCl, 1 m*M* EDTA).
2. Gold microcarriers (1.6 µL).
3. Polyvinylpyrrolidone (PVP), 360,000 MW (molecular weight).
4. Fresh 100% ethanol.
5. Gold-coat tubing.
6. 0.05 *M* Spermidine.
7. 1 *M* $CaCl_2$.
8. Helios Cartridge Kit (Bio-Rad) containing 0.5 g PVP and 50 ft of gold-coat tubing.

2.2. Loading the DNA/Microcarrier Suspension into Gold-Coat Tubing Using the Tubing Prep Station

1. Tubing prep station.
2. Gold-coat tubing.
3. Nitrogen tank.
4. Tube cutter.
5. Tubing prep station and tube cutter contained in the Helios gene-gun system.

3. Methods

3.1. Precipitation of DNA onto Microcarriers

1. Prepare a stock solution of 10 mg/mL PVP in ethanol in a small screw-cap container. In this process, vortexing should not be used; prepare solution daily.

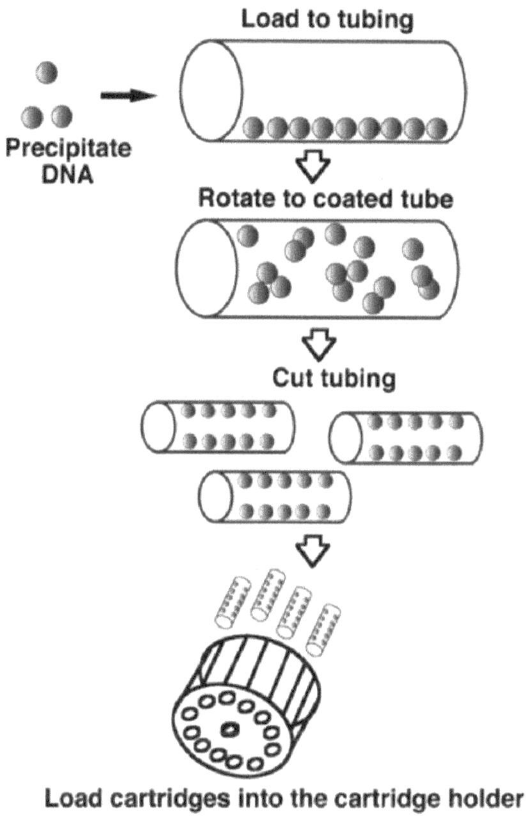

Fig. 1. The process for the preparation of microcarriers. The gene gun introduces DNA into the skin bound into small gold particles. For this purpose, DNA is precipitated onto gold microcarriers, which are then loaded into plastic tubing. The tubing is cut into approx 0.5-in.-long pieces, and the pieces of tubing are loaded into the gene gun, which uses pressurized helium to accelerate the particles into the skin.

2. Weight of gold microcarriers in a 1.5-mL microtube. (Refer to **Table 1** for a detailed description on determining microcarrier loading quantity [MLQ].)
3. To the measured gold, add 100 µL of 0.05 *M* spermidine. (When the volume of the plasmid solution is 200 µL, add 200 µL of 0.05 *M* spermidine.)
4. Vortex the gold and spermidine mixture for a few seconds to break up gold clumps.
5. To the gold and spermidine mixture, add the required volume of plasmid to achieve the desired DNA loading ratio (DLR). For cotransfection of multiple plasmids, add each of the plasmids at this step.
6. Mix the DNA, spermidine, and gold by vortexing for more than 5 s.

Table 1
Microcarriers and DNA Required for Various MLQs and DNA Loading Ratios (DLRs)[a]

	Calculated particle delivery condition		Materials required for selected MLQs and DLRs	
MLQ	DLR			
(mg/shot)	(μg/mg gold)	(μg/shot)	Gold (mg)	DNA (μg)
0.5	2	1	50	100
0.25	4	1	25	100
0.125	8	1	12.5	100

[a]Approximately 50 in. of tubing will usually yield 80–90 cartridges.

7. While vortexing the mixture at a moderate rate on a variable-speed vortexter, add 100 μL of 1 *M* $CaCl_2$ dropwise to the mixture. The volume added should be the same as that of the spermidine in **step 3**.
8. Allow the mixture to precipitate at room temperature for 10 min.
9. Most of the gold will now be in the pellet, but some may be on the sides of the tube. The supernatant should be relatively clear. Spin the mictocarrier solution in a microfuge for 15 s to pellet the gold. Remove the supernatant and discard.
10. Add 100 μL of 100% ethanol and resuspend the pellet by tapping, and then add 1 mL of ethanol. In the first suspension step, the pellet may be difficult to dissolve. Wash the pellet three times with 1 mL of fresh ethanol each time and spin for 5 s in a microfuge between each wash. Discard the supernatants.
11. After the final ethanol wash, resuspended the pellet in 200 μL of ethanol solution containing the appropriate concentration of PVP prepared in **step 1**. Transfer this suspension to a 15-mL disposable polypropylene centrifuge tube with a screw cap. Rinse the microfuge tube once with 200 μL of the same ethanol/PVP solution and add to the centrifuge tube. Add the necessary volume of the ethanol/PVP solution to the centrifuge tube to bring the DNA/microcarrier solution to the desired MLQ.
12. The suspension is now ready for tube preparation. Alternatively, DNA/ microcarrier suspensions can be stored for up to 2 mo at –20°C. Prior to freezing, tighten the cap securely and put Parafilm around the cap of the tube. After storage at –20°C, allow the particle suspension to reach room temperature before breaking the Parafilm seal.

3.2. Loading the DNA/Microcarrier Suspension into Gold-Coat Tubing Using the Tubing Prep Station

1. Prior to preparing cartridge, ensure that the gold-coat tubing is completely dry by purging with nitrogen. Insert an uncut piece of tubing into the opening on the right side of the tubing prep station (*see* **Fig. 2**).

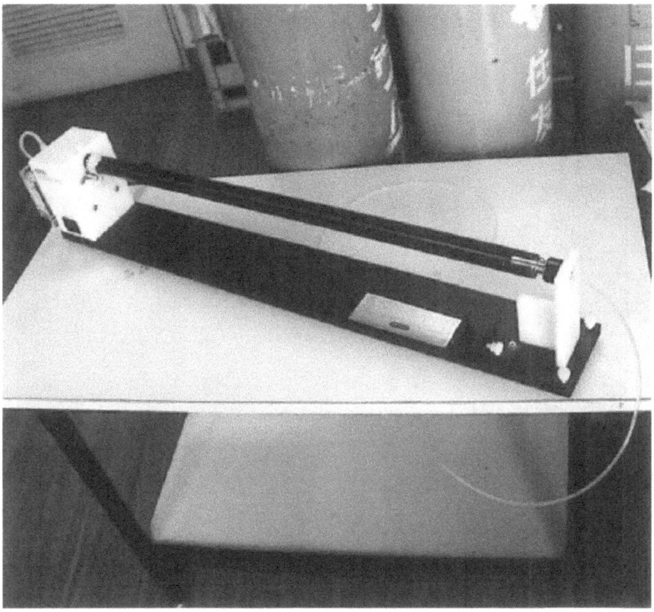

Fig. 2. Tubing prep station and gold-coat tubing.

2. Using the knob on the flow meter, turn on the nitrogen and adjust the flow to 0.3–0.4 liters per minute (LPM) (**Fig. 3**). Allow nitrogen to flow into the gold-coat tubing for at least 15 min immediately prior to using it in the following steps.
3. Remove the gold-coat tubing from the tubing prep station. Turn off the flow of nitrogen to the tubing prep station using the knob on the flow meter.
4. From the dried gold-coat tubing, cut a 75-cm length of tubing for each 3-mL sample of microcarrier/DNA suspension. Insert one end of the gold-coat tubing into the end of the adapter tubing fitted to a 10-mL syringe.
5. Vortex the microcarrier suspension and invert the tube several times to resuspend the gold; immediately remove the cape and quickly draw the gold suspension into the gold-coat tubing approx 60 cm (15 cm from the end). Remove the tubing from the suspension and continue drawing the suspension into the tubing for another 6 cm to leave some space at each end.
6. Immediately bring the gold-coat tubing to a horizontal position and slide the loaded tube, with syringe attached, into the tubing support cylinder in the tubing prep station until the tubing passes through the O-ring.
7. Allow the microcarriers to settle for 3–5 min. Detach the gold-coat tubing from the adapter tubing and attach to the tubing on the 10-mL syringe. Remove the ethanol at constant speed.

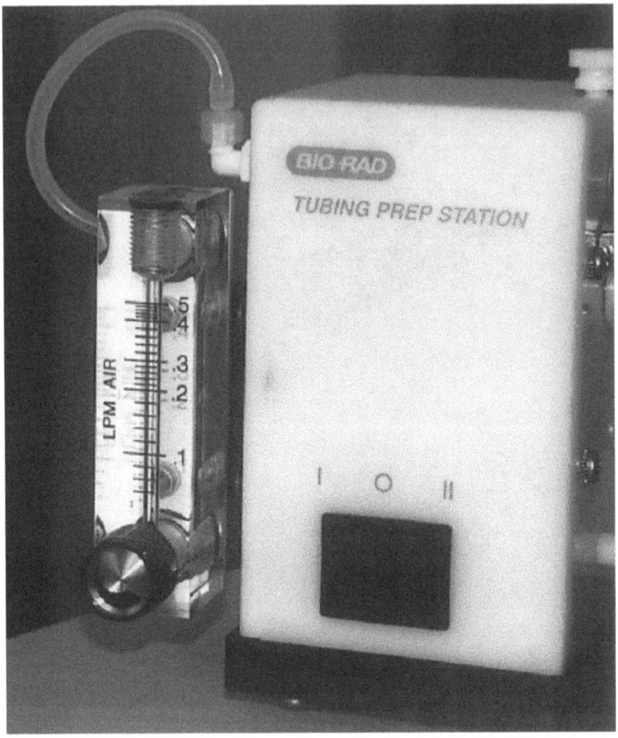

Fig. 3. Nitrogen pressure regulator. The nitrogen regulator is turned on and adjusted to the correct pressure (0.35–0.4 LPM) prior to connecting the nitrogen line to the tubing prep station.

8. Detach the 10-mL syringe from the tubing. Immediately turn the gold-coating tubing 180° while in the groove and allow the gold to begin coating the inside surface of the tubing for 3–4 s.
9. Start rotating the tubing prep station. Allow the gold to smear in the tube for 30 s and then open the valve on the flow meter to allow 0.35–0.4 LPM of nitrogen to dry the gold-coating tubing while it continues to rotate.
10. Continue drying the gold-coat tubing while turning for 5 min.
11. After loading the DNA/microcarrier into the gold-coat tubing, use a tubing cutter to cut the coated tube into 0.5-in-long pieces (*see* **Fig. 4**).
12. Store the pieces of tubing in a desiccated environment. Pieces of tubing stored at 4°C are stable for at least 8 mo.

3.3. Optimization of Gene Gun Parameters

The flexibility of the particle delivery system allows fine-tuning of experimental parameters; however, the optimal parameters for the instrument must be

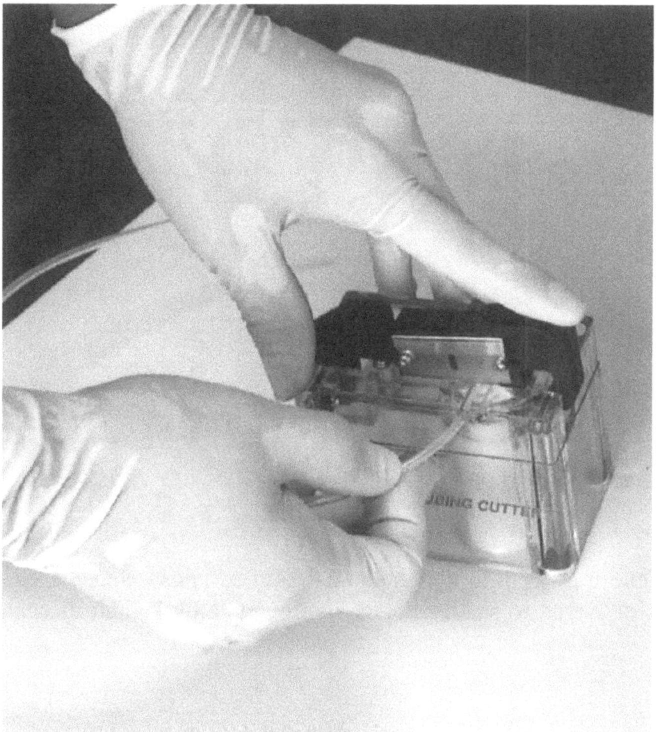

Fig. 4. Tubing cutter. To prepare 0.5-in.-long cartridges, insert the cut ends of tubing into the tubing channels and push down on the handle.

determined in each laboratory. Any quantitative assay may be used to determine the optimum combination of critical parameters for the particular biological system under investigation. Important parameters for evaluation include helium pressure, PVP concentration, the MLQ, and the DLR. It should be noted that absolute transgene expression levels are only a part of the processes leading to immune or other biological response; thus, each researcher must identify those parameters that result in the appropriate level, location, and duration of transgene expression following particle-meditaed delivery. In this section, we describe the process of optimization for gene transfer into the epidermis.

1. A pcDNA3-CAT is an expression plasmid that expresses chlorolamphenicol acetyl transferase (CAT) under the control of a cytomegalovirus immediate–early promoter (Invitrogen Co, Headquarters, CA). The pcDNA3-CAT is purified using a Plasmid Maxi Kit (QIAGEN Inc., Valencia, CA) according to the

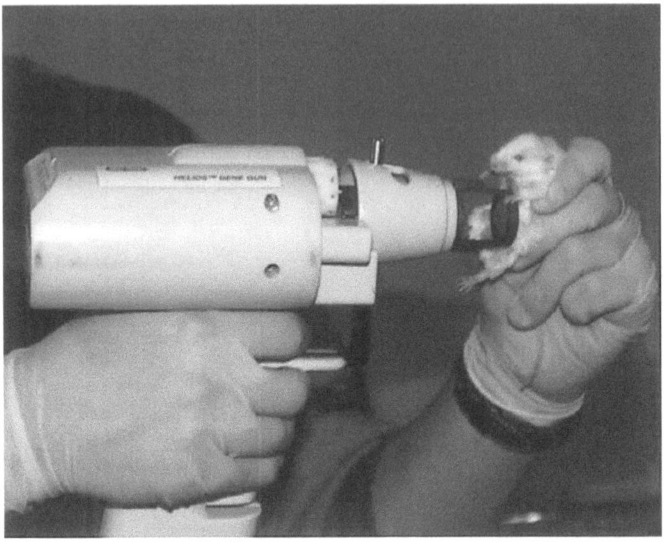

Fig. 5. Gene transfer into the abdominal skin of mice using the gene gun. Plasmid DNA is delivered to shaved dorsal skin of mice by helium discharge using the gene gun.

manufacturer's instructions. Plasmid DNA is coated onto microcarrier of DLR ratios of 2, 4, and 8 as described in **Subheading 2.1.**

2. Abdominal areas of mice are shaved, and each animal's fur is removed with Nair, a commercial depilatory. The skin is carefully rinsed with warm water following depilatory treatment. If the target site is wet or dirty, it is cleaned and dried with 70% ethanol.
3. Each of the cartridges prepared in Procedure 1 is inserted into a cartridge holder, and the cartridge holder is set into the Helios gene gun. Plasmid DNA is delivered to shaved dorsal skin of mice using the gene gun at helium discharge pressures of 200, 300, and 400 psi (*see* **Fig. 5**).
4. Twenty-four hours after the gene transfection, a circular area of target skin of 1.5 cm in diameter is biopsied. Minced skin tissue is homogenized in 2 mL of Triton X-100 in PBS with 1X protease inhibitor mixture (Boehringer-Mannheim, Indianapolis, IN) using a Polytron homogenizer. Homogenates are clarified by microcentrifugation (10,000*g* at 4°C), stored at –70°C, and assayed using a commercially available CAT enzyme-linked immunosorbent assay (ELISA) kit (Roche Diagnostics GmbH, Mannheim, Germany).

Figure 6 shows representative results of experiments to determine the optimal conditions for gene transfer into the epidermis. For another reporter

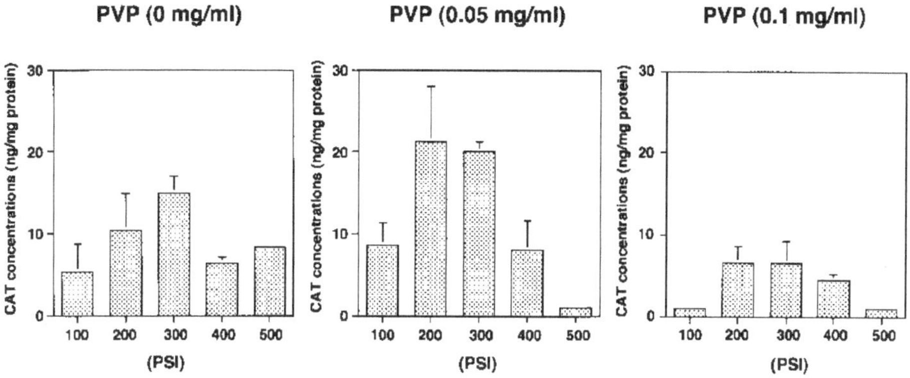

Fig. 6. Optimization of gene gun parameters. To determine the optimal conditions for in vivo gene transfer, several parameters, including PVP concentrations, helium pressure, and DLR, were examined using a pcDNA-CAT expression plasmid. Twenty-four hours after gene transfection, the abdominal skin of mice was removed, and the levels of CAT expression were estimated.

system, an expression plasmid-encoding β-galactosidase or luciferase can also be used in this assay. The duration of transgene expression following particle-mediated delivery can be measured by Western blotting for detection of the target protein or by the reverse transcription–polymerase chain reaction (RT-PCR) technique for detection of target mRNA levels.

3.4. Conditions for DNA Vaccination

The gene-gun system is used for DNA vaccination and gene therapy. In these applications, the gene transfection regime, including transfection times and intervals, is crucial for efficacy. Therefore, each researcher should first determine the optimal regime for DNA immunization.

The 47-kDa *Plasmodium falciparum* serine repeat antigen (SERA) is one of the vaccine candidate antigens for malaria vaccine. We have used this DNA for DNA immunization *(6)*. The number of immunizations is one of the main factors affecting the immune responses. Therefore, we examined the effect of the number of immunizations on the humoral immune response elicited by SERA DNA vaccination. The levels of SERA-specific antibody response to three or four immunizations were the same; however, these levels were higher than those in the case of only two immunizations (*see* **Fig. 7**).

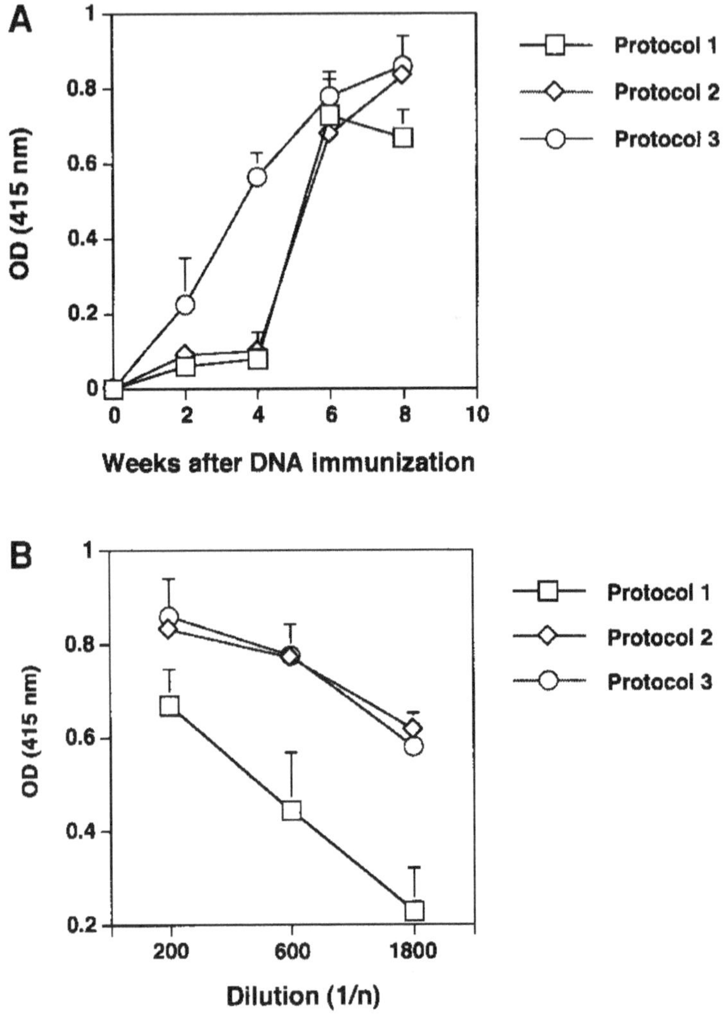

Fig. 7. Effect of number of immunizations on the humoral immune responses elicited by SERA DNA vaccine. Mice were immunized with an expression plasmid encoding the SERA gene by a gene gun according to protocol 1 (immunized at 0 and 6 wk), protocol 2 (immunized at 0, 4, and 6 wk), or protocol 3 (immunized at 0, 2, 4, and 6 wk). Gene transfection occurred at the optimal conditions defined in **Fig. 2** (PVP concentration, 0.05 mg/mL; helium pressure, 300 psi; DLR, 8). (**A**) Sera were collected at 0, 2, 4, 6, and 8 wk after the first DNA immunization. SERA-specific IgG levels in 200-fold diluted sera were determined by ELISA. (**B**) Sera collected at 8 wk after immunization were serially diluted, and SERA-specific IgG levels were determined by ELISA.

References

1. Stanford, J. C., Klein, T. M., Wolf, E. D., and Allen, N. (1987) Delivery of substances into cells and tissue using a particle bombardment process. *Particle Sci. Technol.* **5,** 27–37.
2. Yang, N.-S., Burkholder, J., Roberts, B., Martinell, B., and McCabe, D. (1990) In vivo and in vitro gene transfer to mammalian somatic cells by particle bombardment. *Proc. Natl. Acad. Sci. USA* **87,** 9568–9572.
3. Cheng, L., Ziegelholder, R. P., and Yang, N.-S. (1993) In vivo promotor activity and transgene expression in mammalian somatic tissues evaluated by using particle bombardment. *Proc. Natl. Acad. Sci. USA* **90,** 4455–4459.
4. Donnelly, J. J., Ulmer, J. B., Shiver, J. W., and Liu, M. A. (1997) DNA vaccines. *Annu. Rev. Immunol.* **15,** 617–648.
5. Rakhmilevick, A. L., Turner, J., Ford, M. J., McCabe, D., Sun, W. H., Sondel, P. M., et al. (1996) Gene gun-mediated skin transfection with interleukin 12 gene results in regression of established primary and metastatic tumors. *Proc. Natl. Acad. Sci. USA* **93,** 6291–6296.
6. Sakai, T., Horii, T., Hisaeda, H., Zhang, M., Ishii, K., Nakano, Y., et al. (1999) DNA immunization with *Plasmodium* serine repeat antigen: regulation of humoral immune response by coinoculation of cytokine expression plasmid. *Parasitol. Int.* **48,** 27–33.

14

Purification of the Eucaryotic Heat-Shock Proteins Hsp70 and gp96

Arne von Bonin, Solveig H. Moré, and Minka Breloer

1. Introduction

Heat-shock proteins (HSPs), highly conserved across species, are generally considered as intracellular proteins that have protective functions in situations of cellular stress. A wide variety of stressful stimuli like heat shock, ultraviolet radiation, and viral or bacterial infections induce a substantial increase in intracellular HSP synthesis *(1)*. The main functions ascribed to HSPs (not only restricted to situations of cellular stress) are to act as chaperones of nascent or aberrantly folded proteins. From the immunological point of view, HSPs have obtained significant interest because it could be shown that HSPs like Hsp70 and gp96 purified from tumor and virus-infected cells are capable of eliciting a protective CTL-mediated immunity *(2,3)*. This immunogenicity is based on antigenic peptides that are associated with Hsp70 and gp96 molecules, and peptide-deprived HSP complexes lose their immunization capacity *(4)*.

In addition to the well-established function of HSP to introduce HSP-associated peptides into an alternate major histocompatibility complex (MHC) class I antigen-presenting pathway, recent studies suggested that mammalian HSP preparations are delivering signals to the immune system irrespective of HSP-bound peptides. gp96 and Hsp70 preparations derived from syngenic mouse (liver or spleen) tissue are able to activate long-term established antigen-specific cytotoxic T cell (CTL) clones in vitro and spleen cells in vivo in the absence of antigenic peptides *(5,11)*. Moreover, mouse and human monocytes were found to mount proinflammatory responses when incubated with recombinant Hsp60, Hsp70, or gp96 molecules *(6,7)*. Here, we will

From: *Methods in Molecular Biology, vol. 215:*
Cytokines and Colony Stimulating Factors: Methods and Protocols
Edited by: D. Körholz and W. Kiess © Humana Press Inc., Totowa, NJ

describe the purification of eukaryotic Hsp70 and gp96 molecules from murine tissue and the use of purified HSPs in cellular assays.

1.1. Purification of HSPs and Generation of Antigen-Presenting Cells

In this chapter, we describe the purification of Hsp70 and gp96 from murine tissue (e.g., murine liver tissue). In principle, HSPs can be purified according to the following protocol from all cell lines (e.g., human tumor cell lines or peripheral blood lymphocyte [PBL], in quantities sufficient to perform cellular assays). The purification steps were originally established by Uduno and Srivastava *(3)* and we use a very similar protocol with only minor modifications. As a rule, we obtain 1 mg Hsp70/gp96 from 20–50 g of cellular material. Although Hsp70 and gp96 can be purified from the same sample, we recommend concentrating on the purification of one kind of HSP from a given tissue sample because this way of purifying HSPs, in general, results in higher yields of HSPs.

2. Materials

2.1. Buffers

1. Hypotonic lysis buffer: 10 m*M* $NaHCO_3$, 0.5 m*M* phenyl methyl sulfonyl fluoride (PMSF), pH 7.0.
2. ConA-binding buffer: 125 m*M* NaCl, 20 m*M* Tris-HCl, 1 m*M* $MgCl_2$, 1 m*M* $CaCl_2$, pH 7.2.
3. ConA washing buffer: 5 m*M* sodium phosphate, 300 m*M* NaCl, pH 7.0.
4. ConA elution buffer: 10% α-methyl-Mannoside, 5 m*M* sodium phosphate, 300 m*M* NaCl, pH 7.0.
5. DEAE binding buffer: 20 m*M* sodium phosphate, 20 m*M* NaCl, pH 7.0.
6. DEAE elution buffer: 20 m*M* sodium phosphate, 700 m*M* NaCl, pH 7.0.
7. ATP-binding buffer: 20 m*M* Tris-acetate, 2 m*M* $MgCl_2$, 15 m*M* 2-mercaptoethanol (ME), 20 m*M* NaCl, pH 7.0.
8. ATP-elution buffer: 3 m*M* ATP, 20 m*M* Tris-acetate, 2 m*M* $MgCl_2$, 15 m*M* 2-ME, 20 m*M* NaCl, pH7.0.

2.2. Murine Tissue, Cell Lines, Columns

Thus far, we have used a variety of different murine tumor cell lines (e.g., RMA, EG7, P815, EL4) to purify Hsp70 and gp96 (*see* **Note 1**). Cellular pellets can be stored for months at –70°C before HSPs are extracted from them. In order to reduce protein content, cellular pellets are washed once before freezing the pellets. The same holds true for livers or lungs, which can be stored as whole organs at –20°C for long time periods before purifying Hsp70 or gp96. Mice as a source for the organs were maintained at the animal facility at the Benhard-Nocht-Institute. The age of the mice when sacrificed for preparing the organs seemed of minor importance.

1. ADP or ATP–agarose for purifying Hsp70 was obtained from Fluka (Neu-Ulm, Germany).
2. ConA–Sepharose and DEAE–Sepharose were obtained from Amersham (Germany). DEAE was packed manually in 20-mL columns (Bio-Rad, Munich, Germany).
3. Prepacked PD10 columns were purchased from Amersham-Pharmacia (Freiburg, Germany).

3. Methods

3.1. gp96 Purification

gp96 is purified according to a protocol described by Udono et al. *(8)* with minor modifications. Note that it is extremely important to perform **steps 1–4** at 4°C.

1. Cells were grown in roller bottles to generate at least a 20-mL cell pellet. Mouse liver-derived gp96 was purified from 50 g of, for example, liver tissue (whole organs, disrupted with a conventional kitchen mixer). To inhibit protein degradation, we added Complete® protease inhibitor (Roche Diagnostics, Germany).
2. The pellets were homogenized in 2 vol of hypotonic buffer (*see* **Notes 2** and **3**) and maintained for 20 min at 4°C. Lysates were centrifuged at 100,000*g* for 90 min in a precooled rotor.
3. The supernatant (e.g., 100 mL) from **step 2** was applied to a 50% ammonium sulfate precipitation (add the ammonium sulfate stepwise, 1 h on ice, stirring). The solution was centrifuged at 25,000*g* (e.g., Sorvall centrifuge) and the supernatant was subsequently incubated with 70% ammonium sulfate overnight (stirring!). On the next day, the precipitate was dried (aspirate the supernatant carefully with a vacuum pump) for 30 min and was solubilized in 200 mL ConA-binding buffer. The ConA–Sepharose was packed in a column and the column was washed extensively (10 vol, ConA-binding buffer). Note that loading and washing of the columns takes approx 2 h in general, depending on the purity of the supernatants.
4. ConA-bound material was eluted with 10% α-methyl-Mannoside. Incubate the ConA column with the methyl-Mannoside solution for 20 min at room temperature and collect 3–5 column volumes.
5. The eluate buffer was exchanged to DEAE-binding buffer employing PD10 gel filtration units (Pharmacia, Freiburg, Germany).
6. This partially purified gp96 material was applied to two DEAE–sepharose columns in parallel (Pharmacia, Freiburg, Germany), washed and bound material was eluted with DEAE-binding buffer containing 700 m*M* NaCl. Check the initial 10 fractions (2-mL fractions) in sodium dodecyl sulfate–polyacrylamide gel electrophoresis (SDS-PAGE) (*see* **Fig. 1**). We usually detect the purified

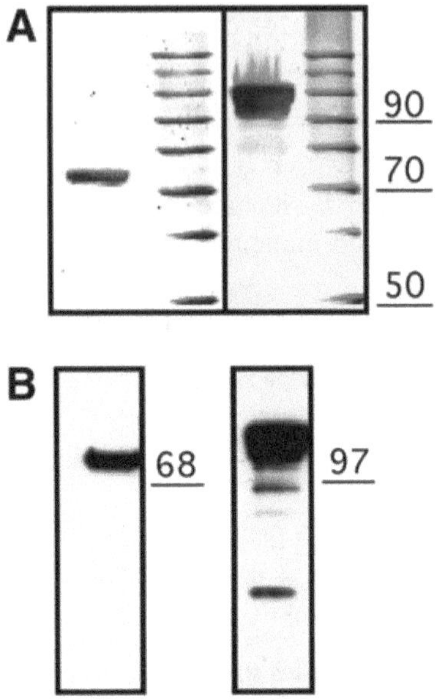

Fig. 1. Silver staining and Western blotting of purified Hsp70 and gp96. **(A)** Shown is the silver staining of 100 ng purified Hsp70 (left) and 200 ng gp96 (right) extracted from murine livers. **(B)** The same preparations shown in **(A)** were analyzed in Western blotting using Hsp70-specific monoclonal antibodies (MAbs) (left) and gp96-specific MAbs (right) obtained from Stressgen. Numbers on the right indicate relative molecular masses.

proteins with a silver-staining procedure. Fractions containing the purified gp96 molecule as the major protein are pooled and concentrated with Amicon 50-kD-spin filters (*see* **Note 4**).

3.2. Hsp70 Purification

Hsp70 purification was performed according to a protocol first described by Peng et al. *(9)*. Employing of ADP–agarose allows purification of Hsp70 in association with endogenously bound peptides, whereas the use of ATP–agarose "strips" the associated peptides from Hsp70.

1. Hsp70 was purified from tumor and liver cell pellets (50–60 g). The cellular pellet was homogenized in ATP-binding buffer. Cellular debris was separated by centrifugation at 100,000*g*.
2. The supernatant was applied to ATP–agarose (Sigma, Deisenhofen, Germany) overnight at 4°C (on a rolling wheel). The ATP–agarose was packed in a column and washed (10 column volumes) with ATP-binding buffer containing 500 m*M* NaCl and, subsequently, with an ATP-binding buffer containing 20 m*M* NaCl (10 column volumes).

3. The ATP-bound material was eluted with ATP-binding buffer containing 3 m*M* ATP. Incubate the column for 30 min at room temperature and collect 7 column volumes. The elution buffer was exchanged to Hsp70–DEAE-binding buffer, again using PD10 columns.
4. The eluate was applied to a DEAE column (*see* **Note 5**) and washed. Hsp70 was coeluted at a concentration of 150 m*M* NaCl (in DEAE-binding buffer). Collect the first eight fractions. Fractions of purified Hsp70 material were tested in SDS-PAGE and Western blotting using MAb specific for Hsp70. Fractions containing Hsp70 as the major proteins (*see* **Fig. 1**) were used for further experiments. Protein concentration was determined employing Coomassi Plus Protein Reagent (Pierce, IL).

3.3. Cellular Assays

The biological activity of the purified Hsp70 and gp96 preparations can be determined in assays containing professional APCs, which produce cytokines or upregulate cell-surface-expressed receptors when coincubated with purified HSPs. In the case of Hsp70 and gp96, it seems that in-vitro-generated APCs like bone-marrow-derived dendritic cells (DCs) and peritoneal exsudate cells (PECs) are less sensitive; e.g., they produce only small amounts of inflammatory cytokines like interleukin (IL)-6, IL-8 or tumor necrosis factor (TNF)-α. In the assays, when another HSP, murine or human Hsp60, was used to activate APCs, the secretion of cytokines was significantly enhanced and clearly distinguishable from contaminating endotoxins like lipopolysaccharide (LPS). However, Hsp60 in our lab could not be purified in sufficient quantities from murine tissue or tumor cell lines and, thus, was obtained in a recombinant form from Stressgen.

As a readout for the activation of APCs, TNF-α in the supernatants was determined with the help of specific enzyme-linked immunosorbent assays (ELISAs) (Pharmingen, Heidelberg, Germany). Because the correct choice of professional APCs is an important step for performing cellular assays, we describe in the following the isolation and generation of PECs and DC cells (*see* **Note 6**), respectively.

1. To induce peritoneal macrophages, mice (e.g., BALB/c or C57BL/6) were injected with 500 μL pristane (Sigma, Deisenhofen, Germany) intraperitoneally.
2. Peritoneal exsudate cells were harvested 5–6 d later by rinsing the peritoneum with medium. Only freshly prepared PECs were used in the experiments described. Fluorescence-associated cell sorting (FACS) analysis of isolated PECs showed a >95% staining for the macrophage surface marker Mac-1 (Caltag, Germany).
3. Bone-marrow-derived dendritic cells were generated as previously described *(12)*. Briefly, bone marrow was collected from tibias and femurs of one to two mice and resuspended in complete RPMI-1640 medium. Two million cells were

placed in round plastic Petri dishes (no tissue culture surface) in 10 mL complete RPMI-1640 containing 20 ng/mL granulocyte–macrophage colony stimulating factor (GM-CSF) (Biomol, Hamburg, Germany).

4. On d 3, 10 mL RPMI-1640 was added containing 20 ng/mL GM-CSF. On d 6 and 8, 10 mL of the culture supernatants were exchanged with 10 mL fresh medium (20 ng/mL GM-CSF).
5. Then 5×10^5 to 1×10^6/mL freshly purified PECs or DCs (on d 7 or 8 of culture) were cocultured with titrated amounts of purified HSP in 24-well or 48-well culture dishes (Greiner, Germany). After 24 (or 48) h of culture, undiluted supernatants were analyzed for cytokines using specific ELISAs (*see* **Note 7**). In the case of Hsp70, up to 200 µg/mL and in the case of gp96, at least 30–50 µg/mL purified HSPs were used to induce cytokine secretion in the APCs (in the case of recombinant Hsp60, even 5 µg/mL Hsp60 showed reproducible effects).

3.4. Concluding Remarks

As a positive control for the induction of cytokines by HSPs, LPS (100 ng/mL) was added to the cultures. Cultures treated with heat-inactivated HSPs or LPS (10 min, 95°C) served as controls. Polymyxin B, an LPS inhibitor (100 U/mL; Sigma, Deisenhofen, Germany), can be added as an internal control, to rule out that contaminating endotoxins are responsible for the observed effects.

In our hands, there was a clear hirachy with respect to the biological potential of the individual HSP to activate professional APCs (*see* **Fig. 2**). Hsp60, even at low protein concentrations, was very effective. Hsp60 is followed by biochemically purified gp96, which showed some reactivity to purified APCs at higher HSP concentrations (>50 µg/mL). Hsp70, however, although being described as acting on APCs, in our hands showed a variable induction of cytokines (and also of costimulatory molecules on APCs), which in many cases could not be clearly separated from endotoxin contaminations. It will be interesting, thus, to analyze whether HSPs belonging to given protein families maintain specific roles within an organism to activate different functions in cells of the immune system.

4. Notes

1. Once established, Hsp70 and gp96 can be purified from tumor cells or cellular tissue in reproducible quantities and qualities.
2. To maintain the yield of the HSPs, care has to be taken during a few critical steps of the purification protocol. Most importantly, freshly made buffers are highly recommended. We routinely prepare fresh buffers for every new purification of HSPs.
3. The buffers should be precooled, and whenever it is possible and not otherwise indicated, individual steps should be carried out at 4°C. When starting to purify

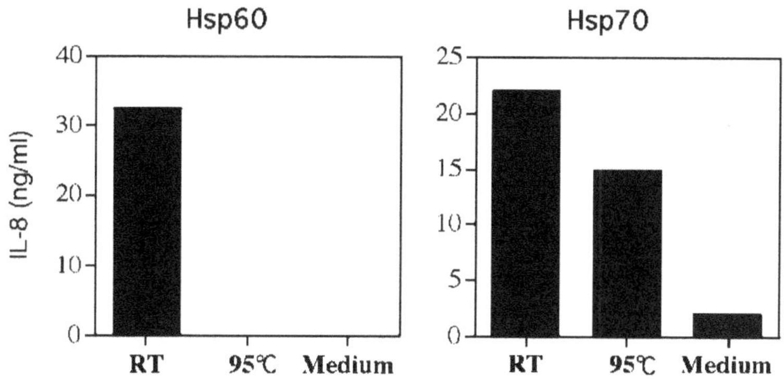

Fig. 2. Cytokine secretion by dendritic cells following the addition of HSPs. Bone-marrow-derived dendritic cells (1×10^6) were incubated alone ("Medium") or together with 10 μg/mL recombinant murine Hsp60 (obtained from Stressgen) or Hsp70 (50 μg/mL), purified according to the described protocol from murine liver tissue. Dendritic cells were incubated overnight at d 6 of culture. HSP were left untreated ("RT") or were boiled for 10 min prior to the addition to the cultures ("95°C"). The reduction of secreted IL-8 to background levels in cultures containing the "boiled" HSP samples as a control indicate that endotoxins (as heat-stable reagents) are not responsible for the observed cytokine secretion. IL-8 (as a typical inflammatory cytokine) in the supernatants of the cultures was determined using an IL-8-specific ELISA. Note the different scales of the *y*-axis in the Hsp60 and Hsp70 diagrams.

HSPs, we try not to introduce large pauses during the individual steps, but, rather, purify the HSP on two subsequent days.

4. Highly pure Hsp70 and gp96 material should be kept in aliquots to minimize contaminations and repeated freeze-and-thaw cycles. Hsp70 can be kept at 4°C for long periods of time (up to several weeks and months), whereas gp96 tends to degrade at 4°C. Therefore, we keep gp96 at –70°C in small, feasible aliquots (e.g., 100-μg aliquots).
5. In the case of gp96, the ConA columns are discarded after one purification and the same holds true for the DEAE columns. In contrast, PD10 columns can be used several times (>20 times).
6. Concerning the release of cytokines from dendritic cells, we observed variable results depending on the batch of dendritic cells. Thus, it seems that individual preparations of DCs respond differentially to the addition of HSPs, whereas PECs from different cultures are more comparable with respect to the pattern and amount of HSP-induced cytokines.
7. For the detection of secreted cytokines, it should be noted that we do not use commercial ELISA kits or even precoated plates but instead buy pairs of cytokine-specific MAbs separately but from one manufacturer.

References

1. Hartl, F. U. (1996) Molecular chaperones in cellular protein folding. *Nature* **381,** 571–580.
2. Blachere, N. E., Udono, H., Janetzki, S., Li, Z., Heike, M., and Srivastava, P. K. (1993) Heat shock protein vaccines against cancer. *J. Immunother.* **14,** 352–356.
3. Udono, H. and Srivastava, P. K. (1994) Comparison of tumor-specific immunogenicities of stress-induced proteins gp96, hsp90, and hsp70. *J. Immunol.* **152,** 5398–5403.
4. Srivastava, P. K., Menoret, A., Basu, S., Binder, R. J., and McQuade, K. L. (1998) Heat shock proteins come of age: primitive functions acquire new roles in an adaptive world. *Immunity* **8,** 657–665.
5. Breloer, M., Fleischer, B., and von Bonin, A. (1999) In vivo and in vitro activation of T cells after administration of Ag-negative heat shock proteins. *J. Immunol.* **162,** 3141–3147.
6. Asea, A., Kraeft, S. K., Kurt-Jones, E. A., Stevenson, M. A., Chen, L. B., Finberg, R. W., et al. (2000) HSP70 stimulates cytokine production through a CD14-dependant pathway, demonstrating its dual role as a chaperone and cytokine. *Nat. Med.* **6,** 435–442.
7. Singh-Jasuja, H., Scherer, H. U., Hilf, N., Arnold-Schild, D., Rammensee, H. G., Toes, R. E., et al. (2000) The heat shock protein gp96 induces maturation of dendritic cells and down-regulation of its receptor. *Eur. J. Immunol.* **30,** 2211–2215.
8. Udono, H., Levey, D. L., and Srivastava, P. K. (1994) Cellular requirements for tumor-specific immunity elicited by heat shock proteins: tumor rejection antigen gp96 primes CD8+ T cells in vivo. *Proc. Natl. Acad. Sci. USA* **91,** 3077–3081.
9. Peng, P., Menoret, A., and Srivastava, P. K. (1997) Purification of immunogenic heat shock protein 70-peptide complexes by ADP-affinity chromatography. *J. Immunol. Methods* **204,** 13–21.
10. Walden, P. R. and Eisen, H. N. (1990) Cognate peptides induce self-destruction of CD8+ cytolytic T lymphocytes. *Proc. Natl. Acad. Sci. USA* **87,** 9015–9019.
11. Singh-Jasuja, H., Toes, R. E., Spee, P., Munz, C., Hilf, N., Schoenberger, S. P., et al. (2000) Cross-presentation of glycoprotein 96-associated antigens on major histocompatibility complex class I molecules requires receptor-mediated endocytosis. *J. Exp. Med.* **191,** 1965–1974.
12. Lutz, M. B., Kukutsch, N. A., Menges, M., Rossner, S., and Schuler, G. (2000) Culture of bone marrow cells in GM-CSF plus high doses of lipopolysaccharide generates exclusively immature dendritic cells which induce alloantigen-specific CD4 T cell anergy in vitro. *Eur. J. Immunol.* **30,** 1048–1052.

15

Genetic Engineering of a Recombinant Fusion Protein Possessing an Antitumor Antibody Fragment and a TNF-α Moiety

Jim Xiang and John R. Gordon

1. Introduction

Tumor necrosis factor-α (TNF-α) is a cytokine (CK) that possesses a wide variety of biological activities, including potent antitumor activities *(1)* and immunomodulatory properties mediated through its binding to two TNF receptors (p55 and p75) *(2)*. Signaling through the p55 receptor is primarily associated with responses such as cytotoxicity *(2,3)* and cytokine secretion *(4)*, whereas the p75 receptor is responsible for lymphoproliferative signals and the activation of T-cells *(5)*. Recently, it has been found that TNF-α has profound effects on dendritic cell (DC) maturation *(6)* and activation *(7)*. In addition, it has also been reported to stimulate T-cell proliferation *(8)* and to activate cytotoxic T-cells *(9)*. Because its systemic administration was shown to mediate the regression of some mouse tumors *(10)*, TNF-α has attracted much attention as a potential antitumor reagent *(11)*. However, the problem of its dose-dependent toxicity has been particularly apparent in human trials, wherein its maximal tolerated dose was 40-fold less than that used in mice *(12,13)*. Systemic administration of TNF-α in treatments of cancer patients has usually resulted in severe and limiting side effects *(11)*, whereas more local delivery (e.g., via isolated perfusion to limbs) has been more effective in mediating tumor regression, indicating that antitumor effects are possible if high local concentrations of TNF-α can be obtained *(14)*. Therefore, an important issue to be addressed is how to achieve a continuously high local concentration of TNF-α within tumors without inducing severe side effects.

From: *Methods in Molecular Biology, vol. 215: Cytokines and Colony Stimulating Factors: Methods and Protocols*
Edited by: D. Körholz and W. Kiess © Humana Press Inc., Totowa, NJ

New developments in modern biotechnology have made new approaches to this problem feasible. One of these approaches is the use of recombinant antibodies (Abs) to target CKs to tumors. Ab molecules are able to bind to various antigens, including the tumor-associated antigen (Ag), present on tumor cells. Antitumor Abs have been used extensively as vehicles to successfully deliver a high concentration of various therapeutic agents, such as antitumor drugs, toxins, and radionuclides, to tumor cells expressing antigens recognized by these Abs ***(15)***. Abs have also been used as delivery vehicles for targeting CKs to which they have been chemically crosslinked (e.g., Ab/IFN-γ; **ref. *16***). However, there can be problems with chemical crosslinking approaches, such as losses of activity and protein aggregation and instability ***(17)***. As an alternate strategy, recombinant fusion proteins possessing both antitumor Ab fragments and CK moieties have been constructed by protein engineering ***(18–20)***. Recently, the therapeutic potentials of recombinant fusion proteins have been evaluated in animal models and proven more effective than equivalent doses of CK alone in suppressing tumor growth and dissemination ***(21)***, in overcoming tumor heterogeneity, and in eradication of established tumors ***(22)***. We have previously used genetic engineering to construct a mouse variable region (V)/human constant region (C) chimeric Ab, ccM4, recognizing the human tumor-associated TAG72 Ag ***(23)***. The TAG72 Ag recognized by the B72.3 Ab is present in the majority of colorectal, gastric, and ovarian adenocarcinomas ***(24)***. Subsequently, we further cloned chimeric heavy- and light-chain genes from a cDNA library of this ccM4 transfectoma cell line ***(25)***; this provided us with cDNA gene fragments, such as the chimeric heavy-chain (M4H2) gene, that are more amenable to genetic manipulation and construction of fused genes than their genomic counterparts.

In this chapter, (1) the design and methods for construction of recombinant fusion protein RM4/TNF-α possessing the anti-TAG72 chimeric Ab fragment RM4 and the TNF-α moiety and (2) the methods for purification and characterization of RM4/TNF-α will be described. The principle and methods are also applicable to the construction of other kinds of recombinant fusion proteins.

2. Material

2.1. Vector Construction

1. Primer (5′ GAATT CAACA TGGAA TGGAG 3′) and primer 2 (5′ GGATCC GGTGG GCATG TGTGA GTTTT GTCAC AAGAT 3′) are complementary to the 5′ end of the V_H region and the 3′ end of the M4H2 hinge region, respectively, in the plasmid pBM4H2 ***(25)***.
2. Primer 3 (5′ ATGGA TCCTA GCTCC TCTCG CACTC CGTCC 3′) and primer 4 (5′ GGTCG ACATT ATTAC AGTGC GATAA TACC 3′) are complementary to

the 5′ and 3′ ends, respectively, of the TNF-α region in the plasmid M13mp18-TNF-α (R&D System, Minneapolis, MN).
3. 10X Reaction buffer: 200 m*M* Tris-HCl, pH 8.8, 100 m*M* KCl, 100 m*M* $(NH_4)_2SO_4$, 20 m*M* $MgSO_4$, 1% Triton X-100, 1 mg/mL nuclease-free bovine serum albumin (BSA).
4. dNTP (Strategene Inc., La Jolla, CA).
5. TwinBlock thermocycler (Ericomp Inc., San Diego, CA).
6. *pfu* DNA polymerase (2.5 U/μL; Strategene Inc.).
7. *Eco*RI/*Bam*HI and *Bam*HI/*Sal*I (Gibco-BRL, Burlington, Ontario, Canada).
8. 1X Digestion buffer solution (Gibco-BRL).

2.2. Expression of Fusion Protein

1. Dulbecco's modified essential medium (DMEM) plus 10% fetal calf serum (FCS) and the selection reagent (mycophenolic acid [6 μg/mL]; Sigma–Aldrich Canada Ltd., Oakville, Ontario, Canada).
2. Electroporation cuvet (Bio-Rad Laboratories, Mississauga, Ontario, Canada).
3. Gene Pulser II System (Bio-Rad Laboratories).

2.3. Purification of RM4/TNF-α by Affinity Chromatography

1. Protein A kappa-lock affinity column (Upstate Biotech, Lake Placid, NY).
2. Centriprep® concentrator (Amicon Canada Ltd., Oakville, Ontario, Canada; molecular-weight cutoff <100 kDa).

2.4. TAG72-Binding ELISA

1. TAG72 epitope-rich mucin from bovine submaxillary glands (Sigma–Aldrich Canada Ltd.).
2. Enzyme-linked immunosorbent assay (ELISA) plate (Corning Laboratory Sciences, Elmira, NY).
3. Peroxidase-conjugated goat anti-kappa chain antibody (Jackson Immuno Research Laboratories, Bar Harbor, ME; 1:5000 in phosphate-buffered saline containing 0.05% Tween-20 [PBST]).
4. ABTS substrate (2-2′-azino-di[3-ethyl-benzthiazoline sulfonate ***(6)***] with H_2O_2; Kirkegaard and Perry Laboratories, Gaithersburg, MD).
5. Stop solution (1% sodium dodecyl sulfate [SDS]; Bio-Rad Laboratories Ltd.).

2.5. Cytotoxicity Assay

1. L929 murine fibroblast cell line (American Type Culture Collection, Rockville, MD).
2. Actinomycin D (stock, 5 mg/mL in 95% ethanol; Sigma–Aldrich Canada Ltd).
3. TNF-α standards (R & D Systems; 1, 5, 25, 100, and 500 pg/mL).
4. 3-[4,5-Dimethylthiazol-2-yl]-2,5-diphenyltetrazolium bromide (MTT, 5 mg/mL in PBS; Sigma–Aldrich Canada Ltd).

3. Methods

3.1. Construction of the Expression Vector mpSV2neo-EP-M4TNF-α-PA

For construction of the fused heavy-chain gene fragment *M4TNF*-α (*see* **Note 1**), four oligonucleotide primers are designed for amplification of the heavy-chain gene fragment *M4* (primers 1 and 2) and the *TNF*-α gene (primers 3 and 4) using the polymerase chain reaction (PCR) method. The *M4* gene fragment includes the mouse variable (V_H), the human constant (C_{H1}), and the hinge regions of the chimeric heavy-chain gene *M4H2* ***(25)***. Primer 1 and primer 2 are used for the introduction of *Eco*RI and *Bam*HI sites into its 5′ and 3′ ends, respectively. Primer 3 and primer 4 are used for introduction of *Bam*HI and *Sal*I sites into its respective ends.

The method for PCR incorporation of new restriction sites into a gene fragment for easy cloning into a vector is as follows:

1. Add 0.1 µg plasmid DNA (pBM4H2 or M13mp18-TNF-α) into a sterile 500-µL microcentrifuge tube.
2. Add 1 µL of the primers (1 mg/mL; primers 1 and 2, or primers 3 and 4) to the tube.
3. Add 10 µL of 10X reaction buffer to the tube.
4. Add 0.8 µL of 100 m*M* dNTP to the tube.
5. Add sterile water to a final volume of 99 µL.
6. Place the tube in a TwinBlock thermocycler. Heat the reaction to 91°C for 5 min and then immediately cool the reaction to 54°C for 5 min.
7. Briefly microcentrifuge the sample and then add 1 µL of *pfu* DNA polymerase. Microcentrifuge the sample again.
8. Carefully overlay the reaction mixture with a drop of mineral oil to prevent evaporation from the reaction during the amplification procedure.
9. Place the tube back in the thermocycler and program the heating block for 30 cycles (91°C for 1 min, 54°C for 1 min, and 72°C for 1 min).
10. After amplification, the *M4* and *TNF*-α gene fragments are visualized on an ethidium bromide-stained 1% (w/v) agarose gel. The *M4* and *TFN*-α gene fragments obtained from the PCR are shown in **Fig. 1**.
11. These two gene fragments are digested with *Eco*RI/*Bam*HI and *Bam*HI/*Sal*I, respectively.
12. Briefly, microgram quantities of DNA are digested with 5–25 U of restriction enzymes in a volume of 10–50 µL 1X digestion buffer solution appropriate for the DNA and endonuclease being used.
13. The reaction mixtures are incubated at 37°C for 1–2 h.
14. The digested M4 and TNF-α gene fragments are then ligated into the *Eco*RI/*Bam*HI and *Bam*HI/*Sal*I sites of PUC18 vector to form the PUC18-M4TNF-α.

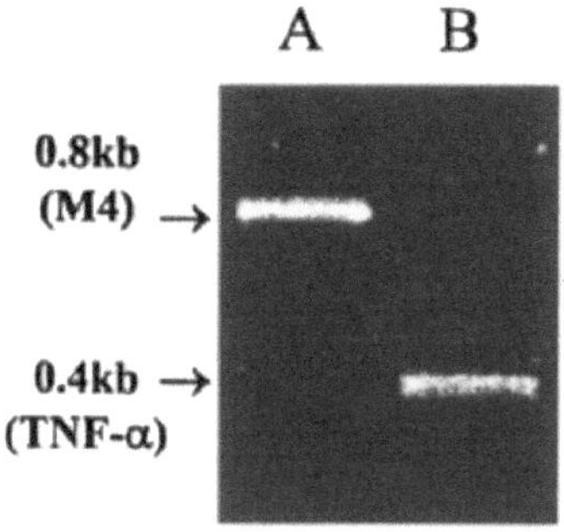

Fig. 1. Ethidium bromide staining of DNA in an agarose gel visualized under ultraviolet light. Note a band of 0.8 kb for the *M4* gene fragment (**A**) and another band of 0.4 kb for the *TNF*-α gene fragment (**B**).

15. The *M4TNF*-α fragment (*Eco*RI/*Sal*I) purified from the PUC18-M4TNF-α by *Eco*RI and *Sal*I is ligated into the *Eco*RI/*Xho*I site of mpSV2neo-EP-PA *(26)* to form the expression vector mpSV2neo-EP-M4TNF-α-PA, as shown in **Fig. 2**.
16. Briefly, the purified M4TNF-α fragment is mixed with the vector mpSV2ne0-EP-PA DNA at a molar ratio of approx 3:1 in 10 μL of solution containing 50 m*M* Tris-HCl, pH 7.6, 10 m*M* $MgCl_2$, 10 m*M* dithiothreitol (DTT), 1 m*M* ATP, and 5 U T4 DNA ligase. The mixture is incubated at 16°C for at least 2 h. Optimal results are obtained by incubating the reaction overnight. In the expression vector, EP and PA stand for the immunoglobulin enhancer (E), promoter (P), and polyadenylation signal regions.

3.2. Expression of the Recombinant Fusion Protein RM4/TNF-α

The expression vector mpSV2neo-EP-M4-TNF-α-PA is transfected into the V_KC_K cell line expressing the chimeric light-chain gene *(23)* for expression of the fusion protein RM4/TNF-α, comprising the anti-TAG72 $F(ab')_2$ (RM4) and the TNF-α moiety. V_KC_K cells are grown in Dulbecco's modified essential medium (DMEM) plus 10% FCS and the selection reagent (mycophenolic acid [6 μg/mL]; Sigma–Aldrich Canada Ltd., Oakville, Ontario, Canada) for the *gpt* gene, in order to keep the light-chain expression vector mpSV2gpt-EP-M4K4-PA *(26)* in V_KC_K cells.

1. Harvest the V_KC_K cells and wash twice with PBS.
2. Resuspend 2×10^7 V_KC_K cells in 0.8 mL phosphate-buffered saline (PBS).
3. Add 10 μg of the expression vector mpSV2neo-EP-M4-TNF-α-PA DNA to the cell suspension and mix gently.
4. Transfer the cell suspension into a chilled 1-mL electroporation cuvet.

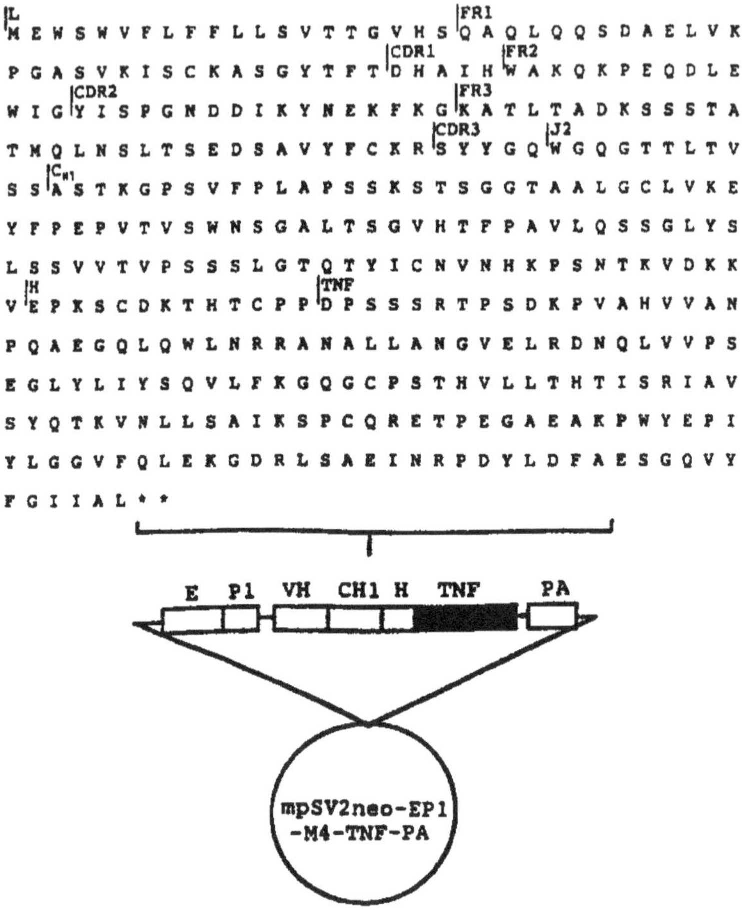

Fig. 2. Expression vector mpSVneo-EP1-M4-TNF-PA and the amino acid sequence of the fused gene fragment *M4/TNF*. Demarcated are the leader peptide region (L), the framework regions (FR), the complementarity-determining regions (CDR), the joining segment (J), the constant region 1 (C_{H1}), the hinge region (H), and the human tumor necrosis factor (TNF). The asterisk represents the stop codon. In the expression vector mpSVneo-EP1-M4-TNF-PA, gene fragments are abbreviated as follows: heavy-chain variable region, V_H; constant region 1, C_{H1}; hinge region, H; human tumor necrosis factor, TNF; immunoglobulin enhancer and promoter, EP1; polyadenylation signal region, PA.

5. Electroporate the cells using 250 V and 125 capacity settings on a Gene Pulser II System.
6. Keep the cuvet on ice for 30 min after electroporation.
7. Transfer the cells into 20 mL DMEM plus 10% FCS and mix gently.

8. Plate 100 μL of the cell suspension into each well of two 96-well plates and incubate at 37°C overnight.
9. Add 100 μL of DMEM plus 10% FCS, G418 (4 mg/mL; Gibco, Burlington, Ontario, Canada) and mycophenolic acid (6 μg/mL) to each well (*see* **Note 2**).
10. Harvest the positive growth clones after 7–10 d and expand them in DMEM plus 10% FCS, G418 (0.5 mg/mL), and mycophenolic acid (6 μg/mL).
11. Grow a positive clone secreting the recombinant fusion protein RM4/TNF-α in a large volume of the medium (*see* **Note 3**).
12. Harvest the culture supernatant for purification of RM4/TNF-α.

3.3. Purification of RM4/TNF-α by Affinity Chromatography

The Protein-A affinity column has been commonly used for purification of immunoglobulin molecules because of its high-affinity binding to the Fc portion of immunoglobulin. However, because the fusion protein RM4/TNF-α contains only the $F(ab)_2$, not the Fc portion of immunoglobulin, we used a kappa-lock affinity column (Upstate Biotech, Lake Placid, NY), which binds the kappa chain of the fusion protein, for purification of RM4/TNF-α.

1. Filter the culture supernatant through a 0.45-μm filter and adjust the pH to 7.2–7.4.
2. Wash the kappa-lock column with several column volumes of PBS.
3. Run the cell culture supernatant through the column matrix several times to saturate the IgG kappa-chain-binding capacity of the column.
4. Wash the column with PBS until the nonspecifically bound fusion protein no longer elutes from the column, using a spectrophotometer or ultraviolet (UV) monitor (OD_{280}) to confirm the end point.
5. Elute the fusion protein from the column by running 10 mL of 0.1 *M* citric acid elution buffer through the column, collecting the elute into 1-mL fractions. In order to minimize the time that the antibodies remain in an acidic environment, elute the column directly into tubes containing 250 μL of 1 *M* Tris-HCl buffer (pH 9.0) for pH neutralization.
6. Analyze the protein content of the eluted fractions by measuring the OD_{280} of the fractions as shown in **Fig. 3** and pool the protein-containing fractions.
7. Dialyze the eluted fusion protein overnight against three changes of PBS (1 L each time). After dialysis, determine the protein concentration of the eluted fusion protein solution as above and, if necessary, concentrate the eluted protein using a Centriprep concentrator.

3.4. Characterization of RM4-TNF-α

In order to check whether the fusion protein RM4/TNF-α contains the functional anti-TAG72 RM4 and TNF-α moiety, we perform two tests, a TAG72-binding ELISA and a cytotoxicity assay, for characterization of RM4 immunoreactivity and TNF-α bioactivity, respectively.

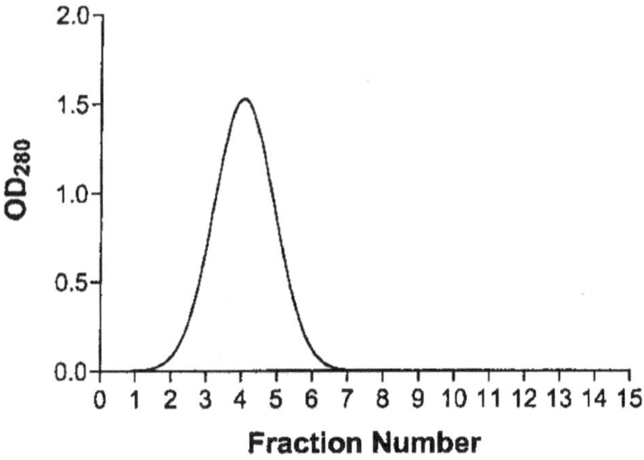

Fig. 3. Analysis of the protein content of eluted fractions. The recombinant fusion protein RM4/TNF-α bound to kappa-lock affinity column was eluted by using 0.1 *M* citric elution buffer. Each fraction of the samples was measured at OD_{280} in a spectrophotometer.

3.4.1. TAG72-Binding ELISA

In the TAG72-binding ELISA, the wells are coated with the TAG72 epitope-rich mucin from bovine submaxillary glands (Sigma–Aldrich Canada Ltd.) ***(23)***, which should capture the TAG72-specific RM4-TNF-α fusion protein from the samples. Subsequently, the wells are blocked with a protein solution that will not bind other assay components (e.g., PBS–10% BSA), and then peroxidase-conjugated goat anti-kappa chain antibody is applied to bind to the kappa-chain of fusion protein RM4/TNF-α as well as to the positive control B72.3 antibody:

1. Coat each well of the ELISA plate with 100 μL of mucin (1 μg/mL) in PBS. Cover the plate and incubate it overnight at 4°C.
2. Wash the wells two times with PBST. Block the plate by adding 200 μL of PBS containing 10% BSA to each well, cover the plate, and incubate at room temperature for 2 h.
3. Wash the wells two times with PBST as in **step 2** and add 100 μL of the fusion protein sample(s). To a separate set of wells, add a control B72.3 antibody standard curve, starting at 2 μg/mL and using doubling dilutions thereafter. Cover the plate and incubate overnight at 4°C.
4. Wash the wells four times with PBST and then add 100 μL of peroxidase-conjugated goat anti-kappa chain antibody to each well and incubate for 90 min at room temperature.

5. Wash the plates four times with PBST, add 100 μL of ABTS substrate to each well, and then place the plates in a dark location for 20–45 min at room temperature to allow the reactions to develop. The plates can be read directly at this point, or 100 μL of stop solution may be added to reduce plate-to-plate variability when reading larger numbers of plates.
6. Read the plates using a microplate reader with a 405-nm wavelength filter and analyze and plot the data.

3.4.2. Cytotoxicity Assay

Tumor necrosis factor-α activity is usually detected using a cytotoxicity assay, often with the L929 murine fibroblast cell line (American Type Culture Collection, Rockville, MD). L929 cells are sensitive to TNF-α, such that this cytokine kills the cells over approx 18 h (*see* **Notes 4** and **5**). These cells are most sensitive to the effects of TNF-α in the presence of low levels of a transcription inhibitor (e.g., actinomycin D).

1. The day before the assay, harvest L929 cells from a flask by trypsinization. To do this, remove the DMEM plus 10% FCS medium and replace it with 10 mL of serum-free DMEM containing 4–5 drops of 1% trypsin (Gibco). Allow the trypsin to digest the cells off of the plastic (*see* **Note 6**).
2. Wash the dislodged cells in DMEM plus 10% FCS medium, resuspend them to 4.5×10^5 cells/mL, and dispense 70 μL to each well of a 96-well plate (*see* **Note 7**), leaving some wells cell free, but containing all other reagents, as assay blanks. Place the plate in the 37°C CO_2 incubator overnight.
3. The next day, just prior to running the assay, remove all of the serum-containing medium from the wells by inverting the plated and vigorously flicking it. Add 180 μL of serum-free DMEM containing 2.5 μg/mL actinomycin D (stock, 5 mg/mL in 95% ethanol; Sigma–Aldrich Canada Ltd).
4. Add recombinant TNF-α standards and the fusion protein samples (at levels within the range of the standard curve) to the appropriate wells, each in a 20 μL total volume, so that the total well volumes equal 200 μL.
5. Return the plate to the 37°C CO_2 incubator overnight.
6. The next day, examine the cells under the inverted microscope to get a feeling for the relative levels of L929 cell death in each well. Add 20 μL of 3-[4,5-dimethylthiazol-2-yl]-2,5-diphenyltetrazolium bromide to all wells, and, again, return the plate to the incubator.
7. After 45–60 min, re-examine the cells visually to confirm the levels of MTT conversion to formazan dye in the mitochondria. Remove 150 μL of medium from each well in the plate and replace it with 100 μL of acidified isopropanol (0.375% HCl in isopropanol).
8. Vortex the plates on an ELISA plate shaker (Lab-Line Instrument Inc., Melrose Park, IL) to solubilize the formazan precipitates within the mitochondria and

read the plates on a microplate reader at 595-nm wavelength. Calculate the cytotoxicity in each of the wells using the formula:

$$\text{Percent cytotoxicity} = \frac{\text{Mean OD}_{590}\text{ medium control wells} - \text{OD}_{590}\text{ experimental well}}{\text{Mean OD}_{590}\text{ medium control wells}} \times 100$$

9. Calculate the mean (±SEM) cytotoxicity for each TNF-α standard (*see* **Note 8**) and fusion protein sample and graph your results. Perform a statistical analysis to confirm that your results are meaningful (*see* **Note 9**).

4. Notes

1. To avoid mutations when amplifying the M4 and TNF-α gene fragments, we use pfu DNA polymerase instead of *Taq* I DNA polymerase. Pfu DNA polymerase, a proofreading DNA polymerase isolated from *Pyrococcus furiosus*, exhibits the lowest error rate of any DNA polymerase studied. This feature makes it the ideal choice for high-fidelity DNA synthesis by PCR.
2. To increase the growth of positive clones in selection medium, we usually reduce the G418 concentration in the selection medium in each well 3 d after adding the selection medium. This can be easily done by careful aspiration of 150 µL of the selection medium in each well and replacement with 150 µL of DMEM plus 10% FCS.
3. To increase the yield of fusion protein, the positive clone should be subcloned once by limiting dilution. The clone secreting the highest amounts of fusion protein can be detected by testing the culture supernatants of the various clones in a TAG72-binding ELISA.
4. L929 cells can lose their sensitivity to TNF-α, particularly if they are not carefully maintained as subconfluent cultures. Use of subconfluent L929 cell monolayers is also critical within the assay, as cellular overgrowth reduces the sensitivity of the assay.
5. The cytotoxic effects of TNF-α are markedly diminished by the presence of other proteins (e.g., FCS) so that, as much as possible, the samples should be free of extraneous proteins.
6. The process of trypsin digestion of subconfluent monolayers of L929 cells off the plates can be exploited by watching the cells and, when they are beginning to lift (i.e., many still remain attached), forcefully slap the flask on your thigh. All cells should be dislodged instantaneously.
7. The L929 target cells for the TNF-α assay should be at about 90% confluency at the time of sample addition to the wells. Because the cells proliferate strongly in DMEM containing 10% FCS, the volume of cells to be added to the wells the night before the assay can be varied, depending on the elapsed time between plating the target cells and actually running the assay.
8. In order to conclusively demonstrate that the cytotoxic activities observed in the L929 cell killing assays are attributable to TNF-α, neutralizing anti-TNF-α Ab can be added to a set of fusion protein control wells. This should block the cytotoxic effects in these wells.

9. It is important to recognize that the MTT assay does not measure cell death *per se*, but rather the mitochondrial activity of the cells. Thus, viable cells that, for whatever reason, display a reduced metabolic activity could incorporate MTT at levels equivalent to populations suffering some degree of TNF-α-mediated cytotoxicity. For this reason, it is prudent to examine the L929 cells microscopically prior to lysing them with the acidified isopropanol.

References

1. Sugarman, B., Aggarwal, B., Hass, P., Figari, I., Palladino, M., and Shepard, H. (1985) Recombinant human tumor necrosis factor-alpha: effects on proliferation of normal and transformed cells in vitro. *Science* **230,** 943–948.
2. Fiers, W. (1991) Tumor necrosis factor: characterization at the molecular, cellular, and in vivo level. *FEBS Lett.* **285,** 199–212.
3. Sarraf, C. E. (1994) Tumour necrosis factor and cell death in tumours. *Int. Oncol.* **5,** 1333–1339.
4. Benigni, F., Faggiioni, R., and Sironi, M. (1996) TNF receptor p55 plays a major role in centrally mediated increases of serum IL-6 and corticosterone after intracerebroventricular injection of TNF. *J. Immunol.* **157,** 5563–5568.
5. Vandenabelle, P., Declerq, W., and Vercammen, D. (1992) Functional characterization of the human tumor necrosis factor receptor p75 in a transfected rat/mouse T cell hybridoma. *J. Exp. Med.* **176,** 1015–1024.
6. Nelson, E., Stroble, S., and Subleski, J. (1999) Cycling of human dendritic cell effector phenotypes in response to TNF: modification of the current maturation paradigm and implications for in vivo immunoregulation. *FASEB J.* **13,** 2021–2030.
7. Rieser, C., Bock, H., Klocker, G., Bartsch, G., and Thumher, M. (1997) Prostaglandin Ea and tumor necrosis factor alpha cooperate to activate human dendritic cells: synergistic activation of interleukin-12 production. *J. Exp. Med.* **186,** 1603–1608.
8. Lahn, M., Kalataradi, H., and Mittelstadt, P. (1998) Early preferential stimulation of γδ T cell by TNF-α. *J. Immunol.* **160,** 5221–5230.
9. Prevost-Blondel, A., Roth, E., Rosenthal, F., and Pircher, H. (2000) Crucial role of TNF-α in CD8 T cell-mediated elimination of 3LL-A9 Lewis lung carcinoma cells in vivo. *J. Immunol.* **164,** 3645–3651.
10. Creasey, A., Reynolds, M., and Laird, W. (1986) Cures and partial regression of murine and human tumors by recombinant human tumor necrosis factor. *Cancer Res.* **46,** 5687–5690.
11. Feinberg, B., Kurzrock, R., Talpaz, M., Blick, M., Saks, S., and Gutterman, J. (1988) A phase I trial of intravenously administered recombinant tumor necrosis factor-alpha in cancer patients. *J. Clin. Oncol.* **6,** 1382–1334.
12. Sherman, M., Spriggs, D., Arthur, K., Immamura, K., Frei, E., and Kufe, D. (1988) Recombinant human tumor necrosis factor administered as a five-day continuous infusion in cancer patients: phase toxicity and effects on lipid metabolism. *J. Clin. Oncol.* **6,** 344–354.

13. Spriggs, D., Sherman, M., Michie, H., Arthur, K., Imamura, K., Wilmore, D., et al. (1988) Recombinant human tumor necrosis factor administered as a 24 hour intravenous infusion a phase I and pharmacologic study. *J. Natl. Cancer Inst.* **80,** 1039–1044.
14. Lienard, D., Ewalenko, P., Delmotte, J., Renard, N., and Lejeune, F. (1992) High-dose recombinant tumor necrosis factor alpha in combination with interferon gamma and melphalan in isolation perfusion of the limbs for melanoma and sarcoma. *J. Clin. Oncol.* **10,** 52–57.
15. Goldenberg, D., Goldenberg, H., Sharkey, R., and Hansen, H. (1990) Clinical studies of cancer radioimmunodetection with carcinoembryonic antigen monoclonal antibody fragments labeled with ^{125}I or ^{99m}Tc. *Cancer Res.* **5,** 149.
16. Alkan, S., Miescher, S., Braun, D., and Hochkeppel, H. (1984) Antiviral and antiproliferative effects of interferons delivered via monoclonal antibodies. *J. Interferon Res.* **4,** 355–363.
17. Brown, B., Drzynski, C., Dearbrom, C., Hadjlan, R., Liberstore, F., and Haber, S. (1988) Conjugation of metallothionein to a murine monoclonal antibody. *Anal. Biochem.* **172,** 22.
18. Gillies, S., Young, D., Lo, K., Foley, S., and Reisfeld, R. (1991) Expression of genetically engineered immunoconjugates of lymphotoxin and a chimeric anti-ganglioside GD2 antibody. *Hybridoma* **10,** 347–356.
19. Gillies, S., Reilly, E., Lo, K., and Reisfeld, R. (1991) Antibody-targeted interleukin 2 stimulates T-cell killing of autologous tumor cells. *Proc. Natl. Acad. Sci. USA* **89,** 1428–1432.
20. Gillies, S., Young, D., Lo, K., and Roberts, S. (1993) Biological activity and in vivo clearance of antitumour antibody/cytokine fusion proteins. *Bioconjugate Chem.* **4,** 230–235.
21. Sabzevari, H., Gillies, S., Mueller, B., Pancook, J., and Reisfeld, R. (1994) An antibody-IL-2 fusion protein overcomes tumor heterogeneity by induction of a cellular immune response. *Proc. Natl. Acad. Sci. USA* **91,** 9626–9630.
22. Becker, J., Varki, N., Gillies, S., Furukawa, K., and Reisfeld, R. (1996) An antibody-IL-2 fusion protein overcomes tumor heterogeneity by induction of a cellular immune response. *Proc. Natl. Acad. Sci. USA* **93,** 7826–7831.
23. Xiang, J., Moyana, T., Kalra, J., Hamilton, T., and Qi, Y. (1992) Construction and characterization of a high affinity chimeric anti-colorectal carcinoma antibody ccM4. *Mol. Biother.* **4,** 174–183.
24. Thor, A., Obuchi, N., Szpak, C., Johnston, W., and Schlom, J. (1986) Distribution of the oncofetal antigen, tumor-associated glycoprotein-72 defined by the mononclonal antibody B72.3. *Cancer Res.* **46,** 3118–3124.
25. Boyd, M. and Xiang, J. (1994) Molecular cloning and expression of functional cDNA genes of a mouse/human chimeric antibody rcM4. *Tumori* **80,** 473–480.
26. Xiang, J., Moyana, T., and Qi, Y. (1997) Genetic engineering of a recombinant fusion protein possessing antitumor Fab and tumor necrosis factor. *J. Biotechnol.* **53,** 3–12.

16

Genetic Engineering of Dendritic Cells by Adenovirus-Mediated TNF-α Gene Transfer

Jim Xiang and Josh Wu

1. Introduction

Dendritic cells (DCs) are one of the most potent antigen-presenting cells (APCs). They migrate as precursors from the bone marrow into various organs, where they usually reside in an inactive state *(1)*. However, during this regional residency, these cells can efficiently endocytose and process antigens *(2)*. Upon activation, they undergo a differentiation process that results in decreased antigen-processing capacity and enhanced expression of major histocompatibility complex (MHC) and costimulatory molecules, after which they migrate to the lymphoid organs to interact with or activate naive T cells *(3,4)*. Because of the critical roles DCs have in the generation of primary immune responses, an important avenue of investigation is their potential for modulating immunologic functions, such as the induction of immune tolerance or tumor immunity. Recently, it has been shown that DCs pulsed with tumor derived MHC class I-restricted peptides or tumor lysates are able to induce significant cytotoxic T-lymphocyte (CTL)-dependent antitumor immune responses in vitro as well as in vivo *(5–7)*. However, the therapeutic efficiency of these DC vaccine strategies has been quite limited, because they have protected against rechallenge with only small numbers of parental tumor cells or inhibited very early-stage-established tumors. Thus, a strategic goal of current cancer vaccine research has become the induction of stronger tumor-specific CTL responses.

The maturational processes of DCs are efficiently regulated, such that these cells can achieve different states of activation/maturation and, thereby, different functional properties *(1)*, depending on the precise nature of the signals they receive from their microenvironment. A number of cytokines that can be

From: *Methods in Molecular Biology, vol. 215: Cytokines and Colony Stimulating Factors: Methods and Protocols*
Edited by: D. Körholz and W. Kiess © Humana Press Inc., Totowa, NJ

produced by DCs themselves (or by other cells within the local microenvironment) can significantly affect DC function at various levels, including their viability, morphology, migration, expression of MHC and accessory molecules, and binding and processing of antigen peptides *(8)*. For example, granulocyte–macrophage colony-stimulating factor (GM-CSF), alpha tumor necrosis factor (TNF)-α, interleukin-4 (IL-4) and gamma interferon (IFN)-γ *(9)*, and inflammatory stimuli such as interleukin (IL)-1, IL-6, and TNF-α *(8,10)* are able to stimulate DCs to mature into cells with a strong T-cell stimulatory potential. Of perhaps more central importance to cancer immunology, Labeur et al. have demonstrated that the induction of antitumor immunity by DC vaccines is correlated with the maturation stage(s) of the DCs *(11)*. In this respect, TNF-α appears to have profound effects on DC function, because it contributes to their maturation *(12)*, activation *(13)*, and migration to, and accumulation within, draining lymph nodes *(14,15)*. Given these effects of TNF-α on DCs, TNF-α becomes a good candidate for induction of DC maturation and enhanced antitumor immunity when these matured DCs are used for cancer vaccines.

Adenoviruses (AdVs) have become popular viral vectors for the delivery of foreign genes into mammalian cells mainly because of their large cloning capacity, their ease of genetic manipulation and growth, and their ability to infect many different tissue types containing both dividing and nondividing cells *(16)*. AdV has a linear double-stranded DNA genome of about 35 kb, flanked by short inverted terminal repeats containing origins of DNA replication *(17)*. The expression of viral genes from Ad DNA genome is highly regulated, which can be divided into early and late phases. There are four regions of early-phase transcription (E1–E4) and five late-transcription regions (L1–L5) in human Ad type 5 (AdV5). Currently, most AdV vectors for gene delivery are based on E1 and E3 deletion mutants *(18)*. Deletion of these regions allows vectors to accommodate larger inserts, removes the region (E1) associated with cellular transformation, and promotes efficient transgene delivery and expression in the absence of significant viral vector gene expression. Several methods have been developed to insert a transgene into an Ad vector; they include (1) direct in vitro ligation of transgene with AdV genomic DNA *(19)*, (2) homologous recombination between AdV genomic DNA and a transfer vector *(20)*, (3) homologous recombination between two transfected plasmids *(21)*, and (4) *Escherichia coli* plasmid, cosmid, and yeast artificial chromosome (YAC) systems *(22–24)*. The recombinant adenoviruses have several biologic characteristics that make them more effective vectors than retroviral ones for somatic gene therapy of tumors *(25)*. Adenovirus vectors have a broad host and cell range and high levels of transgene expression. In addition, there is little possibility of insertion mutation of the host gene. In contrast, this risk may be increased with the use of retroviral vectors, because the insertion is episomal. The ability to produce large quantities of purified virus with relative ease makes this system very attractive in the delivery of transgenes into the cells or tissues for experimental and clinic uses. Recently, the genetic engineering of DCs by adenovirus-mediated cytokine gene transfer has been reported to induce enhanced antitumor immunity when used as cancer vaccines ***(26,27)***.

In this chapter, (1) the design and methods for construction of recombinant adenovirus expressing TNF-α, (2) the methods for generation of bone-marrow-derived DCs, and (3) the methods for genetic engineering of DCs by adenovirus-mediated *TNF*-α gene transfer will be described. The principle and methods are also applicable to construction of other kinds of recombinant adenoviruses and to genetic engineering of other kinds of cells with adenovirus-mediated gene transfer.

2. Methods

2.1. Construction of Recombinant Adenovirus AdV-TNF-α Expressing TNF-α

1. Oligo 1 (5′ GGA ATT CAC TAT AGG GAG ACC CAA GCT GGC TAG 3′).
2. Oligo 2 (5′ GGC TCG AGA TTA TTA CAG TGC GAT AAT ACC GAA GTA CAC 3′).
3. 10X Reaction buffer (200 m*M* Tris-HCl, pH 8.8, 100 m*M* KCl, 100 m*M* $(NH_4)_2SO_4$, 20 m*M* $MgSO_4$, 1% Triton X-100, 1 mg/mL nuclease-free bovine serum albumin [BSA]).
4. dNTP (Stratagene Inc., La Jolla, CA).
5. TwinBlock thermocycler (Ericomp Inc., San Diego, CA).
6. *Taq*I DNA polymerase (2.5 U/μL; Stratagene Inc.).
7. *Eco*RI/*Xho*I (Gibco-BRL, Burlington, Ontario, Canada).
8. pLpA-TNF-α and pJM17 (Microbix Biosystems Inc., Toronto, Ontario, Canada).
9. 293 Cells (Microbix Biosystems Inc.).
10. 1X Citric saline (10X citric saline stock, 50 g potassium chloride, 22 g sodium citrate in 500 mL distilled water [dH_2O]).
11. Lysis buffer (10 m*M* Tris-HCl, pH 7.4, 10 m*M* NaCl, 10 m*M* EDTA, 0.5% sodium dodecyl sulfate [SDS]) containing pronase (0.8 mg/mL; Sigma–Aldrich Canada Ltd, Oakville, Ontario, Canada).
12. Dialysis buffer (10 m*M* Tris-HCl, pH 7.4, 1 m*M* $MgCl_2$, 10% [v/v] glycerol).
13. 1.25 CsCl (36.16 g CsCl dissolved in 100 mL of 1X TD buffer) (*see* **step 15**).
14. 1.40 CsCl (62.0 g CsCl dissolved in 100 mL of 1X TD buffer).
15. 1X TD buffer (170 m*M* NaCl, 6 m*M* KCl, 1 m*M* NaH_2PO_4, pH7.5).
16. 1.33 CsCl (51.2 g CsCl dissolved in 100 mL of 1X TD buffer).

2.2. Genetic Engineering of DC with Adenovirus-Mediated TNF-α Gene Transfer

X-gal staining solution (0.5 mg/mL X-gal, 0.5 m*M* potassium ferricyanide, 0.5 m*M* potassium ferrocyanide, 0.2 m*M* $MgCl_2$).

3. Methods

3.1. Construction of Recombinant Adenovirus AdV-TNF-α Expressing TNF-α

In this section, we describe the construction of recombinant AdV5 expressing TNF-α based on homologous recombination between two plasmids, pLpA-TNF-α and pJM17 *(28)*. In this system, the plasmid pJM17, containing the complete AdV5 *dl*309 genome with an additional 4.3-kb DNA fragment inserted in the E1 region, is used for cotransfection with a transfer vector containing the transgene TNF-α. Because the size of pJM17 exceeds the AdV packaging limit, transfection with pJM17 alone barely generates adenoviruses. Homologous recombination between the flanking AdV sequences in the transfer vector pLpA-TNF-α and pJM17 results in replacement of the additional DNA fragment with the transgene. The reduced size of the construct allows genome packaging and virus formation in 293 cells.

3.1.1. Construction of Transfer Vector pLpA-TNF-α

The plasmid pLpA *(28)*, which contains the left 17% of the AdV5 genome with the deletion of E1 region, is used to make the transfer vector pLpA-TNF-α. To do this, the 0.8-kb DNA fragment encoding the *TNF*-α with an Ig κ-chain leader sequence is obtained by polymerase chain reaction (PCR) amplification from the plasmid pSec-TNF-α *(28)*. Oligo 1 and Oligo 2 are used as primers for PCR. An *Eco*RI and a *Xho*I site are engineered into Oligo 1 and 2, respectively, for easy cloning into a vector. PCR reaction is performed as follows:

1. Add 0.1 μg pSec-TNF-α DNA into a sterile 500-μL microcentrifuge tube.
2. Add 1 μL of the primers (1 mg/mL; primers 1 and 2) to the tube.
3. Add 10 μL of 10X reaction buffer to the tube.
4. Add 0.8 μL of 100 m*M* dNTP to the tube.
5. Add sterile water to a final volume of 99 μL.
6. Place the tube in a TwinBlock thermocycler. Heat the reaction to 91°C for 5 min and then immediately cool the reaction to 54°C for 5 min.
7. Briefly microcentrifuge the sample and then add 1 μL of *Taq*I DNA polymerase. Microfuge the sample again.
8. Carefully overlay the reaction mixture with a drop of mineral oil to prevent evaporation from the reaction during the amplification procedure.

9. Place the tube back in the thermocycler and program the heating block for 30 cycles (91°C for 1 min, 54°C for 30 s, and 72°C for 30 s).
10. After amplification, the 0.4-kb *TNF*-α fragment is visualized on an ethidium bromide-stained 1% (w/v) agarose gel. The *TNF*-α gene fragment obtained from PCR is digested with the restriction enzymes *Eco*RI/*Xho*I. Briefly, microgram quantities of DNA are digested with 5–25 U of restriction enzymes (*Eco*RI and *Xho*I) in a volume of 10–50 µL of 1X buffer solution (Gibco-BRL) appropriate for the DNA and endonuclease being used. The reaction mixtures are incubated at 37°C for 1–2 h. The digested *TNF*-α gene fragment is then ligated with the *Eco*RI-*Sal*I-digested pLpA vector resulting in a plasmid pLpA-TNF-α. Briefly, the digested *TNF*-α gene fragment is mixed with the digested vector pLpA at a molar ratio of 3:1 in 10 µL of solution containing 50 m*M* Tris-HCl, pH 7.6, 10 m*M* $MgCl_2$, 10 m*M* dithiothreitol (DTT), 1 m*M* ATP, and 5 U T4 DNA ligase (Gibco-BRL). The mixture is incubated at 16°C for overnight. Thus, by using this vector, the *TNF*-α expressed from the recombinant AdV-TNF-α is able to be secreted from the infected cells.

3.1.2. Generation of Recombinant AdV-TNF-α Expressing TNF-α

To generate recombinant AdV, calcium phosphate coprecipitation is used to cotransfer plasmids pLpA-TNF-α and pJM17 into 293 cells. The 293 cells are human embryo kidney cells that were transformed by sheared adenovirus type 5 DNA by Dr. Frank Graham. Because 293 cells contain the early region 1 of AdV5, they can complement the growth of E1 defective adenovirus vectors and are used extensively for propagation and titration of adenovirus vectors with deletion of E1 sequences.

3.1.2.1. Cell Culture

1. Thaw one vial of the frozen low passage 293 cells in a 37°C water bath.
2. Transfer cells dropwise into a T75 tissue culture flask containing 15 mL Eagle's modified essential medium (EMEM) (Gibco-BRL, Burlington, Ontario, Canada) medium plus 10% fetal calf serum (FCS).
3. Maintain cells at 37°C in an atmosphere of 5% CO_2. For 5 h after incubation, change media once to remove any traces of the freezing medium from the cells.
4. Remove growth medium from tissue culture flasks when 293 cells are approx 90% confluent.
5. Wash the 293 monolayers two times with 1X citric saline
6. Decant citric saline except for a volume sufficient to cover the cell monolayer.
7. Incubate at 37°C for 3–5 min or until cells begin to loosen from the flask.
8. Complete the cell detachment by tapping the side of the flask.
9. Once the cells are in suspension, dilute the mixture in growth medium and distribute to three flasks for further incubation (*see* **Note 1**).
10. When preparing 293 cells in T75 flasks for transfection, use one flask for each transfection. Cells are ready for transfection when they reach about 80%

confluency. Three hours prior to transfection, replace media with fresh EMEM consisting of 10% FCS.

3.1.2.2. Cotransfection of 293 Cells with Plasmids pLpA-TNF-α and pJM17

The calcium phosphate–DNA complex is made by using the Calcium Phosphate Transfection System from Gibco/BRL (cat. no. 18306-019). One milliliter of calcium phosphate–DNA suspension is needed for transfection of one T75 flask of 293 cells.

1. Prepare 1X HEPES-buffered saline (HBS) by adding 0.885 mL Transfection Qualified Water to 0.1 mL of 10X HBS in a sterile tube and mixing well. Then, add 15 µL of 1 *N* NaOH to the mixture. The amount of 1X HBS prepared in this way is enough for two transfections.
2. Label two sterile polypropylene tubes as A and B. Then, add 0.5 mL of 1X HBS and 10 µL phosphate solution to tube A and mix well. Add 10 µg pLpA DNA and 10 µg pJM17 DNA to tube B in a total volume of 0.43 mL in Transfection Qualified Water, followed by adding dropwise 60 µL of calcium solution. Mix gently the solution in tube B.
3. Precipitate plasmid DNA by adding the solution in tube B dropwise into tube A. As the solutions are combined and mixed in tube A, they should appear milky.
4. Leave the mixed solution at room temperature for 20 min.
5. Add the solution dropwise to 293 cells grown in one T75 flask while gently swirling the medium in the flask.
6. Incubate the cells for 4–6 h. Aspirate the media from the flask, wash cell monolayer once with phosphate-buffered saline (PBS), and replace with the complete media.
7. Maintain the cells by replacing the media twice a week. The initial plaques appear about 2–3 wk after transfection and extensive cytopathic effects (CPE) occur in another 3–5 d.
8. Harvest the cells by gently tapping the flask to dislodge the cells to the medium, pellet cells at 1100 rpm for 10 min, remove supernatant, and resuspend the cell pellet in 6 mL EMEM containing 2% FCS.
9. Lyse cells by three freeze–thaw cycles to release the virus. For each cycle, the cells are completely frozen at –80°C and then thaw at 37°C, followed by vortexing to further disrupt the cells.
10. Store the crude viral lysate (CVL) at –80°C.

3.1.2.3. Plaque Purification of Recombinant AdV-TNF-α

Because of the possibility that the original transfection may generate a mixture of recombinant AdVs with different genotypes, it is necessary to do at least one round of plaque purification of CVL obtained from original transfection.

1. Prepare overlay for plaque assay as follows: Make 400 mL of 2X EMEM containing 100 U/mL penicillin, 100 µg/mL streptomycin, and 8 mL of 5% yeast

extract. The medium can be stored at 4°C for a few weeks. To make 100 mL of complete overlay, prepare 50 mL of 2X EMEM + 5 mL FCS + 0.5 g agarose + 50 mL sterile water. Bring agarose and 2X EMEM to 37°C before mixing and use within about 1 h.

2. Prepare 293 cells in a 6-well plate to be just confluent 1 d after seeding.
3. Prepare 10-fold serial dilutions of CVL using EMEM and do a plaque assay using several different dilutions (e.g., 10^{-5}, 10^{-6}, and 10^{-7}) of the CVL in order to get well-isolated plaques.
4. Remove medium from the plate and inoculate with 0.1 mL of the various dilutions.
5. Incubate at 37°C for 1 h with shaking the plate to disperse the inoculum every 20 min.
6. Add 5 mL agarose overlay to each well. Leave the plate at room temperature for about 0.5 h to let the agar to solidify.
7. Incubate the plate at 37°C until appearance of well-separated plaques (usually takes about 10 d).
8. Isolate virus from well-separated plaques by using a plugged sterile Pasteur pipet to aspirate the infected cells. Place the sample in 1 mL of EMEM in a sterile Eppendorf tube (*see* **Note 2**).
9. Freeze and thaw the plaque isolate three times and store the sample at –70°C.
10. Amplify plaque-purified virus stock by inoculation of a T25 flask of 293 cells with 0.1 mL cell lysate diluted in 0.9 mL EMEM.
11. Incubate at 37°C for 1 h, dispersing the inoculum over the cell monolayer surface every 20 min.
12. Add 10 mL EMEM containing 10% FCS and incubate at 37°C until CPE is extensive.
13. Harvest infected cells and prepare CVL as described in **Subheading 3.1.2.2., step 8**.

3.1.2.4. Analysis of the Potential Recombinant AdV by Restriction Enzyme Digestion

To confirm if the virus generated by cotransfection is the desired recombinant AdV, viral DNA isolated from infected 293 cells is used for restriction digestions. We have used the modified "Hirt" method ***(29)*** to isolate AdV DNA.

1. Culture 293 cells in T75 flasks till the cells are confluent.
2. Aspirate the medium, inoculate with 1 mL CVL prepared from **Subheading 3.1.2.3. (steps 10–13)** diluted with 2 mL EMEM.
3. Incubate the flasks at 37°C for 1 h with dispersing the inoculum over the cell monolayer surface every 20 min.
4. Add 10 mL EMEM containing 10% FCS to each flask and incubate the flasks at 37°C until CPE is extensive (usually takes 24–36 h after inoculation).
5. Transfer infected cells to centrifuge tube and pellet cells at about 200*g* for 10 min at 4°C.

6. Wash cell pellet once with cold PBS and resuspend cells in 1 mL lysis buffer containing pronase.
7. Incubate the sample at 37°C for 20 min.
8. Add 0.25 mL of 5 *M* NaCl and transfer the sample into an Eppendorf tube.
9. Chill on ice for 30 min and centrifuge at 4°C for 10 min.
10. Transfer the supernatant to an Eppendorf tube and add an equal volume of isopropanol to precipitate DNA. Centrifuge the precipitate for 15 min.
11. Wash the pellet with 70% ethanol and air-dry.
12. Dissolve DNA pellet in 50 μL sterile water containing 0.5 mg/mL RNase.
13. Check viral DNA by digestion with appropriated enzyme(s).

3.1.3. Purification of Recombinant AdV-TNF-α by CsCl-Gradient Ultracentrifugation

1. The CVL from 1 T175 flask is used for amplification of 6 T175 flasks of 293 cells.
2. Use 3 mL EMEM for preparation of CVL from every six T175 flasks. Prepare large stocks of CVL from 24–32 T175 flasks for each virus purification.
3. Lay 6 mL CVL onto a CsCl step gradient in a ultracentrifuge tube. The gradient consists of 2.5 mL of 1.25 CsCl layed on top of 2.5 mL of 1.40 CsCl. Fill the tube with 1X TD buffer.
4. Centrifuge the sample at 45,000 rpm for 1 h at 20°C in a Beckman L8-55 Ultracentrifuge using a 80Ti rotor (Beckman Coulter, Mississauga, Ontario, Canada).
5. After centrifugation, two opalescent bands should appear within the gradient. Collect the lower virus containing band using a syringe with a needle.
6. Lay the collected sample on top of 5 mL of 1.33 CsCl in a new ultracentrifuge tube. Centrifuge the sample at 45,000 rpm for 18 h at 20°C.
7. Collect the single heavy opalescent band using a syringe with a needle and add glycerol to the purified virus stock to a final concentration of 10% (v/v).
8. Remove CsCl from purified virus stock by dialysing the sample at 4°C against six changes of 500 mL dialysis buffer (*see* **Note 3**).
9. Aliquot the virus stock as 100 μL to each sterile 500-μL microcentrifuge tubes and store the samples at –80°C (*see* **Note 4**).
10. The absorbency of the virus is read at 260 nm and is used to calculate the titer of the virus stock. The values used to determine the viral titer is an $A_{260} = 1.000$ is about equivalent to 10^{10} plaque-forming units (PFU)/mL (*see* **Note 5**).

3.2. Genetic Engineering of DC with Adenovirus-Mediated TNF-α Gene Transfer

3.2.1. Preparation of Murine Bone Marrow Cells

Although bone marrow cells can be obtained from any of the long bones, the femur and tibia are usually used because they give the best yields.

1. Kill mice by CO_2 inhalation and dip them in 70% ethanol.
2. Place mice on their backs and make a long transverse incision through the skin in the middle of the abdominal area of each mouse. Reflect skin completely from the hindquarters, including the hind legs.
3. Flood the hind legs with 70% ethanol.
4. While grasping the hind legs with the mouse tooth forceps, cut away as much muscle as possible with the scissors.
5. Separate legs from the body at the hip joint and remove the feet. Place the legs in a culture dish containing PBS.
6. Begin the removal of adherent muscle tissue from the femur and tibia by grasping the bones with the mouse tooth forceps and scraping them with the flat forceps. Transfer the partially cleaned bones to another culture dish containing PBS.
7. Separate the tibia and femur with the scissors.
8. Remove the epiphyses with the scissors and puncture the bone ends with a 20-gage needle.
9. Using a syringe with a 25-gage needle attached, expel the marrow by pushing PBS through the center of the bones. Draw the marrow in and out of the needle and syringe to obtain a single-cell suspension.

3.2.2. Generation of Dendritic Cells

1. Deplete red blood cells by resuspending bone marrow cell pellets in 0.84% ammonium chloride, using 10 mL in a 50-mL tube for bone marrow cells obtained from one mouse.
2. Incubate for 5 min at room temperature.
3. Add 10 mL PBS to the tube and spin 10 min at 1000 rpm.
4. After centrifugation, cells from one bone marrow are plated in a six-well plate with 3 mL of DC culture medium (DMEM plus 10% FCS, GM-CSF [10 ng/mL], and IL-4 [10 ng/mL]) in each well.
5. On d 3, the nonadherent granulocytes, T and B cells are gently removed and fresh media were added.
6. On d 5, the loosely adherent proliferating DC aggregates were dislodged and replated by using DC culture medium.
7. On d 8, the immature, nonadherent DCs are harvested and used for in vitro AdV-TNF-α transfection.

The immature DCs generated in this manner should display (1) typical morphologic features of dendritic cells (i.e., numerous dendritic processes) and (2) significant expression of MHC class I and II antigens, costimulatory molecules (CD80 and CD86), and adhesion molecules (ICAM-1, CD11b, CD11c, and CD40).

3.2.3. Transfection of Dendritic Cells with AdV-TNF-α

The susceptibility of murine DCs to adenoviral infection has been previously reported by using AdV-LacZ ***(26)***. Briefly, after the cells had been infected with AdV/LacZ, they were fixed for 5 min at 37°C in PBS containing 2% (v/v) formaldehyde and 0.2% (v/v) glutaraldehyde. The fixative was then removed and the cells were rinsed twice with PBS. The cells were then incubated at 37°C with X-gal staining solution for 20 min to overnight until blue staining was observed. The cells were then counterstained with nuclear fast red. Positive cells had a blue color. The average readings of positive cells from wells in triplicates and from three sections of each tumor were taken as the percentage of transduction. A quantitative assessment of LacZ gene expression (blue cells) was determined using microscopy. A maximal DC transfection (80%) can be reached at "multiplicities of infection" (MOI) of 100. Therefore, MOI of 100 is used for transfection of DCs with AdV-TNF-α.

1. Wash DCs (2×10^6 cells) twice with PBS.
2. Resuspend DCs in 200 μL serum-free DMEM in 1 well of a 24-well plate.
3. Add 2×10^8 PFU AdV-TNF-α to the well and incubate at 37°C for 1 h with gentle agitation every 20 min (*see* **Note 6**).
4. Add 2 mL DMEM plus 10% FCS to the well and incubate DCs at 37°C for 24 h.
5. Harvest AdV-TNF-α-transfected DCs and wash them twice with PBS before use.

4. Notes

1. The 293 cells grow only in monolayers. They are sensitive to the way that they are handled. Growth properties will be altered if cells are allowed to become over confluent and are plated too thinly. Therefore, make no more than 1:5 split when passing the cells from a confluent flask. When passing the cells, warm the growth medium. Early passage cells, up to approximately passage 40, are preferred for use for transfection and construction of recombinant adenovirus.
2. To produce a virus stock after plaque purification of recombinant adenovirus, 293 cells should be infected at a low multiplicity of infection. This allows a selection for the virus capable of supporting productive infection and, thus, helps maintain the integrity of the virus genome. In addition, it is good practice to keep an aliquot of the first plaque-purified virus stock as a master stock.
3. Autoclaved solutions should be used during virus purification to ensure that the virus stock is free of microbial contamination or proteases. The dialysis membranes should be disinfected by boiling in Tris-EDTA buffer immediately before use.
4. Crude viral lysate can be stored in medium plus 2% FCS and 10 m*M* Tris, pH 8.0, at –20°C and –70°C respectively. Purified virus still in CsCl (before dialysis) can be stored at 4°C for 2–3 d. However, purified virus after dialysis must be stored at –70°C.

5. The titer of recombinant adenovirus stock can also be determined by plaque assay, which gives a measurement of the number of PFUs in a given volume of a virus stock when infecting 293 cells. The plaque assay can be done in a method similar to that described in **Subheading 3.1.2.3.**
6. To get efficient transfection, keep the medium volume of cell suspension as low as possible and do not add any FCS, which will absorb the adenovirus and reduce the transfection efficiency.

References

1. Schuler, G., Koch, F., Heufler, C., Kampgen, E., Topar, G., and Romani, N. (1993) Murine epidermal Langerhans cells as a model to study tissue dendritic cells. *Adv. Exp. Med. Biol.* **329,** 243–254.
2. Steinman, R. M. and Bancherear, J. (1988) Dendritic cells and the control of immunity. *Nature* **392,** 145–152.
3. Steinman, R., Witmer-Pack, M., and Inaba, K. (1993) Dendritic cells: antigen presentation, accessory function and clinical relevance. *Adv. Exp. Med. Biol.* **329,** 1–9.
4. Cella, M., Sallusto, F., and Lanzavecchia, A. (1997) Origin, maturation and antigen presenting function of dendritic cells. *Curr. Opin. Immunol.* **9,** 10–16.
5. Porgador, A., Snyder, D., and Gilboa, E. (1996) Induction of antitumor immunity using bone marrow-generated dendritic cells. *J. Immunol.* **156,** 2918–2926.
6. Bellone, M., Lezzi, G., Martin-Fontecha, A., Rivolta, L., Manfredi, A., Protti, M., et al. (1997) Rejection of a nonimmunogeneic melanoma by vaccination with natural melanoma peptides on engineered antigen-presenting cells. *J. Immunol.* **158,** 783–789.
7. Asheley, D., Faiola, B., Nair, S., Hale, L., Bigner, D., and Gilbao, E. (1997) Bone marrow-generated dendritic cells pulsed with tumor extracts or tumor RNA induced antitumor immunity against central nervous system tumors. *J. Exp. Med.* **186,** 1177–1182.
8. Jonuleit, H., Knop, J., and Enk, A. (1996) Cytokines and their effects on maturation, differentiation and migration of dendritic cells. *Arch. Dermatol. Res.* **289,** 1–8.
9. Caux, C., Dezutter-Dambuyant, C., Schmitt, D., and Banchereau, J. (1992) GMCSF and TNF-α cooperate in the generation of dendritic Langerhans cells. *Nature* **360,** 258–269.
10. Jonuleit, H., Kuhn, U., Muller, G., Steinbrink, K., Paragnik, L., Schmitt, E., et al. (1997) Pro-inflammatory cytokines and prostaglandins induce maturation of potent immunostimulatory dendritic cells under fetal calf serum-free conditions. *Eur. J. Immunol.* **27,** 3135–3142.
11. Labeur, M., Roters, B., Pers, B., Mehling, A., Luger, T., Schwarz, T., et al. (1999) Generation of tumor immunity by bone marrow-derived dendritic cells correlates with dendritic cell maturation stage. *J. Immunol.* **162,** 168–175.
12. Sallusto, F., Cella, M., Danieli, C., and Lanzavecchia, A. (1995) Dendritic cells use macropinocytosis and the mannose receptor to concentrate macromolecules

in the major histocompatibility complex class II compartment: downregulation by cytokines and bacterial products. *J. Exp. Med.* **182,** 389–400.
13. Rieser, C., Bock, G., Klocker, H., Bartsch, G., and Thumher, M. (1997) Prostaglandin Ea and tumor necrosis factor alpha cooperate to activate human dendritic cells: synergistic activation of interleukin 12 production. *J. Exp. Med.* **186,** 1603–1608.
14. Cumberbatch, M. and Kimber, I. (1992) Tumor necrosis factor-α induces dendritic cell migration to draining lymph nodes, and possibly provides one stimulation for Langerhans' cell migration. *Immunology* **75,** 257–263.
15. Cumberbatch, M. and Kimber, I. (1995) Tumor necrosis factor-α is required for accumulation of dendritic cells in draining lymph nodes and for optimal contact sensitization. *Immunology* **84,** 31–35.
16. Kay, M. A., Glorioso, J. C., and Naldini, L. (2001) Viral vectors for gene therapy: the art of turning infectious agents into vehicles of therapeutics. *Nat. Med.* **7,** 33–40.
17. Shenk, T. (1996) Adenoviruses: the viruses and their replication, in *Fields Virology*, 3rd ed (Fields, B. N., Knipe, D. M., and Howley, P. M., eds.), Lippincott–Raven, Philadelphia, pp. 2111–2148.
18. Graham, F. and Prevec, L. (1991) Manipulation of adenovirus vectors, in *Methods in Molecular Biology* (Murray, E. J. and Clifton, N. J., eds.), Humana, Totowa, NJ, pp. 109–128.
19. Okada, T., Ramsey, W. J., Munir, J., Wildner, O., and Blaese, R. M. (1998) Efficient directional cloning of recombinant adenovirus vectors using DNA-protein complex. *Nucl. Acids Res.* **26,** 1947–1950.
20. Scarppini, C., Arthur, J., Efstathiou, S., McGrath, Y., and Wilkinson, G. (1999) Herpes simplex virus and adenovirus vectors, in *DNA Viruses: A Practical Approach* (Cann, J., ed.), Oxford University Press, New York, pp. 267–306.
21. McGrory, W. J., Bautista, D. S., and Graham, F. L. (1988) A simple technique for the rescue of early region I mutations into infectious human adenovirus type 5. *Virology* **163,** 614–617.
22. Chartier, C., Degryse, E., Gantzer, M., Dieterle, A., Pavirani, A., and Mehtali, M. (1996) Efficient generation of recombinant adenovirus vectors by homologous recombination in *Escherichia coli*. *J. Virol.* **70,** 4805–4810.
23. Fu, S. and Deisseroth, A.B. (1997) Use of the cosmid adenoviral vector cloning system for the in vitro construction of recombinant adenoviral vectors. *Hum. Gene Ther.* **8,** 1321–1330.
24. Ketner, G., Spencer, F., Tugendreich, S., Connelly, C., and Hieter, P. (1994) Efficient manipulation of the human adenovirus genome as an infectious yeast artificial chromosome clone. *Proc. Natl. Acad. Sci. USA* **91,** 6186–6190.
25. Boviatsis, E., Chase, M., Wei, M., et al. (1994) Gene transfer into experimental brain tumors mediated by adenovirus, herpes simplex virus and retrovirus. *Hum. Gene Ther.* **5,** 183–191.
26. Cao, X., Zhang, W., He, L., Xie, Z., Ma, S., Tao, Q., et al. (1998) Lymphotactin gene-modified bone marrow dendritic cells act as more potent adjuvants for peptide delivery to induce specific antitumor immunity. *J. Immunol.* **161,** 6238–6244.

27. Curiel-Lewandrowski, C., Mahnke, K., Labeur, M., Roters, B., Schmidt, W., Granstein, R., et al. (1999) Transfection of immature murine bone marrow-derived dendritic cells with the granulocyte–macrophage colony-stimulating factor gene potently enhance their in vivo antigen-presenting capacity. *J. Immunol.* **163,** 174–183.
28. Wright, P., Braun, R., Babiuk, L., Hurk, S., Moyana, T., Zheng, C., et al. (1999) Adenovirus-mediated TNF gene transfer induces significant tumor regression in mice. *Cancer Biother. Radiopharm.* **14,** 49–57.
29. Hirt, B. (1967) Selective extraction of polyoma DNA from infected mouse cell cultures. *J. Mol. Biol.* **26,** 365–369.

17

Retroviral Transfer of T-Cell Receptor Genes into Human Peripheral Blood Lymphocytes

Timothy M. Clay and Michael I. Nishimura

1. Introduction

We are developing strategies to target tumor-associated antigens (TAA) and viral antigens by genetically modifying patient peripheral blood lymphocytes to produce antitumor or antiviral reactive cytotoxic T-lymphocyte (CTL). To this end, we have succeeded in redirecting the specificity of peripheral blood T-cells and T-lymphocyte clones by the transfer of T-cell-receptor (TCR) genes from antigen specific T-cell clones into these lymphocytes. The TAA MART-1 is expressed by the majority of human melanoma tumors *(1,2)*, making it an excellent potential target for therapeutic strategies utilizing TCR gene transfer. A single HLA-A2-restricted peptide epitope has been identified in the MART-1 protein (amino acids 27–35: MART-$1_{(27-35)}$). HLA-A2 is present in approx 50% of Caucasian melanoma patients. Therefore, a TCR gene transfer approach using a TCR that recognizes this particular major histocompatibility complex (MHC)–peptide complex could potentially treat 50% of Caucasian melanoma patients. We have previously described the cloning of the TCR genes from a tumor-reactive CTL clone (clone 5) derived from a tumor-infiltrating lymphocyte (TIL) culture from patient 501 *(3,4)*. The clone 5 TCR is HLA-A2 restricted and is specific for the m9–27 peptide epitope (MART-1_{27-35}) of MART-1. Transfection of the α and β TCR genes from clone 5 into Jurkat cells resulted in the expression of a functional TCR on the cell surface *(4)*. Subsequently, we have cloned the full-length clone 5 TCR α- and β-chain cDNAs into the A7 retroviral constructs and successfully used this retrovirus to redirect normal donor PBL to recognize MART-1 *(5)*. These CTLs were capable of lysing MART-1^+, HLA-A^*0201^+ melanoma tumor

From: *Methods in Molecular Biology, vol. 215: Cytokines and Colony Stimulating Factors: Methods and Protocols*
Edited by: D. Körholz and W. Kiess © Humana Press Inc., Totowa, NJ

lines in vitro. These results are proof of principle studies that will lead to the clinical application of this approach.

The following protocols permit initial transduction efficiencies of approx 40%, and this can be increased to >90% following selection for transduced T-cells when an antibiotic resistance gene is present in the retroviral construct. To detect the presence of a functional TCR in transduced PBL, antigen-specific immune assays are necessary and two popular assays of T-cell function are also briefly described. Where applicable, these protocols include the authors' notes and tips to facilitate the implementation of these protocols in your own lab.

2. Materials

2.1. Retroviral Gene Transfer to Human PBL

1. Dulbecco's modified Eagle's medium (DMEM)/10 medium (DMEM medium [Biofluids, Rockville, MD] supplemented with 10% fetal bovine serum (FBS) (Life Technologies, Grand Island, NY), 50 m*M* HEPES (Life Technologies), penicillin (100 U/mL)–streptomycin (100 µg/mL)-Glutamine (2.92 mg/mL) (Life Technologies).
2. DMEM/10 was supplemented with 400 µg/mL G418 (Geneticin; Life Technologies).
3. Very low-protein-binding filter flask (Corning, cat. no. 431097, Corning, NY).
4. Lymphocyte Separation Medium (LDM) (Organon Teknika, Durham, NC).
5. 50-mL Polypropylene centrifugation tubes (Falcon, cat. no. 35-2098, Becton Dickinson, Franklin Lakes, NJ).
6. AIM 'V' serum-free medium containing 600 IU/mL interleukin (IL)-2 (Cetus Oncology Corp., Emmeryville, CA) and 10 µg/mL anti-CD3 monoclonal antibody (OKT3; Ortho Biotech, Raritan, NJ).
7. Polybrene (Hexadimethane Bromide, Sigma Chemical Company, St. Louis, MO).
8. Geneticin (Life Technologies).

2.2. Limiting Dilution Cloning of Transduced PBL Cultures

1. RPMI/11 media (RPMI 1640 [Biofliuds] containing 11% heat inactivated human pooled AB serum (Pel Freez, Brown Deer, WI), penicillin (100 U/mL)–streptomycin (100 µg/mL)–glutamine (2.92 mg/mL) (Life Technologies), 25 m*M* HEPES, 25 µ*M* 2-mercaptoethanol [ME]).
2. Multichannel pipettor.

2.3. Immune Assays to Demonstrate Transgenic TCR Function in Transduced T-Cells

1. Enzyme-linked immunosorbent assay (ELISA) kits (IL-2 and GM-CSF, R & D Systems, Minneapolis, MN; interferon [IFN]-γ, IL-4, IL-10, and tumor necrosis factor [TNF]-α, Endogen, Cambridge, MA, respectively).
2. ^{51}Cr (10–35 mCi/mL; Amersham, Arlington Heights, IL).

3. Methods

3.1. Retroviral Gene Transfer to Human Peripheral Blood Lymphocytes

3.1.1. Routine Culture of Retroviral Producer Cell Lines

The A7/PG13 clone 6 is grown in DMEM/10 medium supplemented with G 418 (*see* **Note 1**). The G418 maintains selection pressure on the cells to ensure continued expression of the A7 retroviral plasmid, which contains the neomycin phosphotransferase gene. Without this selection pressure, the retroviral plasmid promoters can shut down and virus production will cease. HEPES is used to provide buffering against the pH change that occurs during transduction when the T-cells are centrifuged in air and we do not provide the 5% CO_2 atmosphere of an incubator (*see* below):

1. A7/PG13 clone 6 packaging cell line was grown to 80% confluence in DMEM/10 medium supplemented with G418. It is important not to exceed 80% confluence prior to this 18-h culture period, to ensure that the subsequent yield of retroviral supernatant is high. If cell density is too high, cells will die and virus production will be limited.
2. Eighteen hours before supernatant was to be harvested for PBL transductions, the media was exchanged with fresh DMEM/10 media *without G418*.
3. The supernatant containing the retrovirus was harvested and filtered through a 0.22-μm micron, very low-protein-binding filter flask to remove any packaging cells that might otherwise be carried over in to transduced cell cultures. The presence of packaging cell lines in subsequent cultures can give false positives in polymerase chain reaction (PCR) assays for the transgenes.

3.1.2. PBMC Isolation and Retroviral Transduction

Peripheral blood lymphocytes (PBL) are transduced using an adaption of the method described by Bunnell et al. *(6)*. The PBL must be dividing in order to allow integration of the reverse-transcribed retroviral genome into the genome of the host cell. The nuclear membrane dissolves in dividing cells, permitting integration of the retroviral DNA. Soluble anti-CD3 monoclonal antibody and IL-2 (T-cell growth factor) are used to stimulate T-cell proliferation. CD3 is the signalling portion of the TCR–CD3 complex on T-lymphocytes. Ligation of CD3 with the anti-CD3 monoclonal antibody will mimic TCR signaling, and with the addition of IL-2, the T-lymphocytes will proliferate and divide.

1. Human peripheral blood mononuclear cells (PBMC) are isolated from buffy coats from donors (on Institutional Review Board [IRB]-approved protocols) by centrifugation through Lymphocyte Separation Medium (LDM). First, the buffy coat is diluted 1:4 in Hank's buffered salt solution (HBSS) and then overlaid over 10 mL of LSM in 50-mL polypropylene centrifugation tubes. Following

centrifugation at 1000*g* for 20 min at room temperature, the centrifuge is stopped without braking to avoid disturbing the interface.

2. The PBMC layer is harvested and washed twice in HBSS, centrifuging cells at 800*g* for 5 min.
3. Cells are counted and resuspended at a concentration of 1×10^6 cells/mL in AIM 'V' serum-free medium containing 600 IU/mL IL-2 and 10 µg/mL anti-CD3 monoclonal antibody. Cells are cultured for 72 h in 175-cm^2 tissue culture flasks, 1×10^8 cells per flask, in a 37°C humidified incubator with 5% CO_2.
4. On d 3, cells are harvested, counted, and resuspended at 1×10^6 cells/mL in retroviral supernatant containing 600 IU/mL IL-2 and 8 µg/mL polybrene and plated in 24-well tissue culture plates, 2 mL per well (*see* **Note 2**).
5. Plates are centrifuged at 32°C for 60–90 min at 1000*g* (*see* **Notes 3** and **4**). Cultures will be very pink because of the low CO_2 content of air, resulting in pH change. This is countered somewhat by using HEPES-buffered medium to grow the packaging line (*see* **Subheading 3.1.1.**). Plates are then incubated overnight (16–20 h) in a 37°C humidified incubator with 5% CO_2.
6. This transduction procedure (**steps 4** and **5**) is repeated for a further 2 d. These additional rounds of transduction can significantly improve transduction efficiency.
7. On d 7, the PBL are harvested, counted, and resuspended in AIM 'V' medium containing 600 IU/mL IL-2 at 1×10^6 cells/mL and cultured overnight in 175-cm^2 tissue culture flasks, 1×10^8 cells per flask, in a 37°C humidified incubator with 5% CO_2.
8. The following day, cells are harvested and resuspended in AIM 'V' medium containing 600 IU/mL IL-2 and 0.5 mg/mL Geneticin at 1×10^6 cells/mL and cultured overnight in 175-cm^2 tissue culture flasks, 1×10^8 cells per flask, in a 37°C humidified incubator with 5% CO_2. The flask should be stood upright to increase the local cell density, which is optimal for T-cell proliferation. The cells are selected for 5 d in Geneticin, with the concentration of live cells being monitored daily by performing cell counts using trypan blue staining to distinguish live from dead cells. The volume of media is adjusted as necessary to maintain a concentration of 1×10^6 cells/mL. This cell concentration of 1×10^6 cells/mL is optimal for T-cell growth and is a significant factor in recovering transduced cells from these cultures. Significant cell death is observed both during and for several days after this G418 selection step. From the 1×10^8 cells, expect 50–75% of the cells to die off before the proliferation of transduced cells begins to replace the cells that are dying in selection. It is critical to maintain the concentration of live cells at 1×10^6 cells/mL. If the number of live cells falls below 50×10^6, they should be transferred into a 75-cm^2 flask in the appropriate volume of medium to yield a concentration of 1×10^6 cells/mL.
9. On d 12, after 5 d of selection in G418, the cells are harvested and resuspended at 1×10^6 cells/mL in AIM 'V' medium containing 600 IU/mL IL-2.
10. On d 14, the PBL can be tested in functional assays or cloned in limiting dilution to generate T-cell clones.

3.2. Limiting Dilution Cloning of Transduced PBL Cultures

Transduced CTL cultures can be expanded and single-cell cloned using anti-CD3 stimulation, using an adaption of the method of Walter et al. *(7)*. Single-cell cloning allows the reactivities of individual transgenic T-cell clones to be investigated. We have previously described several different functional phenotypes in transduced T-cell clones *(5)*.

1. Transduced PBL are plated at 10, 1, and 0.3 cells per well in 96-well U-bottom microtiter plates in a 0.1 mL volume per well of RPMI/11 media.
2. Then, 2.5×10^5 irradiated allogeneic PBMC (100 Gy) per mL, 5×10^4 irradiated allogeneic Epstein-Barr virus (EBV)-B cells (100 Gy) per mL, and 30 ng/mL anti-CD3 monoclonal antibody are added to the wells in a 0.1 mL volume of RPMI/11.
3. The following day, IL-2 is added in a volume of 20 μL/well to yield a final concentration of 120 IU/mL IL-2.
4. On d 5, the medium is exchanged and fresh RPMI/11 medium containing 120 IU/mL IL-2, *without OKT3*, is added. To exchange the media, remove 180–200 μL of medium with a multichannel pipettor and add back the same volume of RPMI/11 supplemented with 120 IU/mL.
5. On d 8, fresh IL-2 is added to yield a final concentration of 120 IU/mL.
6. Cells are tested for reactivity on d 12 in cytokine release assays and peptide-specific/tumor-specific cloids are restimulated as described above. To further expand the reactive clones, the culture volume is increased and the numbers of PBMC and EBV-B are adjusted accordingly. Twenty-five-milliliter, 75-mL, and 150-mL culture volumes have been used to successfully expand T-cell clones (using 25 cm^2, 75-cm^2, and 175-cm^2 flasks respectively, placed upright).

3.3. Immune Assays to Demonstrate Transgenic TCR Function in Transduced T-Cells

There are a wide variety of immune assays that can be used to test whether the transferred TCR genes are conferring antigen-specific recognition to the transduced T-cells (for reviews on immune assays, *see* **refs. *8–10***). The IFN-γ ELISA assay described here is excellent for "high-throughput" economical screening of large numbers of T-cell clones and provides a sensitive and robust assay for antigen-specific recognition by bulk CTL cultures. In essence, antigen-specific recognition will trigger the secretion of IFN-γ by T-cells, as a result of TCR signaling in response to the ligation of the TCR to the correct peptide–MHC complex. An alternate method is the various forms of lysis assay, where antigen-specific CTLs result in lysis of antigen-expressing target cells. These target cells are labeled with a marker (e.g., radioactive chromium ^{51}Cr, europium, etc.) that is released when the cells are lysed. Using appropriate controls, ^{51}Cr level is a quantitative measure of the percentage of cells lysed by the CTLs. Cytokine-release ELISA and cell lysis assays are described next. Most importantly, before embarking on any TCR gene transfer studies, appropriate target cells for the assessment of antigen recognition must be available.

3.3.1. Assessment of Culture Reactivity by Cytokine Release

Peripheral blood lymphocytes cultures and clones are tested for reactivity in cytokine release assays using commercially available ELISA kits (IL-2 and GM-CSF, IFN-γ, IL-4, IL-10 and TNF-α). These kits are very accurate and straightforward to use, but relatively expensive. Matched antibody pairs are also available from these and other manufacturers, and they allow labs to produce their own ELISA plates much more economically. This requires some optimization for each batch of reagents but can give equally good results at a fraction of the cost. Cytokine release is measured in 24-h culture supernatants. If the TCR in question is HLA-A*0201 restricted, then peptide-pulsed T2-cells make excellent targets for these assays. HLA-matched and HLA-mismatched tumor cell lines can also be used.

1. T2-Cells are pulsed with appropriate peptides for the TCR in question (typically at a concentration of peptide of 1 μg/mL), by incubation in RPMI/11 medium for 2–3 h at 37°C, 5% CO_2. Up to 10×10^6 T2-cells can be pulsed with peptide in a 15-mL conical centrifuge tube at 1×10^6 cells/mL in RPMI/11 medium. The tube should be placed upright in the incubator. To ensure good loading of the peptide onto surface HLA-A*0201 molecules, the T2 cells should be resuspended gently every 30–45 min. Cells can also be tested for reactivity with tumor cell lines and/or fresh tumor cells.
2. The assay consists of 10,000 T-cells as responders and 10,000 stimulator cells (T2-cells pulsed with peptide, or tumor cells), incubated together in a 96-well U-bottom plate in 0.2 mL volume of RPMI/11 medium for 20–24 h at 37°C, 5% CO_2. Cytokine secretion can then be measured in culture supernatants by ELISA, using the manufacturer's protocol.

3.3.2. Assessment of Reactivity by Lysis Assay: 4-h ^{51}Cr-Release Assays

Typical target cells for lysis assays are T2-cells pulsed with peptide and/or appropriate tumor cell lines.

1. Targets are labeled for 90 min with 200 μCi ^{51}Cr per 1×10^7 cells. Typically, 10 million target cells are labeled in a 1 mL volume in a 15-mL conical centrifuge tube.
2. While the targets are labeled with ^{51}Cr, the T-cells can be harvested, counted, and plated at 5000 cells per well in 96-well U-bottom plates in 0.1 mL RPMI/11 medium.
3. After the 90-min labeling step (**step 1**), the targets are washed three times with Hank's buffered salt solution and then plated in triplicate with responders at the following E:T ratios: 80:1, 20:1, 5:1, and 1.25:1, by adding the targets in a

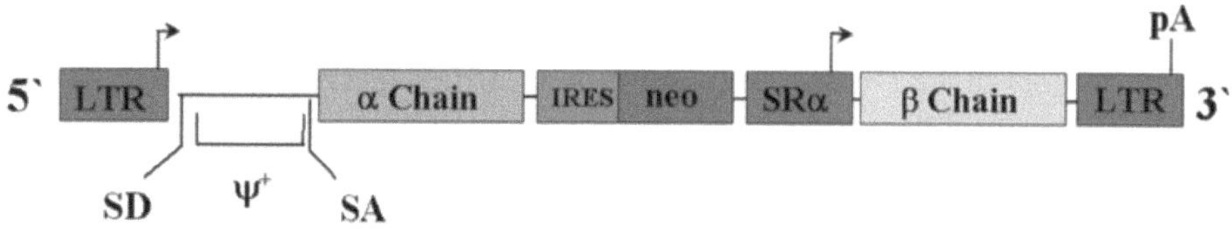

Fig. 1. Structure of the A7 retrovirus plasmid. A modified SAMEN pg1 Moloney murine leukemia virus backbone. Expression of the TCR α-chain is driven by the promoter in 5′ long terminal repeat (LTR), which also drives expression of the neomycin phosphotransferase gene (*neo*) via an internal ribosomal entry site (IRES). β-Chain expression is driven by the hybrid HTLV-I/SV40 SRα promoter. Abbreviations: SD, splice donor site; SA, splice acceptor site; pA, polyadenylation signal; ψ^+, packaging signal. Arrows denote transcription start sites.

0.1 mL volume of complete medium. As a positive control, target cells alone are lysed with 0.1% sodium dodecyl sulfate (SDS) to determine the ^{51}Cr release level when 100% of the cells are lysed.

4. Plates are incubated for 4 h at 37°C without CO_2 and then supernatants are harvested.
5. ^{51}Cr release is measured in supernatants on a suitable automatic γ-counter and percent specific lysis calculated. Specific lysis is calculated using the following equation:

$$\text{Percent specific lysis} = \frac{\text{Experimental cpm} - \text{Spontaneous cpm} \times 100\%}{\text{Maximum cpm} - \text{Spontaneous cpm}}$$

where cpm is counts per minute.

4. Notes

1. This chapter deals only with the retroviral transduction of human PBL and assessment of immune function of the resultant transgenic T-cells. The construction of some of the retroviral vector used in our studies have been described elsewhere *(5)* and vector construction will not be discussed here. The structure of the A7 vector we have used extensively is shown in **Fig. 1**. The PG13 retrovirus producer cell line was transfected with the A7 construct and high titer clones were isolated *(5)*. The A7 virus produced by A7 PG13 clone 6 was used to successfully transduce human T-lymphocytes (**5**; Clay et al., unpublished; Nishimura et al., unpublished).
2. The polybrene improves the retroviral infection efficiency by mediating the association between viral particles and the cell membrane.
3. It is very important that the centrifuge be cleaned thoroughly before use and not be one that is used for centrifuging bacteria. Aerosols of bacteria will result in infection of the T-cell cultures.
4. Centrifugation of the plates increases transduction efficiencies *(6)*.

Acknowledgments

The authors thank Dr. Steven A. Rosenberg, Chief, Surgery Branch, National Cancer Institute, NIH, where we performed our initial studies and developed many of the protocols described herein. We also thank Mary Custer, David Yu, and Jessica Sachs for their technical assistance.

References

1. Kawakami, Y., Eliyahu, S., Delgado, C. H., Robbins, P. F., Rivoltini, L., Topalian, S. L., et al. (1994) Cloning of the gene coding for a shared human melanoma antigen recognized by autologous T cells infiltrating into tumor. *Proc. Natl. Acad. Sci. USA* **91,** 3515–3519.
2. Marincola, F. M., Hijazi, Y. M., Fetsch, P., Salgaller, M. L., Rivoltini, L., Cormier, J., et al. (1996) Analysis of expression of the melanoma-associated antigens MART-1 and gp100 in metastatic melanoma cell lines and in situ lesions. *J. Immunother.* **19,** 192–205.
3. Shilyansky, J, Nishimura, M. I., Yannelli, J. R., Kawakami, Y., Jacknin, L. S., Charmley, P., et al. (1994) T-Cell receptor usage by melanoma-specific clonal and highly oligoclonal tumor-infiltrating lymphocyte lines. *Proc. Natl. Acad. Sci. USA* **91,** 2829–2833.
4. Cole, D. J., Weil, D. P., Shilyansky, J., Custer, M., Kawakami, Y., Rosenberg, S. A., et al. (1995) Characterization of the functional specificity of a cloned T-cell receptor heterodimer recognizing the MART-1 melanoma antigen. *Cancer Res.* **55,** 748–752.
5. Clay, T. M., Custer, M. C., Sachs, J., Hwu, P., Rosenberg, S. A., and Nishimura, M. I. (1999) Efficient transfer of tumor antigen-reactive TCR to human peripheral blood lymphocytes confers anti-tumor reactivity. *J. Immunol.* **163,** 507–513.
6. Bunnell, B. A., Muul, L. M., Donahue, R. E., Blaese, R. M., and Morgan, R. A. (1995) High-efficiency retroviral-mediated gene transfer into human and nonhuman primate peripheral blood lymphocytes. *Proc. Natl. Acad. Sci. USA* **92,** 7739–7743.
7. Walter, E. A., Greenberg, P. D., Gilbert, M. J., Finch, R. J., Watanabe, K. S., Thomas, E. D., et al. (1995) Reconstitution of cellular immunity against cytomegalovirus in recipients of allogeneic bone marrow by transfer of T-cell clones from the donor. *N. Engl. J. Med.* **333,** 1038–1044.
8. Mosca, P. J., Hobeika, A. C., Clay, T. M., Morse, M. A., and Lyerly, H. K. (2001) Direct detection of cellular immune responses to cancer vaccines. *Surgery* **129,** 248–254.
9. Morse, M. A., Clay, T. M., Hobeika, A. C., Mosca, P. J., and Lyerly, H. K. (2001) Monitoring cellular immune responses to cancer immunotherapy. *Curr. Opin. Mol. Ther.* **3,** 45–52.
10. Clay, T. M., Hobeika, A. C., Mosca, P. J., Lyerly, H. K., and Morse, M. A. (2001) Assays for monitoring cellular immune responses to active immunotherapy of cancer. *Clinical Cancer Res.* **7,** 1127–1135.

18

Gene Transfer to Articular Chondrocytes with Recombinant Adenovirus

Glyn D. Palmer, Elvire Gouze, Jean-Noel Gouze, Oliver B. Betz, Christopher H. Evans, and Steven C. Ghivizzani

1. Introduction

Articular cartilage forms the load-bearing surfaces of diarthrodial joints. Its dense extracellular matrix, composed primarily of water, collagen, and proteoglycans is maintained by a resident population of chondrocytes. These all combine to form a highly organized structure that provides the tissue with its unique biomechanical properties. However, partly because of its avascular nature, cartilage has poor regenerative properties, and damage from disease or injury will often lead to an inferior repair tissue that ultimately fails. One approach to promoting cartilage repair is through the administration of biological agents that enhance matrix synthesis or decrease matrix degradation. Several cytokines are known to promote matrix synthesis in chondrocytes, including the bone morphogenic proteins (BMPs), transforming growth factor-βs (TGF-βs) and insulinlike growth factors (IGFs). Improved repair has been reported following in vivo administration of recombinant *IGF-1 (1)* or *BMP-2 (2)* to cartilage lesions. However, the maintainence of effective intralesional cytokine concentrations for extended periods following administration is hindered by the relatively short half-lives of these proteins. The transfer of cDNAs encoding these cytokines represents an alternative treatment strategy. By this approach, local expression of the gene to cells within sites of damage could achieve sustained, biologically relevant levels of protein that are synthesized locally *(3)*.

From: *Methods in Molecular Biology, vol. 215: Cytokines and Colony Stimulating Factors: Methods and Protocols*
Edited by: D. Körholz and W. Kiess © Humana Press Inc., Totowa, NJ

Adenoviral vectors are currently the most commonly used viral gene delivery vectors for in vitro and in vivo studies. They can infect both quiescent and dividing cells and can therefore be used for direct in vivo gene delivery. In addition, their method of production is technically straightforward, and high titers can be consistently obtained. Several studies have demonstrated that direct in vivo delivery of recombinant adenoviral vectors to normal and arthritic rabbit joints generates high levels of transgene expression that persist for several days ***(4,5)***. However, because it is difficult to transduce chondrocytes *in situ*, ex vivo gene delivery may be a more suitable method for generating high levels of localized transgene expression within a cartilage lesion. This procedure first requires the in vitro transduction of chondrocytes with the appropriate genes and then implantation of the genetically modified cells into the experimentally induced defect ***(6,7)***. This approach allows for the standardization of the gene transfer method used, so that optimal transduction efficiency and transgene expression may be achieved prior to implantation.

Adenoviral vectors have been used to successfully deliver chondrogenic growth factors such as IGF-1, BMP-2, and TGF-β_1 ***(1,8–10)*** to articular chondrocytes in vitro. Importantly, in these studies, nanogram amounts of cytokine gene product were produced by transduced chondrocytes; this was accompanied by significant increases in proteoglycan and type II collagen synthesis compared to noninfected controls. Moreover, gene transfer maintained matrix synthesis in the presence of interleukin-1 (IL-1), a strong inhibitory agent ***(8)***. Therefore, infection of chondrocytes with recombinant adenoviral vectors encoding specific cytokines serves as a useful tool to evaluate the ability of expressed proteins, either alone or in combination, to promote matrix synthesis in vitro, as a prelude to evaluating their ability to enhance cartilage repair in vivo. This chapter details the infection of primary articular chondrocytes with recombinant adenoviral vectors, using those encoding IGF-1 (Ad.IGF-1) and BMP-2 (Ad.BMP-2) as examples. The process requires large-scale preparation of chondrocytes from articular cartilage and generation of pure, high-titer recombinant adenovirus by amplification in 293 cells. Following infection of chondrocyte cultures, the production of cytokine gene products from genetically modified chondrocytes can be demonstrated by enzyme-linked immunosorbant assay (ELISA). Although representative Ad.IGF-1 and Ad.BMP-2 infections are described in this chapter, this technology can be used to deliver any cDNA to chondrocytes.

2. Materials

1. Cre8 cells (generous gift of Stephen Hardy; *see* **ref. 11**).
2. pAdlox adenoviral shuttle plasmid (Genebank accession number U62024).
3. *Sfi*I restriction endonuclease (New England Biolabs, Beverly, MA).

4. Phenol:chloroform:isoamyl alcohol (25:24:1 vol/vol/vol).
5. Sodium acetate, 3 *M*, pH 5.2.
6. Ethanol, 95%.
7. TE: 10 m*M* Tris-HCl, pH 8.0, 1 m*M* EDTA.
8. DOC lysis buffer: 20% ethanol; 100 m*M* Tris-HCl, pH 9.0; 0.4% sodium deoxycholate.
9. Spermine-HCl, 0.5 *M*.
10. RNase A (Sigma–Aldrich, St. Louis, MO), 10 mg/mL in TE.
11. Sodium dodecyl sulfate (SDS), 10%.
12. EDTA, 0.5 *M*, pH 8.0.
13. Pronase (Sigma–Aldrich, St. Louis, MO), 50 mg/mL in water.
14. Isopropyl alcohol.
15. Purified ψ 5 adenoviral DNA.
16. 2.5 *M* $CaCl_2$.
17. 2X HEPES-buffered saline (HBS): For 100 mL, dissolve 1.6 g NaCl, 0.074 g KCl, 0.027 g $Na_2HPO_4 \cdot 2H_20$, 0.2 g dextrose, and 1 g HEPES in 90 mL double-distilled water (ddH_2O). Adjust pH to 7.05 with 0.5 *N* NaOH. Adjust to final volume with ddH_2O. Filter sterilize.
18. 293 Cells (American Type Culture Collection, Manassas VA, ATCC cat. no. CRL 1573).
19. Cesium chloride solutions:
    a. 1.4 g/mL CsCl = 53.0 g CsCl (Sigma–Aldrich, St Louis, MO) + 87 mL 10 m*M* Tris-HCl, pH 7.9;
    b. 1.2 g/mL CsCl = 26.8 g CsCl + 92 mL of 10 m*M* Tris-HCl, pH 7.9;
    c. Dissolve CsCl into solution and sterile filter using a 0.2-µm bottle top filter (e.g., NalgeNunc Int., Rochester, NY).
20. Thin-walled ultraspeed centrifuge tubes (e.g., 10 mL polyallomer tubes; Sorvall, Newton, CT).
21. Benzonase (Sigma–Aldrich, St. Louis, MO).
22. Dialysis buffer: 10 m*M* Tris-HCl, pH 7.5; 200 m*M* NaCl; 1 m*M* EDTA; 4% (w/v) sucrose. For 2 L, combine 20 mL of 1 *M* Tris-HCl, pH 7.5, 80 mL of 5 *M* NaCl, 4 mL of 0.5 *M* EDTA; 80 g sucrose, and make up to final volume with ddH_2O. Prepare the same day, and chill at 4°C before use.
23. Dialysis tubing, molecular-weight cutoff (MWCO) 50,000 (e.g., Specta/Por, Spectrum Laboratories, Rancho Dominquez CA).
24. Phosphate-buffered saline (PBS): For 1 L, combine 8 g of NaCl, 0.2 g KCl, 1.15 g Na_2HPO_4, and 0.2 g KH_2PO_4 and make up with ddH_2O. Sterilize by autoclaving.
25. Tissue culture reagents (e.g., Gibco-BRL, Gaithersburg, MD; Sigma–Aldrich, St. Louis, MO):
    a. Dulbecco's modified Eagle's medium (DMEM);
    b. Ham's F-12 medium;
    c. Trypsin–EDTA (0.05% trypsin, 0.53 m*M* EDTA);
    d. Gey's balanced salt solution (Gibco-BRL) (GBSS);

e. Fetal bovine serum (FBS);
f. Powdered, cell-culture-tested trypsin, and type B clostridial collagenase.

26. Spectra/mesh nylon filter (Spectrum Laboratories).
27. Tissue culture vessels (e.g., Becton Dickinson, Franklin Lakes, NJ; NalgeNunc Int., Rochester, NY):
    a. 75-cm^2 tissue culture flasks;
    b. 10-cm^2 tissue culture dishes;
    c. 24-well tissue culture plates.
28. IGF-1 ELISA kit and protocol (Diagnostics Laboratory systems, Webster, TX).
29. Recombinant human BMP-2 and murine anti-human BMP-2 antibody kindly supplied by Genetics Institute, Cambridge, MA.

3. Methods

3.1. Generation of Recombinant Adenovirus

First-generation, *E1*, *E3*-deleted, serotype 5 recombinant adenoviral vectors containing *IGF-1* and *BMP-2* were constructed using *Cre-lox* recombination by the system of Hardy et al. ***(11)***. The resulting vectors were designated Ad.IGF-1 and Ad.BMP-2. For this system, the gene of interest is inserted directionally into an adenoviral shuttle plasmid, pAdlox, containing the 3′ inverted terminal repeat of the virus, a viral packaging signal (ψ), a cDNA expression cassette driven by the cytomegalovirus promoter/enhancer, and, finally, a *loxP* Cre recombinase recognition sequence. Recombinant adenovirus is generated by cotransfection of linearized Adlox shuttle plasmid with ψ5 adenoviral genomic DNA, which has its packaging sequence flanked by *loxP* sites. The transfection is performed in a 293 cell line called Cre8, which constitutively expresses high levels of Cre recombinase ***(11)***. These cells generate recombinant adenoviral particles following Cre-mediated recombination between the *loxP* site in the shuttle vector and the 3′ *loxP* site in the ψ5 adenoviral backbone. Propagation of nonrecombined ψ5 virus is selected against via deletion of the packaging signal by the Cre recombinase. Plaques isolated from the cotransfected plates are almost exclusively recombinants. Any contaminating ψ5 can be eliminated by subsequent propagation of the adenovirus in 293 Cre8 cells, or by plaque purification if necessary. It is important to note that stringent safety procedures should be followed while working with recombinant adenovirus (*see* **Note 1**). It is also strongly advised that the investigator be familiar with the appropriate federal, state, and institutional regulations and guidelines concerning microbiological safety.

1. Plate a 10-cm dish of Cre8 cells to 60% confluence in DMEM/10% FBS/1% penicillin–streptomycin.
2. Digest 3.5 μg of pAdlox vector containing the gene of interest with *Sfi*I (*see* **Note 2**). Use 0.5 μg for electrophoresis in a 1% agarose gel to verify digestion.

This will generate two bands, a 2500-bp fragment and a fragment of 1600 bp plus the cDNA insert.

3. Purify the digested plasmid DNA by extraction in phenol:chloroform:isoamyl alcohol. Recover the top, aqueous phase and add 1/10 vol 3 *M* NaOAc. Precipitate the DNA by addition of 3 vol of ethanol. Recover the DNA by centrifugation for 10 min at 12,000*g* and resuspend the DNA pellet in 25 µL TE.
4. To transfect the Cre8 cells, add 25 µL of 2.5 *M* $CaCl_2$, 3 µg of *Sfi*I-digested pAdlox, 3 µg of purified ψ5 adenoviral DNA, and sterile water to 250 µL. Mix thoroughly. Add 250 µL of 2X HBS to a separate tube. Slowly add the DNA/$CaCl_2$ mixture to the HBS, and gently agitate the solution so that a fine precipitate forms.
5. Incubate the DNA/HBS mixture at room temperature for 30 min. At this time, replace the medium on the cell culture with fresh DMEM. Afterward, add the total transfection volume, dropwise, to the 10-cm dish of cells and incubate overnight. Change the media the following day. Discard the media into a large volume of undiluted bleach.
6. After 5–7 d, small plaques should begin to appear in the cell monolayer (*see* **Note 3**).
7. After 7–8 d posttransfection, plate untransfected Cre8 cells into two 10-cm dishes. One plate will be used to prepare a viral stock; the second will be used to verify that the virus generated is indeed a recombinant (*see* **Note 4**).
8. When prominent cytopathic effects are observed throughout the transfected cells (approx 8–10 d), harvest the cells and media from the dishes using a cell scraper (*see* **Note 5**). Transfer the mixture to a 50-mL capped tube.
9. Freeze and thaw the harvested cell/media mixture three times to lyse the cells and release the recombinant adenovirus. Store the lysate at –80°C.
10. To amplify and purify the adenoviral stock, thaw the viral lysate and mix 0.5 mL with 4 mL of fresh DMEM. Remove the medium from one of the new Cre8 cultures (from **step 7**) and replace with the medium–lysate mixture. Return cells to the incubator for 4 h (*see* **Note 6**).
11. After the incubation, supplement the culture with fresh medium to normal culture conditions. Culture the cells until prominent cytopathic effects are observed throughout the monolayer (approx 1–2 d). Harvest the cells and media, as in **step 9**, and store the lysate at –80°C.

3.2. Preparation of Adenoviral DNA

This procedure is performed to either amplify ψ5 adenovirus for DNA isolation or to purify recombinant viral DNA for analysis by restriction digestion ***(11)***.

1. Remove culture media from 10-cm confluent culture of 293 cells (for ψ5) or Cre8 cells (for recombinant). Mix 50 µL of viral lysate with 4 mL of media and add to cells. Incubate for 2–4 h at 37°C.
2. Following incubation, supplement media to normal culture level.

3. When the cells of the monolayer are rounded and begin to detach, harvest the cells and media using a cell scraper.
4. Pellet the cells using a tabletop centrifuge at 2000*g*, at 4°C for 10 min, and resuspend in 400 µL of TE. Discard the media into a large volume of undiluted bleach.
5. Add 400 µL of DOC lysis buffer to the cell pellet. Lyse the cells by drawing the mixture in and out through a 1-mL pipet 10–15 times. Transfer the lysate to a microcentrifuge tube and add 8 µL of 0.5 *M* spermine–HCl. Incubate on ice for 10 min.
6. Centrifuge at 12,000*g* for 5 min to pellet cellular debris and genomic DNA.
7. Remove the supernatant containing the viral particles and transfer to a new microcentrifuge tube. Add 4 µL of 10 mg/mL RNase A and incubate at 37°C for 10 min.
8. To release the viral DNA from the particles, add 60 µL of 10% SDS, 20 µL of 0.5 *M* EDTA, and 40 µL of 50 mg/mL pronase. Incubate for 60 min at 40°C.
9. Extract once with an equal volume of phenol:chloroform:isoamyl alcohol and transfer the aqueous layer (top) into a new microcentrifuge tube.
10. Add 90 µL of 3 *M* sodium acetate; mix and fill the tube with isopropanol. Vortex and centrifuge at 12,000*g* for 10 min.
11. Remove the liquid and wash the pellet with 70% ethanol. Remove all traces of the ethanol with a pipet, and resuspend the DNA pellet in 25 µL TE. The viral DNA is now suitable for restriction digestion analysis or transfection as needed.

3.3. Amplification of Adenoviral Vectors

To generate recombinant virus for large-scale infections, replication-deficient adenovirus is typically amplified in 293 cells (*see* **Note 1**).

1. Grow 293 cells to confluence in DMEM/10% FBS/1% penicillin–streptomycin in six 75-cm^2 tissue culture flasks.
2. Aspirate media and wash cells two times in prewarmed GBSS. Dilute adenoviral stock in a small volume of GBSS and add to cells. Typically, 5×10^7 plaque-forming units (PFUs) of virus is added to 5 mL of GBSS for each 75-cm^2 flask.
3. Incubate at 37°C, in a 5% CO_2 incubator for 3–4 h.
4. Remove the viral solution and add 15 mL of DMEM with serum and antibiotics to the flasks. Incubate as in **step 3** until prominent cytopathic effects are observed in cells (*see* **Note 5**). Typically, the cells are ready to harvest after 2 d.
5. Harvest cells and media from flasks using a cell scraper and transfer into 50-mL capped centrifuge tubes.
6. Centrifuge at 2000*g*, at 4°C for 10 min using a tabletop centrifuge. Discard the media into a large volume of undiluted bleach. Resuspend the cell pellets in 5–10 mL GBSS.
7. Freeze–thaw the cells three times, alternating with a 37°C water bath and dry-ice/ethanol bath.

8. Add Benzonase (50 U/mL) to the freeze–thaw lysate and incubate at 37°C for 30 min.
9. Centrifuge lysate at 2000*g*, 4°C for 10 min, and aspirate supernatant. Incubate supernatant on ice while preparing cesium chloride (CsCl) gradients.
10. Prepare a CsCl step gradient in prechilled polyallomer tubes containing 1/3 volume of the following:
    a. 1.4 g/mL CsCl (bottom layer);
    b. 1.2 g/mL CsCl (middle layer);
    c. Viral cell lysate (top layer) (*see* **Note 7**).
11. Centrifuge the tubes in a swinging bucket rotor at 40,000*g*, 4°C for 1 h.
12. Harvest the viral band from the tube in a minimal volume using a 16-gage needle and 3-mL syringe. Typically, two bands will be seen near the interface of the 1.2- and 1.4 g/mL CsCl layers. The lower band containing the infectious particles is the band that is collected. Dilute the harvested band at least twofold in 10 m*M* Tris-HCl, pH 8.0, for recentrifugation. Mix thoroughly to ensure that the cesium and virus are evenly mixed with the diluent.
13. Repeat **steps 10–12** two more times.
14. Transfer the collected adenovirus fraction to dialysis bags and dialyze against 500 mL of prechilled dialysis buffer for at least 6 h at 4°C. Repeat two more times. Aliquot the recombinant adenovirus into multiple, sterile Eppendorf tubes and store at –80°C. This will avoid multiple freeze–thawing of large viral stocks, which dramatically reduce the infectivity of the recombinant adenovirus.
15. To measure virus particle concentration, mix a 50-µL aliquot with 950 µL GBSS and determine A_{260} by spectrophotometry; one A_{260} is approximately equal to 10^{12} viral particles/mL. The percentage of infectious virions typically ranges between 1% and 10% of the total number of viral particles. The infectious titer may be determined by performing a plaque assay on confluent cultures of 293 cells.

3.4. Isolation of Articular Chondrocytes

Articular cartilage can be obtained from the knee and shoulder joints of skeletally mature New Zealand white rabbits. The steps involved in chondrocyte isolation must be performed in a sterile environment.

1. Immediately following sacrifice, shave cartilage from joint surfaces using aseptic technique and place shavings in Ham's F-12 medium supplemented with 1% penicillin–steptomycin in a 10-cm^2 tissue culture dish. Wash twice in medium to remove excess synovial fluid.
2. Using a scalpel, mince cartilage into small (approx 1 mm) pieces.
3. Aspirate medium and transfer cartilage pieces into a flask containing a sterile-filtered solution of trypsin (2% w/v) in serum-free medium (or a 2% w/v of pronase in 0.15 *M* NaCl). Incubate for 30 min in a humidified atmosphere of 5% CO_2 at 37°C.

4. Aspirate solution after 30 min and rinse cartilage pieces with sterile GBSS to remove residual trypsin.
5. Incubate overnight with a sterile-filtered solution of collagenase (2% w/v) in Ham's F-12 medium supplemented with 10% FBS and 1% penicillin–streptomycin.
6. Sterile filter the digestion mixture through a nylon mesh to separate chondrocytes from the undigested matrix components.
7. Pellet cells by centrifugation at 2000*g* for 10 min in a tabletop centrifuge and resuspend in a suitable volume of Ham's F-12 with serum/antibiotics. Count cells using a hemocytometer, and determine cell viability by trypan blue exclusion.

3.5. Adenoviral Transduction of Chondrocytes

The steps involved in chondrocyte transduction must be performed under sterile conditions.

1. Seed chondrocytes at a density of 2×10^4 cells/well in monolayer cultures on 24-well plates in Ham's F-12 nutrient medium supplemented with 10% FBS and 1% penicillin–streptomycin. Change medium every 3 d until cells become fully confluent (7 –10 d, approx 2×10^5 cells/well).
2. Rinse confluent cultures twice with sterile GBSS to remove traces of serum. Label an Eppendorf tube for each culture to be transduced.
3. Calculate the appropriate amount of adenoviral stock solution (product of **Subheading 3.3.**) to achieve the desired MOI (the mutiplicity of infection defined as the number of infectious viral particles per cell) and add to GBSS to a final volume of 300 µL in the appropriate Eppendorf tube (*see* **Note 8**). Using a 200-µL pipet, carefully mix the solution by pipetting up and down.
4. Add the viral solution to the cultures and incubate at 37°C for 2–4 h.
5. Following transduction, aspirate the viral solution and add 0.5 mL Ham's F-12 medium supplemented with FBS and antibiotics.

3.6. Cytokine Measurements

The concentration of expressed gene products in the media are determined 48 h after transduction (*see* **Note 9**). IGF-1 concentrations were determined using a commercially available ELISA kit and protocol (Diagnostics Laboratory Systems Webster, TX). BMP-2 concentrations were also determined by ELISA, using recombinant human *BMP-2* and the murine anti-human BMP-2 antibody (both supplied by Genetics Institute, Cambridge, MA). Typical synthesis levels of IGF-1 or BMP-2 following transduction of articular chondrocytes with Ad.IGF-1 and Ad.BMP-2 are shown in **Fig. 1**.

4. Notes

1. Wild-type adenovirus has been associated with upper respiratory disease in humans, and recombinant virus is highly infectious for numerous human tissues.

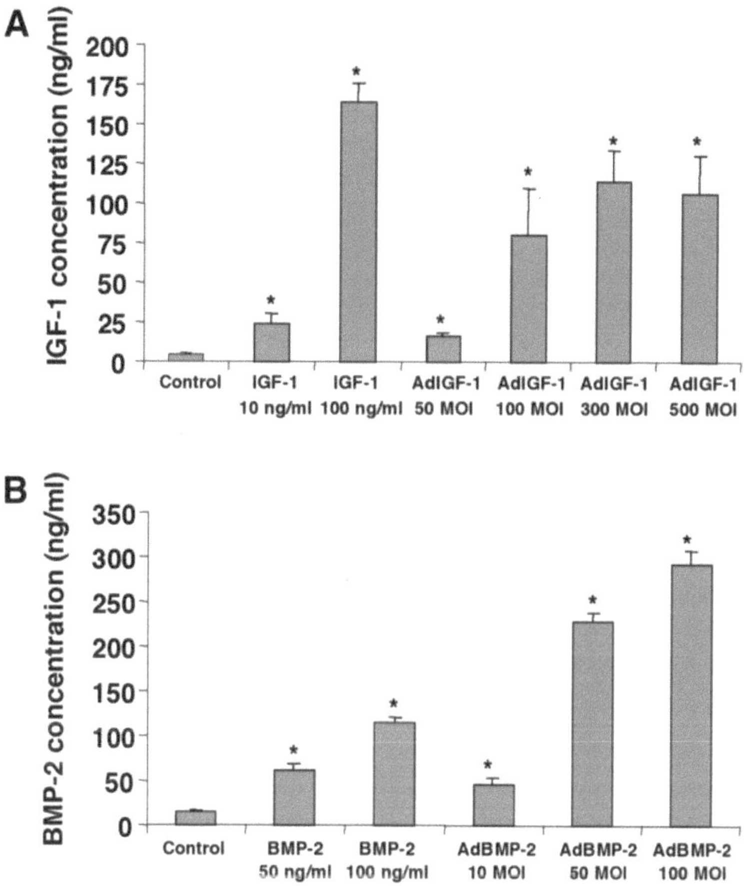

Fig 1. Production of (**A**) IGF-1 or (**B**) BMP-2 by transduced chondrocytes. Chondrocyte cultures were infected with Ad.IGF-1 or Ad.BMP-2 at the indicated MOI or were supplemented with exogenous cytokine at the indicated doses. After 48 h, all conditioned media were assayed for IGF-1 or BMP-2. The controls represent cytokine production in resting cultures of rabbit articular chondrocytes after 48 h. * = $p < 0.05$ vs control. Infection with AdIGF-1 or AdBMP-2 increased respective cytokine production in a dose-dependent manner. Note that IGF-1 production after exogenous IGF-1 addition is much greater than 100%; this may be the result of the ability of IGF-1 to induce its own synthesis in chondrocytes.

First-generation adenoviral vectors are replication defective and are only able to replicate in permissive cells. The 293 cell lines used here contain the 5′ 11% of the wild-type adenoviral genome, which contains the *E1* locus. Because these proteins are provided in trans, they can complement the *E1* deletion in the adenoviral vector. Recombinant adenovirus is a biosafety level 2 hazardous agent. Wherever possible, work should be performed in a class II biological safety cabinet. In addition, proper protective clothing and eyewear should be worn at all times. Solid waste should be rinsed with 10% bleach solution and disposed in biohazard containers. Liquid waste should be decanted or aspirated into a large container of bleach. Likewise, surfaces on which all virus work was performed should be decontaminated with 10% bleach. Extra caution should be used during the handling of sharp objects.

2. *Sfi*I digestion cuts pAdlox at the upstream and downstream boundaries of the adenoviral sequences contained in the plasmid. This fragment contains the 5′ inverted terminal repeat of the adenovirus, the ψ packaging signal, the expression cassette, and *loxP* site for recombination with the 3′ portion of the ψ5 adenoviral backbone. Omit this linearization step if the gene of interest contains an internal *Sfi*I recognition sequence; efficient recombination will still occur with a circular plasmid.
3. Plaques in the replication permissive cells form when a productive recombination event has occurred in a single cell. Within this cell, a functional adenoviral genome has formed, and the cell produces adenoviral particles intracellularly. Because adenovirus is not enveloped, particles are released from the host cell by lysis of the cellular membrane. This releases virus into the media to infect, primarily, cells in the immediate vicinity. These cells will, in turn, undergo viral synthesis and subsequent lysis. This local viral production will generate a small zone of clearing in the monolayer of cells. Cellular debris will accumulate around the growing perimeter of the plaque. When plaques begin to form, it is important not to change the culture medium, because this will remove the virus. At this point of the procedure, the intent is to have the viral infection spread to the majority of cells in the dish for amplification. During this process, if the medium becomes yellow, supplement the existing medium with a small volume of fresh medium.
4. When generating novel recombinants, it is essential to verify that the adenoviral genome contains the cDNA of interest. It is also necessary to determine that the new vector expresses a functional gene product. Depending on the gene product, this can be achieved by infecting cells of interest with different dilutions of the adenovirus and measuring the protein synthesized by ELISA, Western blot, or a functional assay.
5. When the cells become cytopathic, they appear granular and begin to round and detach from the plate. At the time of harvest, small plaques should also be visible with the naked eye. Because adenovirus is a lytic infection, it is important to follow 293-based cells closely, as they begin to become cytopathic. High yields will be obtained if the cells are full of virus at the time of harvest, but have

not yet burst to release viral particles into the liquid medium. Thus, the initial centrifugation of the cells serves to concentrate the virus.

6. This step is necessary to increase the titer of the adenoviral stock and reduce contamination of the ψ5 helper virus. The stock generated from this procedure should be used to infect cultures for subsequent large-scale adenoviral preparations. To reduce the chance of generating replication-competent virus, it is recommended that viral preparations not be continually passaged but propagated from this original stock.
7. Each viral supernatant should be diluted in 10 m*M* Tris-HCl, pH 8.0, and divided among multiple (two to four) CsCl gradients. This ensures that the gradient does not become overloaded with a large amount of virus and cellular debris, which could form aggregates and lead to preparations of low purity. Impurities are often a source of the inflammation noted when using adenoviral preparations in vivo.
8. A minimal volume of viral solution is recommended for efficient transduction. However, the cultures should be monitored during the infection period to ensure that the cells remain completely covered by the viral solution and do not dry out. Infection of chondrocytes is performed over a wide range of MOI to optimize the levels of expressed gene product.
9. Following transduction, chondrocytes will continue to express the transgene product for a week or more. During this time, the cultures can be monitored for various markers of chondrocyte metabolism, including proteoglycan and type II collagen synthesis. However, it should be noted that this procedure cannot be used to assess the effects of long-term transgene expression, as the viral genome remains episomal and will not be maintained upon serial passage.

References

1. Nixon, A. J., Saxer, R. A., and Brower-Toland, B. D. (2001) Exogenous insulin-like growth factor-I stimulates an autoinductive IGF-I autocrine/paracrine response in chondrocytes. *J. Orthop. Res.* **19,** 26–32.
2. Sellers, R. S., Peluso, D., and Morris, E. A. (1997) The effect of recombinant human bone morphogenetic protein-2 (rhBMP-2) on the healing of full-thickness defects of articular cartilage. *J. Bone Joint Surg. Am.* **79,** 1452–1463.
3. Evans, C. H., Ghivizzani, S. C., Smith, P., Shuler, F. D., Mi, Z., and Robbins, P. D. (2000) Using gene therapy to protect and restore cartilage. *Clin. Orthop.* S214–S219.
4. Roessler, B. J., Allen, E. D., Wilson, J. M., Hartman, J. W., and Davidson, B. L. (1993) Adenoviral-mediated gene transfer to rabbit synovium in vivo. *J. Clin. Invest.* **92,** 1085–1092.
5. Nita, I., Ghivizzani, S. C., Galea-Lauri, J., Bandara, G., Georgescu, H. I., Robbins, P. D., et al. (1996) Direct gene delivery to synovium. An evaluation of potential vectors in vitro and in vivo. *Arthritis Rheum.* **39,** 820–828.
6. Kang, R., Marui, T., Ghivizzani, S. C., Nita, I. M., Georgescu, H. I., Suh, J. K., et al. (1997) Ex vivo gene transfer to chondrocytes in full-thickness articular cartilage defects: a feasibility study. *Osteoarthritis Cartilage* **5,** 139–143.

7. Baragi, V. M., Renkiewicz, R. R., Qiu, L., Brammer, D., Riley, J. M., Sigler, R. E., et al. (1997) Transplantation of adenovirally transduced allogeneic chondrocytes into articular cartilage defects in vivo. *Osteoarthritis Cartilage* **5,** 275–282.
8. Smith, P., Shuler, F. D., Georgescu, H. I., Ghivizzani, S. C., Johnstone, B., Niyibizi, C., et al. (2000) Genetic enhancement of matrix synthesis by articular chondrocytes: comparison of different growth factor genes in the presence and absence of interleukin-1. *Arthritis Rheum.* **43,** 1156–1164.
9. Shuler, F. D., Georgescu, H. I., Niyibizi, C., Studer, R. K., Mi, Z., Johnstone, B., et al. (2000) Increased matrix synthesis following adenoviral transfer of a transforming growth factor beta1 gene into articular chondrocytes. *J. Orthop. Res.* **18,** 585–592.
10. Mi, Z., Ghivizzani, S. C., Lechman, E. R., Jaffurs, D., Glorioso, J. C., Evans, C. H., et al. (2000) Adenovirus-mediated gene transfer of insulin-like growth factor 1 stimulates proteoglycan synthesis in rabbit joints. *Arthritis Rheum.* **43,** 2563–2570.
11. Hardy, S., Kitamura, M., Harris-Stansil, T., Dai, Y., and Phipps, M. L. (1997) Construction of adenovirus vectors through Cre-lox recombination. *J. Virol.* **71,** 1842–1849.

19

Gene Therapy of X-Linked Severe Combined Immunodeficiency

Salima Hacein-Bey-Abina, Genevieve de Saint Basile, and Marina Cavazzana-Calvo

1. Introduction

This review describes the main steps which led to carrying out the gene therapy clinical trial for X-linked severe combined immune deficiency (SCID-X1) patients, from the preparation of the retroviral vector up to the design of GMP conditions to transduce ex vivo the CD34+ cells of patients. This gene therapy protocol is currently being applied and the encouraging preliminary results published *(1)*. The success of this protocol could be mainly attributed to the physiopathology of the SCID-X1 disease (for review, *see* **ref.** ***2***), which is a good model for a gene therapy application. However, progress in the technology of gene transfer into hematopoietic progenitor cells (HPCs) may also have contributed to this success. The most important improvement concerns the use of fibronectin, which increases the transduction efficiency by colocalization of the retroviral particles and the HPC. The other improvement is related to the use of the early-acting cytokines such as stem cell factor (SCF), Flt3 ligand, and megakaryocyte growth and differentiation factor (M-GDF), which promote immature $CD34^{+}CD38^{low}$ cycle induction, making them permissive for transgene integration with minimal loss of their lymphoid potential *(3)*.

1.1. Preparation of Defective Retroviral Vector Containing the γc Gene

Our standard packaging cell line for virus production is ψ CRIP and MFG/B2, the defective retrovirus vector, but different packaging cells or other

From: *Methods in Molecular Biology, vol. 215:*
Cytokines and Colony Stimulating Factors: Methods and Protocols
Edited by: D. Körholz and W. Kiess © Humana Press Inc., Totowa, NJ

retroviral vector constructs can be used. The MFG vector is derived from the Moloney murine leukemia virus (MLV). The *gag*, *pol*, and *env* genes, which encode the virus proteins, were removed and sequentially introduced into NIH 3T3 murine fibroblasts to create the ψ CRIP packaging cell line *(4)*. The cis retroviral sequences necessary to the production and maturation of the recombinant viral genome transcripts, their reverse transcription, and their integration into the genome of the infected cell were retained. The sequence of the recombinant provirus and the flanking fragments of the cellular DNA are carried by the bacterial plasmid pBR322. The MFG-B2 vector contains the B2 mutation (G-A) in position 1005 *(5)*. Sequence expression is controlled by the viral long terminal repeat (LTR).

1.2. Obtaining of the Recombinant Virus Producer Cell Clone

- A cDNA encoding the γ-chain common (γc) to the interleukin (IL)-2, IL-4, IL-7, IL-9, IL-15 and IL-21 receptor is prepared from total RNA isolated from the lymphocytes of a healthy subject. The prepared fragment contains the entire coding sequence, from the initial methionine to the stop codon, and sequencing analysis confirms that it is identical to the published sequence.
- The 1.13-kb fragment (*AFL*III-*Bam*HI) of the γc cDNA is inserted by ligation at the unique *Nco*I site of the vector, which should coincide with the first methionine of the γc cDNA and with the unique *Bam*HI vector site. The MFG/B2-γc vector constructed in this way is completely sequenced.
- The ψ CRIP packaging cell line is transfected by a coprecipitation in calcium phosphate (2 *M*) of the retroviral construction and the pSV2-neo plasmid at a 10 to 1 molar ratio. Cells that are integrated and stable and those expressing exogenous DNA are selected by exposure to 1 mg/mL G418. Individual clones of G418-resistant cells are tested for their capacity to produce retroviral vectors (*see* **Subheading 2.1.**).

2. Materials

2.1. Virus Titration of Producer Cell Clones and GMP-Produced Supernatants

This chapter deals with the different procedures used to determine the best producer cell clone (the clone producing the highest virus titer supernatant) and to test the titer of each subsequent GMP-produced supernatant from the selected clone.

2.1.1. Transduction Protocol of the NIH-3T3 Cell Line

1. 10-cm-Diameter culture Petri dishes.
2. Polybrene (Sigma Chemical, CO).

3. Dulbecco's modified Eagle's medium (DMEM) growth medium.
4. Fetal calf serum (FCS) (Stem Cell Technologies).
5. Trypsine–EDTA 1X (Gibco-Life Technologies).
6. Phosphate-buffered saline (PBS) 1X.

2.1.2. Southern Blot Analysis

1. Ethanol 100%.
2. Agarose.
3. Electrophoresis buffer, TBE 1X (89 m*M* Tris-borate, 2 m*M* EDTA, pH 8.0).
4. Whatman 3MM paper.
5. Nylon membrane for capillary transfer.
6. NaOH, 0.5 *N*, for DNA denaturation.
7. Prehybridization, hybridization, and posthybridization solutions are as described classically (Molecular Cloning, Laboratory Manual ***[6]***).

2.1.3. Transgene Expression Study

1. Anti-γc chain specific antibody phycoerythrin labeled (Tugh 4, rat anti-human IgG2, Pharmingen, San Diego, CA).
2. PBS 1X.
3. FCS.
4. FACScalibur (Becton–Dickinson Immunocytometry systems).

2.1.4. Titration by TAQMAN Quantitative Polymerase Chain Reaction (PCR) Analysis

2.1.4.1. Transduction of the HCT-116 Human Cell Line

1. HCT-116 human cell line (provided by Genethon III, Evry, France).
2. Six-well flat-bottom tissue culture plates.
3. DMEM growth medium.
5. FCS.
6. Trypsine–EDTA 1X (Gibco-Life Technologies).
7. X-vivo 10 (Biowhitaker).
8. Polybrene.
9. PBS 1X.

2.1.4.2. PCR Analysis

1. DNA extraction kit (Promega, GenomicWizard).
2. Two sets of primers and probe (target and reference):

   One of them positioned in the U5-R region of the MLV LTR:

   Forward primer: ATT GAC TGA GTC GCC CGG

   Reverse primer: AGC GAG ACC ACA AGT CGG AT

   Taqman probe: 5′FAM-labeled TAC CCG TGT ATC CAA TAA ACC CTC TTG CAG TT

The second is specific to one part of the human albumin sequence (used to normalize the amount of DNA):

Forward primer: GCT GTC ATC TCT TGT GGG CTG T
Reverse primer: ACT CAT GGG AGC TGC TGG TTC
Taqman probe: 5′VIC labeled CCT GTC ATG CCC ACA CAA ATC TCT CC

3. Plasmid PMOI Alb (provided by Genethon III, Evry, France).
4. Universal master mix 1X (Perkin Helmer).
5. Multiwell 96 plates for Taqman analysis (Perkin-Elmer).
6. ABIPRISM 7700 sequence detector (Applied Biosystem).

2.2. Clinical Protocol: Selection and Transduction of Patients' CD34+ Cells

2.2.1. Selection of CD34+ Cells from Patients' Bone Marrow

1. PBS–EDTA working buffer (Miltenyi Biotech).
2. Human serum albumin 20% (HSA).
3. Human serum albumin 4%.
4. Anti-CD34 antibodies (Miltenyi Biotech).
5. X-vivo-10 (Biowhitaker).
6. Fetal calf serum (Stem Cell Technologies).
7. CliniMACS tubing set (Miltenyi Biotech).
8. 150/300-mL transfer bags (MacoPharma).
9. 500-mL Triple-layer bags (MacoPharma).
10. Luer connector (Braun).

2.2.2. Preactivation Step

1. X-vivo-10 (Biowhitaker).
2. Fetal Calf Serum (Stem Cell Technologies).
3. SCF (Amgen).
4. Flt3-Ligand (Immunex).
5. IL-3 (Novartis).
6. M-GDF(Amgen).
7. Sterile containers of 85, 180, 390, and 600 cm2 (Nexell).

2.2.3. Coating of Culture Sterile Containers with the Human Recombinant Fragment of Fibronectin (CH296)

1. Tube containing 4 mL of concentrated CH-296 human fibronectin (2.5 mg per vial; Takara-Shuzo, Japan).
2. PBS 1X (Gibco-Life Technologies).
3. Human serum albumin 4 % (HSA).
4. Culture bags of 85, 180, 390, or 600 cm^2 (Nexell), depending on the cell number.

2.2.4. Transduction Step

1. The appropriate number of cryopreserved bags of retrovirus containing supernatant (MFG-96B2) (150 mL/bag).
2. The same cytokines as used in the preactivation step (i.e., SCF, FLT3-L, M-GDF and IL-3) (they are used at the same concentration established for the preactivation step)
3. Protamine sulfate (CHOAY).
4. The culture bags precoated with fibronectin CH-296 as described above.
5. Braun connections.
6. Stock bottles.

2.2.5. Preparation of the Transduced Cells for Administration to the Patient

1. Human serum albumin 4%.
2. 70-mL "Transfer" bag (Cryocyte) or a 150/300-mL (MacoPharma) bag, depending on the number of cells.
3. Braun connection.

3. Methods

3.1. In Vitro Analysis: Virus Titration

3.1.1. Transduction Protocol of the NIH-3T3 Cell Line

One of our standard cell lines for retroviral vector titration is the murine cell line NIH-3T3.

1. Day 1: Culture 5×10^5 NIH-3T3 cells per 10-cm-diameter culture Petri dish for 24 h.
2. Day 2: Incubate the NIH-3T3 cells for 2 h with 4 mL of virus supernatant per Petri dish, in the presence of 8 µg/mL of polybrene. Then, add 6 mL of DMEM containing 10% FCS.
3. Day 3: Replace the medium with 10 mL of DMEM containing 10% FCS.
4. Day 4: Discard the culture medium and remove the adherent cells by applying 500 µL of trypsine–EDTA in each Petri dish. Incubate for 2 min at 37°C, then remove the cells using 10 mL culture medium. Centrifuge for 5 min at 1500 rpm. Wash the pellet with PBS 1X and count the cells. An aliquot is removed for immunofluorescence analysis (*see* below). DNA extraction is performed with the rest of the cells for Southern blot analysis.

3.1.2. Southern Blot Analysis

1. In a final volume of 10 µL, digest the following cells with an enzyme able to liberate the ligated insert (γc) from the retroviral vector:
   10 µg of DNA from each producer cell clone to be tested;
   10 µg of DNA from noninfected NIH-3T3 fibroblasts;

20 pg of γc vector plasmid (corresponding to 1 vector copy);
2 pg of γc vector plasmid (corresponding to 0.1 vector copy).

2. Digested DNAs are precipitated with ethanol, loaded on a 0.8% agarose gel, and electrophoresed overnight at 55 V.
3. The gel is then transferred onto a nylon membrane and hybridized to a P^{32}-labeled γc cDNA probe. Prehybridization, hybridization, and posthybridization washes were performed according to standard procedures ***(6)***.
4. Filters were exposed to X-ray film, in the presence of an intensifying screen at –80°C, 24–48 h. Quantification of the band intensity of the producer cell clones is compared to the intensity of copies 1 and 0.1 of the vector using a PhosphorImager.

   A (MFG-96B2 γc) clone capable of transferring approx 0.5 unrearranged copies of the MFG-B2γc provirus into the genome of the NIH-3T3 murine fibroblasts under these conditions was selected, amplified, and then frozen at 6×10^6 cells/mL in liquid nitrogen to be stored.

3.1.3. Transgene Expression Study (Determination of Transduction Efficiency)

1. Distribute 5×10^5 infected NIH-3T3 diluted into PBS 1X + T2% FCS into appropriate FACS tubes.
2. Add 10 μL of anti-γc PE-specific antibody, incubate 30 min at 4°C, and then wash with PBS
3. Resuspend the cell pellet in 500 μL of PBS 1X.

The percentages of γc–positive cells are assessed by FACS analysis. In addition to the frequency of transduced cells that reflect the transduction efficiency, the titer could also be calculated as follows (when different dilutions of the supernatant were used in the transduction protocol):

$$\text{Infectious particle per mL} = \text{\% of transduced cells} \times \frac{\text{Number of transduced cells}}{\text{Volume of supernatant (mL)}}$$

The Titration curve (% of transduced cells) = f (volume of supernatant) is plotted, and the titer is calculated in the linear part of the curve.

3.1.4. Titration by TAQMAN Quantitative PCR Analysis

The principle is to transduce a human cell line with serial dilutions of MFG-γc-clone-derived supernatant and to analyze the transgene integration by real-time quantitative PCR.

3.1.4.1. Transduction Protocol of the HCT-116 Human Cell Line

Our standard cell line is HCT-116 cells, but other human cell lines permissive for amphotropic Moloney derived vector can be used.

1. Day 1: Split 1 × 10^5 HCT-116 cells/well on a six-well plate in DMEM +10% FCS corresponding to the number of points tested in duplicate. Do not forget two wells for cell numeration on the day of transduction. The cells are incubated 24 h at 37°C, 5% CO_2.
2. Day 2: Control the adherence of the cells and their state of mitosis (the cells are infected when they reach 50–60% confluence). Add 1 mL of trypsin–EDTA to one or two wells reserved for the cell count. Thaw the viral supernatant at 37°C and dilute it in X-vivo + 4% FCS to obtain a volume of pure supernatant in the wells as follows: 1000, 800, 600, 400, 200, 150, 100, 75, 50, 25, 10 μL, and 0. Put 1 mL of each supernatant dilution on the cells in the appropriate well together with 8 μg/mL of polybrene. The cells are incubated at 37°C in 5% CO_2 for 2 h, at which point 1 mL of DMEM +10% FCS is added and the incubation continued at 37°C for 24 h.
3. Day 3: The medium in each well is removed and replaced by fresh medium.
4. Day 4: The infected cells are harvested by trypsin–EDTA exposure, washed with PBS 1X, pelleted, and stored at –20°C until DNA extraction.

3.1.4.2. PCR Analysis

Real-time quantitative PCR is performed to determine infectious titer of MFG-γc-derived supernatant by determining the number of integrated vectors in a population of infected cells.

1. Genomic DNA from infected cells (HCT-116) is extracted using the Promega kit and then used for the quantitative PCR template.
2. Prepare serial dilutions of the plasmid PMOI Alb ranging from 2 × 10^6 to 2 × 10^1 copies. The plasmid contains two copies of the LTR and one copy of the human albumin sequence and is used as the reference for viral copy number quantification.
3. About 100 ng /10 μL of DNA from infected cells or from the plasmid sample, diluted in H_2O were added to 15 μL of duplex mix containing 1X of the universal master mix; 0.2 μ*M* of each primer and 0.1 μ*M* of each probe. All samples are run in duplicate.

The amplification conditions are 50°C for 2 min, 95°C for 10 min, and 50 cycles at 95°C, 15 s and 60°C, 1 min. Amplification, data acquisition, and analysis are performed on the ABIPRISM 7700.

A standard curve [Ct = *f*(log DNA concentration)] is plotted for each PCR and enables quantitation of the number of copies. The sample DNA duplicate is extrapolated on the standard curve for quantitation. Then, for each sample in duplicate, the ratio (number of LTR copies/number of Alb copies) is calculated and the infectious titer determined as follows:

Infectious particle per mL = (Number of LTR copies/number of Alb copies) × (Number of transduced cells/volume of supernatant (mL).

The titration curve (LTR/Alb) = *f*(volume of supernatant) is then plotted, and the titer is calculated in the linear part of the curve.

3.2. Selection and Transduction of Patients' CD34+ Cells

3.2.1. CD34+ Cell Selection from Patients' Bone Marrow

The mononuclear cells (MNCs) were first isolated from the total bone marrow (BM) by Ficoll gradient density separation. Hematopoietic progenitor cells bearing the CD34 antigen are then selected by an immunomagnetic procedure using the CliniMACS system from Miltenyi Biotech.

3.2.1.1. Preparation of the Bags

1. Prepare two bags of buffer PBS–EDTA with 0.4% HSA (20-mL sample of the 20% albumin is injected in each bag of 1 L PBS–EDTA).
2. Prepare two triple-layer bags:
   - Detach a bag after having made three seals and clamp the tube from each bag and the central tubing.
   - Bag 1: Use the sterile connection equipment to transfer 500 mL of PBS–EDTA–0.4% HSA into one of the two bags. Record the volume and the contents on the bag. Write "MNC+Abs" on the other bag and stick on a bar-code label. Seal the central tubing.
   - Bag 2: Use the sterile connection equipment to transfer 300 mL of PBS–EDTA–0.4% HSA into one of the two bags. Record the volume and the contents on the bag. Write "Abs wash supernatant" on the other bag and stick on a bar-code label. Seal the central tubing.

3.2.1.2. Incubation with Anti-CD34 Antibodies

1. Take a sample of the anti-CD34 antibodies (7.5 mL) and transfer this sample into the cell bag via the coupler.
2. Incubate the cells with the antibodies for 30 min at room temperature (22°C) on the orbital shaker.

3.2.1.3. Washing the Anti-CD34 Antibodies After Incubation

1. Once the 30-min incubation is complete, establish a sterile connection with **bag 1**.
2. Transfer the labeled MNC to the bag labeled "MNC + Abs."
3. Expel the air from the bag and rinse the bag used for the incubation three times.
4. Centrifuge at 200*g* for 10 min at 20°C.
5. Connect **bag 2** aseptically to the centrifuged cell bag without removing the bag from the bucket.
6. Place the cell bag on the plasma extractor then slowly release the press. Unclamp the tubing from **bag 2** labeled "Ab wash supernatant" and unclamp the central tubing.

7. Decant as much of the supernatant as possible and then clamp the tubing. Adjust the volume to 300 mL (largest volume that can be deposited on the column) with PBS–EDTA–0.4% HSA solution on the precision balance. Clamp the tubing and seal. Place the bag under the hood.

3.2.1.4. Selection of the CD34+ Cells

1. While the centrifuge is running, prepare the tubing set under the hood and install it on the separator to prime the column by following the instructions displayed on the control screen.
2. Once the priming is complete, connect a PALL filter to the bag containing the cells under the laminar-flow hood.
3. Connect the bag with the filter to the tubing set and follow the instructions on the control screen; then, start the automated selection of labeled cells.
4. At the end of the procedure, clamp the tubing above the collection bag and then seal and place the bag under the hood.
5. Transfer the positive fraction into a 50-mL tube.
6. Rinse the bag with PBS–EDTA–0.4% HSA using a 50-mL syringe fitted with a pump needle.
7. Transfer the rinsing solution into another 50-mL tube.
8. Centrifuge both tubes at 200*g* for 10 min at 20°C.
9. Decant the supernatant into two 50-mL tubes.
10. Resuspend the cells in X-vivo 10 culture medium containing 4% of FCS (the volume depends on the number of cells selected).
11. Remove an aliquot for quality controls:
    - cell count and viability by trypan blue exclusion method;
    - immunofluorescence (CD34, CD3, CD14, and CD19 positive cells);
    - bacteriological control: take a 1-mL sample of the negative fraction (*see* **Notes 1** and **2**).

3.2.2. Preactivation Step

Early-acting cytokines were used for cell cycle induction and retroviral gene transfer. We submitted selected hematopoietic progenitor cells to the action of SCF (300 ng/mL) , Flt3 ligand (300 ng/mL), M-GDF (100 ng/mL), and IL-3 (60 ng/mL) during 24 h before the transduction step.

1. An appropriate culture bag (85, 180, 360, or 600 cm^2) will be used depending on the number of CD34+ selected cells. Total volumes of the medium (X-vivo 10 containing 4% FCS) and each cytokine (the CD34+ cell concentration should be 0.5×10^6 cells /mL) are calculated.
2. Prepare the mixture and introduce the cell suspension into the bag(s) via a Braun connection and a syringe body. Put the bags in the 5% CO_2 incubator at 37°C for 24 h.

3.2.3. Procedure for the Coating of Culture Sterile Containers with the Human Recombinant Fragment of Fibronectin (CH296)

The extracellular matrix protein, fibronectin, has been reported to be an effective support layer for gene transfer into human hematopoietic progenitor cells because it can replace stromal cells *(7,8)*. The gene transfer efficiency is significantly increased when an adhesion protocol on the human recombinant fibronectin fragment CH296 is used.

Procedure (for one culture bag of 85 cm^2):

1. Remove from each bag the yellow covering protecting the injection site and attach a Braun connection.
2. Take off the lid and insert a 20-mL syringe body.
3. Put into the syringe 11 mL PBS 1X plus 1 mL of concentrated fibronectin solution (final concentration 50 μg/mL).
4. Replace the lid on the Braun connection and incubate for 2 h at room temperature under the culture hood.
5. At the end of the incubation period, aspirate the whole of the fibronectin solution, and using a new syringe, distribute into the bag 20 mL of HSA 4% for nonspecific sites saturation.
6. Incubate for 30 min at room temperature.
7. Aspirate the saturation solution and wash the bag twice with 24 mL of PBS IX.
8. Leave the bag with the final PBS 1X rinse solution under the hood until use or at 4°C if it has to be kept for several days before being used.

3.2.4. Transduction Step

1. At the end of the preactivation step the CD34+ cells are harvested and pelleted by spinning 10 min at 350*g*.
2. After the centrifugation, the supernatant is decanted and the cells resuspended in a small volume of X-vivo 10.
3. The cells are counted following the trypan blue exclusion method. (A sample is removed for immunofluorescence control [*see* **Note 2**].)
4. Calculate the number of CH296-coated bags necessary for seeding the cells, knowing that the capacity of the bags are as follows:
   (5–8) × 10^6 cells in the 85-cm^2 bags (volume: 25–30 mL);
   (10–16) × 10^6 cells in the 180-cm^2 bags (volume: 53–63 mL);
   (27–33) × 10^6 cells in the 360-cm^2 bags (volume: 115–140 mL);
   (42–50) × 10^6 cells in the 600-cm^2 bags (volume: 170–210 mL).
5. Prepare the concentrated cellular suspension in a stock bottle (250 or 500 mL), adding the desired quantities of cytokines and protamine sulfate (used at 4 μg/mL).
6. Thaw the bag(s) of MFG-96B2 supernatant in a 37°C water bath. After thawing, remove the protective plastic overbag(s) and place it/them under the hood.

7. Homogenize the content of the bag and introduce a Braun connector into one of the two openings.
8. Pour out the desired volume of supernatant in the stock bottle.
9. Fill the fibronectin coated bags as follows:
   - Clamp the tube of the bag, remove the lid, and connect a 50-mL syringe body.
   - Distribute the predefined volume of prepared cellular suspension.
   - Replace the lid.
   - Repeat the operation according to the number of bags there are to seed.
   - Place the bag in the incubator at 37°C with 5% CO2 for 24 h.

   The transduction step is repeated three times at 24-h intervals.

3.2.5. Preparation of the Cells for Administration to Patient

1. Remove the culture bags from the incubator and place them under the culture hood.
2. Put the cells back in suspension by gently shaking the bag in a rotating movement.
3. Aspirate the cellular suspension using a 50-mL syringe.
4. Empty the content into Falcon 50-mL tube(s) and wash the bag once in an adjusted volume of 0.9% NaCl, corresponding to the capacity of the culture bag. Rinse the surroundings of the bags carefully to detach the cells that have adhered to the fibronectin.
5. Centrifuge the cells for 10 min at 350*g* at room temperature.
6. The supernatant is carefully decanted in all the tubes (avoid discarding it until you have counted the cells). The pellets are pooled in two or four tubes, depending on the cell density to facilitate cell washing.
7. Wash the cells once by filling the tubes with 0.9% NaCl solution and twice with albumin solution (HSA 4%) in order to eliminate as much as possible the free viral particles and reagents.
8. At the end of the last centrifugation, resuspend the two or four pellets with a small volume of HSA 4%, Pool them and take out a small aliquot of the resuspended cells for cell counting (following the trypan blue exclusion method) and for quality control testing (*see* **Notes 1–3**).
9. Transfer the cells into a 70-mL "transfer" bag (Cryocyte) or a 150/300-mL bag depending on the number of cells. Adjust the final volume to 50 mL or 100 mL depending on the weight of the recipient. (The cellular suspension must be at a minimum concentration of 10×10^6 cells/mL/kg.)
10. Label the bag and send to the clinical department.

4. Notes

1. Every day, during the clinical protocol, an aliquot of the removed culture supernatant is submitted to a bacteriological control.
2. In addition to the cell count, a phenotype analysis by direct immunofluorescence is performed to evaluate transduction rate and CD34 maintenance on a daily basis

until the day of injection. This analysis specifically identifies the total number of CD34+ cells and double positive CD34+ γc + cells contained in the graft. The preclinical studies allow us to predict that approx 30–40% of the CD34+ cells will express the transduced γc-chain at the end of the three transduction cycles *(3,9)*.

3. The natural killer (NK) cell differentiation potential of transduced and untransduced CD34+, assessed for the final product on the last day, is studied in vitro in the continuous presence of SCF and IL-15. After 3 wk, transduced cultured cells should be differentiated into NK cells (CD56+) able to kill the K562 target cells ***(10)***.

References

1. Cavazzana-Calvo, M., Hacein-Bey, S., de Saint Basile, G., Gross, F., Yvon, E., Nusbaum, P., et al. (2000) Gene therapy of human severe combined immunodeficiency (SCID)-X1 disease. *Science* **288,** 669–672.
2. Fischer, A. (2000) Severe combined immunodeficiencies (SCID). *Clin. Exp. Immunol.* **122,** 143–149.
3. Hacein-Bey, S., Gross, F., Nusbaum, P., Hue, C., Hamel, Y., Fischer, A., et al. (2001) Optimization of retroviral gene transfer protocol to maintain the lymphoid potential of progenitor cells. *Hum. Gene Ther.* **12,** 291–301.
4. Danos, O. and Mulligan, R. C. (1988) Safe and efficient generation of recombinant retroviruses with amphotropic and ecotropic host ranges. *Proc. Natl. Acad. Sci. USA* **85,** 6460–6464.
5. Riviere, I., Brose, K., and Mulligan, R. C. (1995) Effects of retroviral vector design on expression of human adenosine deaminase in murine bone marrow transplant recipients engrafted with genetically modified cells. *Proc. Natl. Acad. Sci. USA* **92,** 6733–6737.
6. Sambrook, J., Fritsch, E. F., and Maniatis, T. (1989) *Molecular Cloning, A Laboratory Manual*, 2nd ed., Cold Spring Harbor Laboratory Press, Cold Spring Harbor, NY.
7. Moritz, T., Dutt, P., Xiao, X., Carstanjen, D., Vik, T., Hanenberg, H., et al. (1996) Fibronectin improves transduction of reconstituting hematopoietic stem cells by retroviral vectors: evidence of direct viral binding to chymotryptic carboxy-terminal fragments. *Blood* **88,** 855–862.
8. Hanenberg, H., Xiao, X. L., Dilloo, D., Hashino. K., Kato, I., and Williams, D. A. (1996) Colocalization of retrovirus and target cells on specific fibronectin fragments increases genetic transduction of mammalian cells. *Nat. Med.* **2,** 876–882.
9. Hacein-Bey, S., Basile, G. D., Lemerle, J., Fischer, A., and Cavazzana-Calvo, M. (1998) Gamma c gene transfer in the presence of stem cell factor, FLT-3L, interleukin-7 (IL-7), IL-1, and IL-15 cytokines restores T-cell differentiation from gammac(-) X-linked severe combined immunodeficiency hematopoietic progenitor cells in murine fetal thymic organ cultures. *Blood* **92,** 4090–4097.

10. Cavazzana-Calvo, M., Hacein-Bey, S., de Saint Basile, G., De Coene, C., Selz, F., Le Deist, F., et al. (1996) Role of interleukin-2 (IL-2), IL-7, and IL-15 in natural killer cell differentiation from cord blood hematopoietic progenitor cells and from gamma c transduced severe combined immunodeficiency X1 bone marrow cells. *Blood* **88,** 3901–3909.

20

Gene Transfer for Generation of Tumor and Leukemia Vaccines

Dagmar Dilloo and Andree Zibert

1. Introduction

Tumor vaccine strategies aim to compensate for the reduced immunogenicity of tumor cells *(1–4)*. Neoplastic cells fail to induce an effective antineoplastic immune response for a number of reasons, such as defective antigen-processing mechanisms, decreased levels of MHC expression, or lack of costimulatory molecules. In tumor vaccine generation, these deficits may be overcome either by introducing genes encoding for immunomodulatory molecules into the tumor vaccine cells or by employing professional antigen-presenting cells such as dendritic cells (DCs) that have been primed with tumor antigens. For preparation of tumor vaccine cells, choices have to be made with regard to autologous versus allogeneic target cells, the immunomodulatory genes used for transduction, and the appropriate gene transfer system used for the introduction of immunomodulatory molecules.

In the tumor vaccine setting, the efficacy of different cytokines *(5–7)*, chemokines *(8,9)*, as well as costimulatory surface molecules *(10–12)* have been demonstrated. There is currently no indication that any single cytokine or costimulatory molecule may be optimal for all different tumor entities targeted by the tumor vaccine approach. Choices on a particular cytokine/chemokine combination or costimulatory surface molecule are based on preclinical studies, primarily murine tumor vaccine models. Yet, the observed efficacy of the tumor vaccine may not always transfer to the clinical setting. The choice of autologous versus allogeneic tumor vaccine cells also has major implications on the gene transfer system employed. Autologous tumor vaccines are prepared from primary tumors or malignant hematopoietic cells obtained from individual

From: *Methods in Molecular Biology, vol. 215: Cytokines and Colony Stimulating Factors: Methods and Protocols*
Edited by: D. Körholz and W. Kiess © Humana Press Inc., Totowa, NJ

patients. Although such autologous vaccines have the definite advantage of presenting an antigen profile that is identical to the original neoplastic cell, isolation and expansion of the primary malignant cells are often problematic. Thus, for genetic modification of autologous tumor cells, gene transfer systems that allow for highly efficient target cell transduction even in the absence of cell proliferation are warranted. Here, adenovirus and herpesvirus vectors have proven particularly useful providing transient transgene expression in primary human neoplastic cells ***(13)***. In contrast, allogeneic tumor vaccines are often prepared from stably transduced, partially HLA-matched tumor cell lines ***(14)***. For stable transduction, retroviral vectors are most commonly used because of their ability to integrate into the host cell genome. Retroviral vectors have found widespread application for gene transfer into hematopoietic cells ***(15)***. However, as retroviral vectors require target cells to be in cycle, they are of limited value for transduction of primary human neoplastic cells, acute leukemia cells in particular. In contrast, lentiviral vectors transduce nondividing cells and have been shown to exhibit moderate gene transfer efficiency in leukemic target cells ***(16)***.

Ex vivo cultured DCs may be derived either from peripheral blood monocytes or from CD34-positive hematopoietic progenitor cells, and retroviruses have been successfully used for their genetic modification ***(17)***. Also, adenoviral vectors have been shown to efficiently transduce immature and mature dendritic cells ***(18,19)***. For use in tumor vaccine strategies, dendritic cells have to be provided with the tumor-associated antigens (TAA) for stimulation of an effective antineoplastic immune response ***(20)***. If a TAA has already been identified, the encoding cDNA may be transfered into DCs utilizing viral or nonviral vector systems. Alternatively, DCs can be pulsed with protein or peptides derived from TAA. In the absence of defined TAA, however, tumor cell lysates, apoptotic bodies, peptides, or total tumor mRNA may be loaded onto DCs for tumor-antigen presentation ***(17)***.

Thus, depending on the tumor vaccine strategy pursued and the target cells in question, one may opt for different vector systems to achieve optimal expression of cytokines or other costimulatory molecules. An important consideration when aiming for bulk transduction of high numbers of vaccine cells is the fact that viral vectors differ in their ability to grow to high titers, in their transduction efficiency of primary cells of different origin, and in their capability of transducing nondividing vs dividing primary cells. There are also differences in viral replication intermediates that governs either stable or transient transgene expression in target cells. The protocols detailed below focus on gene transfer with vector systems, based on adenovirus and herpes simplex (HSV) viruses for tumor and leukemia vaccine generation, respectively. As retroviral vectors have been successfully employed for transduction of dendritic cells, a protocol for retroviral gene transfer into hematopoietic cells is provided below emphasizing the importance of colocalization of virus and target cells for retroviral transduction ***(21,22)***.

2. Materials

1. All transductions of target cells are performed in biological safety cabinets certified for use of respective viral vectors, providing protection for the product as well as the personal.
2. Outside the biological safety cabinets viral vectors and genetically modified target cells are handled in closed systems.
3. For the generation of single cells from tumor material, sterile scalpels and tube-shaped glass mortar and pestle and cell strainers (70 μm nylon, Falcon, Becton Dickinson Labware, Le Pont De Claix, France) are required.
4. For digestion of tumor material, trypsin–EDTA (Gibco-BRL, Paisly, Scotland) or Accutase (PAA Laboratories GmBH, Linz, Austria) is employed.
5. For cell culture, general plastic laboratoryware and 15-mL as well as 50-mL conical plastic tubes with lids are needed. Cell culture and most gene transfer experiments are carried out in RPMI supplemented with heat-inactivated 10% (v/v) fetal bovine serum (FCS), penicillin–streptomycin (100 μg/mL), and glutamine (2 m*M*) if not otherwise specified.
6. Neuroblastoma cells are cultured in EHS matrix six-well plates which have been coated with a naturally deposited extracellular matrix obtained from an Engelbreth-Holm-Swarm tumor cell line (Becton Dickinson Labware, Bedford, MA).
7. IMDM (Sigma) supplemented with 20% FCS, penicillin–streptomycin, and Retronectin® (BioWhittaker Europe, Taufkirchen, Germany) is used for retroviral transductions. Growth factors required to stimulate proliferation of hematopoietic target cells may include stem cell factor (rhSCF), interleukin-3 (rhIL-3) interleukin-6 (rhIL-6), flt-3 ligand (flt3-L), thrombopoietin, and granulocyte colony-stimulating factor (rhG-CSF).
8. For differentiation of progenitor cells to DCs, rhSCF, flt-3L, interleukin-4 (IL-4), granulocyte-macrophage colony-stimulating factor (rhGM-CSF) (CellGenix, Freiburg, Germany) were used.
9. Enzyme-linked immunosorbent assay (ELISA) kits and fluorescein-coupled monoclonal antibodies for determination of secreted or cell-associated transgenes are purchased from commercial distributors.

3. Methods

The following protocols describe small- and large-scale experiments. Small-scale experiments supply useful information with regard to the multiplicity of infection (MOI) (*see* **Note 1**) required to achieve adequate gene transfer efficiency. Small-scale experiments also allow assessment of transgene expression levels over time, which is of particular interest for vector systems mediating only transient gene expression. In the preclinical setting, large-scale experiments may be performed to generate sufficient vaccine cells for murine in vivo studies. Large-scale experiments are also useful to assess upscaling strategies and as a feasibility study for subsequent clinical-grade vaccine production.

3.1. Adenoviral Vectors for Gene Transfer

Because of their versatility, adenoviral vectors are attractive candidates for tumor vaccine generation and considerable experience with the use of adenoviral vectors in clinical tumor vaccine studies has by now been collected. The principle advantages of adenoviral vectors are the accommodation of large-size foreign genes of up to 35 kb, the availability of a high-titer virus that can be further concentrated by centrifugation, the broad range of susceptible target cells, as well as the ability to efficiently infect nondividing cells achieving high levels of—albeit transient—transgene expression *(23)*. One can exploit these qualities of adenoviral vectors for high-efficiency transduction of solid tumor vaccines. Hematopoietic and blood cells—with the exception of monocytes and dendritic cells— are relatively resistant to adenoviral infection. A potential problem is the strong immunogenicity of viral proteins that may impede multiple applications of adenovirally transduced target cells. Clinical toxicity of viral protein expression—although proven hazardous following direct in vivo application *(24)*—is of lesser concern in the tumor vaccine setting where genetic modification of tumor cells is generally performed ex vivo.

Here, adenoviral protein expression may even enhance the immunogenicity of adenovirally transduced tumor vaccine cells. Safety-modified viruses have been designed to counteract the potential disadvantages of viral protein expression as well as the risk of recombination events with wild-type virus. Deletions range from E1/E3 deletions on the prototype *(23)* to second-generation adenovirus vectors that lack all viral coding sequences. These latter vectors are generated by cotransfection of linearized plasmids encoding the transgene that was ligated to the viral encapsidation and replication signals, together with DNA of mutant adenovirus lacking the encapsidation signal *(25)*.

Following are two representative adenoviral transduction protocols for primary isolated solid tumor cells and dendritic cells.

3.1.1. Adenoviral Transduction of Primary Human Solid Tumor Cells

The solid tumor gene transfer protocol has been validated for primary human neuroblastoma cells; however, it should translate with minor modifications to a number of other adherently growing solid tumors as well (*see* **Notes 2** and **3**). The following protocols are for preclinical use only and require additional modifications to meet the standards of good manufacturing practice if intended for clinical use.

3.1.1.1. Tumor Cell Isolation and Establishment of Short-Term Culture

1. For isolation of primary tumor cells, transfer the tumor specimen obtained from surgery or by biopsy to a 150-mm Petri dish and chop it into small pieces max. 4 mm^2 in size using two scalpels.
2. Place tumor pieces into a 50-mL conical tube containing 20 mL of trypsin–EDTA. Incubate at room temperature (RT) for 30 min. (*See* **Note 2**.)
3. Add 20 mL RPMI/10% FCS to stop the digestive process and pellet cells and remaining undigested tumor pieces by centrifugation for 10 min at 400*g*.
4. Remove supernatant and resuspend cells in RPMI/10% FCS and pass them through a Falcon cell strainer (70 μm) into a fresh 50-mL conical tube. Set tube with strained single-cell suspension aside.
5. Transfer the nondigested pieces remaining in the strainer into a funnel-shaped glass mortar and add fresh RPMI/10% FCS.
6. Apply turning motion to mortar and pestle to dissociate the remaining cells from connective fibers.
7. Use a pipet to pass the cellular suspension from the mortar through a fresh Falcon cell strainer into the 50-mL conical tube set aside in **step 4** and pellet the combined cells by centrifugation for 10 min at 400*g*.
8. For short-term cell culture, seed $(1–2) \times 10^6$ tumor cells/mL in a six-well culture dish with a maximal volume of 3 mL/well and incubate at 37°C in 5% CO_2 (*see* **Note 3**).
9. After 24–72 h, a good portion of tumor cells has settled. Gently remove floating cells and transfer into a second six-well plate. Feed the adherent cells in the first plate with fresh media.
10. For large-scale transductions, place up to 1×10^6 tumor cells/mL into T75 flasks with a maximal volume of 12 mL. Otherwise, follow **steps 8** and **9**. For assessment of a number of adenoviral vectors at the same time or for comparison of different experimental conditions, one may choose to downscale the cell culture volume using a 24-well plate with 2×10^5 cells/mL in RPMI/10% FCS and follow **steps 8** and **9**.

3.1.1.2. Adenoviral Gene Transfer into Primary Human Solid Tumor Cells

1. For analytical transductions, use the 24-well cell cultures once foci of adherent tumor cells or an adherent cell layer has developed. Take care to not let cultures overgrow by fibroblasts prior to adenoviral transductions. Remove complete media and add in 0.2 mL RPMI/2% FCS per well instead.
2. Add adenoviral supernatant. Start pilot studies with an MOI of 2–5.
3. Incubate for 2 h at 37°C and 5% CO_2. Shake gently every 15 min during this period.
4. Add 0.3 mL RPMI/10% FCS and incubate overnight at 37°C in 5% CO_2.
5. The next morning, carefully aspirate the medium from the cells and place into a 15-mL conical tube. Spin down any cells floating for 10 min at 400*g* and place pelleted cells back to the adherent cell culture.
6. Add 1 mL of RPMI/10% FCS and place cell culture dish into the incubator for the desired time until the cells are analyzed for transgene expression.

7. For large-scale transduction of solid tumor cells, grow cells in a T75 cell culture flask until foci of adherent tumor cells or an adherent cell layer has developed.
8. Remove cell culture medium and spin down any floating cells for 10 min at 400*g* and return pelleted cells to the plate of adherently growing cells.
9. Add 5 mL RPMI/2% FCS and the adenoviral vector at a suitable MOI determined by small-scale transduction experiments as described in **steps 1–6**. As removal of adherent cells for cell counting prior to transduction is problematic with regard to reduction of gene transfer efficiency, estimate cell numbers per flasks following short-term culture based on preliminary large-scale cell culture experiments with the respective tumor (*see* **Note 3**).
10. Incubate for 2 h at 37°C and 5% CO_2. Shake gently every 15 min during this period.
11. Add 10 mL RPMI/10% FCS per flask.
12. Following the overnight incubation at 37°C in 5% CO_2, carefully aspirate the medium from the cells. Spin down floating cells for 10 min at 400*g* and return pelleted cells to the plate of adherently growing cells.
13. Add 15 mL RPMI/10% FCS per 10^6 cells and incubate at 37°C in 5% CO_2 for the desired time until analysis.

*3.1.2. Adenoviral Transduction of Human DCs (*see **Note 4***)*

1. For analytical transduction, resuspend 10^5 DCs in 0.2 mL RPMI/2% FCS in a 24-well tissue culture plate.
2. Add adenovirus vector. Start pilot studies with an MOI of 30–250.
3. Incubate for 2 h at 37°C in 5% CO_2. Shake gently every 15 min during this period.
4. Add 0.3 mL RPMI/10% FCS.
5. Following the overnight incubation at 37°C in 5% CO_2, carefully aspirate the medium from the cells. Spin down floating cells for 10 min at 400*g* and place pelleted cells back into DC culture.
6. Add 0.5 mL complete medium and incubate at 37°C in 5% CO_2 for the desired time until analysis.

3.1.3. Analysis of Transgene Expression in Transduced Target Cells

For vaccine generation, genes encoding for different immunomodulatory molecules such as cytokines, chemokines, and costimulatory surface molecules are introduced into the tumor cells. Assessment of secreted immunomodulatory molecules analyzed over time in serial samples of cellular supernatant can be readily performed by ELISA and indicates levels as well as duration of transgenic expression. Assessment of immunomodulatory surface molecules expressed in trans on the genetically modified tumor cells can also serve to determine transduction efficiency. Alternatively, an adenoviral vector encoding for the green fluorescent protein or equivalent molecules can be used to determine gene transfer rates. The principles of analysis as outlined below pertain to tumor cells as well as dendritic cells and to small- and large-scale cultures alike.

3.1.3.1. Analysis of Cytokine and Chemokine Expression

To analyze levels and kinetics of cytokine/chemokine expression in adenovirally transduced tumor or dendritic cells by ELISA, cellular supernatant is sampled at different time-points posttransduction. For adenovirally transduced target cells, 24-h intervals over 4 d have proven useful.

1. For sampling, remove the entire supernatant from the cell culture dish also when studying large-scale cultures in T75 flasks.
2. Spin down any floating cells for 10 min at 400*g*; remove and aliquot the supernatant for immediate analysis. Aliquots of supernatant may also be stored by –20°C or –80°C depending on the storage conditions for the secreted protein and analyzed at a later point in time.
3. Place pelleted cells back into the cell culture dish, add 2 mL (6-well tissue culture plate) or 10 mL (T75 tissue culture flask) RPMI/10% FCS and incubate for the next 24 h at 37 °C in 5% CO_2.
4. Repeat **steps 2** and **3** several times to determine the duration of transgene expression at levels required for efficient immune stimulation.
5. Aliquots of supernatant obtained from the adenovirally transduced target cells may also be used for safety analysis such as sterility, mycoplasma, and endotoxin testing.

3.1.3.2. Analysis of Transgenic Surface Molecule Expression

1. To analyze surface molecule expression in adenovirally transduced tumor or dendritic cells over time, it is best to seed sufficient aliquots of target cells in separate wells prior to transduction to allow for sampling of the entire cell population/well at each time-point of the analysis. By providing separate aliquots for each time-point, disturbance of the cell culture as well as sampling errors by removing only subpopulations from the wells resulting in skewing of the expression analysis are avoided.
2. Remove the cells from the cell culture dish by pipetting or gentle scraping. Try to avoid trypsin–EDTA treatment for cell removal.
3. Pellet the cells by centrifugation for 10 min at 400*g* and resuspend the cell pellet in buffer or medium as required by the respective protocols for immune fluorescent staining and analysis by microscopy or flow cytometry.

3.2. HSV-Based Vectors for Gene Transfer

Herpes simplex virus-based vectors, although originally designed for gene therapy of nervous system disease ***(26)***, have gained broader application, including the transduction of hematopoietic cells ***(27,28)***. Transgene expression mediated by these vector systems is transient; however, new developments of HSV/AAV hybrid vectors have resulted in stable transgene expression ***(29)***.

HSV-based vectors for gene therapy are mostly derived from HSV type 1 viruses and share a number of benefits described for adenoviral vectors, such as high-titer vector preparations and the broad range of susceptible target cells. In contrast to adenoviruses, HSV vectors also transduce leukemia cells with high efficiency.

A major disadvantage is the cellular toxicity resulting from cytopatic effects (CPE) of HSV infection, which does, however, vary considerably between target cells. As with adenoviral vectors, there is also the inherent risk of recombination events with wild-type virus present in the host and the possibility of unfavorable immune responses directed against HSV proteins. Expression of the HSV protein ICP47 results in downregulation of MHC I molecules ***(30)***, a particular problem for tumor vaccine generation in some target cells. Generation of deletion mutants of HSV vectors were designed to cirumvent the above problems. Deletion of the gene coding for the glycoprotein gH that is essential for viral infectivity results in the HSV type-2-derived DISC (disabled infectious single cycle) vector. This virus is capable of only one round of infection, as the viral particles generated in infected noncomplementing target cells do not express *gH* ***(27,31)***. Other replication-defective vectors have been generated by deleting one or more of the viral immediate–early genes (e.g., *ICP4* and *ICP27*) necessary for viral replication ***(32)***. These vectors do not replicate in infected target cells and are propagated in complementing cell lines that stably express the deleted genes in trans. Depending on the extent of the viral genome deletion that leads to a reduced synthesis of viral proteins, these viral mutants also exhibit reduced levels of cytopathic effects.

For preclinical studies, a versatile HSV mini-vector systems called amplicon has been developed. While in recombinant HSV vectors, the transgene is integrated into the viral genome, amplicon vectors encapsidate a plasmid (amplicon–plasmid) that encodes the transgene, the HSV origin of replication and the HSV packaging signal ***(33)***. The name amplicon has been coined based on the rolling mechanism of HSV amplicon replication resulting in multiple copies of the transgene packaged per viral particle. Amplicon vectors can carry up to 150 kb of DNA, which allows for insertion of large-size genes or multiple genes, which is a useful feature when designing vaccination strategies with combinations of various immunomodulatory genes. HSV helper-virus-dependent amplicon vectors are produced by transfection of the amplicon–plasmid and coinfection with a deleted HSV-helper virus into the complementing HSV packaging cell line. The downside of helper-virus-dependent amplicon generation is that "empty" helper virus particles are generated together the amplicon particles and contribute to target cell toxicity. This may be overcome by a helper-virus-independent bacterial artificial chromosome BAC amplicon vector. The latter vector is generated by cotransfection of the transgene carrying amplicon–plasmid and a BAC vector that contains the complete HSV genome except for the HSV packaging signal. Helper-virus-independent BAC amplicon vectors were recently improved by cloning the ICP27-deleted HSV-1 genome into BAC ***(34)*** as a safety measure. Principally, BAC amplicons should be less toxic to target cells, as helper-virus-independent BAC amplicon vectors do not support *de novo* synthesis of HSV genes. Because HSV has been recently reported to efficiently transduce hematopietic cells ***(27,28)***, we provide a protocol for HSV-mediated transduction of human leukemia cells.

3.2.1. Determination of Vector Dose Required for HSV Gene Transfer by Titration of Transgene Expression

Recombinant HSV vectors are titrated by plaque assay using a complementing cell line. In this assay, viral replication results in the formation of plaques on the adherent cell layer. When calculating the multiplicity of infection (MOI) needed for transduction of target cells, the number of plaque-forming units (PFUs) is taken as representative of the number of infectious viral particles carrying the transgene. When titrating helper-virus-dependent amplicons, the plaque assay is a measure for the helper virus particles that are also present in the amplicon preparation, yet it does not indicate the concentration of amplicon particles carrying the transgene. The amplicon/helper virus ratio can be determined by comparative analysis of amplicon/herpesvirus DNA restriction fragment intensity or by polymerase chain reaction (PCR). Alternatively, the concentration of transducing viral particles (transducing "units") can be estimated by titration of transgene expression after transduction of a reference cell line with serial dilutions of viral stock.

Determination of transducing "units" in a reference cell line is also useful for titrating helper-virus-independent BAC amplicons, as these viruses are replication defective and cannot be assessed by plaque assay. Transgenes encoding the green fluorescent protein or a cell surface molecule allow for ready assessment of gene transfer efficiency by flow cytometry. Although any cell line (e.g., mouse 3T3 fibroblasts) may be used as a reference cell line, ideally a cell line is chosen that is closely related to the definitive target cell. When comparing different viral preparations on a reference cell line, serial dilutions of the viral stocks may be used to determine the volume required for transgene expression in 50% of a defined number of target cells. As gene transfer efficiency of herpesviral vectors is often associated with cellular toxicity, one may wish to use a reference cell line also sensitive to the cytopathic effect (CPE) of HSV vectors to determine the viral dose that mediates optimal transduction at acceptable cytotoxicity levels. This may be of particular interest when evaluating amplicon stocks where the helper virus content may significantly influence the extent of CPE. For leukemic cells, the murine B-cell lympoma cell line A20 may be used for evaluation of transgene expression and CPE of viral stocks. The following respective protocol for HSV-based vectors encoding a cell surface molecule may also be adapted to nonhematopoietic target cells.

1. Incubate serial dilutions of the HSV vector preparation with 10^5 A20 cells in a total volume of 0.3 mL medium (RPMI/10% FCS/0.05 m*M* β-mercaptoethanol) in a 24-well tissue culture plate. Plate several aliquots per tested vector dilution for determination of transgene expression at different time-points (24, 48, 72 h) post-transduction.
2. Shake cells gently every 15 min for 2 h (*see* **Note 5**).
3. Incubate for 24 h at 37 °C in 5% CO_2.
4. Determine transgene expression by flow cytometry for detection of green fluorescent protein or surface marker expression. Also, determine viable cells by trypan blue staining or by flow cytometry using propidium iodide.
5. Repeat **steps 3** and **4** at later time-points.
6. Plot transgene expression against virus dilution and against viable cells to characterize individual preparations of viral stock.
7. Determine dilution of virus that efficiently transduces target cells with minimal cytopathic effect (*see* **Note 6**).

3.2.2. Transduction of Leukemia Cells Using HSV-Based Vectors

1. Human leukemic cells derived from bone marrow aspirates or peripheral blood are obtained by Ficoll® density gradient centrifugation. The mononuclear interphase is collected and washed twice in PBS to remove any remaining Ficoll solution. Resuspend mononuclear cells in RPMI/10% FCS at 4×10^5 cells/mL.
2. For small-scale transductions, transfer 2×10^5 cells in 0.5 mL RPMI/10% FCS in a 12-well tissue culture plate. (*See* **Note 7**.)
3. Add HSV vector. For recombinant HSV stocks, start with MOI 1–10. For helper-virus-dependent amplicon viral stocks, start with MOI of 5–20 based on the titer obtained by plaque assay. For helper-virus-independent BAC amplicon, base your choice of viral dose on the determination of transduction efficiency obtained in the reference cell line.
4. Incubate plate at 37°C in 5% CO_2.
5. Shake tissue culture plate gently every 15 min for the next 2 h.
6. Incubate plate overnight at 37 °C in 5% CO_2. Addition of 0.5 mL RPMI/10% FCS are sufficient for a 24-h incubation.
7. Determine transgene expression 24 h post-transduction (*see* **Subheading 3.1.3.**).
8. Medium-scale transductions may be required for preparation of leukemia vaccines to be assessed in a preclinical animal model.
9. Plate up to 3×10^6 leukemic cells per well of a 6-well culture plate in 1 mL RPMI/10% FCS (*see* **Note 7**).
10. Add sufficient amount of HSV virus stock as determined in the analytical experiments.

11. Incubate plate at 37°C in 5% CO_2. Shake tissue culture plate every 15 min for the next 2 h.
12. Add 2 mL of complete medium. Incubate plate at 37°C in 5% CO_2 overnight.
13. At 18 h posttransduction, transfer cells into 15-mL tube, top up with 10 mL PBS, and spin down cells for 10 min at 400*g*. Resuspend cells in 10 mL PBS and repeat centrifugation step. Adjust cells to the desired concentration in PBS for vaccination.

3.3. Retrovirus for Gene Transfer

Although transient expression of immunomodulatory genes is sufficient for most tumor vaccine applications, the generation of allogeneic vaccines from tumor cell lines affords a stable integration of the transgene to allow for large-scale expansion of the vaccination cell line. Retroviral gene transfer results in integration of the provirus into the host genome with stable expression of the transgene. At present, retroviral vectors are the only vectors for which substantial clinical experience with transduction of hematopoietic cell is available ***(35)***. Retroviral transduction of leukemic cells can be achieved in acute and chronic myeloid leukemia; however, it does require the cells to be in cycle ***(36,37)***. Stably transduced cells after retroviral gene transfer have importance as a bystander cell vaccine, where nontumor cells like fibroblasts are transduced and admixed to the tumor vaccine cells, which leads to the expression of immunomodulatory molecules at the vaccination. Potential disadvantages are the low titers of retroviral stock preparations and the comparatively small insert size of only 6–7 kb that can be accommodated into the retroviral backbone. Nonetheless, multiple gene expression, which may be desirable for the generation of combination vaccines, can be achieved utilizing an alternative promoter or a viral internal ribosome entry sites (IRES) in addition to the retroviral 5′ long terminal repeats for gene transcription ***(38)***.

Most retroviral vectors are based on the genome of murine retroviruses, usually the moloney leukemia virus, a member of the group C oncoviruses ***(39)***. As human cells do not express the murine ecotropic receptor required for viral binding, the retroviral backbone is packaged in an envelope carrying a heterotypical viral envelope protein, such as the envelope protein of amphotropic virus ***(40)***, the gibbon-ape leukemia virus (GALV) ***(41)***, or the vesicular stomatitis virus (VSV) ***(42)***. Such pseudotyped viruses enhance the versatility of available retroviral vectors. Yet, for receptor–ligand interaction to occur, frequency and duration of contact between virus and target cell needs to be enhanced. Several techniques have been developed to improve colocalization of virus and target cell such as centrifugation flowthrough procedure ***(43–45)***, and the use of tissue culture plates coated with fibronectin or fibronectin fragments ***(21,22)***. As gene transfer efficiency critically depends on cell proliferation, exposure of target cells to growth factors prior to transduction significantly improves gene transfer rates into hematopoietic cells. Various growth factor combinations have been assessed in different settings including SCF, thrombopoietin, flt-3 ligand, IL-3, IL-6, and G-CSF. For expansion of hematopoietic progenitor cells for DC differentiation, SCF, flt-3 ligand, IL-4, and GM-CSF have proven effective. Myeloid leukemic cell isolates may be brought into cycle by various growth factor combinations allowing for subsequent retroviral transduction ***(36,37)***.

3.3.1. Retroviral Gene Transfer into Human Hematopoietic Cells Using Fibronectin

1. Incubate hematopoietic progenitors with a growth factor combination as required to induce cell cycling.
2. At the day of transduction, dilute Retronectin in PBS as suggested by the manufacturer to yield 4 $\mu g/cm^2$. For coating 24-well or 6-well plates, assume an area of 2.0 cm^2 and 9.6 cm^2 per well, respectively. Use nontissue culture dishes.
3. Add Retronectin solution to the dish.
4. Incubate plate at least for 2 h at RT.
5. Decant solution and block with 2% bovine serum albumin (BSA)/PBS for 30 min at RT.
6. Wash coated dish with PBS.
7. Resuspend 10^6 cells in 1–4 mL of retroviral supernatant aiming for an MOI of 2–5. If retroviral titers are low, you may have to reduce cell number/volume. (*See* **Note 8.**)
8. Transfer cells suspended in retroviral supernatant onto Retronectin-coated plates. Add growth factors as used for prestimulation of target cells.
9. Incubate for 2 h at 37°C in 5% CO_2.
10. Carefully remove supernatant and pellet any floating cells by centrifugation for 10 min at 400*g*. Resuspend cell pellet in fresh retroviral supernatant supplemented with growth factors and place back into the Retronectin-coated plates.
11. After 2 h, replace retroviral supernatant by fresh complete medium containing growth factors as above. To this end, remove retroviral supernatant and pellet any floating cells by centrifugation for 10 min at 400*g*. Resuspend cell pellet in fresh complete medium containing growth factors and place back into the dishes.
12. Incubate at 37°C at 5% CO_2 until next day.
13. Repeat incubation with viral supernatants to cells after 24 h and 48 h proceeding through **steps 7–12**.
14. Analyze transgene expression starting 24–48 h posttransduction (*see* **Subheading 3.1.3.**).

4. Notes

1. Multiplicity of infection refers to the number of infectious viral particles per target cell. For quantification of viral concentration in different vector preparations, refer also to **Subheading 3.2.1.**

2. Trypsin–EDTA is a standard approach for tissue digestion and has proven useful for neuroblastoma and other tumor cells. Accutase is an alternative for digestion of neuroblastoma specimens exhibiting lower cellular toxicity and may be found useful also for other tumor entities. In some cases, one may have to refrain from the use of digestive enzymes and use physical homogenization of tumor specimens only.
3. Isolated neuroblastoma cells should be plated on EHS-coated tissue culture plates or flasks. For some tumor entities, collagen-coated tissue culture plates may be useful, whereas untreated tissue culture plates may be sufficient for others. Cell numbers/well are indicated in the protocol for neuroblastoma cells and may need to be adjusted for other tumor entities.
4. For the generation of human DCs from CD14+ or CD34+ cells, refer to respective current protocols. Some authors report that mature DCs are more sensitive toward adenoviral transductions ***(18)***. However, this issue is controversial ***(17)***. Also, lipid formulations like Lipofectamine have been shown to enhance adenoviral transduction ***(46)***. Our protocol involves transduction of DCs while growing in tissue culture plates; however, DCs can also be harvested and infected in suspension.
5. Repeated mixing of viral supernatant during the first 2 h of infection as well as small volumes of cell/virus mixture can improve transduction efficiency. You may have to adjust the cell/volume ratio depending on your target cells. Some authors suggest using lower amounts of FCS during the first 2 h of infection to enhance infectivity; however, we did not observe any difference in the target cell lines tested.
6. In addition to a CPE that is mediated by viral proteins, cytotoxicity can also be caused by impurities of the viral preparation itself. If possible, use highly purified viral stocks that have been purified by sucrose gradient centrifugation. You may test for cytotoxicity caused by impurities of the viral stock by boiling the virus at 95°C for 30 min prior to incubation with cells. If cytotoxicty in target cells is still observed under these denaturing conditions, the effect is most likely caused by impurities in the supernatant of the viral stock, such as remnants of cellular debris and not the result of *de novo* synthesis of viral proteins.
7. If transducing adherent tumor cells instead of leukemia cells, you may want to wait until tumor cells have settled in the tissue culture plate. Alternatively, transductions of adherent cells can also be performed in suspension to allow for infection in a small volume. We have found that short trypsinization of the target cell does not affect subsequent HSV transduction. However, depending on the target cells used, you may decide to remove the adherent cells from the plates by treatment with EDTA alone or simply by scraping. For the first 2 h of incubation with HSV vectors resuspend target cells tightly packed in a small volume in a 15-mL conical tube and flick the bottom of the tube every 15 min. Alternatively resuspend target cells in 1–2 mL of medium in a 15-mL conical tube and rotate the tube on a rolling device at 37°C for the entire transduction period.
8. When working with low-titer retroviral supernatants or supernatants that are toxic to target cells, preload the Retronectin-coated dishes with retroviral supernatants

by first incubating the dishes with retroviral supernatant alone. After 2 h at 37°C in 5% CO_2, remove supernatant and add fresh retroviral supernatant. Repeat for several rounds as seems necessary to compensate for the low titer of the supernatant used. After preloading of the Retronectin-coated dish with retrovirus, add target cells in complete medium containing the appropriate growth factors in a volume of 1.5 mL/well for a 6-well plate and 0.5 mL/well for a 24-well plate. Incubate for 2 h at 37°C and 5% CO_2 and top up with growth factor containing medium for overnight incubation. After 24 h and 48 h, you may wish to repeat the whole process by transferring the cells onto a new dish that has again been preloaded with retrovirus.

References

1. Pardoll, D. M. (1998) Cancer vaccines. *Nat. Med.* **4,** 525–531.
2. Brenner, M. K. (2001) Gene transfer and the treatment of haematological malignancy. *J. Intern. Med.* **249,** 345–358.
3. Wierda, W. G. and Kipps, T. J. (2000) Gene therapy of hematologic malignancies. *Semin. Oncol.* **27,** 502–511.
4. Van Tendeloo, V. F., Van Broeckhoven, C., and Berneman, Z. N. (2001) Gene-based cancer vaccines: an ex vivo approach. *Leukemia* **15,** 545–558.
5. Stripecke, R., Skelton, D. C., Gruber, T., Afar, D., Pattengale, P. K., Witte, O. N., et al. (1998) Immune response to Philadelphia chromosome-positive acute lymphoblastic leukemia induced by expression of CD80, interleukin 2, and granulocyte-macrophage colony-stimulating factor. *Hum. Gene Ther.* **9,** 2049–2062.
6. Golumbek, P. T., Lazenby, A. J., Levitsky, H. I., Jaffee, L. M., Karasuyama, H., Baker, M., et al. (1991) Treatment of established renal cancer by tumor cells engineered to secrete interleukin-4. *Science* **254,** 713–716.
7. Schadendorf, D., Czarnetzki, B. M., and Wittig, B. (1995) Interleukin-7, interleukin-12, and GM-CSF gene transfer in patients with metastatic melanoma. *J. Mol. Med.* **73,** 473–477.
8. Dilloo, D., Bacon, K., Holden, W., Zhong, W., Burdach, S., Zlotnik, A., et al. (1996) Combined chemokine and cytokine gene transfer enhances antitumor immunity. *Nat. Med.* **2,** 1090–1095.
9. Zhang, W., He, L., Yuan, Z., Xie, Z., Wang, J,. Hamada, H., et al. (1999) Enhanced therapeutic efficacy of tumor RNA-pulsed dendritic cells after genetic modification with lymphotactin. *Hum. Gene Ther.* **10,** 1151–1161.
10. Heuer, J. G., Tucker-McClung, C., Gonin, R., and Hock, R. A. (1996) Retrovirus-mediated gene transfer of B7-1 and MHC class II converts a poorly immunogenic neuroblastoma into a highly immunogenic one. *Hum. Gene Ther.* **7,** 2059–2068.
11. Dunussi-Joannopoulos, K., Weinstein, H. J., Nickerson, P. W., Strom, T. B., Burakoff, S. J., Croop, J. M., et al. (1996) Irradiated B7-1 transduced primary acute myelogenous leukemia (AML) cells can be used as therapeutic vaccines in murine AML. *Blood* **87,** 2938–2946.

12. Dilloo, D., Brown, M., Roskrow, M., Zhong, W., Holladay, M., Holden, W., et al. (1997) CD40 ligand induces an antileukemia immune response in vivo. *Blood* **90,** 1927–1933.
13. Bonnet, M. C., Tartaglia, J., Verdier, F., Kourilsky, P., Lindberg , A., Klein, M., et al. (2000) Recombinant viruses as a tool for therapeutic vaccination against human cancer. *Immunol. Lett.* **74,** 11–25.
14. Todryk, S. M., Birchall, L. J., Erlich, R., Halanek, N., Orleans-Lindsay, J. K., and Dalgleish, A. G. (2001) Efficacy of cytokine gene transfection may differ for autologous and allogeneic tumour cell vaccines. *Immunology* **102,** 190–198.
15. Kohn, D. B. (2001) Gene therapy for genetic haematological disorders and immunodeficiencies. *J. Intern. Med.* **249,** 379–390.
16. Stripecke, R., Cardoso, A. A., Pepper, K. A., Skelton, D. C., Yu, X. J., Mascarenhas, L., et al. (2000) Lentiviral vectors for efficient delivery of CD80 and granulocyte–macrophage- colony-stimulating factor in human acute lymphoblastic leukemia and acute myeloid leukemia cells to induce antileukemic immune responses. *Blood* **96,** 1317–1326.
17. Bubenik, J. (2001) Genetically engineered dendritic cell-based cancer vaccines. [review]. *Int. J. Oncol.* **18,** 475–478.
18. Dietz, A. B., Bulur, P. A., Brown, C. A., Pankratz, V. S., and Vuk-Pavlovic, S. (2001) Maturation of dendritic cells infected by recombinant adenovirus can be delayed without impact on transgene expression. *Gene Ther.* **8,** 419–423.
19. Regn, S., Raffegerst, S., Chen, X., Schendel, D., Kolb, H. J., and Roskrow, M. (2001) Ex vivo generation of cytotoxic T lymphocytes specific for one or two distinct viruses for the prophylaxis of patients receiving an allogeneic bone marrow transplant. *Bone Marrow Transplant.* **27,** 53–64.
20. Esche, C., Shurin, M. R., and Lotze, M. T. (1999) The use of dendritic cells for cancer vaccination. *Curr. Opin. Mol. Ther.* **1,** 72–81.
21. Hanenberg, H., Xiao, X. L., Dilloo, D., Hashino, K., Kato, I., and Williams, D. A. (1996) Colocalization of retrovirus and target cells on specific fibronectin fragments increases genetic transduction of mammalian cells. *Nat. Med.* **2,** 876–882.
22. Moritz, T., Patel, V. P., and Williams, D. A. (1994) Bone marrow extracellular matrix molecules improve gene transfer into human hematopoietic cells via retroviral vectors. *J. Clin. Invest.* **93,** 1451–1457.
23. Zhang, W. W. (1999) Development and application of adenoviral vectors for gene therapy of cancer. *Cancer Gene Ther.* **6,** 113–138.
24. Greenberg, D. S. (2000) Stricter regulation proposed for US gene-therapy trials. *Lancet* **355,** 1977.
25. Kochanek, S., Clemens, P. R., Mitani, K., Chen, H. H., Chan, S., and Caskey, C. T. (1996) A new adenoviral vector: replacement of all viral coding sequences with 28 kb of DNA independently expressing both full-length dystrophin and beta-galactosidase. *Proc. Natl. Acad. Sci. USA* **93,** 5731–5736.

26. Wolfe, D., Goins, W. F., Yamada, M., Moriuchi, S., Krisky, D. M., Oligino, T. J., et al. (1999) Engineering herpes simplex virus vectors for CNS applications. *Exp. Neurol.* **159,** 34–46.
27. Dilloo, D., Rill, D., Entwistle, C., Boursnell, M., Zhong, W., Holden, W., et al. (1997) A novel herpes vector for the high-efficiency transduction of normal and malignant human hematopoietic cells. *Blood* **89,** 119–127.
28. Tolba, K. A., Bowers, W. J., Hilchey, S. P., Halterman, M. W., Howard, D. F., Giuliano, R. E., et al. (2001) Development of herpes simplex virus-1 amplicon-based immunotherapy for chronic lymphocytic leukemia. *Blood* **98,** 287–295.
29. Jacobs, A., Breakefield, X. O., and Fraefel, C. (1999) HSV-1-based vectors for gene therapy of neurological diseases and brain tumors: part II. Vector systems and applications. *Neoplasia* **1,** 402–416.
30. Hill, A., Jugovic, P., York, I., Russ, G., Bennink, J., Yewdell, J., et al. (1995) Herpes simplex virus turns off the TAP to evade host immunity. *Nature* **375,** 411–415.
31. Forrester, A., Farrell, H., Wilkinson, G., Kaye, J., Davis-Poynter, N., and Minson, T. (1992) Construction and properties of a mutant of herpes simplex virus type 1 with glycoprotein H coding sequences deleted. *J. Virol.* **66,** 341–348.
32. Krisky, D. M., Wolfe, D., Goins, W. F., Marconi, P. C., Ramakrishnan, R., Mata, M., et al. (1998) Deletion of multiple immediate–early genes from herpes simplex virus reduces cytotoxicity and permits long-term gene expression in neurons. *Gene Ther.* **5,** 1593–1603.
33. Spaete, R. R. and Frenkel, N. (1982) The herpes simplex virus amplicon: a new eucaryotic defective-virus cloning-amplifying vector. *Cell* **30,** 295–304.
34. Saeki, Y., Ichikawa, T., Saeki, A., Chiocca, E. A., Tobler, K., Ackermann, M., et al. (1998) Herpes simplex virus type 1 DNA amplified as bacterial artificial chromosome in Escherichia coli: rescue of replication-competent virus progeny and packaging of amplicon vectors. *Hum. Gene Ther.* **9,** 2787–2794.
35. Williams, D. A. and Smith, F. O. (2000) Progress in the use of gene transfer methods to treat genetic blood diseases. *Hum. Gene Ther.* **11,** 2059–2066.
36. Cornetta, K., Moore, A., Leemhuis, T., Moen, R. C., Tricot, G., Leibowitz, D., et al. (1994) Retroviral mediated gene transfer in chronic myelogenous leukaemia. *Br. J. Haematol.* **87,** 308–316.
37. Roddie, P. H., Paterson, T., and Turner, M. L. (2000) Gene transfer to primary acute myeloid leukaemia blasts and myeloid leukaemia cell lines. *Cytokines Cell Mol. Ther.* **6,** 127–134.
38. Baum, C., Richters, A., and Ostertag, W. (1999) Retroviral vector-mediated gene expression in hematopoietic cells. *Curr. Opin. Mol. Ther.* **1,** 605–612.
39. Miller, A. D. (1992) Retroviral vectors. *Curr. Topics Microbiol. Immunol.* **158,** 1–24.
40. Orlic, D., Girard, L. J., Jordan, C. T., Anderson, S. M., Cline, A. P., and Bodine, D. M. (1996) The level of mRNA encoding the amphotropic retrovirus receptor in mouse and human hematopoietic stem cells is low and correlates with the efficiency of retrovirus transduction. *Proc. Natl. Acad. Sci. USA* **93,** 11,097–11,102.

41. Kiem, H. P., Heyward, S., Winkler, A., Potter, J., Allen, J. M., Miller, A. D., et al. (1997) Gene transfer into marrow repopulating cells: comparison between amphotropic and gibbon ape leukemia virus pseudotyped retroviral vectors in a competitive repopulation assay in baboons. *Blood* **90,** 4638–4645.
42. Burns, J. C., Friedmann, T., Driever, W., Burrascano, M., and Yee, J. K. (1993) Vesicular stomatitis virus G glycoprotein pseudotyped retroviral vectors: concentration to very high titer and efficient gene transfer into mammalian and nonmammalian cells. *Proc. Natl. Acad. Sci. USA* **90,** 8033–8037.
43. Palsson, B. and Andreadis, S. (1997) The physico-chemical factors that govern retrovirus-mediated gene transfer. *Exp. Hematol.* **25,** 94–102.
44. Liu, H., Hung, Y., Wissink, S. D., and Verfaillie, C. M. (2000) Improved retroviral transduction of hematopoietic progenitors by combining methods to enhance virus–cell interaction. *Leukemia* **14,** 307–311.
45. Hutchings, M., Moriwaki, K., Dilloo, D., Hoffmann, T., Kimbrough, S., Johnsen, H. E., et al. (1998) Increased transduction efficiency of primary hematopoietic cells by physical colocalization of retrovirus and target cells. *J. Hematother.* **7,** 217–224.
46. Lee, S. G., Yoon, S. J., Kim, C. D., Kim, K., Lim, D. S., Yeom, Y. I., et al. (2000) Enhancement of adenoviral transduction with polycationic liposomes in vivo. *Cancer Gene Ther.* **7,** 1329–1335.

21

Cytokine Gene Delivery into the Central Nervous System Using Intrathecally Injected Nonreplicative Viral Vectors

Roberto Furlan, Stefano Pluchino, Peggy C. Marconi, and Gianvito Martino

1. Introduction

The delivery of drugs through the bloodstream in patients affected by central nervous system (CNS)-confined multifocal diseases can be therapeutically ineffective because of the presence of the blood–brain barrier (BBB), which forms an inaccessible wall to the majority of CNS-targeting molecules. The BBB is a specialized endothelial structure formed by the interaction between endothelial cells and astrocytes. It can be distinguished from the normal endothelium for the presence of tight junctions between endothelial cells, which are impermeable to macromolecules and even ions, and for the reduced endocytic activity, which considerably decreases the number of molecules that can cross the BBB in a nonspecific fashion *(1)*. Only the presence of specific transport mechanisms assures that molecules essential for the brain metabolism (e.g., amino acids and glucose) reach the brain parenchyma.

1.1. CNS Drug Delivery

Chronic inflammatory demyelinating diseases of the CNS, such as multiple sclerosis (MS), might benefit from anti-inflammatory therapies *(2)*. However, promising treatments such as those based on the systemic administration of anti-inflammatory cytokines did not result in a consistent therapeutic effect in MS patients *(3)*. The scarce capacity of cytokines to cross the BBB, along with their short half-life and autocrine/paracrine activity, might render necessary

From: *Methods in Molecular Biology, vol. 215:*
Cytokines and Colony Stimulating Factors: Methods and Protocols
Edited by: D. Körholz and W. Kiess © Humana Press Inc., Totowa, NJ

the delivery of these molecules directly into the CNS. Biological and physical vectors engineered with heterologous genes coding for anti-inflammatory cytokines might represent the appropriate tool to deliver cytokines into the CNS *(4)*. Results obtained in rodents affected by experimental autoimmune encephalomyelitis (EAE), the animal models of MS, support this working hypothesis *(5)*.

1.2. The Ependymal Way to Access the CNS

Here, we summarize the technical procedure and the troubleshooting of a novel strategy we recently established to access the CNS using viral vectors engineered with heterologous genes coding for anti-inflammatory cytokines. This approach is based on the injection into the cerebrospinal fluid (CSF) space trough the cisterna magna (i.c.) of nonreplicative viral vectors (**Fig. 1A**). Injected vectors, in turn, infect exclusively neuroectodermal cells lining the CSF space (including the Virchow–Robin spaces) and forming the blood–CSF barrier surrounding both the brain and the spinal cord (i.e., ependymal, choroidal, and leptomeningeal cells) (**Fig. 1b**). The viral genome enters into the nucleus of infected cells and dictates heterologous gene transcription (**Fig. 1c**). The protein coded by the transgene is then translated into the cell cytoplasm and secreted into the CSF (**Fig. 1d**). Secreted proteins diffuse, via the ependymal layer or the pia mater, into the CNS parenchyma, where they are still biologically active and can exert therapeutic activity.

Only vectors fulfilling the following criteria can be used: (1) vectors able to infect nondividing cells because ependymal and leptomeningeal cells cycle at a very slow rate; (2) vectors that can be obtained at very high titers because only very small volumes (up to 10 µL in mice) can be injected; (3) vectors with very low or no immunogenicity because the protocol is designed to interfere with an already ongoing immune reaction; (4) vectors expressing the transgene long term because repeated intrathecal injection of the vectors is a not feasible approach in a routine clinical setting.

Several viral vectors fulfill the above-mentioned criteria, among which herpes simplex virus (HSV) type-1-derived vectors, adenoviral vectors (AD), adeno-associated viral vectors (AAV), and lentiviral vectors. However, only HSV-1 and AD vectors, to our knowledge, have been intrathecally delivered, so far.

1.2.1. Herpes Simplex Type-1-Derived Vectors

Several features make HSV-1 a likely candidate as a vector for gene transfer: (1) At least one-third of the 152 kb of the HSV-1 genome is made of genes nonessential for replication. These genes can be deleted and substituted by exogenous genes without any detrimental effect for in vitro viral growth. (2) HSV-1 can be easily propagated in several different cell lines (e.g.,

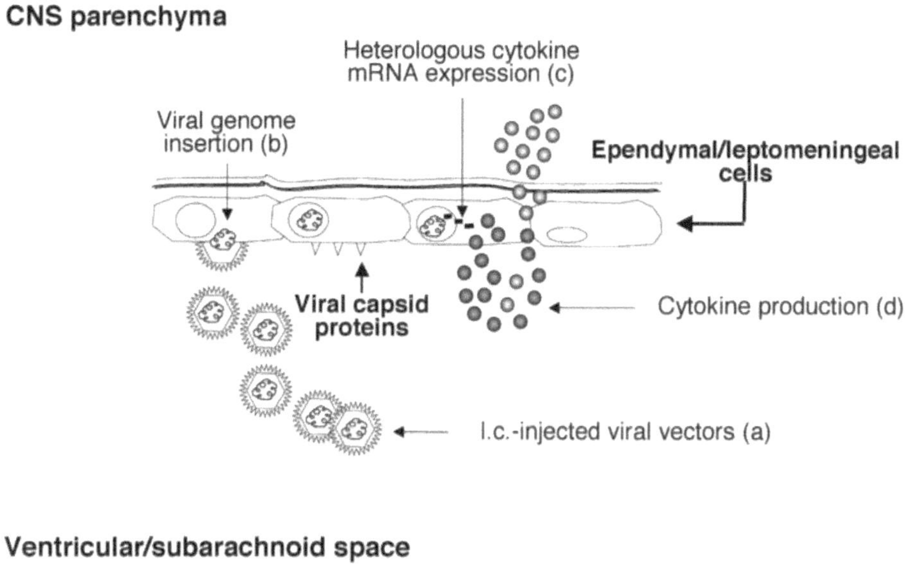

Fig. 1. The ependymal route. Nonreplicative viral vectors engineered to contain a cytokine gene are injected intracisternally in cerebrospinal fluid spaces (a); these vectors insert their genome into the cells lining ventricles and sub-arachnoid spaces (ependymal and leptomeningeal cells) (b) and induce them to transcribe (c) and translate the cytokine gene which is then released (d) into the cerebrospinal fluid. From there the exogenously produced cytokine can travel through the ependymal cell layer into the brain parenchima and exert there its potentially beneficial effect.

complementing cell lines), allowing the generation of high-titer viral stocks. (3) HSV-1 is able to infect several different cell types, regardless of the cell cycle, with high efficiency, making it an ideal candidate vector for several applications (i.e., infection of postmitotic neurons). (4) During the lytic cycle, many HSV-1 genes are expressed with high efficiency. A heterologous gene driven by a viral promoter can, therefore, produce large amounts of protein. (5) HSV-1 is able to persist in a state of latency for the whole life of its host. During latency, the viral genome is circularized and remains as an episome in the cell nucleus. Lytic genes are silent and only latency-specific transcripts are present. The introduction of foreign sequences under the control of latency-specific promoters may allow long-term transcription of transgenes.

We have been working with nonreplicative deletion mutant HSV-1 vectors. These deletions lead to the inability to replicate in normal conditions and lower their cytotoxicity. Immediate–early genes are the main target of the mutations because these genes are both essential for viral replication and responsible for most of the cytopathicity. Immediate–early genes are ICP infected cell polypeptides ICP0, ICP22, ICP4, ICP27, and ICP47, in order of decreasing toxicity. Among those, only ICP4 and ICP27 are essential for replication, although the lack of some of the others (ICP0, ICP22) produces a marked decrease in viral titers. HSV-1 deletion mutants have been generated lacking three immediate–early genes (ICP4, ICP27, and ICP22) ***(6,7)*** and are propagated on a corresponding complementary cell line producing ICP4 and ICP27 (ICP22 is nonessential for in vitro viral growth). Using these herpetic vectors engineered to express interleukin (IL)-4, interferon (IFN)-γ, and fibroblast growth factor (FGF)-II, we have obtained encouraging results in both mice and nonhuman primates affected by EAE ***(5,8–13)***.

1.2.2. Adenoviral Vectors

Adenoviruses have a double-stranded DNA genome of about 35–40 kb and lack an envelope. About 50 serotypes have been described, most of which causing, in humans, benign diseases of the respiratory tract. Serotypes 2 and 5 are the most studied for the obtainment of gene therapy vectors. In the adenoviral genome, between two inverted repeats (ITR) functioning as origin of replication, there are complex transcriptional units that can be divided into four early (E) and five late (L) regions. Of those, region E1A, the first expressed after infection, is essential for viral replication. First-generation vectors have been obtained deleting the E1 gene and replacing it with a 5- to 8-kb transgene expression cassette. The E1 gene product was provided in trans by a complementing cell line. Because of the high toxicity and immunogenicity of these vectors, high-capacity (HC), also named helper-dependent (HD) vectors, almost completely devoid of viral sequences, have been developed. HD vectors, which depend on an helper virus for in vitro growth, have only the two ITRs and the packaging signal and are, therefore, able to accomodate up to 30–35 kb of exogenous DNA. Contaminating helper virus is eliminated both by Cre-mediated excision of the packaging signal from its genome and, subsequently, by gradient purification. Main features of these vectors are (1) the ability to infect many different cell types regardless of the cell cycle (including postmytotic cells), (2) reduced toxicity and immunogenicity, and (3) long-term expression of the transgene (up to 6 mo). The genome can be entirely manipulated as a plasmid and grown in bacteria, making its genetic engineering easier. Adenoviral vectors have been used to infect ependymal–leptomeningeal cells, also in nonhuman primates ***(14–17)***.

1.3. Injection of HSV-1 Vectors into the Cisterna Magna to Access the CNS

To access the CSF space of mice, the most common procedure is an intraventricular injection performed using a stereotactic apparatus. This

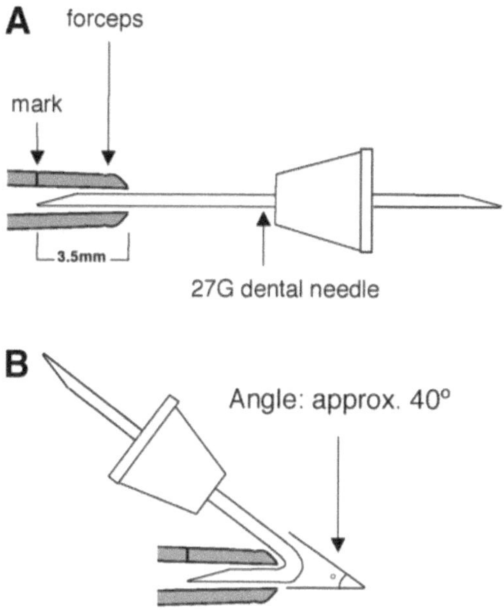

Fig. 2. Needle preparation. Hold the needle tightly with the tip at 3.5 mm from the end of the forceps (**A**); bend the needle with the forceps at an angle of approx 40°, keeping the cutting edge inside (**B**).

procedure is, however, time-consuming and limits the number of mice that can be treated in a single experiment. Here, we describe the application of a quick and simple intracisternal (i.c.) injection technique that can be used to deliver cytokine genes within the CNS. We also show a method to sample CSF from mice that represents an essential corollary technique to verify the efficiency of heterologous protein production within the CNS.

2. Materials

2.1. Injection Procedure

1. Flat forceps.
2. Dental needle 27G × 13/16 in. (0.40 × 21 mm) (*see* **Fig. 2**).
3. Bunsen burner.
4. Polyethylene tubing; inner diameter = 0.38 mm (0.015 in.); outer diameter = 1.09 mm (0.043 in.) (Becton Dickinson, cat. no. 427406).
5. 10-µL Hamilton syringe.
6. Diethyl ether.
7. Multichannel pipet reservoir.
8. Pipet tip box.

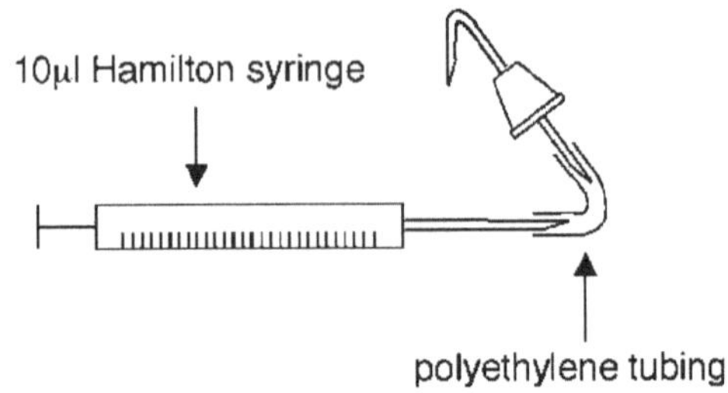

Fig. 3. The injecting device. The injecting device is obtained connecting the bent dental needle to the 10 μL Hamilton syringe through a small (approx 1 cm) piece of polyethylene tubing.

2.2. Cerebrospinal Fluid Sampling

1. 100-μL Pyrex disposable microsampling pipets (Corning).
2. Bunsen burner.
3. Microdissecting scissors.
4. Cloralium hydrate (store at 4°C for up to 2 mo).
5. Disposable scalpel.
6. Cotton swabs.
7. Pipet tip box.
8. Butterfly needle.
9. 10-mL syringe.

3. Methods

3.1. Intracisternal Injection of Vectors in Mice

1. Needle preparation. Needles are prepared as described by Ueda et al. *(18)*. Measure 3.5 mm from the tip of a large forceps and tag the point with a marker (**Fig. 2A**).
2. Used the marked forceps as a tool to bend a 27-gage dental needle, keeping the cutting edge toward the inside of the loop (**Fig. 2A,B**). The needle should be J shaped, with an angle of approx 40° (**Fig. 2B**).
3. Before recapping, the needle should be briefly flamed.
4. Dental needles have a short needle at the opposite end as well. Connect a 1-cm-long polyethylene tubing (internal diameter 0.38 mm) to the short end. Several needles can be prepared and kept for further use.
5. Connect the dental needle, through the polyethylene tubing, to the needle of a 10-μL Hamilton syringe (**Fig. 3**).
6. Fill the resulting injecting device, aspirating the viral vector containing solution from a reservoir for multichannel pipets.
7. Briefly anesthetize the mouse with ether, lean it on a small box (e.g., a pipet tip box), and bend the head slightly forward (**Fig. 4**). Run the needle along the external surface of the occiput and insert it into the cleft between the occiput and

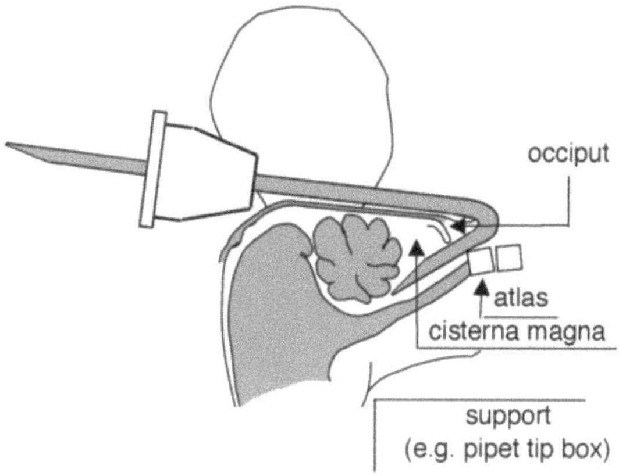

Fig. 4. The anesthetized mouse is put on a support and the head bent forwards. The needle is inserted as shown at the midline in the cleft between atlas and occiput.

the atlas vertebra through the intact skin, muscles, and ligaments in the midline at the back of the neck (**Fig. 4**). The bent part of the needle is kept in close contact with the internal surface of the occiput for the entire length (**Fig. 4**). The operator should learn to recognize the feeling of the needle being "hooked" to the mouse skull.

8. Inject the vector-containing solution in approx 10 s (best done by a second operator) and keep the needle in place a few more seconds before extracting it (*see* **Notes 1** and **2**). The same needle can be used for several mice to be injected with the same vector. Mice recover rapidly from the procedure, showing no evident adverse effects of the injection (*see* **Notes 3–6**).

The entire injection procedure takes less than 1 min. If the procedure has been performed correctly, your vector should have been distributed throughout the ventricular and CSF space (*see* **Note 7**). If a reporter gene-containing vector (e.g., β-galactosidase) is employed, the brain of an injected mouse should appear, upon specific staining, as in **Fig. 5B** (the solid arrow indicates the injection site; dashed arrows indicate viral vector-infected leptomeningeal cells). A sham-injected mouse brain is shown in **Fig. 5A** for comparison.

3.2. Cerebrospinal Fluid Sampling

1. Cerebrospinal fluid can be sampled only at sacrifice, because the procedure permanently lesions neck muscles and the cisterna magna.
2. Glass needle preparation. Keep the center of a 100-μL Pyrex disposable micro-sampling pipet on the flame of a Bunsen burner until it starts to melt. Take the two extremities apart quickly, so that the pipet is divided into two parts with two short terminal cones (*see* **Fig. 6**).

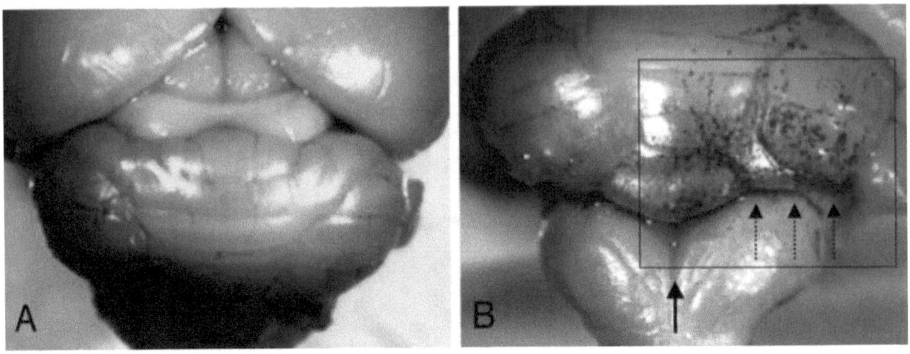

Fig. 5. Dissemination of the d120-IFNγ-lacZ herpetic vector within the CNS of a representative Balb/C mouse after injection of 10^9 plaque forming unit (PFU) of the vector in the mouse' cisterna magna (I. C.). Dark spots in panel **B** indicate the vector-infected cells around the injection site in a representative animal injected I. C. with the vector and sacrificed 72 h later. In panel **A** the same brain area obtained from a sham-injected animal is shown.

3. Using microdissecting scissors, cut the terminal of the cone to obtain a cutting edge as indicated by the dashed lines in **Fig. 6**.
4. Cut off the needle from a butterfly needle and connect the glass needle to its tubing.
5. Anesthetize the mouse with an intraperitoneal injection of cloralium hydrate (400 mg/kg of weight) and check for the absence of corneal and deep pain reflexes.
6. Lean the mouse on a support where you can bend the mouse head forward.
7. With a scalpel, cut the skin of the mouse twice horizontally, above the occiput (above the first vertebras of the column, and once vertically to connect the two previous incisions on the midline).
8. Open the skin, remove the muscles of the neck with a cotton swab and wait for the small hemorrhages to stop (*see* **Note 8**). You should see exposed the dura mater above the cisterna magna as a transparent sierosa (*see* **Fig. 7**).
9. Using the glass needle, punch the dura mater, slightly lateral to the midline (*see* **Note 9**).
10. The CSF enters the pipet by capillarity, usually 5–10 μL.
11. Empty the pipet in an Eppendorf tube by connecting it to a syringe through the tubing of a butterfly needle (after removing the needle) and blowing air into it.
12. Sacrifice the mouse.

4. Notes

1. Applying excessive pressure will cause hydrocephalus.
2. Ten microliters is the highest injectable volume. Injection of more fluid will result in reflux along the needle track.

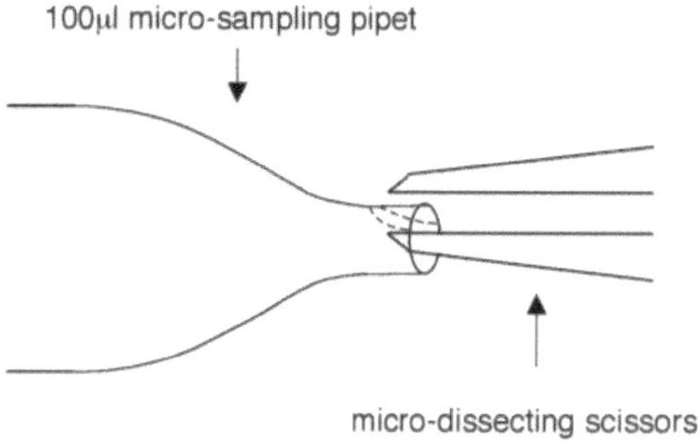

Fig. 6. Preparation of the glasss needle for the collection of the cerebrospinal fluid. The needle is obtained tearing apart a 100 µL micro-sampling pipet on an open flame to obtain a short conical end. A sharp edge is obtained as shown with micro-dissecting scissors.

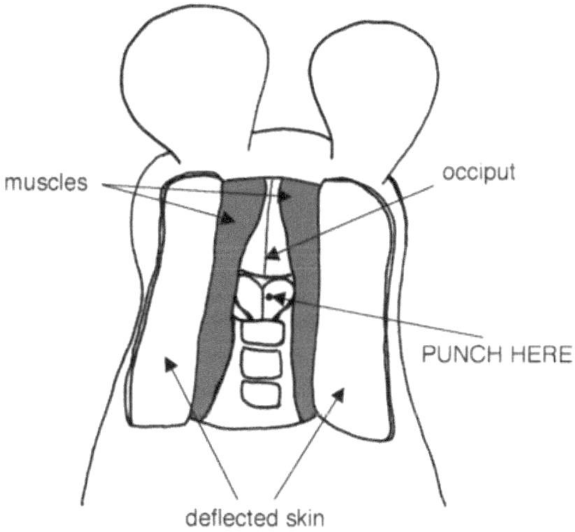

Fig. 7. Cerebrospinal fluid sampling. The anesthetized mouse is put on a support where the mouse head can be bent forward. After skin and the muscles of the neck have been removed and the small hemorragies have stopped, use the glass needle to punch the exposed dura mater above the cisterna magna slightly lateral to the midline, as shown.

3. Mice sometimes become ataxic because of cerebellar lesions. The length of 3.5 mm indicated for the bent part of the needle works with most mice strains at different ages. If, however, this becomes a recurrent problem or if the injection technique is applied to small or very young mice, shortening the needle length has to be considered.
4. Mice that underwent intracisternal injection will lose some weight during the forthcoming 2 d, probably because of reduced food intake.

5. After intracisternal injection, we were able to demonstrate a transient local inflammatory reaction, with an increase of the blood–brain barrier permeability, lasting less than a week. Neuropathological signs of inflammation are, however, absent.
6. Intracisternal injection can be repeated, but, in our experience, at least 5 d apart.
7. Although liquoral circulation goes from the choroid plexi, where it is produced, to the cauda, the intracisternal injection will transiently revert the flux direction allowing viral vector particles to reach the whole ventricular system.
8. Usually, several cotton swabs have to be used in order to tear muscles apart and wipe the blood until it stops. Because avoiding blood contamination is a crucial issue, be very careful in this step.
9. A blood vessel runs exactly along the midline of the dura mater above the cisterna magna. Insertion of the glass syringe at the midline will therefore result, almost invariably, in a blood contamination of the CSF sample.

References

1. Saunders, N. R., Habgood, M. D., and Dziegielewska, K. M. (1999) Barrier mechanisms in the brain, I. Adult brain. *Clin. Exp. Pharmacol. Physiol.* **26,** 11–19.
2. Martino, G. and Hartung, H. P. (1999) Immunopathogenesis of multiple sclerosis: the role of T cells. *Curr. Opin. Neurol.* **12,** 309–321.
3. Hohlfeld, R. and Wiendl, H. (2001) The ups and downs of multiple sclerosis therapeutics. *Ann. Neurol.* **49,** 281–284.
4. Martino, G., Poliani, P. L., Marconi, P., and Furlan, R. (2000) Cytokine gene therapy of autoimmune demyelination revisited using herpes simplex virus type-1-derived vectors. *Gene Ther.* **7,** 1087–1093.
5. Martino, G., Furlan, R., Comi, G., and Adorini, L. (2001) The ependymal route to the CNS: an emerging gene therapy approach for MS. *Trends Immunol.* **22,** 483–490.
6. Wu, N., Watkins, S. C., Schaffer, P. A., and DeLuca, N. A. (1996) Prolonged gene expression and cell survival after infection by a herpes simplex virus mutant defective in the immediate-early genes encoding ICP4, ICP27, and ICP22. *J. Virol.* **70,** 6358–6369.
7. Marconi, P., Krisky, D., Oligino, T., Poliani, P. L., Ramakrishnan, R., Goins, W. F., et al. (1996) Replication-defective herpes simplex virus vectors for gene transfer in vivo. *Proc. Natl. Acad. Sci. USA* **93,** 11,319–11,320.
8. Martino, G., Furlan, R., Galbiati, F., Poliani, P. L., Bergami, A., Grimaldi, L. M. E., et al. (1998) A gene therapy approach to treat demyelinating diseases using nonreplicative herpetic vectors engineered to produce cytokines. *Multiple Sclerosis* **4,** 222–227.
9. Furlan, R., Poliani, P. L., Galbiati, F., Bergami, A., Grimaldi, L. M. E., Comi, G., et al. (1998) Central nervous system delivery of interleukin-4 by a non replicative herpes simplex type I viral vector ameliorates autoimmune demyelination. *Hum. Gene Ther.* **9,** 2605–2617.

10. Furlan, R., Poliani P. L., Marconi, P. C., Bergami, A., Ruffini, F., Adorini, L., et al. (2001) Interleukin-4 gene delivery in the central nervous system at the time of disease onset inhibits progression of autoimmune demyelination. *Gen. Ther.* **8,** 13–19.
11. Poliani, P. L., Brok, H., Furlan, R., Ruffini, F., Bergami, A., Desina, G., et al. (2001) Delivery of a non-replicative herpes simplex type-1 vector engineered with the IL-4 gene to the central nervous system protects rhesus monkeys from hyperacute autoimmune encephalomyelitis. *Hum. Gene Ther.* **12,** 905–920.
12. Furlan, R., Brambilla, E., Ruffini, F., Poliani, P. L., Bergami, A., Marconi, P. C., et al. (2001) Intrathecal delivery of IFNγ protects C57BL/6 mice from chronic-progressive experimental autoimmune encephalomyelitis by increasing apoptosis of CNS-infiltrating lymphocytes. *J. Immunol.* **167,** 1821–1829.
13. Ruffini, F., Poliani, P. L., Brambilla, E., Marconi, P. C., Furlan, R., Bergami, A., et al. Short-term gene therapy using fibroblast growth factor-II induces remyelination and protects mice from chronic experimental autoimmune encephalomyelitis. *Gene Ther.*, in press.
14. Akli, S., Caillaud, C., Vigne, E., Stratford-Perricaudet, L. D., Poenaru, L., Perricaudet, M., et al. (1993) Transfer of a foreign gene into the brain using adenovirus vectors. *Nat. Genet.* **3,** 224–228.
15. Bajocchi, G., Feldman, S. H., Crystal, R. G., and Mastrangeli, A. (1993) Direct in vivo gene transfer to ependymal cells in the central nervous system using recombinant adenovirus vectors. *Nat. Genet.* **3,** 229–234.
16. Ohashi, T., Watabe, K., Uehara, K., Sly, W. S., Vogler, C., and Eto, Y. (1997) Adenovirus-mediated gene transfer and expression of human β-glucuronidase gene in the liver, spleen, and central nervous system in mucopolysaccharidosis type VII mice. *Proc. Natl. Acad. Sci. USA* **94,** 1287–1292.
17. Driesse, M. J., Kros, J. M., Avezaat, C. J., Valerio, D., Vecht, C. J., Bout, A., et al. (1999) Distribution of recombinant adenovirus in the cerebrospinal fluid of nonhuman primates. *Hum. Gene Ther.* **10,** 2347–2354.
18. Ueda, H., Amano, H., Shiomi, H., and Takagi, H. (1979) Comparison of the analgesic effects of various opioid peptides by a newly devised intracisternal injection technique in conscious mice. *Eur. J. Pharmacol.* **56,** 265–268.

IV

Ex Vivo Cell Expansion

22

In Vitro Regulation of Colony Stimulating Factor-Mediated Hematopoiesis in Healthy Individuals and Patients with Different Types of Myeloproliferative Disease

Thomas Vraetz, Peter D. Emanuel, and Charlotte M. Niemeyer

1. Introduction

1.1. Hematopoiesis

Human hematopoiesis is initiated by hematopoietic stem cells, the most undifferentiated cell type in the bone marrow. Although hematopoietic stem cells are not yet fully identified, their function is well characterized. They are able to replicate by cell division producing daughter cells with identical properties (self-renewal), differentiate into all myeloid and lymphoid cell lineages, and have a long life-span.

Hematopoiesis is organized in a hierarchical manner. Undifferentiated cells proliferate and thereby differentiate into more mature cells with reduced differentiation capacity. Pluripotent stem cells differentiate into progenitor cells, which have lost the capacity of self-renewal. Progenitor cells themselves give rise to precursor cells, which can be recognized by their morphology. Immature precursor cells have a high proliferative potential, whereas more mature precursor cells have lost the property of cell division and, instead, mature to fully determined peripheral blood cells *(1)*.

Assays for progenitor cells have first been described by in 1965 by Pluznik and Sachs *(2)* and in 1966 by Metcalf and Bradley *(3)*. When mononuclear cells (MNC) of blood or marrow are cultured with growth factors in semisolid media like methylcellulose or agar, progenitor cells will proliferate and form colonies of mature blood cells. The various colonies formed are identified by

From: *Methods in Molecular Biology, vol. 215:*
Cytokines and Colony Stimulating Factors: Methods and Protocols
Edited by: D. Körholz and W. Kiess © Humana Press Inc., Totowa, NJ

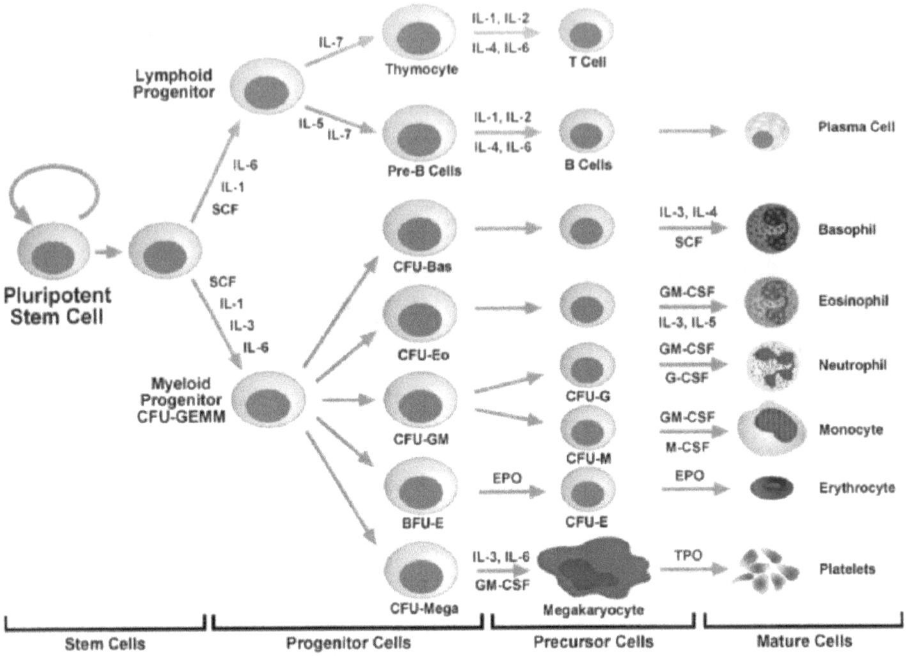

Fig. 1. Schematic illustration of differentiation and maturation of human hematopoietic cells. After differentiation of the pluripotent stem cell, multipotent myeloid and lymphoid progenitors develop, and they subsequently give rise to precursor cells. The progenitor cells can be assigned by their colonies formed; therefore, they are named as CFU or BFU. The specific differentiation process influenced by growth factors are indicated.

their characteristic appearance when studied with an inverted microscope or by morphology of single cells plugged from the colony. Because one progenitor cell was shown to give rise to one colony, the progenitor cell can be referred to as colony-forming unit (CFU). CFU-GEMM are multipotential progenitors giving rise to mixed colonies with granulocytes, erythroid cells, monocytes, and megakaryocytes (*see* **Fig. 1**). CFU-GM are more mature progenitors still capable of differentiating into either the granulocytic or macrophage lineage, whereas CFU-G give rise to granulocyte colonies only, and CFU-M to macrophage colonies only. Cells derived from immature erythroid progenitors grow in grouped colonies of bright red hemoglobinized nucleated erythrocytes, so-called bursts. The progenitors, therefore, have been named burst-forming-units erythroid (BFU-E). In contrast, more mature erythroid progenitors, CFU-E, proliferate into much smaller colonies without a burst appearance.

Proliferation and differentiation of normal progenitors in culture is dependent on the addition of growth factors, also named colony stimulating factors (CSFs). Historically, media conditioned by mononuclear blood cells or tumor cell lines were utilized as a source of CSFs. In the 1980s, CSFs were purified and various factors like erythropoietin (EPO), granulocyte–macrophage CSF (GM-CSF), granulocyte CSF (G-CSF), marcrophage CSF (M-CSF), and interleukin-3 (IL-3) were cloned. Subsequently, stem cell factor (SCF) and thrombopoietin (TPO) became known. In concert with other cytokines, SCF, IL-3, and GM-CSF act on early hematopoietic progenitors, whereas G-CSF, M-CSF, TPO, and EPO are more lineage-specific (*see* **Fig. 1**).

Hematopoietic growth factors are produced by stromal cells or mature blood cells. Most of them bind to extracellular domains of their specific receptors, thereby initiating a signaling cascade. Intracellular second messengers transduce the signal into the nucleus, where specific genes encoding for proliferation or differentiation are activated. These signaling pathways are controlled by a complex system of inhibitors and enhancers associated with the intracellular signal transducing proteins.

In vitro culture systems can be used for the quantification of hematopoietic progenitor cells, when it is useful to know the frequency of repopulating cells, such as in grafts for stem cell transplantation. In addition, they may be helpful for studying the pathogenesis of various blood diseases. In some diseases, like aplastic anemia, the frequency of CFUs is reduced, whereas in some neoplastic diseases, CFU-derived colonies are markedly increased. In specifically considering the myeloproliferative disorders, the knowledge about hematopoiesis and its regulation can be used to characterize the origin of the proliferating cells and to study which CFUs may play a role in the pathogenesis. In addition, progenitor assays can facilitate the diagnosis of hematopoietic disorders.

1.2. Myeloproliferative Disorders

Chronic myeloproliferative disorders are clonal diseases generally arising from a defective stem or early progenitor cell. They are characterized by an overwhelming production of differentiated and undifferentiated nonlymphoid cells with excessive cell numbers in the peripheral blood. In traditional classification schemes, myeloproliferative disorders included chronic myelogenous leukemia (CML), polycythemia vera (PV), and essential thrombocythemia (ET) *(4)*. Each of these diseases has a characteristic overproduction of cells of one or more hematopoietic cell lineages with often observed transformations into acute leukemia. The recent WHO classification added additional diseases to this myeloproliferative category.

In addition to the above group of chronic myeloproliferative disorders, there are diseases that have been placed by the recent WHO classification in a new category of disorders with myelodysplastic and myeloproliferative features. This class of hematopoietic malignancies includes juvenile myelomonocytic leukemia (JMML), a unique disorder of early childhood, chronic myelomonocytic leukemia, and atypical CML.

Clonal myeloproliferative disorders have to be distinguished from nonclonal disorders with primary or secondary overproduction of mature blood cells. One such example is the distinction between primary erythrocytosis, which can be noted in patients with congenital defects in the EPO signal transduction pathway, versus secondary erythrocytosis caused by spurious EPO production from a renal cell carcinoma.

1.2.1. Chronic Myelogenous Leukemia

Chronic myelogenous leukemia has an incidence of 1 to 1.5 cases per 100,000. Although the neoplasia can affect all age groups, it is most often diagnosed in older individuals. The clinical course of CML is typically divided into three phases: (1) a chronic phase, which can be stable over years, (2) an accelerated phase with increased myeloid proliferation, lasting approx 6–18 mo, and (3) a final blast phase, exhibiting symptoms and a course similar to acute leukemia.

The initial chronic phase is characterized by myeloid hyperplasia and leukocytosis with hematopoietic precursors and mature cells in peripheral blood. Basophilia can be recognized in many cases. In contrast to acute leukemia, the excessively produced myeloid precursors are able to differentiate into functional granulocytes. About half of the patients are diagnosed incidentally by a routine blood test. The accelerated phase can be viewed as a transitional period between the chronic and blast phases, during which there is an increase in the total white blood cell count as well as blast count, development of anemia and worsening splenomegaly, and worsening clinical symptoms. In two-thirds of all cases of CML, the disease terminates in a blast phase with a progressive fatal course of acute leukemia.

Chronic myelogenous leukemia is characterized by a reciprocal translocation between the long arms of chromosomes 9 and 22. The resulting shortened chromosome 22 can be detected cytogenetically and has been named Philadelphia chromosome (Ph). The Ph translocation fuses the *bcr* (breakpoint cluster region) gene on chromosome 22 and the c-*abl* proto-oncogene on chromosome 9. The new gene, *bcr-abl*, a cytoplasmic tyrosine kinase, is transcriptionally controlled by the *bcr* promoter, is constitutively active, and phosphorylates several substrates responsible for cell growth, differentiation, and apoptosis.

Whereas the Ph chromosome can be demonstrated by standard banding cytogenetics, *bcr-abl* can be detected with high sensitivity by techniques such as fluorescent *in situ* hybridization (FISH) or reverse transcriptase–polymerase chain reaction (RT-PCR). These techniques are useful tools for diagnosis and clinical follow-up, including the detection of minimal residual disease.

In in vitro culture, CML mononuclear cells from peripheral blood or bone marrow show a high proliferative potential. This increased number of CFU-derived colonies is the result of a high number of CD34+ cells in the CML bone marrow and blood. The in vitro proliferation and differentiation of CML progenitor cells is, however, dependent on supplemented growth factors like SCF, IL-3, IL-6, or GM-CSF. In CML, neither growth-factor-independent proliferation nor hypersensitivity to small amounts of growth factors has been described.

1.2.2. Polycythemia Vera

Polycythemia Vera is an acquired hematopoietic stem cell disorder with clonal expansion of erythroid, myeloid, and megakaryocyte lineages. It is the most common primary polycythemia, with a median age at diagnosis of 60 yr. A familial incidence and appearance in childhood have been reported. The predominant feature of this disease is increased red cell production. Erythrocytosis with high hematocrit results in a hyperviscosity syndrome associated with thrombotic or bleeding episodes. Clinically, hepatosplenomegaly, plethora, erythromelalgia, and pruritus are often present. In the later "spent" phase, ineffective erythropoiesis, marrow fibrosis, and extramedullary hematopoiesis develop. Eventually, transformation into acute leukemia may occur.

In in vitro progenitor culture systems, MNCs of patients with PV give rise to an increased number of erythroid, myeloid, and mixed colonies ***(5)***. In addition, PV erythroid progenitors show the ability to form BFU-E- and CFU-E-derived colonies in the absence from EPO ***(6,7)***. These so-called "endogenous erythroid colonies" have first been described in serum-containing culture systems ***(5,8)***. Results with serum-deprived culture methods are conflicting with respect to hypersensitivity for EPO or true EPO independence ***(9–11)***. Nevertheless, the formation of these endogenous erythroid colonies is the hallmark of PV, and in vitro cultures with/without EPO are important diagnostic tools to differentiate PV from others forms of erythrocytosis. In addition to EPO, hypersensitivity of PV erythroid progenitors toward other growth factors like insulin-like growth factor-1 (IGF-1) has been described ***(12)***. It seems likely that a final common signaling pathway is involved in the hypersensitivity for the various cytokines. The molecular pathogenesis of PV still needs to be elucidated.

1.2.3. Essential Thrombocythemia

Essential thrombocythemia is characterized by an autonomous acceleration of clonal thrombopoiesis. ET is a rare disease, generally affecting adults older than 50 yr of age. The clinical features consist of thrombocytosis (>600,000 platelets/μL) with dysfunctional platelets resulting in thromboembolic complications and a hemorrhagic diathesis. There is an increased proliferation of megakaryocytes (MK), giving rise to platelets with abnormal structure and an intrinsic hypersensitivity for excessive activation and aggregation. With subsequent endothelial damage, microvascular circulation disturbances and occlusion may lead to transient ischemic attacks, acute cardiac symptoms, or venous thromboses.

The pathogenesis of ET is not elucidated yet. Several investigators noted spontaneous growth of CFU-Mega-derived colonies in different serum-containing in vitro culture systems. Molecular analysis of thrombopoiesis-stimulating factors showed aberrant facts that may be involved in the phenomenon. As in the pathogenesis of PV, it is likely that ET is caused by an intrinsic stimulation of proliferation signals in progenitor cells. There is currently no evidence for paracrine stimulation, overexpression, or mutation of thrombopoietin or its receptor ***(13)***. In some instances, endogenous erythroid colony formation has been described in patients diagnosed of ET ***(14)***. These patients have been shown, however, to develop a PV ***(15,16)***, suggesting that the presence of endogenous formation of red cell colonies in such patients indicates the diagnosis of PV. Hypersensitivity of ET progenitor cells to TPO has been described recently, but, as with PV, the molecular pathogenesis of ET still needs to be further elucidated ***(17)***.

1.2.4. Juvenile Myelomonocytic Leukemia

Juvenile Myelomonocytic Leukemia is a clonal myeloproliferative disorder of early childhood ***(18,19)***. The disease is characterized by leukocytosis with monocytosis, the presence of immature hematopoietic precursor cells in the peripheral blood, and leukemic infiltration of various organs. Massive hepatosplenomegaly and pulmonary infiltrates rapidly lead to death unless stem cell transplantation is performed.

In in vitro culture, two interesting observations can be made. First, JMML myeloid progenitor cells exhibit so-called "spontaneous growth" of CFU-GM. This colony formation is dependent on growth factors secreted by leukemic monocytes in a paracrine fashion. Adherence depletion of monocytes before culture abrogates spontaneous growth. Second, myeloid progenitor cells show hypersensitivity for hematopoietic growth factors like GM-CSF ***(20)*** and SCF ***(21)***. These growth characteristics can be utilized as a diagnostic tool. When isolated MNCs from JMML patients are incubated with various amounts of GM-CSF, the GM-CSF dose-response curve of JMML myeloid progenitor cells is left shifted compared to controls. In addition, in some patients abnormally huge BFU-E-derived colonies are noted. Clonality studies have confirmed that JMML derives from a clonal origin of the pluripotent stem cell and that the erythroid lineage is unequivocally involved in the JMML clonal process. Further, JMML BFU-E are also hypersensitive to GM-CSF ***(20)***.

The molecular mechanisms involved in the pathogenesis of JMML are not totally clarified yet, but the detection of abnormalities and mutations in the GM-CSF signaling pathway support the importance of GM-CSF in the pathogenesis of the disease. Signal transduction of GM-CSF from the cell surface to the nucleus in normal cells involves at least two distinct pathways: the activation of the Jak–STAT pathway and the Ras signaling pathway (*see* **Fig. 2**). Members of the Ras family of signaling proteins regulate cellular proliferation by cycling between an active guanosin triphosphate (GTP)-bound state (Ras-GTP) and an inactive guanosine diphosphate (Ras-GDP)-bound state. *RAS* gene point mutations causing constitutively high Ras-GTP levels were noted in up to 25% of JMML patients ***(23)***. The conversion from Ras-GTP to the Ras-GDP is facilitated by GTP-ase-activating proteins (GAP), which act as negative regulators. One such GAP is neurofibromin, the protein encoded by the gene for neurofibromatosis (NF1). About 15% of children with JMML carry the clinical diagnosis of NF1 ***(18)***. In addition, *NF1* inactivating gene mutations have been detected in JMML patients in the absence of the clinical diagnosis of NF1 ***(22)***. The evidence that GM-CSF hypersensitivity in JMML is Ras-mediated is provided by experiments with homozygous mice lacking the *Nf1* gene. Hematopoietic cells of these *Nf1–/–* animals are hypersensitive to GM-CSF ***(24–26)***. By generating mice whose hematopoietic system was reconstituted with *Nf1–/–* hematopoietic stem cells, it could be demonstrated that loss of the *Nf1* gene by itself is sufficient to produce a myeloproliferative disease similar to JMML ***(26)***.

1.2.5. Primary Familial Erythrocytosis

Primary familial erythrocytosis, also named primary familial polycythemia, is a disorder typically inherited in an autosomal manner. The major characteristics are an elevated red cell mass and hemoglobin concentration in association with a low serum level of EPO. MNCs of patients with primary familial erythrocytosis form BFU-E-derived colonies in the absence of EPO. The EPO hypersensitivity of BFU-Es in in vitro culture differentiates primary familial erythrocytosis from all forms of secondary erythrocytosis. In contrast to PV, an acquired form of polycythemia, leukemic progression has not been described

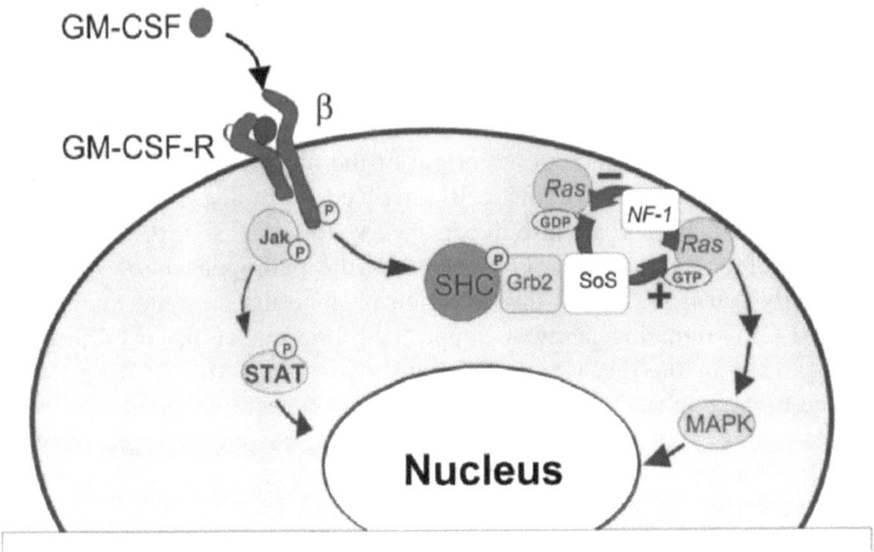

Fig. 2. GM-CSF–Ras signal transduction pathway. After binding of GM-CSF to its receptor (GM-CSF-R), the signal is transduced by two distinct pathways into the nucleus: the Jak–STAT and the Ras pathway. Abnormalities in Ras signaling were found in many JMML patients. Ras signal transduction is directly linked to an association with guanosin triphosphate (GTP). *RAS* gene point mutations cause constitutively active and high *Ras*-GTP levels and are found in up to 25% of JMML patients ***(21)***. The conversion of the active Ras complex to inactive GDP-bound Ras is facilitated by GTP-ase-activating proteins (GAP). Neurofibromin (NF1) is such a GAP. Defective NF1 has been detected in 30% of JMML patients ***(18,22)***. Both defects potentially may lead to accelerated GM-CSF signal transduction and, therefore, to the observed GM-CSF hypersensitivity ***(20)***.

in primary familial erythrocytosis. Although EPO hypersensitivity of erythroid progenitors in vitro can be observed in both primary familial erythrocytosis and PV, an additional increase in number and size of myeloid colonies would suggest PV.

To understand the molecular mechanism for the EPO hypersensitivity in primary familial erythrocytosis, the EPO receptor (EPO-R) was investigated. Several authors have described mutations in the EPO-R, leading to a truncation of the EPO-R cytoplasmic carboxyl terminal end. The truncated EPO-R lacks the negative regulation through the SHP-1 phosphatase, which normally dephosphorylates Jak-2, the transducer of the proliferative signal to the nucleus. In other families with primary familial erythrocytosis, the EPO-R gene was intact, suggesting that other components of the EPO-R signaling pathway might be responsible for the EPO hypersensitivity *(27–29)*.

2. Materials

1. Iscolve's modified Dulbecco's medium (IMDM), stored at 4°C.
2. Phosphate-buffered saline (PBS) w/o magnesium and calcium.
3. Ficoll–Paque™ (Pharmacia Biotech).
4. Fetal calf serum (FCS), mycoplasma tested, complement inactivated, stored at –20°C.
5. Methylcellulose culture mix (*see* **Subheading 3.1.1.**), stored at –20°C.
6. Cytokines like recombinant human GM-CSF, EPO, IL-1β, IL-3, IL-6, SCF, and G-CSF.
7. 50-mL and 14-mL test tubes with round bottom (e.g., Falcon).
8. Tissue culture dishes (33- and 100-mm tissue culture dishes, 24-well tissue culture plates).

3. Methods

3.1. Progenitor Cell Assay in Methylcellulose Culture Media

In this assay, methylcellulose is used as a gelling agent to produce a semisolid medium allowing the progeny of one progenitor cell to stay together and be recognized as one colony. Compared to other semisolid support media like agar, methylcellulose-based media permits better growth of erythroid colonies while still allowing optimal growth of myeloid colonies. Differentiation of myeloid colonies into colonies of granulocytes, monocytes, or mixed colonies relies, however, on the skilled view of the experienced observer through an inverted microscope. For staining of cells, colonies have to be picked from the methylcellulose media and cytospin preparations have to be prepared. In contrast, colonies in an agar-based system can be stained.

3.1.1. Preparation of Methylcellulose Culture Media

It is less expensive but more troublesome to prepare one's own methylcellulose culture media. (*see* **Note 1**). The following gives a recipe for making methylcellulose and preparing a culture media with 0.9% methylcellulose, 30% FCS, 1% bovine serum albumin (BSA), and 10^{-4} *M* β-mercaptoethanol. Preparation of a stock of 2.7% methylcellulose:

1. Weigh out 27 g methylcellulose (e.g., from Gibco) in a 2-L flask.
2. Add a large magnetic stirring bar; autoclave for 30 min and dry.
3. When the flask is cool, add 490 mL sterile double distilled H_2O.
4. Boil for several minutes while stirring.

5. Let cool on hot plate to room temperature while stirring all the time (takes about 3 h). Stir until smooth.
6. Add 490 mL sterile 2X IMDM (17.6 g IMDM + 3.14 g $NaHCO_3$ for 500 mL) and 10 mL penicillin–streptomycin (100X).
7. Stir in cold room overnight. Aliquot in 50-mL tubes and store at –20°C. Leave one tube in the incubator for 2 wk to test for sterility.
8. Preparation of 24 mL methylcellulose culture mix: Add the following to a 50-mL tube: 10 mL of 2.7% methylcellulose (let drip in tube, do not attempt to use pipet), 9.0 mL of FCS, 2.0 mL of IMDM with β-mercaptoethanol (30 mL IMDM + 5 μL β-mercaptoethanol), 2.7 mL BSA, 0.3 mL $NaHCO_3$ 7%. Mix by vortexing.

3.1.2. Preparation of Cells

Cells for plating in the methylcellulose progenitor assay can be prepared by ammonium chloride lysis only (removes erythroid cells only), by density gradient separation (to obtain mononuclear cells), or by CD34+ enrichment (e.g., with magnetic beads or FACSorting). MNCs are most conveniently prepared by Ficoll–Paque density gradient.

1. For optimal separation, blood (1+1) or bone marrow (1+2) are diluted with media like IMDM.
2. Then, 12.5 mL of the diluted cell suspension is carefully layered over 12 mL of Ficoll in a 50-mL tube.
3. The tube is centrifuged at 400*g* at 18°C for 40 min.
4. After careful aspiration of the interphase, MNCs are washed three times with 20–40 mL of media.
5. Cells are subsequently counted after trypan blue staining.
6. In most instances, the adherence depletion of monocytes is advisable. Adherent monocytes are depleted by incubation (37°C, 5% CO_2, humidified air) for 1–2 h or overnight at a concentration of 1×10^7 cells in 10 mL IMDM/2% FCS on a 100-mm tissue culture dish. Nonadherent MNCs are removed and washed.
7. For progenitor culture of peripheral blood and bone marrow cells of normal donors MNCs are adjusted to 5×10^7/mL and 1×10^6/mL, respectively. With the addition of 125 μL cell suspension to 1 mL of methylcellulose culture mix and 125 μL of growth factors (total volume 125 μL), the final plating density is 1×10^6/mL for peripheral blood and 5×10^4/mL for bone marrow MNCs.

3.1.3. Addition of Growth Factors

Hematopoietic growth factors can be added in form of conditioned media or recombinant cytokines. Commercially available cytokines generally have to be diluted in PBS/0.1% human serum albumin or bovine serum albumin to prepare stock solutions that are aliquoted and stored at –20°C to 70°C. Repeated freezing and thawing should be avoided. After thawing, the tubes

Table 1
Example of Pipetting Scheme to Test for GM-CSF Hypersensitivity in JMML

	GM-CSF			Media
Tube	Final conc.	Stock conc.	Vol. of stock added	(Vol. added) (μL)
1		—		125
2	10 pg/mL	0.4 ng/mL	31.7	94.2
3	20 pg/mL	0.4 ng/mL	62.5	62.5
4	40 pg/mL	0.4 ng/mL	125	0
5	80 pg/mL	3.2 ng/mL	31.7	94.2
6	160 pg/mL	3.2 ng/mL	62.5	62.5
7	320 pg/mL	3.2 ng/mL	125	0
8	1 ng/mL	40 ng/mL	31.7	94.2
9	2 ng/mL	40 ng/mL	62.5	62.5
10	4 ng/mL	40 ng/mL	125	0
11	8 ng/mL	80 ng/mL	125	125

Note: Adherence-depleted mononuclear cells of peripheral blood are cultured at a final concentration of 5×10^5/mL. In each 15-mL round-bottom tube, 1 mL of methylcellulose culture mix and 125 μL of the cell suspension at 5×10^6/mL are added. GM-CSF and media is added according to the pipetting schema.

should be always kept on ice. In the assay system proposed below, cytokines are added in a total volume of 125 μL for a total volume of 1250 μL. The concentrations of the cytokines have to be adjusted accordingly (*see* **Table 1**).

3.1.4. Plating and Incubation of Methylcellulose Culture

1. One milliliter of prewarmed methylcellulose culture media is pipetted into a 15-mL round-bottom test tube.
2. Cells and cytokines in 125 μL each are added to a total volume of 1250 μL.
3. A pipetting schema to test for GM-CSF hypersensitivity in JMML is given in **Table 1**.
4. Tubes are vortexed and allowed to stand for 5–15 min for bubbles to rise to the top.
5. With a 1-mL pipet, 0.5 mL of the culture is pipetted into 2 duplicate wells of a sterile 24-well tissue culture plate each.
6. To guarantee optimal humidity, 4 wells of the 24-well tissue culture plate are filled with 0.5 mL of sterile water.
7. Cultures are incubated at 37°C, 5% CO_2, in >95% humidified air for 14 d. If it is intended to pick colonies, 1 mL of culture can also be placed in a 35-mm covered tissue culture dish.

3.1.5. Evaluation of Colonies

Cultures are monitored daily with an inverted microscope. Erythroid, granulocyte, monocyte, and mixed colonies are identified by their growth characteristics. At d 14 of culture, the number of colonies are counted. Colonies with less than 40 cells are neglected. The number of colonies is given per 10^5 cells plated.

Individual colonies can be picked from the culture under an inverted microscope with an Pasteur glass pipet. Over an open flame, the tip of the pipet can be stretched to form a very fine opening. A small amount of media is aspirated in the pipet before a colony and the surrounding media is sucked up. The aspirated media is then smeared directly on a glass slide or placed in an microspin tube to prepare a centrifuge preparation. Pappenheim staining of the slide is performed (*see* **Note 2**).

3.2. Progenitor Culture in Agar

In contrast to methylcellulose-based culture systems, soft agarose culture systems predominantly support the growth of granulocyte and monocyte colonies, with erythroid and/or CFU-GEMM colonies rarely being seen. This can be advantageous in certain circumstances, such as investigating CFU-GM colony hypersensitivity to growth factors in JMML, wherein a background of other colonies (e.g., BFU-E, CFU-E, CFU-GEMM) may only serve to complicate the analysis and enumeration of colonies. In addition to not supporting erythroid development, the other primary difference from agar systems is that colonies and/or cells cannot be easily plucked from the media. Rather, the entire agarose gel must be removed from the tissue culture dish, dried, and then stained.

3.2.1. Preparation of Soft Agarose Gel and Culture Media

The culture media to be used is generally prepared in a concentrated state (two times normal) and then added 1:1 to agarose prepared (at a two times normal concentration) in water such that the final concentration is the normal working concentration. The following is our recipe for use in evaluating hypersensitive CFU-GM colony growth patterns in myeloproliferative diseases. Bacto-agar (Difco Labs, Detroit, MI) is mixed in water at 0.6% concentration and boiled for 4 min, then kept at 42°C.

1. The base culture media, McCoy's 5A (Gibco), is prepared at a two times normal concentration.
2. To make 100 mL of media with nutrients, the following are added: 61.8 mL of McCoy's 5A 2X; 1.6 mL minimum essential amino acids solution 50X without L-glutamine (Cellgro); 0.8 mL nonessential amino acid solution 100X (Cellgro); 2.0 mL of 100 m*M* sodium pyruvate solution (Cellgro); 1.0 mL of 200 m*M*

L-glutamine (100X) (Gibco); 1.6 mL of L-asparagine from a stock solution prepared at 2 mg/mL (Gibco); 0.2 mL of L-serine from a stock solution prepared at 8 mg/mL (Gibco); 1.0 mL of penicillin–streptomycin; 30 mL defined FBS (HyClone). This is a 2X solution and it is warmed to 37°C.

3. For plating, the 2X McCoy's media with added nutrients (at 37°C) is mixed with the 0.6% agar solution (at 42°C) so that the final culture media is 1X McCoy's in 0.3% agar. After mixing, the cells are then added (generally at a final concentration of 5×10^4/mL to 1×10^5/mL), and the mixture is quickly plated in 1-mL aliquots into 35 × 10-mm tissue culture dishes. Generally, dishes are established in triplicate.

3.2.2. Preparation of Cells

Isolation of MNCs (*see* **Subheading 3.1.2.**). To assess for "spontaneous" CFU-GM colony growth, whole MNC preparations from either bone marrow or peripheral blood can be combined with media and agar as in **Subheading 3.2.1.**

To assess for hypersensitive growth factor responses, generally the monocytes must be removed from the MNC prep prior to plating, as these cells are responsible for significant amounts of cytokine production that could interfere with the assessment of added growth factors. Especially in JMML, rigorous depletion of monocytes by plastic adherence must be obtained in order to abolish "spontaneous" CFU-GM colony growth. We have found that for agar-based assays, the best adherence depletion media is Hanks' balanced salt solution (containing calcium chloride, magnesium chloride, and magnesium sulfate) with 13% human AB serum (heat inactivated at 56°C for 30 min, then stored in frozen aliquots). MNCs are resuspended in this media at a concentration of 2×10^6/mL on a 100-mm tissue culture dish. Cells are incubated (37°C, 5% CO_2, humidified air) for a minimum of 90 min. Three successive depletions are performed (one of the incubations can be an overnight incubation). It is important that the tissue culture dishes are not swirled or otherwise manipulated. Simply tilt gently, aspirate the liquid, and plate onto a new 100-mm tissue culture dish and then reincubate.

3.2.3. Addition of Growth Factors or Other Additives to Agar-Based Systems

Hematopoietic growth factors (obtained and prepared as in **Subheading 3.1.3.**) can be added to agar culture in one of two ways. Prior to establishment of 1-mL agar cultures, small aliquots (e.g., 10 µL) can be placed into the bottom of the dish before the agar/nutrient media is plated. Alternatively, aliquots (generally 100 µL) can be uniformly spread over the agar surface after it has gelled to a semisolid state. The cytokine will diffuse down through the agar. Other agents can also be added in similar fashion.

Similar to the scheme depicted in **Table 1**, assessment of GM-CSF hypersensitivity in an agar-based system will include the following GM-CSF concentrations: 0, 0.01 ng/mL 0.02 ng/mL, 0.04 ng/mL, 0.08 ng/mL, 0.16 ng/mL, 0.32 ng/mL, and 2.0 ng/mL agar cultures. A range of standard GM-CSF responsiveness from normal hematopoietic cells will need to be established ***(19)***.

3.2.4. Evaluation of Colonies

Colonies are enumerated in agar as in **Subheading 3.1.5.** To remove agar from plates in order to perform staining, the entire tissue culture plate is carefully immersed in water to loosen the agar from the plate so that it will float freely. The loosened 1-mL agar button is then floated onto a large glass slide, carefully removed from the water, excess water drawn off by touching the agar edges with paper, and then the agar is allowed to dry onto the glass side prior to staining with the desired stain (*see* **Notes 3–5**).

4. Notes

1. Methylcellulose culture media as serum-containing or serum-deprived ready-to-use mix can be purchased from different companies. In some media preparations, growth factors are already added. For the purpose of the experiments described here, we suggest purchasing a serum-deprived media without the addition of colony stimulating factors. We currently use MethoCult™ SF^{BIT}H436, a serum-deprived media from StemCell Technologies (www.stemcell.com).
2. Colony formation in the methylcellulose assay can easily be disturbed. Failure of growth is most often the result of the presence of toxic substances in the double distilled water used to prepare the different media. To establish the assay, it may be advisable to use commercially available tissue-culture-tested methylcellulose mix and media. To prepare the culture media oneself, different batches of methylcellulose and serum have to be screened for optimal growth. The incubator has to be maintained at a stable CO_2 concentration and humidity. Routine cleaning of the incubator is not recommended. If there is bacterial or fungal overgrowth, the water used for cleaning should be free of disinfectant.
3. When preparing the mixture of agar/nutrient media/cells, it is very important that the cells be added last and that the mixture be close to 37°C before the cells are added. However, the agar will soon start to gel below this temperature, so plating must be done efficiently. Prior to placing into the incubator, it is sometimes helpful to put the dishes at 4°C for 10 min to ensure that the agar is completely gelled. Contrary to the methylcellulose, the cells will not grow well in agar that is not sufficiently semisolid.
4. Tissue culture water (used in media and agar preparation) and FBS can both drastically affect cultures and their reproducibility. It is advisable to use the purest water obtainable. FBS lots should be carefully screened and not routinely interchanged.

5. In order to prevent the 35 × 10-mm tissue culture dishes containing the 1-mL agar cultures from drying out during the 14-d incubation for CFU-GM colony growth, it is advisable to place six of the dishes into a 150 × 25-mm tissue culture dish, and then place an additional 35 × 10-mm dish in the center with the top off and fill that additional dish with water for humidity. The 150 × 25-mm dishes should not be stacked in the incubator, but rather each allowed to lay flat on the rack.

References

1. Dexter, T. M., Allen, T. D., and Lajtha, L. G. (1977) Conditions controlling the proliferation of haemopoietic stem cells in vitro. *J. Cell. Physiol.* **91,** 335–344.
2. Pluznik, D. H. and Sachs, L. (1965) The cloning of normal "mast" cells in tissue culture. *J. Cell. Physiol.* **66,** 319–324.
3. Bradley, T. R. and Metcalf, D. (1966) The growth of mouse bone marrow cells in vitro. *Aust. J. Exp. Biol. Med. Sci.* **44,** 287–299.
4. Harris, N. L., Jaffe, E. S., Diebold, J., et al. (2000) The World Health Organization classification of hematological malignancies report of the Clinical Advisory Committee Meeting, Airlie House, Virginia, November 1997. *Mod. Pathol.* **13,** 193–207.
5. Prchal, J. F. and Axelrad, A. A. (1974) Bone-marrow responses in polycythemia vera. [letter]. *N. Engl. J. Med.* **290,** 1382.
6. Weinberg, R. S. (1997) In vitro erythropoiesis in polycythemia vera and other myeloproliferative disorders. *Semin. Hematol.* **34,** 64–69.
7. Pahl, H. L. (2000) Towards a molecular understanding of polycythemia rubra vera. *Eur. J. Biochem.* **267,** 3395–3401.
8. Eaves, C. J. and Eaves, A. C. (1978) Erythropoietin (Ep) dose-response curves for three classes of erythroid progenitors in normal human marrow and in patients with polycythemia vera. *Blood* **52,** 1196–1210.
9. Correa, P. N. and Axelrad, A. A. (1991) Production of erythropoietic bursts by progenitor cells from adult human peripheral blood in an improved serum-free medium: role of insulinlike growth factor 1. *Blood* **78,** 2823–2833.
10. Fisher, M. J., Prchal, J. F., Prchal, J. T., and D'Andrea, A. D. (1994) Anti-erythropoietin (EPO) receptor monoclonal antibodies distinguish EPO-dependent and EPO-independent erythroid progenitors in polycythemia vera. *Blood* **84,** 1982–1991.
11. Casadevall, N., Vainchenker, W., Lacombe, C., et al. (1982) Erythroid progenitors in polycythemia vera: demonstration of their hypersensitivity to erythropoietin using serum free cultures. *Blood* **59,** 447–451.
12. Correa, P. N., Eskinazi, D., and Axelrad, A. A. (1994) Circulating erythroid progenitors in polycythemia vera are hypersensitive to insulin-like growth factor-1 in vitro: studies in an improved serum-free medium. *Blood* **83,** 99–112.
13. Bellucci, S. and Michiels, J. J. (2000) Spontaneous proliferative megakaryocytopoiesis and platelet hyperreactivity in essential thrombocythemia: is thrombopoietin the link? *Ann. Hematol.* **79,** 51–58.

14. Eridani, S., Pearson, T. C., Sawyer, B., Batten, E., and Wetherley-Mein, G. (1983) Erythroid colony formation in primary proliferative polycythaemia, idiopathic erythrocytosis and secondary polycythaemia: sensitivity to erythropoietic stimulating factors. *Clin. Lab. Haematol.* **5,** 121–129.
15. Eridani, S., Sawyer, B., and Pearson, T. C. (1987) Patterns of in vitro BFU-E proliferation in different forms of polycythaemia and in thrombocythaemia. *Eur. J. Haematol.* **38,** 363–369.
16. Shih, L. Y. and Lee, C. T. (1994) Identification of masked polycythemia vera from patients with idiopathic marked thrombocytosis by endogenous erythroid colony assay. *Blood* **83,** 744–748.
17. Axelrad, A. A., Eskinazi, D., Correa, P. N., and Amato, D. (2000) Hypersensitivity of circulating progenitor cells to megakaryocyte growth and development factor (PEG-rHuMGDF) in essential thrombocythemia. *Blood* **96,** 3310–3321.
18. Niemeyer, C. M., Arico, M., Basso, G., et al. (1997) Chronic myelomonocytic leukemia in childhood: a retrospective analysis of 110 cases. European Working Group on Myelodysplastic Syndromes in Childhood (EWOG-MDS). *Blood* **89,** 3534–3543.
19. Pinkel, D. (1998) Differentiating juvenile myelomonocytic leukemia from infectious disease. *Blood* **91,** 365–367.
20. Emanuel, P. D., Bates, L. J., Castleberry, R. P., Gualtieri, R. J., and Zuckerman, K. S. (1991) Selective hypersensitivity to granulocyte-macrophage colony-stimulating factor by juvenile chronic myeloid leukemia hematopoietic progenitors. *Blood* **77,** 925–929.
21. Sawai, N., Koike, K., Ito, S., et al. (1996) Aberrant growth of granulocyte–macrophage progenitors in juvenile chronic myelogenous leukemia in serum-free culture. *Exp. Hematol.* **24,** 116–122.
22. Side, L. E., Emanuel, P. D., Taylor, B., et al. (1998) Mutations of the NF1 gene in children with juvenile myelomonocytic leukemia without clinical evidence of neurofibromatosis, type 1. *Blood* **92,** 267–272.
23. Flotho, C., Valcamonica, S., Mach-Pascual, S., et al. (1999) RAS mutations and clonality analysis in children with juvenile myelomonocytic leukemia (JMML). *Leukemia* **13,** 32–37.
24. Bollag, G., Clapp, D. W., Shih, S., et al. (1996) Loss of NF1 results in activation of the Ras signaling pathway and leads to aberrant growth in haematopoietic cells. *Nat. Genet.* **12,** 144–148.
25. Largaespada, D. A., Brannan, C. I., Shaughnessy, J. D., Jenkins, N. A., and Copeland, N. G. (1996) The neurofibromatosis type 1 (NF1) tumor suppressor gene and myeloid leukemia. *Curr. Topics Microbiol. Immunol.* **211,** 233–239.
26. Largaespada, D. A., Brannan, C. I., Jenkins, N. A., and Copeland, N. G. (1996) Nf1 deficiency causes Ras-mediated granulocyte/macrophage colony stimulating factor hypersensitivity and chronic myeloid leukaemia. *Nat. Genet.* **12,** 137–143.
27. Kralovics, R. and Prchal, J. T. (2000) Congenital and inherited polycythemia. *Curr. Opin. Pediatr.* **12,** 29–34.

28. Kralovics, R., Indrak, K., Stopka, T., Berman, B. W., Prchal, J. F., and Prchal, J. T. (1997) Two new EPO receptor mutations: truncated EPO receptors are most frequently associated with primary familial and congenital polycythemias. *Blood* **90,** 2057–2061.
29. Kralovics, R. and Prchal, J. T. (2001) Genetic heterogeneity of primary familial and congenital polycythemia. *Am. J. Hematol.* **68,** 115–121.

23

In Vitro Generation of Dendritic Cells from Cord Blood CD34⁺ Hematopoietic Progenitors Cells

Bruno Canque, Michele Rosenzwajg, and Jean Claude Gluckman

1. Introduction

This review deals with the principles and methods used to generate dendritic cells (DCs) from cord blood $CD34^+$ hematopoietic progenitors cells (HPC). Inasmuch as this culture system does not raise any particular technical difficulty, we will first thoroughly expose the principal characteristics of this method and discuss its advantages and limitations as compared with the other methods currently used to differentiate DCs in vitro, with particular emphasis on monocyte-derived DCs (MDDCs). We will then discuss the influence of $CD34^+$ HPC origin (cord blood, bone marrow, thymus) and phenotype (pluripotent primitive HPC vs lymphoid- or myeloid-committed HPCs) on DC differentiation pathways. We will finally present the protocol used in our laboratory to generate DCs from cord blood CD34+ HPCs, as adapted from the original method developed by Caux et al. *(1)* and optimized for the analysis of DC developmental pathways *(2–7)*. Serum-free culture conditions required for the generation of DCs for therapeutical use will not be discussed here.

It is now widely recognized that DCs are the major antigen-presenting cells (APCs) to T-lymphocytes, which not only initiate but also control cellular immune responses (for review, *see* **ref.** *8*). Depending on their origin and activation conditions, DCs have the ability to drive T-helper-cells toward either a T_H1 or T_H2 cytokine production profile, or induce differentiation of $CCR7^+$ central memory T-lymphocytes *(9–11)*. In vivo, DCs display an astonishing diversity because no fewer than nine populations have been reported in humans. This includes Langerhans cells (LCs) from the epidermis and mucosae; dermal

From: *Methods in Molecular Biology, vol. 215: Cytokines and Colony Stimulating Factors: Methods and Protocols*
Edited by: D. Körholz and W. Kiess © Humana Press Inc., Totowa, NJ

and submucosal interstitial DCs; marginal zone splenic DCs; interdigitated DCs from the T areas of lymphoid organs; germinal center DCs (GCDCs); as well as DCs in the thymus, liver, and blood. The signification of this heterogeneity is still a matter of discussion, but it probably reflects a variety of origins and/or differentiation pathways as well as different activation/maturation stages. In this respect, one may consider that the phenotype of DCs mainly reflects the microenvironmental conditions they have encountered during their lifetime.

Dendritic cells are characterized by a unique capacity to respond rapidly (i.e., in 4–6 h) to a wide range of stimuli (e.g., viruses, microbial products, cell–cell interactions, cytokines). Upon a process called *activation/maturation*, they undergo an extensive and highly coordinated phenotypic and functional switch, in terms of (1) antigen captation, processing, and presentation capacity, (2) actin polymerization and motility, (3) chimiokine-receptor expression pattern and migratory pathways, (4) upregulation of costimulatory and adhesion molecules, as well as (5) cytokine production profile. From a synthetical point of view, one may consider that they are recruited as immature DCs to inflammatory foci where they take up antigens and become activated before migrating to T-cell areas of lymphoid organs. *In situ*, they undergo final maturation upon cognate interaction with T-lymphocytes and activate both primary and secondary cellular immune responses.

1.1. In Vitro Systems for the Differentiation of DCs

Upon appropriate culture conditions, DCs can differentiate from diverse progenitors and precursors, ranging from fetal liver $CD34^{+}CD38^{-}$ primitive HPCs to granulocyte-committed precursors and monocytes from adult blood *(12–17)* and even from leukemia blasts *(14,18)*. Although the physiological relevance of these differentiation pathways still remains elusive, this set of observations clearly indicates that, upon appropriate stimulation, a wide array of cells have the capacity to differentiate into professional APCs, in a manner which is at least partly independent of their initial lineage commitment and differentiation stage. From a developmental point of view, this raises the question of a unified ontogenic definition of the cell population referred to as DCs, the identification of which is most commonly based on morphologic, phenotypic, and functional criteria, but still awaits precise characterization at the molecular level. In practice, DCs are generated from two major cell types: $CD34^{+}$ HPCs and monocytes. When cultured under appropriate conditions, both allow for the rapid (in 5–10 d) differentiation of DCs that can be used for fundamental and clinical studies. Although these systems display some similarities, they are far from being equivalent. When choosing an in vitro differentiation system, one must keep in mind that DCs that derive from $CD34^{+}$

HPCs or monocytes differ in their differentiation pathways, phenotype, and function *(5–7,19–22)*.

From a technical standpoint, monocytes can be obtained in great numbers from adult blood; they do not require extensive purification procedures and differentiate into DCs after 4–6 d in culture in the only presence of interleukin (IL)-4 and granulocyte–macrophage colony stimulating factor (GM-CSF). For this reason, monocytes represent the system most commonly used for large-scale in vitro generation of DCs, known as MDDCs, for clinical trials (for review, *see* **ref.** ***23***). This system displays two major intrinsic limitations that need to be considered: (1) It results in the differentiation of a single population of so-called myeloid DCs with a defined phenotype and function and (2) immature CD1a⁺CD83⁻ MDDCs have a spontaneous tendency to dedifferentiate into macrophages and, thus, cannot be maintained in culture for more than 9–10 d. In addition, this in vitro DC differentiation system exhibits only limited flexibility relative to $CD34^+$ HPCs. For example, culturing monocytes with GM-CSF, IL-4, and transforming growth factor (TGF)-β1, leads to DCs that express several LC markers, such as E-cadherin or Lag/Langerin, but lack Birbeck granules (BGs) and remain $DC\text{-}SIGN^+$, the expression of which specifies dermal DCs. Conversely, supplementing cultures of $CD34^+$ HPC with TGF-β1, leads to differentiation of typical $Lag/Langerin^+$ $DC\text{-}SIGN^-BG^+$ LC. Note also that, at variance with MDDCs, immature DCs derived from $CD34^+$ HPCs are phenotypically stable and can be maintained for up to 3 wk in culture. Differentiating DCs from $CD34^+$ HPCs is, however, technically more demanding than from monocytes. Indeed, the recent diffusion of commercial kits based on anti-CD34 monoclonal antibody (MAb)-coated magnetic beads has greatly facilitated the initial purification of HPCs, but the major drawback of this system resides in the limited availability of cord blood, bone marrow, or granulocyte (G)-CSF-mobilized blood samples. At variance with MDDCs, whose differentiation proceeds without mitosis, DCs differentiate from HPCs through multiple rounds of cell divisions, which may be of interest in the view of using them as vehicles for gene therapy. The second major characteristic of this system is that it allows for the simultaneous differentiation of two DC populations *(7,24–26)* that differ at both the phenotypic and functional levels *(5,6,21,22)*.

The first reports describing the experimental conditions for efficient in vitro generation of human DCs from HPCs were published independently in 1992 *(1,27)*. In both reports, DCs were obtained from cord blood $CD34^+$ HPCs upon culture in RPMI 1640 medium supplemented with 10% fetal calf serum (FCS) in the presence of GM-CSF and tumor necrosis factor (TNF)-α. Multiple cytokines combinations have since been tested in order to optimize DCs recovery from CD34$^+$ HPCs. The work undertaken in our laboratory over the last 6 yr has led to the definition of a so-called "standard condition" for DC differentiation, based on the association of early-acting factors (stem cell factor [SCF] and Flt3 [fetal liver tyrosine kinase 3]-ligand [FL]), which promote early expansion of CD34$^+$ HPCs, to factors known to induce DC differentiation (TNF-α, GM-CSF). However, one has to keep in mind that this is an oversimplification, because both GM-CSF and TNF-α also interfere with early expansion of CD34$^+$ HPCs. Many cytokines (IL-1, IL-3, IL-6, IL-7, erythropoietin [EPO], thrombopoietin [TPO]) or even chimiokines (RANTES, MIP-1α/β) have also been tested in this system, but most failed to increase DC recovery or displayed an inhibitory effect on their differentiation. For example, compared to the standard condition, addition of IL-3 at culture initiation potentiates CD34$^+$ HPC expansion, but results in lower DC percentages *(6)*. In addition, this model displays a high degree of flexibility that needs to be emphasized. DC differentiation as well as maturation can both be manipulated in order to obtain populations greatly enriched in at least three well-defined DC subtypes. Adding IL-4 or IL-13 between culture d 5 and 8 leads to the differentiation of homogeneously immature CD1a^{bright}CD83$^-$ DCs, which represent up to 80% of the nonadherent cells recovered on d 12. Conversely, adding trimeric CD40 ligand (CD40L) or monocyte-conditioned medium (MCM) instead of IL-4 or IL-13 leads to a mature CD1a^{low}CD83$^+$ DC population that transiently adhere to plastic *(2,3,5,28)*. Also, as mentioned earlier, DC differentiation can be biased toward LC. When cultures are conducted in the continuous presence of TGF-β1 in addition to SCF, FL, GM-CSF and TNF-α, Lag/Langerin$^+$ DC-SIGN$^-$BG$^+$ LCs represent >90% of DCs recovered after 8 d *(29)*.

1.2. The Differentiation Pathways of DC from CD34$^+$ HPCs

Dendritic cell differentiation pathways from CD34$^+$ HPCs are now relatively well characterized. DCs arise from two major populations of CD1a$^+$CD14$^-$ and CD1a$^-$CD14$^+$ precursors that differ in many aspects *(5–7,21,22)*. Bipotent CD1a$^-$CD14$^+$ macrophage/DC precursors express the macrophage-CSF receptor (M-CSFR/CD115) and strongly proliferate in response to M-CSF, which also elicits their differentiation into macrophages; they require the continuous presence of IL-4, TNF-α, or TGF-β1 to unambiguously differentiate into either DCs phenotypically close to dermal DC/GCDCs or LCs. At variance with the former precursors, CD1a$^+$CD14$^-$ DC precursors do not express CD115/M-CSFR, which accounts for their lack of differentiation potential into macrophages, and they depend on TNF-α for differentiating into DCs. This cell population also displays an intrinsically high susceptibility to apoptosis *(6)* and differentiates into CD11b$^-$CD11c$^+$ DCs that express LC markers in the absence of exogenous TGF-β1. We have shown that $CD1a^+CD14^-$ DC precursors mainly derive from lymphoid-committed HPCs, whereas their $CD1a^-CD14^+$ homologs are of myeloid origin *(7,26)*. These phenotypic and ontogenic differences between $CD1a^+CD14^-$ and $CD1a^-CD14^+$ precursor-derived DCs are apparently also relevant at the functional level because, when differentiated under the standard culture condition, $CD1a^+CD14^-$ precursor-derived DCs have a greater capacity to stimulate allogeneic T-lymphocytes than their $CD1a^-CD14^+$ precursor-derived counterparts in vitro *(5)*, the latter retaining higher antigen uptake capacity *(21)*. Although these observations support the hypothesis of an ontogenic control of DC function, one may keep in mind that, at least in humans, their in vivo relevance is far from being established.

A third population of DC precursors of lymphoid origin, plasmacytoid cells, or pDC2, identified as the principal type I interferon-producing cells of the immune system, has been recently described *(30,31)*. These precursors are characterized by high-level expression of CD123/IL-3R and of pre-TCR-α *(32)* and dependence on IL-3 for survival. They also require CD40 ligation or virus stimulation to differentiate into DCs, now referred to as DC2s *(13,33–35)*. In addition to the unique capacity to produce high amounts of type I interferon, DC2s display additional features that distinguish them from $CD1a^+CD14^-$ and $CD1a^-CD14^+$ precursor-derived DCs. They have an apparently more limited capacity to stimulate allogeneic T-lymphocytes than the former *(34)*, do not produce IL-12, and require virus stimulation to drive T-helper-lymphocytes to produce T_H1 cytokines *(11,30)*. pDC2s have recently been differentiated in vitro from certain $CD34^+$ HPC populations upon culture with FL only or coculture with the bone marrow murine S17 stromal cell line *(24,25)*. However, whereas culture of $CD34^+$ HPCs with FLs only results in delayed and poorly efficient pDC2 differentiation (<5% of cells recovered after 3–4 wk of culture) *(24)*, co-culture of fetal liver $CD34^+CD38^-$ or thymic $CD34^+CD1a^-$ HPCs with S17 cells, without exogenous growth factors, appears more efficient. This method allows the generation of 5–15% $CD123/IL\text{-}3R^{bright}pre\text{-}TCR\text{-}\alpha^+$ pDC2s after only 5 d *(25)*. Of note, these data are in line with the theoretical assumption that all DC populations found in vivo can, in principle, differentiate from $CD34^+$ HPCs in vitro.

1.3. Influence of CD34⁺ HPC Origin and Lineage Commitment on DC Differentiation

It is well established that the ability of $CD34^+$ HPCs to self-renew and expand is ontogenetically regulated *(36)*. Compared to $CD34^+$ HPCs from the adult bone marrow, $CD34^+$ HPCs from the fetal liver or cord blood are highly enriched in pluripotent primitive progenitors with long-term repopulating capacity and in highly proliferative potential colony-forming cells (HPP-CFCs) with remarkable clonogenic as well as expansion capacity ***(37–40)***. CD34$^+$ HPC populations also differ with regard to the balance between immature pluripotent and lineage-committed progenitors and, among lineage-committed progenitors, in the relative proportion of erythroid, megakaryocytic, lymphoid, and myeloid progenitors. For example, B-lineage-committed CD34$^+$CD19$^+$ HPCs, which are absent or barely detectable in the cord blood, represent up to 50% of bone marrow CD34$^+$ HPCs. This is also true for thymic CD34$^+$ HPCs, which comprise >90% of unipotent T-lineage-committed CD34$^+$CD1a$^+$ progenitors ***(26)***.

CD34$^+$ HPCs from the fetal liver or bone marrow, neonatal cord blood or from adult bone marrow or G-CSF-mobilized blood, as well as from the thymus, have been successfully used to generate DCs in vitro ***(25,26,41–45)***. However, recent reports have shown that these populations cannot be considered as equivalent in terms of DC differentiation potential, the estimation of which is primarily based on quantitative parameters such as percentages and absolute numbers of CD1a$^+$ DCs. There is strong evidence that DC recovery closely parallels CD34$^+$ HPC expansion capacity: upon 7–8 d culture with SCF, FL, GM-CSF, and TNF-α, the mean expansion of cord blood CD34$^+$ HPCs ranges from 100-fold to 150-fold relative to 20-fold to 30-fold for adult bone marrow CD34$^+$ HPCs and about 10-fold for thymic CD34$^+$CD1a$^-$ HPCs. Thus, as compared with cord blood CD34$^+$ HPCs, thymic CD34$^+$CD1a$^-$ HPCs generate only limited numbers of DCs ***(26)***. There is also evidence that limited expansion of CD34$^+$ HPCs correlates with accelerated kinetics of DC differentiation, because CD1a$^+$ DC percentages peak as early as culture d 4–6 in thymic CD34$^+$CD1a$^-$ HPC cultures relative to 10–12 d for cord blood CD34$^+$ HPCs.

Such remarkable heterogeneity of CD34$^+$ HPC populations prompted us to analyze the relationship between lineage commitment and DC differentiation among cord blood CD34$^+$ HPC populations. Based on CD7 and CD45RA expression, we have characterized two novel CD34$^+$ HPC populations that differ with regard to lymphoid differentiation potential ***(7)***: Whereas CD7$^-$CD45RA$^+$ single positive (SP) HPCs behave as typical myeloid progenitors, double positive (DP) CD7$^+$CD45RA$^+$ HPCs have the capacity to generate CD56$^+$ natural killer (NK) cells in culture with SCF, IL-2, IL-7, and IL-15, an indication of their lymphoid potential. When cultured under our standard condition for DC differentiation, both populations generate DCs, albeit with different efficiency: CD1a$^+$ DC represent 78 ± 8% of nonadherent cells on d 8 of culture of DP HPCs vs 37 ± 15% for their SP counterparts; however, because of limited expansion, the DC yield is threefold lower from DP HPCs than from SP HPCs.

Our major finding was that CD34$^+$ HPC commitment strongly influences the DC differentiation pathways, in that only DP progenitors have the capacity to generate M-CSF-resistant CD1a$^+$CD14$^-$ unipotent DC precursors that differentiate into Lag$^+$S100$^+$ LCs in the absence of exogenous TGF-β1 *(7,26)*. Conversely, SP HPCs differentiate along a unique CD1a$^{+/-}$CD14$^+$ M-CSFR/CD115$^+$ M-CSF-sensitive pathway and require the continuous presence of TGF-β1 to acquire LC markers. These differences are also functionally relevant, because DP HPC-derived DCs produce twofold to fourfold more IL-6, IL-12, and TNF-α upon CD40 ligation and elicit threefold to sixfold greater allogeneic T-lymphocyte reactivity than their SP HPC-differentiated homologs.

2. Materials

2.1. Cell Purification

1. Anti-CD34 MAb-coated magnetic beads (M-450 Dynabeads [Dynal®, Oslo, Norway]).
2. Anti-mouse IgG1-coated magnetic beads: RAM IgG1 CELLectionTM Kit (Dynal®).
3. Anti-CD1a MAb (Coulter Clone, Margency, France).
4. Phosphate-buffered saline (PBS), 2% fetal calf serum (FCS).
5. DNase I from bovine pancreas cell culture grade (Boehringer-Mannheim, Germany).

2.2. Culture Media, Cytokines, and Growth Factors

1. RPMI 1640, 10% FCS, 1% glutamine, 1% antibiotics.
2. GM-CSF (Schering Plough, Kenilworth, NJ) (store at –20°C).
3. SCF (Amgen, Thousand Oaks, CA) (stable at 4°C).
4. Flt3 ligand (Immunex, Seattle, WA) (stable at 4°C).
5. TNF-α (Genzyme, Cambridge, MA) (store at –20°C).
6. TGF-β1 (Genzyme) (store at –20°C).
7. IL-4 (Schering Plough, Kenilworth, NJ) (store at –20°C).
8. Trimeric CD40L (Immunex) (store at –20°C).

2.3. Monoclonal Antibodies

1. Anti-CD34 MAb (clone 581; Immunotech, Marseille, France).
2. CD1a MAb (clone HI149, Becton Dickinson, San Jose, CA).
3. CD83 MAb (clone HB15A, Immunotech).

3. Methods

3.1. Isolation of CD34$^+$ Cells from Cord Blood

Although, diverse methods such as adherence onto plastic-immobilized CD34 MAb or fluorescence-activated cell sorter (FACS) sorting of CD34-MAb-immunolabeled cells have been tested, we recommend use of commercial CD34-MAb-coated magnetic beads for purification of CD34$^+$ HPCs. In spite of the relatively high cost of magnetic beads, this method allows for the rapid

(2–3 h) and efficient isolation of CD34$^+$ HPCs with excellent viability. Here, we describe the protocol we use for purification of CD34$^+$ HPCs from cord blood with M-450 Dynabeads.

1. Mononuclear cells (MNCs) are isolated by conventional Ficoll–Paque (Pharmacia, Uppsala, Sweden) centrifugation from normal cord blood collected according to institutional guidelines.
2. The MNCs are then washed twice in RPMI 1640 medium and resuspended in cold (4°C) PBS, 2% FCS (10^8 cells/mL).
3. Cord blood can be stored overnight at 4°C, but it has to be processed no later than 24–36 h after collection.
4. Magnetic beads are harvested (4×10^7 beads/10^8 MNCs) and washed four to five times in cold PBS, 2% FCS, before being added to the cells in the presence of DNase I (50–100 µg/mL), the latter to prevent cell clusterization.
5. The cell beads suspension is then incubated and rotated (10–20 rpm) at 4°C for 30 min, which optimizes interactions between CD34-MAb-coated beads and the cells.
6. Rosetted CD34$^+$ cells are then magnetically separated from CD34$^-$ cells by three to four passages through the Dynal MPCs (magnetic particles concentrator).
7. Cell beads conjugates are resuspended in PBS, 2% FCS (100 µL/10^8 MNCs) and diluted further in the same volume of DETACHaBEAD.
8. Cell suspensions are incubated for 30 min at 37°C with gentle vortexing every 10 min to allow the release of the rosetted cells.
9. CD34$^+$ cells are finally separated from M-450 Dynabeads by three to five passages through Dynal MPCs.
10. The CD34$^+$ cells are then pooled, pelleted, and resuspended in a small volume of RPMI 1640, 10% FCS, 1% glutamine, 1% antibiotics.
11. The purity of the preparation is determined by labeling about 5×10^3 cells with either an irrelevant or relevant CD34 MAb (clone 581 diluted 1:100) in PBS, 2% FCS. After another 30 min incubation at 4°C, cells are washed and FACS analyzed. Purity of the resulting CD34$^+$ HPCs should be >85%.

3.2. Standard Procedures for DC Differentiation

1. CD34$^+$ cells are seeded at $1–2 \times 10^4$/mL in six-well plates and cultured for 5 d in complete culture medium (RPMI 1640, 10% FCS, 1% glutamine, 1% antibiotics) supplemented with the following human recombinant cytokines: 20 ng/mL GM-CSF, 50 ng/mL SCF, 50 ng/mL FL, and 50 U/mL TNF-α. Cultures are conducted at 37°C in humidified 5% CO_2 (*see* **Notes 1–3**).
2. Medium and cytokines are renewed on d 3. When the experimental design requires differential sorting of CD1a$^+$CD14$^-$ and CD1a$^-$CD14$^+$ precursors, the medium can be supplemented by 2.5% human AB serum, which prevents acquisition of CD1a expression by bipotent CD1a$^-$CD14+ DC precursors *(16)*.
3. Nonadherent cells are harvested on d 5, counted, and cultured for another 3 d with fresh medium and cytokines. DCs can be maintained in cultures for up

Table 1
Kinetics of Cord Blood $CD34^+$ HPC Growth and DC Differentiation Under the Standard Culture Condition

Days in culture	NAC numbers ($\times 10^{-6}$)	Total $CD1a^+$ DCs (%)	DC precursors (%): $CD1a^+CD14^-$ (%)	DC precursors (%): $CD1a^-CD14^+$ (%)
5	3.5 ± 1.5	18 ± 3	8 ± 2	15 ± 3
8	12 ± 4	27 ± 9	—	—
12	19 ± 5	36 ± 14	—	—

Note: $CD34^+$ HPCs were cultured in the presence of SCF, FL, GM-CSF, and TNF-α, referred to as the standard condition. Viable nonadherent cell (NACs) numbers were normalized relative to 1×10^5 seeded $CD34^+$ HPCs. Percentages of $CD1a^+$ DCs, and $CD1a^+CD14^+$ and $CD1a^-CD14^+$ DC precursors were determined by FACS analysis of cells labeled with the corresponding MAbs; they represent mean ± SD value of three to eight independent experiments.

to 3 wk, provided that fresh medium and cytokines are added every 4–5 d. Optimal results in terms of both DC recovery and viability are obtained on culture d 10–12.

4. Dendritic cell differentiation is followed by sequential labeling with fluoresceine isothiocyanate (FITC) or phycoerythrin (PE) conjugated CD1a MAb (clone HI149, dilution 1:100). The maturation stage is assessed starting from culture d 8 by labeling with a CD83 MAb (clone HB15A, dilution 1:100). An example of cell growth and DC differentiation kinetics is shown in **Table 1**.

Starting from 1×10^5 cord blood $CD34^+$ HPCs, this basic protocol leads, in principle, to the generation of $5–12 \times 10^6$ $CD1a^+$ DCs on culture d 10–12 and results in DC populations that appear highly heterogeneous with regard to their phenotype (Lag$^{+/-}$, Langerin$^{+/-}$, BG$^{+/-}$), their differentiation pathways (via $CD1a^+CD14^-$ or $CD1a^-CD14^+$ precursors) and maturation stages (CD83$^{+/-}$ DCs). Therefore, this system can be readily manipulated to obtain homogeneous populations of DCs with defined phenotypes. For example, when cultures are conducted in the continuous presence of TGF-β1 (1-10 ng/mL), typical Langerin$^+$Lag$^+$BG$^+$ LCs represent >90% of the $CD1a^+$ DCs recovered on culture d 8–10 ***(29,46)***. Also, the addition of IL-4 or IL-13 (50 ng/mL) from culture d 5 results in homogeneously immature $CD1a^{bright}CD83^-$ DCs 72 and 96 h later, whereas adding trimeric CD40L (250–500 ng/mL; kind gift from Immunex) results in $CD1a^{+/-}CD83^+$ mature DCs.

3.3. Isolation of Immature and Mature DC from Cultures of $CD34^+$ HPCs

After 10–12 d, $CD34^+$ HPC-derived DCs represent 40–80% of nonadherent cells, depending on the culture condition. For this reason, and at variance with

MDDCs, analysis of their phenotype or function usually requires a secondary purification step. In our hands, both types of DC precursors and mature $CD83^+$ DCs, which are relatively resistant to transient growth factor deprivation as well as to mechanical trauma, can be FACS sorted from d 5–6 and d 10–12 cultures, respectively. However, purification of immature $CD1a^{bright}CD83^-$ DCs, especially those obtained from cultures supplemented with IL-4, is more delicate. These cells display exquisite susceptibility to mechanical stress, which often results in high mortality and precludes conventional FACS sorting.

1. As an alternative for isolation of immature DCs, we have successfully used beads with a rat MAb against mouse IgG1 attached to the beads via a DNA linker (Dynal RAM IgG1 CELLection™ Kit), so that immunoselected cells can be released from the beads by incubation with releasing buffer containing DNase I *(47)*.
2. RAM IgG1 beads are harvested and washed four to five times in PBS, 0.1% BSA, before being incubated for 30 min with CD1a T6 MAb (3–4 μg/10^7 beads).
3. CD1a-MAb–coated beads are washed again to remove unbound MAb, and added to the cells (10^7 cells/mL), at a ratio of 5 beads per target cell.
4. The mix is incubated and rotated (10–20 rpm) for 15 min at 4°C.
5. Rosetted $CD1a^+$ cells are then separated from $CD1a^-$ cells by three to four passages through the Dynal MPCs, and resuspended in RPMI 1640, 2% FCS (200 μL/5 × 10^7 beads).
6. Cell beads conjugates are finally incubated for 15 min at 37°C with DNase I (200 U/10^7 beads) and gently vortexed every 5 min, and $CD1a^+$ cells are separated by four to five passages through the Dynal MPCs.
7. Immature DCs are then pooled, pelleted, and resuspended in a small volume of RPMI 1640, 10% FCS, 1% glutamine, 1% antibiotics. Because this purification procedure may also alter DC viability, we recommend culturing them for another 24 h under the standard condition plus IL-4 before use.

4. Notes

1. As mentioned previously, differentiating DCs from $CD34^+$ HPCs does not raise any special technical difficulty. Nonetheless, optimal DC recovery depends on both the quality of $CD34^+$ HPC purification, which must preserve their viability, and the use of optimal cytokine combinations and dosages. With regard to isolation of $CD34^+$ HPCs, we recommend discarding $CD34^+$ HPC preparations, the purity of which is <80%. Although the presence of contaminant T- or B-lymphocytes or monocytes does not significantly interfere with $CD34^+$ HPC growth and DC differentiation, this may represent a major impediment for studies of their differentiation pathways as well as function.
2. We have performed systematic dose–effect studies for all cytokines and growth factors used in this system. With regard to early-acting factors, we have shown that, provided they are used above a 50-ng/mL baseline, SCF and FL synergisti-

cally support efficient expansion of $CD34^+$ HPCs without any adverse effect and do not interfere with DC differentiation or function. However, GM-SCF and TNF-α display a more complex activity spectrum than the former. GM-CSF not only potentiates the growth of myeloid HPCs, but it also represents a key survival factor for both $CD1a^+CD14^-$ DC precursors and immature DCs. When cultures are conducted with suboptimal doses of GM-CSF (<20 ng/mL), this unequivocally leads to apoptosis of as much as 50–75% of $CD1a^+$ DCs, as noted on culture d 10–12 and, thereby, in a considerable drop in DC recovery ***(6)***. For these reasons, we recommend using this cytokine at 20–50 ng/mL.

3. Tumor necrosis factor-α is known to exhibit pleiotropic and sometimes dual dose-dependent effects on human hematopoiesis as well as to interfere with all steps of DC differentiation ***(6)***. During the first 5 d of culture, this cytokine synergizes with early-acting factors to potentiate $CD34^+$ HPC cycling and expansion, which results in a more than a twofold increase in nonadherent cell recovery by d 8. From a developmental standpoint, TNF-α also represents a key factor that controls differentiation of $CD1a^+CD14^-$ DC precursors from early HPCs. During the second week of culture, TNF-α promotes the differentiation of bipotent $CD1a^-CD14^+$ precursors into DCs at the expense of macrophages while inducing subsequent maturation of immature DCs. For these reasons, the addition of suboptimal doses of TNF-α (10–25 U/mL) results in only limited $CD34^+$ HPC expansion and inefficient DC differentiation. Conversely, when used at high doses (250–1000 U/mL), TNF-α inhibits $CD34^+$ HPC growth and induces early maturation of DCs that acquire CD83 expression as early as culture d 6–7. Thus, the concentration of TNF-α in $CD34^+$ HPC cultures must be carefully determined (50–100 U/mL) and every new batch of the cytokine needs to be repeatedly tested in dose-response experiments (range: 10–1000 U/mL) before use in routine experiments.

References

1. Caux, C., Dezutter-Dambuyant, C., Schmitt, D., and Banchereau, J. (1992) GM-CSF and TNF-alpha cooperate in the generation of dendritic Langerhans cells. *Nature* **360,** 258–261.
2. Rosenzwajg, M., Canque, B., and Gluckman, J. C. (1996) Human dendritic cell differentiation pathway from $CD34^+$ hematopoietic precursor cells. *Blood* **87,** 535–544.
3. Rosenzwajg, M., Camus, S., Guigon, M., and Gluckman, J. C. (1998) The influence of interleukin (IL)-4, IL-13, and Flt3 ligand on human dendritic cell differentiation from cord blood $CD34^+$ progenitor cells. *Exp. Hematol.* **26,** 63–72.
4. Rosenzwajg, M., Tailleux, L., and Gluckman, J. C. (2000) CD13/N-Aminopeptidase is involved in the development of dendritic cells and macrophages from cord blood CD34(+) cells. *Blood* **95,** 453–460.
5. Canque, B., Camus, S., Yagello, M., and Gluckman, J. C. (1998) IL-4 and CD40 ligation differently affect the differentiation, maturation, and function of human

CD34$^+$ cell-derived CD1a$^+$CD14$^-$ and CD1a$^-$CD14$^+$ dendritic cell precursors in vitro. *J. Leuk. Biol.* **64,** 235–244.
6. Canque, B., Camus, S., Yagello, M., and Gluckman, J. C. (1998) Special susceptibility to apoptosis of CD1a$^+$ dendritic cell precursors differentiating in vitro from cord blood CD34$^+$ progenitors. *Stem Cells* **16,** 218–228.
7. Canque B., Camus S., Dalloul, A., Kahn, E., Yagello, M., Dezutter-Dambuyant, C., et al. (2000) Characterization of dendritic cell differentiation pathways from cord blood CD34(+)CD7(+)CD45RA(+) hematopoietic progenitor cells. *Blood* **96,** 3748–3756.
8. Banchereau, J., Briere, F., Caux, C., Davoust, J., Lebecque, S., Liu, Y. J., et al. (2000) Immunobiology of dendritic cells. *Annu. Rev. Immunol.* **18,** 767–811.
9. Langenkamp, A., Messi, M., Lanzavecchia, A., and Sallusto, F. (2000) Kinetics of dendritic cell activation: impact on priming of T_H1, T_H2 and non polarized T cells. *Nat. Immunol.* **1,** 311–316.
10. Moser, M. and Murphy, K. M. (2000) Dendritic cell and the regulation of TH1-TH2 development. *Nat. Immunol.* **1,** 199–205.
11. Cella, M., Facchetti, F., Lanzavecchia, A., and Colonna, M. (2000) Plasmacytoid dendritic cells activated by influenza virus and CD40L drive a potent TH1 polarization. *Nat. Immunol.* **1,** 305–310.
12. Cella, M., Jarrossay, D., Facchetti, F., Alebardi, O., Nakajima, H., Lanzavecchia, A., et al. (1999) Plasmacytoid monocytes migrate to inflamed lymph nodes and produce large amounts of type I interferon. *Nat. Med.* **5,** 919–923.
13. Grouard, G., Rissoan, M. C., Filgueira, L., Durand, I., Banchereau, J., and Liu, Y. J. (1997) The enigmatic plasmacytoid T cells develop into dendritic cells with IL-3 and CD40-ligand. *J. Exp. Med.* **185,** 1101–1111.
14. Oehler, L., Berer, A., Keil, F., Weinlander, G., Konig, M., Haas, O. A., et al. (2000) Generation of dendritic cells from human chronic myelomonocytic leukemia cells in fetal calf serum-free medium. *Leuk. Lymphoma* **38,** 577–586.
15. Strobl, H., Scheinecker, C., Riedl, E., Csmarits, B., Bello-Fernandez, C., Pikl, W., et al. (1998) Identification of CD68$^+$Lin$^-$ peripheral blood cells with dendritic precursor characteristics. *J. Immunol.* **161,** 740–748.
16. Takamizawa, M., Rivas, A., Fagnoni, F., Benike, C., Kosek, J., Hyakawa, H., et al. (1997) Dendritic cell that process and present nominal antigen to naive T lymphocytes are derived from CD2+ precursors *J. Immunol.* **158,** 2134–2142.
17. Ito, T., Inaba, M., Inaba, K., Toki, J., Sogo, S., Iguchi, T., et al. (1999) A CD1a$^+$/CD11c$^+$ subset of human blood dendritic cells is a direct precursor of Langerhans cells. *J. Immunol.***163,** 1409–1419.
18. Kohler, T., Plettig, R., Wetzstein, W., Schmitz, M., Ritter, M., Mohr, B., et al. (2000) Cytokine-driven differentiation of blasts from patients with acute myelogenous and lymphoblastic leukemia into dendritic cells. *Stem Cells* **18,** 139–147.
19. Ferlazzo, G., Klein, J., Paliard, X., Wei, W. Z., and Galy, A. (2000) Dendritic cells generated from CD34$^+$ progenitor cells with flt3 ligand, c-kit ligand, GM-CSF, IL-4, and TNF-alpha are functional antigen-presenting cells resembling mature monocyte-derived dendritic cells. *J. Immunother.* **23,** 48–58.

20. Ferlazzo, G., Wesa, A., Wei, W. Z., and Galy, A. (1999) Dendritic cells generated either from CD34+ progenitor cells or from monocytes differ in their ability to activate antigen-specific CD8+ T cells. *J. Immunol.* **163,** 3597–3604.
21. Caux, C., Massacrier, C., Vanbervliet, B., Dubois, B., Durand, I., Cella, M., et al. (1997) CD34+ hematopoietic progenitors from human cord blood differentiate along two independent dendritic cell pathways in response to granulocyte-macrophage colony-stimulating factor plus tumor necrosis factor alpha: II. Functional analysis. *Blood* **90,** 1458–1470.
22. Caux, C., Vanbervliet, B., Massacrier, C., Dezutter-Dambuyant, C., de Saint-Vis, B., Jacquet, C., et al. (1996) CD34+ hematopoietic progenitors from human cord blood differentiate along two independent dendritic cell pathways in response to GM-CSF+TNF alpha. *J. Exp. Med.* **184,** 695–706.
23. Gluckman, J. C., Canque, B., Chapuis, F., and Rosenzwajg, M. (1997) In vitro generation of human dendritic cells and cell therapy. *Cytok. Cell. Mol. Ther.* **3,** 187–196.
24. Blom, B., Ho, S., Antonenko, S., and Liu, Y. J. (2000) Generation of interferon alpha-producing predendritic cell (Pre-DC)2 from human CD34(+) hematopoietic stem cells. *J. Exp. Med.* **192,** 1785–1796.
25. Spits, H., Couwenberg, F., Bakker, A. Q., Weijer, K., and Uittenbogaart, C. H. (2000) Id2 and Id3 inhibit development of CD34(+) stem cells into predendritic cell (pre-DC)2 but not into pre-DC1. Evidence for a lymphoid origin of pre-DC2. *J. Exp. Med.* **192,** 1775–1784.
26. Dalloul, A., Patry, C., Salamero, J., Canque, B., Grassi, F., and Schmitt, C. (1999) Functional and phenotypic characterization of thymic CD34+CD1a− progenitors-derived dendritic cells: predominance of CD1a+ differentiation pathway. *J. Immunol.* **162,** 5821–5828.
27. Santiago-Schwarz, F., Belilos, E., Diamond, B., and Carsons, S. E. (1992) TNF in combination with GM-CSF enhances the differentiation of neonatal cord blood stem cells into dendritic cells and macrophages. *J. Leuk. Biol.* **52,** 274–281.
28. Caux, C., Massacrier, C., Vanbervliet, B., Dubois, B., Van Kooten, C., Durand, I., et al. (1994) Activation of human dendritic cells through CD40 cross-linking. *J. Exp. Med.* **180,** 1263–1272.
29. Jaksits, S., Kriehuber, E., Charbonnier, A. S., Rappersberger, K., Stingl, G., and Maurer, D. (1999) CD34+ cell-derived CD14+ precursor cells develop into Langerhans cells in a TGF-beta1-dependent manner. [in process citation]. *J. Immunol.* **163,** 4869–4877.
30. Kadowaki, N., Antonenko, S., Lau, J. Y., and Liu, Y. J. (2000) Natural interferon alpha/beta-producing cells link innate and adaptive immunity. *J. Exp. Med.* **192,** 219–226.
31. Siegal, F. P., Kadowaki, N., Shodell, M., Fitzgerald-Bocarsly, P. A., Shah, K., Ho, S., et al. (1999) The nature of the principal type 1 interferon-producing cells in human blood. *Science* **284,** 1835–1837.
32. Res, P. C., Couwenberg, F., Vyth-Dreese, F. A., and Spits, H. (1999) Expression of pTalpha mRNA in a committed dendritic cell precursor in the human thymus. *Blood* **94,** 2647–2657.

33. Rissoan, M. C., Soumelis, V., Kadowaki, N., Grouard, G., Briere, F., de Waal Malefyt, R., et al. (1999) Reciprocal control of T helper cell and dendritic cell differentiation. *Science* **283,** 1183–1186.
34. Pulendran, B., Banchereau, J., Burkeholder, S., Kraus, E., Guinet, E., Chalouni, C., et al. (2000) Flt3-Ligand and granulocyte colony-stimulating factor mobilize distinct human dendritic cell subsets in vivo. *J. Immunol.***165,** 566–572.
35. Vandenabeeleh S., Hochreinh H., Mavaddat, N., Winkel, K., and Shortman, K. (2001) Human thymus contains 2 distinct dendritic cell populations. *Blood* **97,** 1733–1741.
36. Lansdorp, P. M., Dragowska, W., and Mayani, H. (1993) Ontogeny-related changes in proliferative potential of human hematopoietic cells. *J. Exp. Med.* **178,** 787–791.
37. Hows, J. M., Bradley, B. A., Marsh, J. C., Luft, T., Coutinho, L., Testa, N. G., et al. (1992) Growth of human umbilical-cord blood in longterm haemopoietic cultures. *Lancet* **340,** 73–76.
38. Hao, Q. L., Shah, A. J., Thiemann, F. T., Smogorzewska, E. M., and Crooks, G. M. (1995) A functional comparison of $CD34^+$ $CD38^-$ cells in cord blood and bone marrow. *Blood* **86,** 3745–3753.
39. Broxmeyer, H. E., Douglas, G. W., Hangoc, G., Cooper, S., Bard, J., English, D., et al. (1989) Human umbilical cord blood as a potential source of transplantable hematopoietic stem/progenitor cells. *Proc. Natl. Acad. Sci. USA* **86,** 3828–3832.
40. Gluckman, E., Broxmeyer, H. A., Auerbach, A. D., Friedman, H. S., Douglas, G. W., Devergie, A., et al. (1989) Hematopoietic reconstitution in a patient with Fanconi's anemia by means of umbilical-cord blood from an HLA-identical sibling. *N. Engl. J. Med.* **321,** 1174–1178.
41. Res, P., Martinez-Caceres, E., Jaleco, A., Staal, F., Noteboom, E., Weijer, K., et al. (1996) $CD34^+CD38^{dim}$ cells in the human thymus can differentiate into T, natural killer, and dendritic cells but are distinct from pluripotent stem cells. *Blood* **12,** 5196–5206.
42. Herbst, B., Köhler, G., Mackensen, A., Veelken, H., Kulmburg, P., Rosenthal, F. M., et al. (1996) In vitro differentiation of $CD34^+$ hematopoietic progenitor cells toward distinct dendritic cell subsets and MIIC-positive Langerhans cells and the interdigitating dendritic cell type. *Blood* **88,** 2541–2548.
43. Herbst, B., Kohler, G., Mackensen, A., Veelken, H., Mertelsmann, R., and Lindemann, A. (1997) $CD34^+$ peripheral blood progenitor cell and monocyte derived dendritic cells: a comparative analysis. *Br. J. Haematol.* **99,** 490–499.
44. Galy, A., Travis, M., Cen, D., and Chen, B. (1995) Human T, B, natural killer, and dendritic cells arise from a common bone marrow progenitor cell subset. *Immunity* **3,** 459–473.
45. Marquez, C., Trigueros, C., Franco, J., Ramiro, A., Carrasco, Y., Lopez-Botet, M., et al. (1998) Identification of a common developmental pathway for thymic natural killer cells and dendritic cells. *Blood* **91,** 2760–2771.
46. Caux, C., Massacrier, C., Dubois, B., Valladeau, J., Dezutter-Dambuyant, C., Durand, I., et al. (1999) Respective involvement of TGF-beta and IL-4 in the

development of Langerhans cells and non-Langerhans dendritic cells from CD34+ progenitors. *J. Leuk. Biol.* **66,** 781–791.

47. Canque, B., Bakri, Y., Camus ,S., Yagello, M., Benjouad, A., and Gluckman, J. C. (1999) The susceptibility to X4 and R5 human immunodeficiency virus-1 strains of dendritic cells derived in vitro from CD34(+) hematopoietic progenitor cells is primarily determined by their maturation stage. *Blood* **93,** 3866–3875.

24

Generation of Transgenic T Cells from Human CD34+ Cord Blood Cells

Bruno Verhasselt, Evelien Naessens, and Veronique Stove

1. Introduction

Hematopoietic stem cells (HSCs), progenitor cells capable of generating all peripheral blood cell lineages, are believed not to express lineage markers (Lin) found on the latter, nor CD38, but may or may not express CD34. Besides HSCs, more committed $CD34^+$ progenitors expressing CD38, and lineage markers like CD7 or IL-7R can generate T cells in the thymus *(1)*. Cord blood, fetal liver, bone marrow, and peripheral blood $CD34^+$ cells can generate T cells. These progenitors differ in their cytokine responsiveness and mitogenic potential, as exemplified by retroviral transduceability. Gene-marked and cultured progenitors can be studied for their ability to generate T cells, either in vivo or in vitro *(2,3)*. Human fetal thymus fragments are transplanted in severe combined immune deficiency (SCID) mice to establish an in vivo model for T cell generation from these manipulated progenitors (SCID-hu). Alternatively, an in vitro model can be used in which a cultured fetal thymic lobe of non-obese diabetic (NOD)-SCID mice serves as the three-dimensional cellular network that supports human T-cell generation (fetal thymic organ culture [FTOC]). In contrast to the SCID-hu in vivo model, FTOC allows comparing numerous experimental conditions, such as cytokine cocktail titrations, with a manageable work load *(3)*. In the FTOC, progenitor cells generate CD4/CD8 double-positive immature and ultimately single-CD4-positive and single-CD8-positive mature thymocytes. In this chapter, we will detail 1° isolation of hematopoietic progenitor cells from human cord blood, 2° cytokine-stimulated retroviral transduction of these progenitors, and 3° assay of T-cell generation from gene-marked progenitors in FTOC.

From: *Methods in Molecular Biology, vol. 215: Cytokines and Colony Stimulating Factors: Methods and Protocols*
Edited by: D. Körholz and W. Kiess © Humana Press Inc., Totowa, NJ

2. Materials

2.1. Isolation of Precursor Cells

1. Phosphate-buffered saline (PBS; dissolve $Na_2HPO_4 \cdot 12H_2O$ [3.008 g/L or 8.4 m*M*), $NaH_2PO_4 \cdot 1H_2O$ (0.255 g/L or 1.85 m*M*], NaCl [8.182 g/L or 140 m*M*] in 0.8 L milli-Q water, adjust pH to 7.2–7.3; adjust volume to 1 L; osmolarity should be 275–285 mOsm, filter through 0.22-μm filter, store at 4°C)
2. Fetal calf serum (FCS; Invitrogen, Paisley, UK).
3. PBS with 2% (v/v) FCS.
4. Lymphoprep (Nyegaard, Oslo, Norway).
5. Dimethyl sulfoxide (DMSO; Serva, Heidelberg, Germany).
6. Hemocytometer counting chamber.

2.1.1. For Isolation of Highly Purified $CD34^+$ Lineage $Marker^-CD38^{+/-}$ Cells

1. Mouse anti-human monoclonal antibodies (MAbs): glycophorin-A, CD19; fluorescein isothiocyanate (FITC)-labeled CD7-FITC, CD1-FITC, CD3-FITC, CD4-FITC, CD8-FITC; phycoerythrin (PE)-labeled CD34, biotin-labeled CD38. Conjugate: streptavidin, labeled with allophycocyanin (APC) or tricolor (TC). These can be obtained from many different suppliers.
2. Sheep anti-mouse immunoglobulin-coated Dynabeads (Dynal AS, Oslo, Norway).
3. Magnetic particle concentrator (Dynal AS).
4. Cell sorter equipped with an argon-ion laser (488 nm) and helium–neon diode laser (635 nm).

2.1.2. Isolation of $CD34^+$ Cells

1. MACS Direct CD34 Progenitor Cell Isolation Kit (large scale; Miltenyi Biotec, Bergisch Gladbach, Germany).
2. Magnetic cell separator MidiMACS (Miltenyi Biotec).
3. MACS MultiStand (Miltenyi Biotec).
4. LS separation columns (Miltenyi Biotec). Capacity: maximum 2×10^9 total cells and 10^8 magnetically labeled cells.
5. Cell strainer, 40 μm (Becton Dickinson, Mountain View, CA).

2.2. Culture and Transduction

1. Cell culture incubator providing a humidified atmosphere containing 7.5% (v/v) CO_2 in air.
2. Iscove's modified Dulbecco's medium, supplemented with penicillin (100 IU/mL), streptomycin (100 μg/mL), and 10% heat-inactivated FCS (complete IMDM, all products from Invitrogen).
3. All cytokines are used at a fixed concentration during culture: 100 ng/mL recombinant human c-kit ligand (stem cell factor [SCF]), 100 ng/mL recombinant human flt3/flk-2 ligand (FL), 20 ng/mL thrombopoietin (TPO), 10 ng/mL

recombinant human interleukin-3 (IL-3), 100 U/mL recombinant human IL-6 (cytokines from multiple suppliers).

4. Retrovirus (e.g., MFG-GFP virus produced with Phoenix-NA amphotropic packaging cell line). The retrovirus used encoded the marker gene enhanced green fluorescent protein (EGFP; Clontech, Palo Alto, CA), constructed and produced as described previously *(2)*.
   a. A MFG retroviral vector encoding EGFP is constructed by inserting a *Nco*I–*Bam*HI fragment containing EGFP between the unique corresponding restriction sites of the MFG retroviral vector (kindly provided by R. C. Mulligan, Harvard Medical School, Boston, MA).
   b. A plasmid preparation from DH5α bacteria (Invitrogen) is prepared using standard techniques.
   c. Twenty micrograms of plasmid is transfected using calcium–phosphate precipitation (Invitrogen) into 1.5×10^6 Phoenix-A retroviral packaging cells (kindly provided by Dr. G. P. Nolan, Stanford University School of Medicine, Stanford, CA), grown to semiconfluency on 6-cm-diameter Petri dishes (Becton Dickinson).
   d. After 2 d of culture, cells are detached from dish with Cell Dissociation Buffer (Invitrogen).
   e. Green fluorescent protein strong positive cells are sorted using a cell sorter (expect more than 50% GFP positive cells).
   f. Sorted cells are seeded at least $2 \times 10^3/cm^2$ in a tissue culture flask (e.g., 25-cm^2 flask).
   g. Another two rounds of sorting of the GFP strongest cells are done upon confluency of the cell culture, to establish a stable producer cell line in about 4 wk time.
   h. Once established, the producer cell line is seeded in tissue culture flasks. Medium is refreshed 1 d before harvest of the supernatants. Highest titers are obtained at about 80–90% confluency.
   i. Conditioned medium (i.e., retroviral supernatants) is pooled, chilled, and spun (500*g*, 10 min at 4°C) and aliquots are stored at – 70°C.
5. RetroNectin™ (Takara Biomedicals, Otsu Shiga, Japan). Non-tissue-culture-treated plates are coated following instructions of Takara Biomedicals. Quencing solution (2% [m/v] BSA [bovine serum albumin] in PBS) can be left on, allowing storage of coated wells for at least 1 wk at 4°C.
6. Cell Dissociation Buffer (Invitrogen).

2.3. Fetal Thymus Organ Culture

1. Microdissection tools (e.g., curved pincet).
2. Pregnant (d 14–15) NOD-LtSz-*scid*/*scid* (NOD-SCID) mice.
3. Dulbecco's phosphate buffer (DPBS):
   a. Dissolve 0.132 g $CaCl_2 \cdot 2H_2O$ in 0.1 L milli-Q water (for 0.9 m*M* final concentration).
   b. Dissolve 7.60 g NaCl (130 m*M* final concentration), 0.22 g KCl (3 m*M* final concentration), 2.87 g $Na_2HPO_4 \cdot 12H_2O$ (8 m*M* final concentration), 0.20 g

KH_2PO_4 (1.5 m*M* final concentration), 0.10 g $MgCl_2$•$6H_2O$ (0.5 m*M* final concentration), in that order in 0.8 L milli-Q water.
   c. Add 0.1 L $CaCl_2$•$2H_2O$ solution slowly and while stirring continuously to 0.8 L salt solution.
   d. Adjust pH to 7.2–7.3, adjust volume to 1 L (osmolarity should be 275–285 mOsm), filter through 0.22-µm filter; store at 4°C.
4. Fetal thymus organ culture is done in complete IMDM containing 10% heat-inactivated human AB serum (BioWhittaker, Walkersville, MD) instead of FCS.
5. Terasaki plates (Invitrogen).
6. Nuclepore filter (Nuclepore, Costar, Cambridge, MA).
7. Gelfoam sponge (Upjohn, Kalamazoo, MI).
8. Small tissue grinder (Potter–Elvehjem, polytetrafluoroethylene (PTFE) pestle and glass tube, 3 mL working capacity; KIMBLE / KONTES, Vineland, NJ).
9. Tris buffer:
   a. Solution A: dissolve 16.36 g NaCl (final concentration 140 m*M*), 6.06 g Tris (final concentration: 25 m*M*), 5 mg phenol red in 0.5 L milli-Q water.
   b. Solution B: dissolve 0.206 g $CaCl_2$•2 H_2O (final concentration: 0.7 m*M*) in 0.1 L milli-Q water. In 1 L milli-Q water, dissolve 0.820 g KCl (final concentration: 5.5 m*M*), 0.430 g Na_2HPO_4•$12H_2O$ or 0.170 g Na_2HPO_4 (final concentration: 0.6 m*M*), 0.246 g $MgSO_4$•$7H_2O$ (final concentration: 0.5 m*M*). Add 0.1 L $CaCl_2$ solution slowly and while stirring continuously to 1 L salt solution.
   c. Add solution A to solution B, adjust pH to 7.2, adjust volume to 2 L (osmolarity should be 290–310 mOsm).
   d. Autoclave aliquots 20 min, 121°C, 2 atm.
10. Trypan blue.
11. PBS containing 1% (w/v) BSA and 0.1% (w/v) NaN_3 (PBS/BSA/NaN_3).
12. Anti-FcγRII/III MAb to avoid nonspecific binding of MAbs by the murine cells (e.g., clone 2.4.G2, Dr. J. Unkeless, Mount Sinai School of Medicine, New York).
13. Anti-mouse CD45 cychrome (e.g., PharMingen, San Diego, CA).
14. Propidium iodide (1 mg/mL).
15. Flow cytometer equipped with an argon-ion laser (488 nm) and helium–neon diode laser (635 nm).

3. Methods

3.1. Isolation of Precursor Cells

Cord blood donations are stored at room temperature and are to be processed within 18 h after collection. *x* mL cord blood is diluted with an equal volume of PBS, and layered above *x* mL Lymphoprep; all reagents at room temperature (*see* **Note 1**).

1. Pipet 15 mL of Lymphoprep in a sterile 50-mL polypropylene tube.
2. Gently pipet 30 mL of diluted cord blood on top of the Lymphoprep.

3. Spin tubes immediately after filling: 1000 g for 15 min at room temperature, no brake at end of run.
4. Aspirate mononuclear cell fraction carefully and slowly, including clusters attached to the wall of the tube (a total volume of no more than 7.5 mL is aspirated).
5. Wash twice in 30 mL cold PBS (500 g, 10 min at 4°C). The mononuclear cell fraction (median 1.5×10^6 $CD45^+$ cells/mL cord blood) can now be frozen in 9 vol of FCS and 1 vol of DMSO for later use (up to 2 yr and more) after storage in liquid nitrogen.
6. If either used directly or after thawing, cells are to be kept at 4°C at all times until initiation of culture. Cells are washed once (thawed cells twice) in 10 mL PBS/2% (v/v) FCS (500 g, 10 min at 4°C) and counted.

To study T-cell development from well-defined progenitor populations, these have to be sorted on the basis of expression levels of CD34, CD38, Lin, and, possibly, other markers (*see* **Subheading 3.1.1.**). Alternatively, if less stringent demands are put forward on the progenitor phenotype, bulk $CD34^+$ cells can be isolated (*see* **Subheading 3.1.2.**).

3.1.1. Isolation of Highly Purified $CD34^+$ Lineage $Marker^-CD38^{+/-}$ Cells

1. Resuspend up to 5×10^8 cells in 500 µL PBS/2% (v/v) FCS.
2. Stain with glycophorin-A, CD19, and CD7–FITC, 45 min at 4°C (*see* **Note 2**).
3. Wash labeled cells twice in 10 mL PBS/2% (v/v) FCS (500*g*, 10 min, 4°C).
4. Resuspend in 1 mL PBS/2% (v/v) FCS.
5. Wash sheep anti-mouse immunoglobulin coated Dynabeads twice in 2 mL PBS and once in 2 mL PBS/2% (v/v) FCS using the magnetic particle concentrator (take four beads per cell).
6. Resuspend beads in 1 mL PBS/2% (v/v) FCS and mix with 1 mL cell suspension. Incubate 30 min (with gentle shaking every 5 min) at 4°C.
7. Place the cell/beads mix in the magnetic particle concentrator for 5 min; cells coated with beads are drawn from suspension.
8. Aspirate the unlabeled cells from the tube, still in place in the magnetic particle concentrator.
9. Count the cells; expect 10% of the initial number. To make sure that almost all of the beads are removed, 1 µL of the cell suspension is checked microscopically in a hemocytometer. If there are still beads found, it is necessary to re-elute the cell suspension in the magnetic particle concentrator.
10. Cells are washed once in 10 mL PBS/2% (v/v) FCS, resuspended in 200 µL, and stained with CD1–FITC, CD3–FITC, CD4–FITC, CD8–FITC, CD34–PE, and CD38–BIO for 45 min, 4°C.
11. Cells are washed twice in 10 mL PBS/2% (v/v) FCS, resuspended in 200 µL, and further stained with SA–TC or SA–APC for 45 min, 4°C.
12. After a final wash in PBS/2% (v/v) FCS, cells are resuspended at 10^6 cells/mL PBS/2% (v/v) FCS. Cells that are $FITC^-$, $CD34^+$, and either CD38 positive or

negative are sorted on a cell sorter. On the average, 1200 $CD34^+Lin^-CD38^+$ and 60 $CD34^+Lin^-CD38^-$ cells (purity always at least 99.5%) can be obtained per 10^6 mononuclear cord blood cells.

3.1.2. Isolation of $CD34^+$ Cells

Alternatively, precursor cells can be isolated with colloidal superparamagnetic MicroBeads conjugated to CD34 antibodies to obtain >95% pure total CD34-positive cells (*see* **Note 3**).

1. Resuspend cells at a density of 10^8 cells/0.3 mL PBS/2% (v/v) FCS.
2. Add 100 µL FcR Blocking Reagent (per 10^8 total cells) to the cell suspension to inhibit unspecific or Fc-receptor-mediated binding of CD34 MicroBeads to nontarget cells.
3. Incubate for 30 min at 4°C with 100 µL CD34 MicroBeads per 10^8 total cells.
4. Wash labeled cells in 10 mL PBS/2% (v/v) FCS (500*g*, 10 min, 4°C).
5. Resuspend in 10 mL PBS/2% (v/v) FCS and pass through a 40-µm Nylon cell strainer to remove clumps.
6. Centrifuge (500*g*, 10 min, 4°C) and resuspend in 3 mL PBS/2% (v/v) FCS.
7. Place the LS separation column in the magnetic field of the MACS separator (attached to MACS MultiStand).
8. Fill and rinse three times with 3 mL PBS/2% (v/v) FCS.
9. Apply cells to the column; after they have passed through, wash the column three times with 3 mL PBS/2% (v/v) FCS.
10. Remove the column from the separator and place in a suitable tube.
11. Pipet 3 mL PBS/2% (v/v) FCS on top of the column; elute the retained cells (with pressure) by using the supplied plunger.
12. To increase purity, repeat the magnetic separation step with a new prefilled MACS column (**steps 7–9**).
13. Examine the last drop of the third washing solution microscopically to make sure that the $CD34^-$ cells are washed out completely. (No cells should be found in 1 µL filling a hemocytometer counting chamber.)
14. Elute the retained cells using the plunger as described above (**step 11**). Purity is evaluated by flow cytometry (stain with CD34 MAb) and is about 90% after the first column and 98% after the second column.

3.2. Culture and Transduction

For retroviral transduction of $CD34^+$ cord blood cells, cytokine-stimulated culture is essential *(2)*. Numerous studies have analyzed optimal cytokine cocktails, cytokine concentrations, and transduction protocols. The following is an example of a protocol that works best in our hands for cord blood progenitors *(4)*.

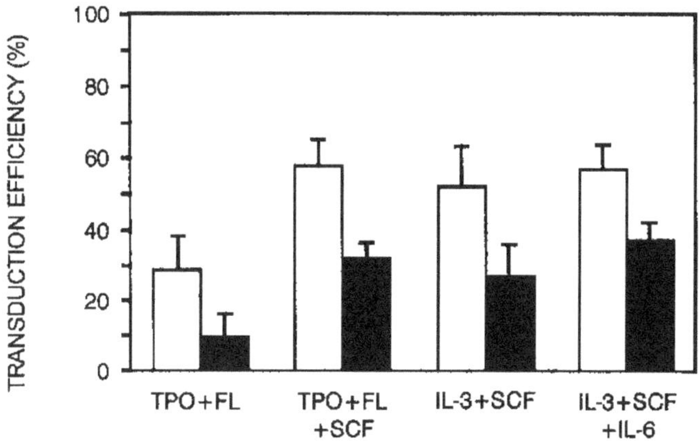

Fig. 1. Transduction efficiencies of CD34⁺CD38⁺Lin⁻ and CD34⁺CD38⁻Lin⁻ cord blood cells. Sorted cells were cultured in medium supplemented with cytokines and transduced with a retrovirus encoding GFP after 1 d of culture. Columns represent average percentage of GFP⁺ cells at d 3 of culture of CD34⁺CD38⁺Lin⁻ (□) and CD34⁺CD38⁻Lin⁻ (■) CB cells, from experiments with eight different donors. Error bars indicate standard deviation. (Reprinted with permission from **ref. *4***.)

1. Resuspend $CD34^+$ cord blood cells in complete IMDM supplemented with (combinations of) SCF, FL, TPO, IL-3, and IL-6 and seed in 96-well round-bottom tissue culture plates at ($2.5–100 \times 10^3$ cells in 150 µL medium per well) (*see* **Note 4**).
2. Culture for 24–48 h.
3. Remove half of the medium volume and replace with viral supernatant, supplemented with cytokines (to keep final cytokine concentration unchanged).
4. Resuspend cells and seed the complete well on RetroNectin™-coated 96-well flat-bottom plates (*see* **Note 5**).
5. Culture for 24–48 h. After this period, wells are resuspended; attached cells are recovered by washing well in PBS, followed by incubation with Cell Dissociation Buffer for 10 min at room temperature.
6. A fraction of the cells can be used to assay transduction efficiencies (e.g., by flow cytometry), and the remainder can be cultured on or used in FTOC. If desired, cells can be sorted for GFP expression before transfer to FTOC.

If the culture is to be continued, cytokine-supplemented medium is refreshed at least twice a week *(3)*. To avoid medium exhaustion, cell density must not exceed 2×10^6 cells/mL. In the second week of culture, cells are transferred to 24-well plates, after resuspension and incubation of the well in Cell Dissociation Buffer to remove adherent cells from the well surface. For a third week of culture, the cells are transferred to six-well tissue culture plates or to 25-cm^2 tissue culture flasks. **Figure 1** shows transduction results for $CD34^+CD38^-Lin^-$ cells and $CD34^+CD38^+Lin^-$ cord blood cells (*see* **Note 6**).

3.3. Fetal Thymic Organ Culture

NOD-LtSz-*scid/scid* (NOD–SCID) mice are bred in a pathogen-free breeding facility. After overnight mating, the appearance of vaginal plugs is noted as d 0 of pregnancy, and these females are pooled in separate cages (*see* **Note 7**).

1. At d 14–15 of pregnancy, NOD–SCID mice are killed by cervical dislocation.
2. The abdomen is sprayed with ethanol, and the uterus is dissected free in a horizontal laminar flow. Transfer the complete uterus to a Petri dish filled with cold DPBS.
3. Release embryos from the uterus by cutting close to the uterine constrictions between embryos. Care is taken not to damage the head region of the embryo (**Fig. 2A**).
4. For dissection of the thymic lobes, the embryo is punched dorsally on a clean tissue covering a polystyrene foam block. The needles used to punch the embryo are placed through the face and the abdomen in such a way that the embryo is stretched, with dorsoflexion of the head (**Fig. 2B**).
5. Using curved pincets, both thymic lobes (one lobe is indicated with an asteriks in **Fig. 2C,D**) residing in the upper thorax region are microdissected free under a macroscope. The lobes are placed in a Petri dish containing cold complete IMDM. Further microdissection is frequently necessary to free lobes from surrounding tissues. Lobes can be stored in a cell culture incubator several hours before use.
6. Cultured $CD34^+$ cord blood cells are harvested in a biohazard flow as described in **Subheading 3.2.5.** and washed in 5 mL complete IMDM, not supplemented with cytokines (500*g*, 10 min, at 4°C).
7. Resuspend pellet in a small-volume complete IMDM, so that 25 µL contains the equivalent of 10,000 cultured $CD34^+$ cord blood cells.
8. Transfer 25 µL cell suspension to a well of a Terasaki plate. The outer wells of the plate are filled with 25 µL PBS. The wells are now ready to receive a thymic lobe. As work with transduced cells is performed in a biohazard flow, a video-monitored marcoscope is very useful for observing the manipulations described in the following **steps 9**, **10**, **13**, and **14**.
9. With a 1000-µL plastic tip and pipettor, an individual thymic lobe is picked in a volume of about 100 µL IMDM. The lobe is allowed to sink to the tip orifice by gravity and is transferred to a Terasaki well by touching the 25-µL drop with the tip (no fluid is to be pipetted out). In this way, the lobe will fall into the Terasaki well.

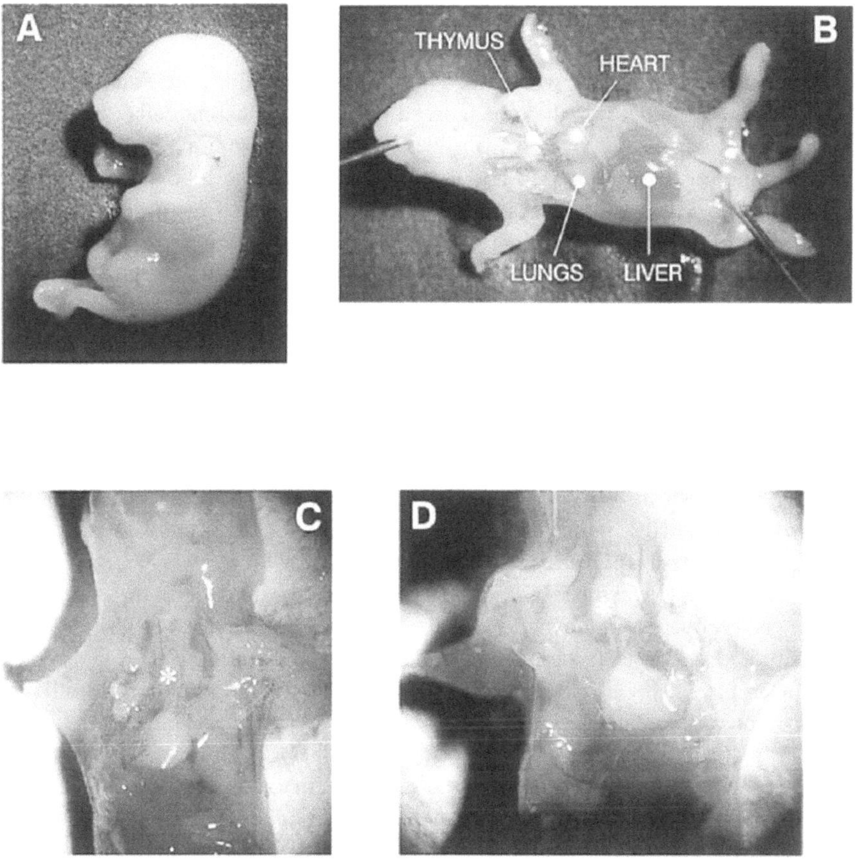

Fig. 2. Microdissection of a d-15 NOD–SCID embryo (for explanation, *see* the text).

10. After all lobes are transferred, the Terasaki plate is closed and inverted in a fast, gentle movement. The well bottoms are now facing up; the 25-µL drop is "hanging."
11. The Terasaki plate is placed in a large Petri dish, also containing small dishes filled with PBS. The latter serve to maximize humidity. The closed large Petri dish is placed in a cell culture incubator.
12. After about 42–44 h of culture, cut Gelfoam sponges (1 cm^2) are soaked in 3 mL IMDM with 10 % human serum not supplemented with cytokines, in a six-well tissue culture plate, for at least 4 h. After the Gelfoam has absorbed medium, residual air is squeezed out with sterile pincets. A nuclepore filter is placed on the Gelfoam sponge, taking care not to trap air bubbles.

13. After 48 h of culture, at d 0 of FTOC, the Terasaki plate is inverted to bottom down and lobes are removed by aspirating the complete well with a 1000-µL tip. The tip is blown out in a well of a 24-well plate filled with 1 mL complete IMDM. By extensive pipetting, the lobe is freed from most adherent cells.
14. Thymic lobes are recovered from the 24-well plate with a 1000-µL plastic tip and pipettor, in a volume of about 100 µL IMDM. The lobe is allowed to sink to the tip orifice by gravity, and the lobe is transferred to the nuclepore filter by touching (again, no fluid is to be pipetted out). Two to four lobes can be transferred to one filter; annotation of the position on the six-well plate cover can allow different experimental lobes to be placed on one filter.
15. The plate, with each well filled with either FTOC cultures or PBS, is put in a large Petri dish and placed in the cell culture incubator. For 14 d, the culture can be left untouched, but inspected regularly for air bubbles that can rise from the gelfoam toward the filter (*see* **Note 7**).
16. After 14 d of FTOC, the Gelfoam sponges are replaced by fresh ones. To this end, in a fresh six-well plate, cut Gelfoam sponges (1 cm^2) are soaked in 1.5 mL IMDM with 10% human serum not supplemented with cytokines, and again squeezed free from excessive air (as in **step 12**).
17. Used medium from the FTOC is aspirated beside the used Gelfoam (about 1.5 mL can be recovered) and expelled on top of the fresh Gelfoam.
18. With a sterile pincet, the nuclepore filter is picked from the used Gelfoam and put on the fresh Gelfoam. (If important, the filter should be placed in the same position as annotated).
19. The plate, with each well filled with either FTOC cultures or PBS, is put in a large Petri dish and placed in the cell culture incubator. The FTOC can be continued for another 14 d, with extra surveillance for air bubble formation. Sometimes, degradation of the Gelfoam necessitates an extra refreshment of the sponge in this second period of culture (done as in **steps 16–18**; *see* **Note 8**).
20. At the end of the culture, lobes are picked carefully from the filter and put in a small tissue grinder containing 150 µL Tris buffer. The lobe is mechanically disrupted to obtain a single-cell suspension. Thymocytes can be stained with trypan blue and counted with a hemocytometer.
21. Thymocytes are washed in 5 mL cold PBS/BSA/NaN_3 (500*g*, 10 min, at 4°C).
22. Cells are resuspended in 250 µL cold PBS/BSA/NaN_3 and preincubated on ice for 15 min with saturating amounts of anti-FcγRII/III MAb.
23. Cells are stained with anti-mouse CD45 cychrome, distributed over multiple tubes and further stained for 45 min with panels of FITC-, PE-, and APC-labeled mAbs, depending on the reporter gene present in the virus.
24. Staining tubes are washed with 3 mL of cold PBS/BSA/NaN_3.
25. Resuspended cells (150 µL) are kept on ice. Just before flow cytometry, 5 µL of propidium iodide is added to the tubes.
26. The cells are analyzed on a flow cytometer with an argon-ion laser tuned at 488 nm and helium–neon diode at 635 nm. Murine and dead cells are gated out by mouse

CD45 cychrome and propidium iodide, respectively. Negative controls included isotype MAbs conjugated with the corresponding fluorochrome. **Figure 3** shows an example of FTOC phenotype, initiated with $CD34^+CD38^-Lin^-$ cells and $CD34^+CD38^+Lin^-$ cord blood cells.

4. Notes

1. Umbilical cord blood can be collected in regular EDTA tubes, filled with a syringe containing umbilical cord blood after *V. umbilicalis* aspiration. Alternatively, a blood transfusion bag, containing a reduced amount of anticoagulant compared to blood donation bags, can be filled with cord blood by gravity after catheterization of the *V. umbilicalis*. The volume yield is highly variable; however, it is higher when blood is taken before compared to after birth of the placenta. An experienced staff obtains on the average 70 mL with syringe aspiration, and more than 100 mL with catheterization. The mononuclear cell fraction is after centrifugation over Lymphoprep visible as a fuzzy interface with cell clusters, sometimes contaminated with red blood cell (RBC), usually in aggregates. Reisolation over Lymphoprep will only marginally reduce RBC contamination.
2. Antibodies should be titrated to determine the optimal amount. However, 1 μg of monoclonal per 10^6 cells can be used as a rule of thumb.
3. Take care not to introduce air bubbles into the column in **step 8**. Elution of the cells (**step 11**) is enhanced by forceful and fast plunger movement.
4. A concentrated stock solution in medium of each cytokine cocktail (e.g., 20 times concentrated) is very useful to compensate for substitutions of medium with cytokine-free medium such as retroviral supernatants.
5. Transduction efficiencies are higher after 48 h, compared to 24 h, of culture before the addition of virus, especially for $CD34^+CD38^-$ cord blood cells. Transduction efficiency may be increased if RetroNectin-coated plate is spun immediately after seeding of the cells (1400*g*, 90 min at 32°C).
6. Lentiviral transduction protocols may vary, such as polybrene-assisted spinoculation. However, addition of cytokines also facilitates lentiviral transduction. Depending on the virus used, transduction efficiency can be assayed by flow cytometry, selective culture, immunocytochemistry, and many other techniques.
7. Sometimes, the lobe remains attached to the bottom of the well after inverting the Terasaki plate (**step 10**). By gentle ticking on the well bottom with a pencil, the lobe is released to sink, together with the cells, to the apex of the drop, thus to the gas–liquid interface. We suspect that the appearance of air bubbles near the filter is the result of degradation of the Gelfoam by agents released from the FTOC. It is generally observed after more than 14 d of FTOC, especially those initiated with cytokine-stimulated cord blood precursors. Not surprisingly, these bubbles typically appear beneath a thymic lobe. When the bubbles touch the filter beneath a lobe, nutrient access is hampered and the lobe concerned is lost for analysis. To avoid this, the filter can be shuffled away from the bubbles over

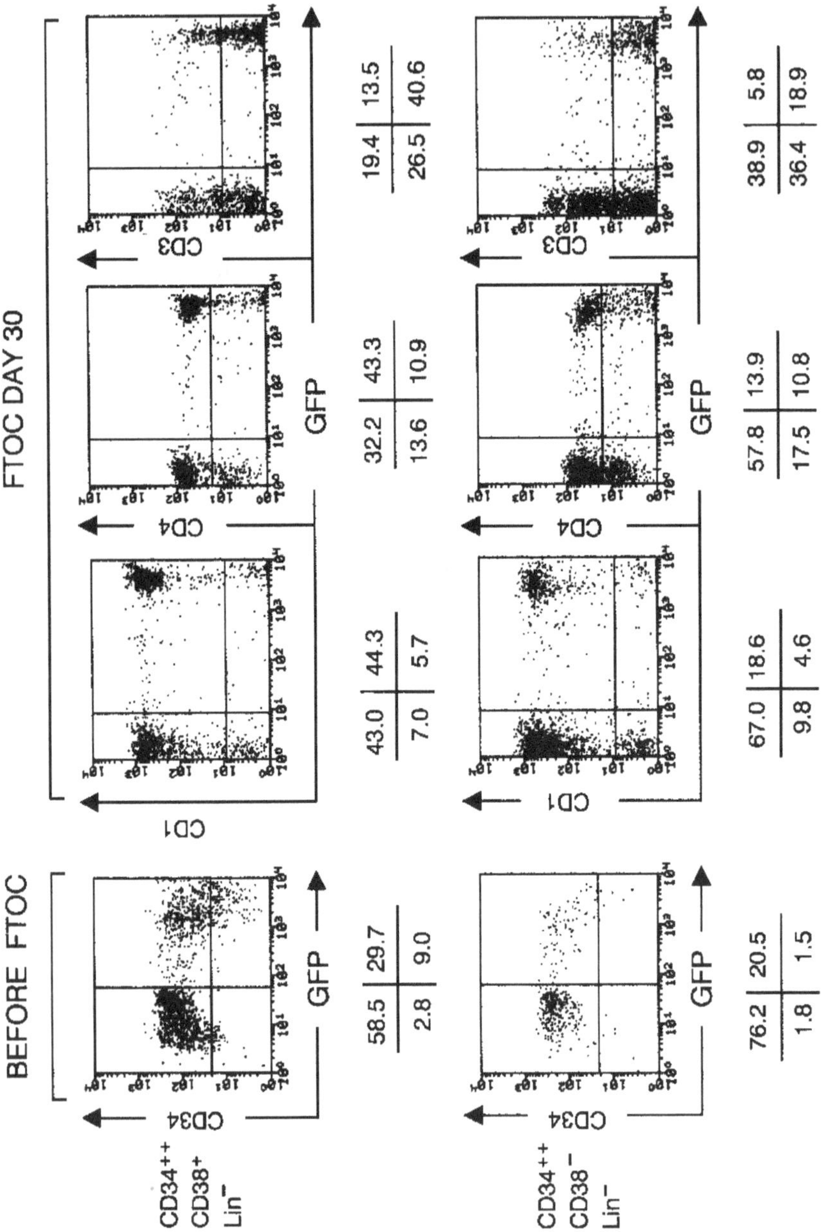
BEFORE FTOC
FTOC DAY 30
CD34++
CD38+
Lin−
CD34++
CD38−
Lin−
CD34
CD1
CD4
CD3
GFP
58.5 | 29.7
2.8 | 9.0
43.0 | 44.3
7.0 | 5.7
32.2 | 43.3
13.6 | 10.9
19.4 | 13.5
26.5 | 40.6
76.2 | 20.5
1.8 | 1.5
67.0 | 18.6
9.8 | 4.6
57.8 | 13.9
17.5 | 10.8
38.9 | 5.8
36.4 | 18.9

Fig. 3.

the surface of the sponge. This is, of course, done in a biohazard flow, using a sterile pincet.

8. To obtain more mature thymocytes, FTOC can be prolonged for a third and final period of 10–14 d.

References

1. MacDonald, H. R., Radtke, F., and Wilson, A. (2001) T cell fate specification and alpha beta/gamma delta lineage commitment. *Curr. Opin. Immunol.* **13,** 219–224.
2. Verhasselt, B., De Smedt, M., Verhelst, R., Naessens, E., and Plum, J. (1998) Retrovirally transduced CD34^{++} human cord blood cells generate T cells expressing high levels of the retroviral encoded Green Fluorescent Protein marker in vitro. *Blood* **91,** 431–440.
3. Verhasselt, B., Kerre, T., Naessens, E., Vanhecke, D., De Smedt, M., Vandekerckhove, B., et al. (1999) Thymic repopulation by CD34$^+$ human cord blood cells after expansion in stroma-free culture. *Blood* **94,** 3644–3652.
4. Verhasselt, B., Naessens, E., De Smedt, M., and Plum, J. (2000) Efficiency of transgenic T cell generation from gene-marked cultured human CD34$^+$ cord blood cells is determined by their maturity and the cytokines present in the culture medium. *Gene Ther.* **7,** 830–836.

Fig. 3. *(opposite page)* Both CD38$^+$CD34$^+$Lin$^-$ and CD38$^-$CD34$^+$Lin$^-$ cord blood cells generate T cells expressing high levels of GFP in FTOC. "Before FTOC" indicates flow-cytometric analysis of the progeny of CD38$^+$ and CD38 CD34$^+$Lin$^-$ cord blood cells after culture for 3 d in medium with SCF and IL-3 supplemented with retroviral supernatant after 1 d of culture. Dot plots shows CD34 PE staining vs GFP expression gated on live cells. The values in the crosses indicate the percentage of cells present in the corresponding quadrant. Quadrants were set arbitrarily. FTOC d 30 indicates flow-cytometric analysis of thymocytes recovered from FTOC initiated with transduced CD38$^+$CD34$^+$Lin$^-$ and CD38$^-$CD34$^+$Lin$^-$ cord blood cells shown before FTOC. After 30 d of culture, thymocytes recovered from FTOC were stained with IgG1–PE (not shown), CD1–PE, CD4–TC, and CD3–PE and analyzed by flow cytometry. Dot plots show staining vs GFP expression of live human cells recovered. The values in the crosses indicate the percentage of cells present in the corresponding quadrant. Quadrants were set to include 99.5% of the cells stained, with isotypic control antibody in the lower quadrants. The data shown are representative of three independent experiments (reprinted with permission from **ref.** ***2***).

25

Ex Vivo Expansion of Umbilical Cord Blood Cells on Feeder Layers

Martin Bornhäuser

1. Introduction

This chapter deals with the culture of hematopoietic stem cells contained in human umbilical cord blood in order to amplify the number of progenitor cells. The aim of expanding human umbilical cord blood cells (HUCBC) is to use a larger inoculum for hematopoietic stem cell (HSC) transplantation of allogeneic recipients *(1)*. It is hoped that the recovery of blood counts can be significantly fastened by the use of amplified HUCBC, especially in adult patients. Until now, successful transplantation of HUCBC has mainly been performed in children where enough cells can be infused to guarantee fast and stable engraftment *(2)*. In these transplants, the decreased risk of acute graft-versus-host disease (GvHD) has been one of the major steps forward in the field of pediatric hematopoietic stem cell transplantation. Retrospective studies have proven that HUCBC may be used as an alternative source of HSC in the related and unrelated donor setting *(2,3)*.

During the last years, it has been convincingly shown in preclinical and clinical trials that HUCBC can be expanded in vivo without losing their long-term repopulating ability, which had been feared during the start of these investigations *(4,5)*.

Different factors of culture conditions and methods influence the results of ex vivo expansion:

1. The separation method of cord blood cells may be crucial to obtain optimum culture results.

From: *Methods in Molecular Biology, vol. 215: Cytokines and Colony Stimulating Factors: Methods and Protocols*
Edited by: D. Körholz and W. Kiess © Humana Press Inc., Totowa, NJ

2. Different cytokines might either help to preserve the immature phenotype of the cultured cells or induce terminal differentiation.
3. Feeder layers like umbilical cord blood endothelial cells or marrow stroma cells mimic the physiological surroundings and may support the expansion of primitive hematopoietic stem cells.

The above-mentioned points will be addressed in detail in the following subheadings, with special emphasis on methods of culture and analysis.

2. Materials

2.1. Cell Separation

1. Dulbecco's phosphate-buffered saline (DPBS) w/o magnesium and calcium, pH 7.2 (Gibco, Paisley, Scotland) (Storage temperature, 4°C).
2. Immuflot (Immucor GmbH, Rödermark, Germany) (storage temperature, 4°C).
3. Human serum albumin (HSA) 5% (Immuno GmbH, Heidelberg, Germany) (storage temperature, 4°C).
4. CellGro® SCGM (stem cell growth medium, CellGenix, Freiburg, Germany) (storage temperature, 4 °C).
5. Cell counter (Technicon H3 RTC™, Bayer Dignostics GmbH, München, Germany).
6. Lysis buffer: 8.29 g ammonium chloride, 1.0 g potassium hydrogen carbonate, 0.037 g sodium–EDTA in 1 L aqua, pH 7.4 (storage temperature, 4°C).

2.2. Establishment of Stromal Layers

1. Stroma medium: RPMI 1640, Dutch modification, 10% fetal calf serum (FCS) (Gibco, Paisley, Scotland) (*see* **Note 1**), 2 m*M* L-glutamine (Sigma–Aldrich, Steinheim, Germany), 100 IU/mL penicillin, mg/mL streptomycin, (penicillin–streptomycin solution: 10.000 U penicillin and 10 mg streptomycin per milliliter in 0.9% sodium chloride [Sigma–Aldrich GmbH, Steinheim, Germany]), 1.0 µmol/L hydrocortisone (Aventis, Frankfurt, Germany) (storage temperature, 4°C).
2. All solutions sterile filtered and cell culture tested.
3. 25-cm² Flasks (NUNC, Wiesbaden, Germany) (8–10 mL medium, 5 mL trypsin for incubation).
4. 80-cm² flasks (NUNC) (20 mL medium, 10 mL trypsin for incubation).
5. 175-cm² flasks (NUNC) (50 mL medium, 20 mL trypsin for incubation).
6. Trypsin–EDTA solution: 0.5 g porcine trypsin and 0.2 g EDTA × 4 Na per liter of HBSS (Hank's balanced salt solution) (Sigma–Aldrich, Steinheim, Germany) (aliquots stored at –20°C)
7. Gamma source (15 Gy).

2.3. Establishment of HUVEC

1. DPBS.
2. 10 mg/mL Gentamycin sulfate (aliquots stored at –20°C).
3. 250 µg/mL amphotericin B (aliquots stored at –20°C).

4. Medium for the transport of the umbilical cord (250 mL DPBS, 1.25 mL gentamycin sulfate solution, 2.5 mL amphotericin B solution).
5. 5000 U/mL Sodium–heparin, sterile filtered (Sigma-Aldrich, Steinheim, Germany) (storage temperature, 4°C).
6. Collagenase solution 0.1% (Sigma-Aldrich, Steinheim, Germany) (aliquots stored at –20°C in a 50-mL tube filled with 25 mL).
7. Endothelial cell growth medium (ECGM; PromoCell, Heidelberg, Germany) (storage temperature, 4°C).
8. Fibronectin (Sigma; 20 µg/mL) (solution storage temperature, 4°C; lyophylisate storage temperature –20°C).
9. Trypsin inhibitor solution Type I-S (Sigma–Aldrich, Steinheim, Germany). From soybean, chromatographically prepared, 1 mg will inhibit 1–3 mg of trypsin.

2.4. CD34+ Progenitor Cell Purification

CD34 Selection Kit, Miltenyi Biotec GmbH (Bergisch-Gladbach, Germany) (storage temperature, 4°C).

2.5. Methylcellulose Progenitor Culture

1. Complete methycellulose medium containing Isocove's Modified Dulbecco's Medium (IMDM) with 30% fetal bovine serum (FBS), 3 U/mL erythropoietin, 50 ng/mL stem cell factor (SCF), 20 ng/mL granulocyte–macrophage colony stimulating factor (GM-CSF), 20 ng/mL interleukin (IL)-3, 20 ng/mL IL-6, and 20 ng/mL G-CSF (Methocult GF H4431; Stem Cell Technologies, Vancouver, Canada) (aliquots stored at –20°C).

2.6. Expansion Cultures

1. SCF (Amgen, Munich, Germany) (aliquots stored at –80°C).
2. Flt3-ligand (R&D, Mannheim, Germany) (aliquots stored at –80°C).
3. IL-3 (R&D, Mannheim, Germany) (aliquots stored at –80°C).
4. G-CSF (Amgen) (aliquots stored at –80°C).
5. IL-6 (R&D) (aliquots stored at –80°C).
6. Erythropoietin (R&D) (aliquots stored at –80°C).
7. Megakaryocyte growth and development factor (MGDF) (Amgen) (aliquots stored at 4°C).

2.7. Flow Cytometry

1. FACS SCAN (Becton Dickinson, San Jose, CA).
2. Class III CD34 antibody HPCA-2-PE, IgG1 (Becton Dickinson) (storage temperature, 4°C).
3. Anti-CD41–FITC (fluorescein isothiocyanate, P2 clone, IgG1 (Coulter-Immunotech Diagnostics, Miami, FL) (storage temperature, 4°C).
4. Anti-CD38–FITC, T16 clone, IgG1 (Coulter-Immunotech Diagnostics, Miami, FL) (storage temperature,4°C).

5. Anti-CD117–FITC (c-kit), 104D2 clone, IgG1 (Serotec-Biozol Diagnostica, München, Germany) (storage temperature, 4°C).
6. Anti HLA–DR–FITC, B-F1 clone, IgG1 (Serotec-Biozol Diagnostica).
7. Annexin V Kit (with PI) (R&D) (storage temperature, 4°C).

2.8. Transwell Cultures

Six-well plates (NUNC, Wiesbaden, Germany) (Anapore membranes, 0.2 µm) (*see* **Note 2**).

3. Methods

3.1. Cell Separation

Cord blood is collected under sterile conditions after puncture of the umbilical cord blood vein, diluted with Dulbecco's PBS w/o magnesium and calcium, pH 7.2, to obtain a cell concentration of (1–2) × 10^6/mL.

Density gradient centrifugation is performed as follows:

1. Cells are layered carefully on Immuflot and centrifuged at 800*g* for 20 min.
2. The mononuclear cells of the interphase are collected carefully and washed twice with PBS at 600*g* for 10 min, supplemented with 0.5% human serum albumin (HSA 5%; Immuno GmbH, Heidelberg, Germany).
3. The pellet of the second wash is resuspended in 5 mL of CellGro SCGM (Cell Genix, Freiburg, Germany). Cells are counted automatically (e.g., by Technicon H 3 RTC™, Bayer Diagnostics GmbH, München, Germany).

An alternative method to density gradient centrifugation is chemical lysis:

1. About 50 mL of cord blood is mixed with 200 mL of ice-cold lysis buffer, consisting of 8.29 g ammonium chloride, 1.00 g potassium hydrogen carbonate, and 0.037 g sodium–EDTA in 1 L aqua with a pH of 7.4.
2. After 15 min of incubation, the nucleated cells are separated by centrifugation, and the supernatant is removed.
3. The pellet is resuspended in 5 mL of CellGro SCGM.

3.2. Establishment of Stromal Layers

1. Bone marrow aspirates are obtained by aspiration at the posterior iliac crest from healthy donors. Marrow usually is filtered using a 500-µm filter and 200-µm filter, respectively.
2. Afterward, the marrow is mixed with the same volume of PBS, and mononuclear cells are separated by density gradient centrifugation as mentioned earlier. Washed cells are resuspended in stroma medium.
3. Then, 7 × 10^7 Cells are placed in 80-cm^2 flasks (NUNC) in 20 mL of medium and maintained at 37°C and 5% CO_2. The medium volumina for different flask sizes were provided in **Subheading 2.2.**

4. Half of the medium is exchanged twice a week.
5. After 2 wk, adherent layers with 80–90% confluency are passaged with trypsin–EDTA solution 1X (Sigma–Aldrich, Steinheim, Germany).
6. Therefore, old medium is removed. All solutions (trypsin, PBS) are prewarmed to 37°C.
7. The culture flask is carefully rinsed with 10 mL of PBS twice without touching the cell layer and removing the PBS thereafter.
8. Ten milliliters of trypsin solution are pipetted into the culture flask and then sat at 37°C for 3–5 min (*see* **Note 3**).
9. The flask is the checked under the inverted microscope to make sure that cells are lifting off of the flask (*see* **Note 4**).
10. Trypsin is quickly inactivated with 20 mL stroma medium and cells are then transferred into a 50-mL conical tube and centrifuged with 200*g* for 5 min.
11. The pellet is resuspended in stromal medium and the cells of one 80-cm^2 culture vessel are transferred in six 25-cm^2 flasks or six 6-well plates for further expansion experiments.
12. Confluent layers are irradiated with 15 Gy with a gamma source before initiating culture experiments to inhibit further proliferation and immunological interaction with HUCBC.

3.3. Establishment of HUVEC

The method used was adapted from **ref. *6***.

1. Umbilical cords are placed in the transport medium.
2. The umbilical cord vein is cut on both sides in order to realize sterile conditions. A tube of a butterfly canula with a luer-lock adapter is fixed to one side of the umbilical cord vein. The other end stays clamped during the whole procedure.
3. The vessel is then flushed several times with a 20-cm^3 syringe of PBS until no more blood or clots are detectable in the rinsing solution.
4. Prewarmed collagenase solution 0.05% (Sigma–Aldrich) is reconstituted and filled into the vein (the one with the luer-lock adapter) and incubated for 4 min at 37°C.
5. To stop the enzymatic reaction, trypsin inhibitor solution from soybean is used. The inhibitor solution is dissolved in serum-free medium (DPBS, w/o Ca^{2+}, Mg^{2+}) at a concentration of 1 mg/mL. Sterile filtration of the solution should be performed beforehand using a 0.2-µm cellulose acetate membrane. The luer-lock adapter is used once again to instillate trypsin inhibitor solution.
6. The vein is reopened allowing the HUVEC solution to be collected in a 50-mL Falcon. The vein is flushed approx 3 times using the inhibitor solution until the 50-mL tube is filled completely.
7. The solution is then centrifuged, and after discarding the supernatant, the cells are resuspended in endothelial cell growth medium (ECGM; PromoCell, Heidelberg, Germany) with a low content of FCS (2 %).

8. Before use, the culture flasks are coated with fibronectin (1 mg/50 mL) at 37°C, 5% CO_2 for 30 min.
9. The HUVECs are placed in a culture flask (either 25 cm^2 or 80 cm^2) coated with fibronectin. Half of the medium is exchanged three times a week until confluence is reached.
10. After trypsinizing cells, they are resuspended in 1 mL trypsin inhibitor solution per milliliter of trypsin (*see* **Note 5**).
11. Cells are then centrifuged at 400*g* for 5 min. As much of the trypsin inhibitor solution as possible is removed and the pellet is resuspended in serum-free medium (ECGM).

3.4. $CD34^+$ Progenitor Cell Purification

The HUCBCs are enriched using the MACS system (CD34 Selection Kit, Miltenyi Biotec GmbH, Bergisch-Gladbach, Germany) according to the manufacturer's instructions (*see* **Notes 6** and **7**).

1. Briefly, HUCBCs are washed and resuspended in PBS, 0.5% BSA, and 5 mmol/L EDTA.
2. Cells are first incubated with QBEND-10 antibody (mouse anti-human CD34) in the presence of human IgG as blocking reagent to minimize unspecific antibody binding.
3. After one cell wash, another incubation step with 100 µL MACS microbeads per 10^8 cells follows. Labeled cells are loaded onto a column installed in a magnetic field.
4. Trapped cells are eluted after removal of the column from the magnet.

3.5. Methylcellulose Progenitor Culture

1. As described, 1×10^5 HUCBCs or 1×10^3 CD34-enriched HUCBCs are plated in a complete methycellulose medium containing IMDM with 30% FBS, 3 U/mL erythropoietin, 50 ng/mL SCF, 20 ng/mL GM-CSF, 20 ng/mL IL-3, 20 ng/mL IL-6, and 20 ng/mL G-CSF (Methocult GF H4435; Stem Cell Technologies, Vancouver, Canada).
2. Cultures are incubated at 37°C and 5% CO_2.
3. The cultures are assessed at d 12–14 for the presence of burst-forming unit-erythroid, colony-forming unit-granulocyte-macrophage and mixed colony-forming units (*see* **Note 8**).

3.6. Expansion Cultures

1. The HUCBCs are cultured in 25-cm^2 flasks containing 10 mL serum-free medium (CellGro SCGM; Cell Genix, Freiburg, Germany) at 37°C and 5% CO_2 for 8 d.
2. Preformed stroma or HUVECs are used as the feeder layer. Growth factor cocktail 1 contains 300 ng/mL SCF (Amgen, Munich, Germany), 300 ng/mL FLT3-ligand, ng/mL IL-3 50 (R&D, Mannheim, Germany).

3. An alternative cocktail contains the cytokines of cocktail 1 supplemented with G-CSF 450 ng/L (Amgen), 100 ng/mL IL-6 (R&D) and 1 U/mL erythropoietin (R&D). MGDF (Amgen) is added at concentrations of 25 and 100 ng/mL.
4. Unselected and CD34-enriched HUCBCs are cultured at 5×10^5 and 1×10^4/mL, respectively.
5. After the 8-d culture period, cells are washed twice with PBS/0.5% HSA at 600*g* for 10 min and are then stained using the antibody combinations mentioned in **Subheading 3.8.**

3.7. Transwell Cultures

To test the influence of stroma contact, six-well-plates (NUNC, Wiesbaden, Germany) can be used. The culture dishes are separated by a microporous 0.2-µm Anapore membrane. After establishment of a stroma layer in the lower culture compartment, the purified HUCBCs can be injected into the upper chamber to test for the role of stroma contact or soluble factors.

3.8. Flow-Cytometric Analysis

Analysis of the CD34 content of all samples before and after culture is performed with FACS SCAN (Becton Dickinson, San Jose, CA) using a class III CD34 antibody HPCA-2-PE (Becton Dickinson) and standard software LYSIS II. Double color staining and analysis are performed for all samples using the following antibodies: anti-CD41–FITC, anti-CD38–FITC (Coulter-Immunotech Diagnostics, Miami, FL) and anti-CD117–PE (c-kit) (Serotec, Eching/München, Germany). Double staining of CD34/CD38, CD34/CD117, and CD34/CD41 is performed to quantitate progenitor subsets. Early hematopoietic stem cells are defined as CD34 positive and CD38 dim. Megakaryocytic progenitor should coexpress CD34 and CD41.

To test for viability, propidiumiodide (PI) staining is performed after all cultures. Annexin is used to determine the percentage of apoptotic cells. PI–/Annexin+ events define early apoptosis, whereas PI+/Annexin+ signals are counted as late apoptosis.

4. Notes

1. Two to three charges of heat-inactivated (30 min in a water bath at 56°C) FCS have to be screened for support of adherent and suspension-cell cultures before being used in serial experiments with stroma cells or HUVECs.
2. The Anapore membrane used in Transwell cultures does not allow cell migration but diffusion of soluble factors like chemokines. The company also offers a 3-µm polycarbonate membrane, which allows cell migration from one to the other chamber.

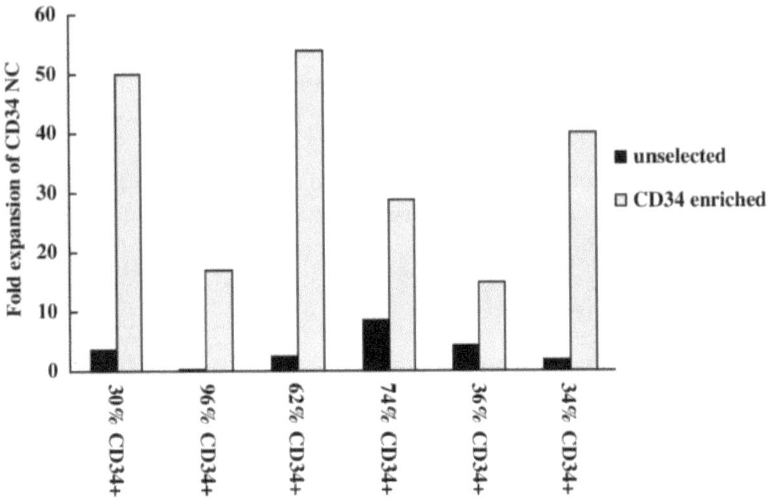

Fig. 1. The bars represent the expansion factors reached in six experiments comparing either CD34-enriched or unselected HUCBCs of the same origin. CD34 purity of each starting sample is provided on the *x*-axis.

3. If cells do not lift off within 5 min the flask can be agitated (rapped) physically to make the cells detaching. Longer incubation with trypsin will deteriorate cell viability.
4. Only single stroma cells should detach from the layer after incubation with trypsin; if the whole layer detaches, the viability of the stroma cells will be reduced.
5. Treatment with trypsin, EDTA, and further maintenance of the HUVEC cultures is performed as mentioned for stroma cells, but the incubation time for trypsin should not be more than 1 min for HUVEC compared to 3–5 min in the case of marrow stroma cells.
6. The first run on the immunomagnetic column leads to an enrichment of CD34+ cells up to 20–40%. A second run increases the purity to more than 95%.
7. CD34 enrichment significantly enhances the expansion potential of HUCBC. **Figure 1** compares the expansion factors of CD34+ cells with and without immunomagnetic pre-enrichment.
8. Colony-forming assay of cord blood cells should be scored after 12–14 d.

References

1. McNiece, I., Jones, R., Bearman, S. I., et al. (2000) Ex vivo expanded peripheral blood progenitor cells provide rapid neutrophil recovery after high-dose chemotherapy in patients with breast cancer. *Blood* **96,** 3001–3007.
2. Rocha, V., Wagner, J. E. J., Sobocinski, K. A., et al. (2000) Graft-versus-host disease in children who have received a cord-blood or bone marrow transplant

from an HLA-identical sibling. Eurocord and International Bone Marrow Transplant Registry Working Committee on Alternative Donor and Stem Cell Sources. *N. Engl. J. Med.* **342,** 1846–1854.

3. Rocha, V., Cornish, J., Sievers, E. L., et al. (2001) Comparison of outcomes of unrelated bone marrow and umbilical cord blood transplants in children with acute leukemia. *Blood* **97,** 2962–2971.
4. Piacibello, W., Sanavio, F., Garetto, L., et al. (1997) Extensive amplification and self-renewal of human primitive hematopoietic stem cells from cord blood. *Blood* **89,** 2644–2653.
5. Kusadasi, N., van Soest, P. L., Mayen, A. E., Koevoet, J. L., and Ploemacher, R. E. (2000) Successful short-term ex vivo expansion of NOD/SCID repopulating ability and CAFC week 6 from umbilical cord blood. *Leukemia* **14,** 1944–1953.
6. Jaffe, E. A. (1980) Culture of human endothelial cells. *Transplant. Proc.* **12,** 49–53.

26

Ex Vivo Production of Cord Blood CD34+ Derived Myeloid Precursors After Serum-Free Static Culture

Sergio Querol

1. Introduction

Cord blood (CB) transplantation is an increasingly used source of hematopoietic progenitor cells for allogeneic stem cell transplantation *(1)*. At birth, blood sequestered into placental vessels can be recovered by draining the cord vein into a closed system. A mean recovery of 90 mL of placental blood is obtained, containing an average of 4×10^6 cells expressing the CD34 antigen. The clinical series published analyzing this transplant modality showed potential advantages in comparison with other sources *(2,3)*. CB has demonstrated higher allogeneic tolerance across HLA barriers, with lower incidence of graft-versus-host disease maintaining the antitumoral effects. Moreover, the easy procurement of the blood without donor risk and the storage in banks of large quantities of products, fully characterized and ready to use, make it attractive for clinical application. However, the appearance of early events related to the delay of myeloid engraftment has resulted in a higher transplant-related mortality during the first month after infusion. This delayed engraftment is related to the limit number of myeloid progenitors contained in a common CB donation. Nevertheless, the overall survival in the global series is comparable between CB transplants and those using postnatal bone marrow as a source of stem cells *(4)*.

A cell therapy approach to increase the probability of early engraftment is the ex vivo production of multilineage myeloid precursors through expansion cultures of a CB unit fraction *(5–7)*. This chapter deals with the methodology for the generation of transplantable myeloid precursors in a serum-free static

From: *Methods in Molecular Biology, vol. 215:*
Cytokines and Colony Stimulating Factors: Methods and Protocols
Edited by: D. Körholz and W. Kiess © Humana Press Inc., Totowa, NJ

culture using clinical-grade reagents. In vitro maintenance of hemopoiesis has been achieved in Dexter-type long-term cultures. Cell-to-cell interaction within the stroma niches and cytokine production by accessory cells promotes stem cell survival, self-renewal, and proliferation/differentiation waves during several months of cultures ***(8)***. The definition of the hematopoietic pathways and the cytokine-mediated proliferation and differentiation events has suggested the definition of expansion cultures intended for generation of large quantities of progenitor cells ***(9)***. Several authors have published the cytokine's ability to promote the stem cell maintenance and expansion in stroma-free cultures ***(10,11)***. Based in these findings, protocols avoiding the use of proteins of animal origin manufactured using GMP rules and approved for ex vivo use need to be developed to test the benefit of the expanded cells in reducing the aplasia period. In this chapter, our experience in developing a clinical protocol on CB expansion will be presented.

2. Reagents for an Expansion Culture

2.1. CB Products

1. Collection: Cord blood samples were collected from term deliveries; previous informed consent of the mothers was obtained. Immediately after delivery and while the placenta was still in the uterus, the umbilical cord was clamped and sectioned and the umbilical vein punctured. Blood was drained by gravity into a blood collection bag and processed within 24 h of collection ***(12)***.
2. Cryopresevation: Cord blood cells were cryopreserved in a controlled-rate freezer. The cryoprotectant solution contained a final volume of 10% dimethyl sulfoxide (DMSO) and 1% dextran (molecular weight [MW] 40,000).
3. Storage: Samples were stored in PL269 plastic bags (Cryocyte-50 mL; Baxter Healthcare, Deerfield, IL) and placed in liquid nitrogen for long-term storage.

2.2. CD34+ Selection Regents

1. Recombinant human deoxyribonuclease I (Pulmozyme®, F. Hoffmann–La Roche Ltd, Basel, Switzerland) ***(13)***.
2. Magnesium chloride (Braun Medical AG, Emmenbrücke, Switzerland).
3. Dextran-40 (Rheomacrodex®; Antibióticos Farma, Madrid, Spain).
4. Human albumin (Instituto Grifols, Parets del Vallès, Spain).
5. Ca^{2+}/Mg^{2+}-free phosphate-buffered saline (PBS) (Braun, France).
6. Sodium citrate (Baxter Deutchsland GmbH, Unterschleissheim, Germany).
7. CD34+ selection kit (Isolex® 300 Stem Cell Reagent Kit, Baxter Healthcare).

2.3. Culture Reagents

1. CellGro® product line (CellGenix, Friburg, Germany) comprised a standardized serum-free medium for optimal growth of human hematopoietic progenitor cells

(SCGM) and GMP-grade cytokines for the optimal growth support: interleukin (IL)-6, fetal tyrosine kinase-3 ligand, thrombopoietin stem cell factor.
2. Semipermeable Teflon bag (American Fluoroseal Corp., Columbia, MD).

2.4. Materials for Progenitor Assessment

1. Methylcellulose-based media (Methocult GF H4434; Stem Cell Technologies, Vancouver, Canada).
2. MegaCult ™ – C (Stem Cell Technologies, Vancouver, Canada).
3. Monoclonal antibodies are used for immunofluorescence analysis: phycoerytrin (PE)-labeled anti-CD34, clone 8G12 (HPCA-2; Becton-Dickinson, San Jose, CA); fluorescein isothyocyanate (FITC)-labeled anti-CD45, clone Kh56 (Cyto-stat; Coulter Corporation, Miami, FL), and an isotypic control (PE–IgG1, Becton-Dickinson). 7-Aminoactinomycin D (7-AAD) is used as a viability marker.

3. Methods for an Expansion Culture

First, we need to define the objective of our culture. Expansion means amplification, the generation of a defined type of cell through culture of starting cells obtained from a progenitor source. In our case, cells required are those involved in the early production of granulocyte and platelets. These cells are monitored by their expression of CD34 antigen and their ability to generate colonies in semisolid media.

The strategy is to use a part (i.e., one-quarter) of a CB unit to obtain a high amount of committed progenitors, to coinfuse with the rest (three-quarters) of the CB without manipulation remaining responsible for the permanent engraftment and alloreactivity.

Hemopoiesis in vitro has been possible by culturing bone marrow in appropriate media (horse and fetal calf serum, corticoids, and enriched media as Iscove's modified Dulbecco's medium [IMDM]). In this type of culture, a feeder layer (stroma) is developed, adhered to the plastic, and hemopoietic stem cells divide, proliferate, and differentiate during several months. Using appropriate cytokines, several authors have published the long-term maintenance of hemopoietic stem cells (HSCs) in stroma-free media. This observation makes possible the definition of an ex vivo strategy for the generation of cells, easy to scale-up, and standardized in different cell therapy laboratories.

3.1. CD34 Selection (see *Note 1*)

1. Cord blood units are thawed by direct immersion in a water bath at 37°C.
2. Immediately after thawing, 2500 UI recombinant human deoxyribonuclease I (rhu-DNase), and 300 μL of 0.5 *M* magnesium chloride ($MgCl_2$) are added.
3. Dimethyl sulfate is washed, adding an equal volume of dextran/albumin solution (5% [v/v] dextran and 2.5% [v/v] human albumin) and then centrifuging at 4°C and 400*g* for 15 min.

4. The supernatant is removed using a plasma extractor, and cells are resuspended in a Ca^{2+}/Mg^{2+}-free PBS containing 1% (v/v) human albumin, 0.5% (v/v) human immunoglobulin, 75 U rhu-DNase/mL, 10 µL of 0.5 M/mL $MgCl_2$, and 0.5% sodium citrate (selection buffer).
5. CD34+ cells are then selected by a direct immunomagnetic approach using a currently available commercial kit (Isolex 300 Stem Cell Reagent Kit; Baxter Healthcare). (*See* **Note 2**).
6. Immunomagnetic beads (IBs) coated with a sheep anti-mouse IgG polyclonal antibody (SAM-M450) are incubated with the 9C5 anti-CD34 IgG monoclonal antibody, for 30 min, at a ratio of 10^7 IB/µg of 9C5 MoAb.
7. Excess MAb is then washed off and the sensitized IBs are resuspended in 4 mL of the selection buffer and incubated with the thawed cells for 30 min in the Isolex-50 column (Baxter Healthcare) in slow agitation (4 cycles/min). The ratio of cells to sensitized beads is around 2×10^9 IB/CB unit.
8. Unbound cells are eluted in the Isolex-300-SA device using adapters for the Isolex-50 column after three washes with the selection buffer.
9. Cells are released from beads by incubation with a releasing peptide for 30 min (PR34+ Stem Cell Releasing Agent; Baxter).
10. The CD34-enriched cell fraction thus obtained is eluted (two washes) using serum-free culture medium (SFM) and collected in the expansion bags. The whole process is performed at room temperature in a closed system.

3.2. Expansion Culture (see *Note 3*)

1. Cells from the CD34+-cell enriched fraction were adjusted at 25,000 cells/mL and cultured at 37°C and 5% CO_2 up to 2 wk in serum-free media (SCGM) supplemented with recombinant human cytokines (*see* **Note 4**). SCGM has been tested previously to use and compared with the culture performance using fetal calf serum. **Table 1** shows the expansion rate of CD34+ cells using both media (*see* **Note 5**).
2. Cytokines used depend on the objective proposed. The need to generate high quantities of hemopoietic progenitors make it necessary to use of cytokine cocktails with synergistic effects, acting at the early and late stages of the hemopoiesis pathway. Common used cytokines in expansion culture are stem cell factor (SCF), fetal tyrosine kinase-3 ligand (Flt3-L), thrombopoietin (TPO), interleukin-6 (IL-6), and interleukin-3 (IL-3). Also, addition of granulocyte–macrophage colony-stimulating factor (GM-CSF), granulocyte colony-stimulating factor (G-CSF), and erythropoietin (EPO) can be considered. We have finally accepted SCF, FLT3-L, IL-6, and TPO as the best option (*see* **Note 6**). This combination promotes a maintenance of immature cells assessed in long-term cultures (cobblestone-area-forming cells) and multilineage expansion of progenitor cells. The use of IL-3 remains controversial. These cytokines are also provided by CellGenix; after testing they were used in clinical protocols.

Table 1
Fold Expansion of Nucleated Cells (NCs), CD34+ Cells, CD34^{bright} Cells, and Colony-Forming Cells (CFCs) in the Presence of Serum (IMDM + FCS) or in its Absence (SCGM), After 6 d of Ex Vivo Expansion of CB-Derived CD34+ Cells in the Presence of SCF, IL-3, and IL-6 (50 ng/mL Each)

	Fold expansion			
Culture medium	NCs	Total CD34+ cells	CD34^{bright} cells	CFCs
IMDM+ FCS	18 (± 10)	9 (± 7)	4 (± 4)	13 (± 15)
CellGro SCGM	17 (± 7)	11 (± 6)	4 (± 2)	14 (± 9)

Note: Mean and standard deviation of 10 experiments.

3. Cells were expanded either in wells, flask, or cell culture bags, at a defined environmental conditions (*see* **Note 7**). Bags used are semipermeable Teflon bags (American Fluoroseal Corp, Columbia, MD) (*see* **Note 8**).
4. Expansion cultures were monitored using cell counts, viability, colony-forming unit (CFU) assays and flow cytometry analysis for antigen expression (*see* **Note 9**).

3.3. Progenitor Assessment

3.3.1. CFU Assay

1. Cells are seeded in a methylcellulose-based media (Methocult GF H4434; Stem Cell Technologies, Vancouver, Canada) at 0.5×10^3 cells/mL for the CD34+ fraction and at $(1–5)\times10^3$ cells/mL after 1–2 wk expansion.
2. After culture at 37°C and 5% CO_2 for 14 d, cells are scored for the presence of CFU-GM, and erythroid and immature myeloerythroid colonies (BFU-E/CFU-Mix) (*see* **Note 10**).

3.3.2. BFU/CFU-Meg Assay

1. Progenitor-derived megakaryocyte (CFU-Meg) colonies are grown in MegaCult-C (Stem Cell Technologies, Vancouver, Canada). MegaCult-C is a collagen-based system. The collagen has a final concentration of 1–2% at 37°C. The medium contains the human recombinant cytokines: 10 ng/mL IL-3, 10 ng/mL IL-6, and 50 ng/mL TPO, 1% bovine serum albumin, 10 ng/mL bovine pancreatic insulin, 200 μg/mL human transferrin, 2 m*M* L-glutamine, 10^{-4} *M*, 2-mercaptoethanol in IMDM.
2. The samples are grown in double-chamber slides and incubated at 37°C with 5% CO_2 in humidified air for 10–12 d.

3. After that, the chamber slides are dehydrated and fixed with methanol/acetone (1:3) for immunocytochemical staining. CFU-Meg colonies are identified by a primary antibody to the megakaryocytic-specific antigen GPIIb/IIIa (CD41) linked to a secondary biotinylated antibody–alkaline phosphatase avidine conjugated detection system.

3.3.3. Flow Cytometry Analysis (see **Note 11**)

1. The following monoclonal antibodies are used for immunofluorescence analysis: PE-labeled anti-CD34, clone 8G12 (HPCA-2; Becton-Dickinson, San Jose, CA); FITC- labeled anti-CD45, clone Kh56 (Cyto-stat, Coulter Corporation, Miami, FL), and an isotypic control (PE–IgG1; Becton Dickinson). 7-Aminoactinomycin D (7-AAD) is used as a viability marker.
2. Aliquots of (1–5) $\times 10^5$ cells/tube are stained for 15 min, at room temperature in the dark, with MAb and washed in PBS with 0.1% bovine serum albumin (Sigma) and 0.01% sodium azide, and analyzed using a cytometer.
3. For analysis, an initial gate is used to exclude 7-AAD+ dead cells and CD45-negative events. The number of viable events CD34+/CD45 low/side scatter low are considered as CD34+ cells (*see* **Note 12**).
4. The percentage of CD34+ cells is calculated dividing the CD34+ by the total CD45+ events. An isotypic negative control is used to determine CD34+ positivity.

3.4. Clinical Experience

Following this protocol, we have selected and expanded five CB units for clinical purposes using Teflon bags for culture. Five patients have received a CB unit, ex vivo expanded using 20 ng/mL IL-3, 100 ng/mL SCF, 100 ng/mL TPO, and 300 ng/mL Flt-3 ligand. Mean purity obtained was 65 ± 16% and recovery of CD34+ cells from the fresh CB unit was 65 ± 14%. Nonspecific loses of cells were 14 ± 17%, and total CD34+ cells were recovered in the positive and negative fraction of 78 ± 17%. This represents a loss of about 22% of CD34+ cells in tubing and clumping during the procedure. Overall expansion after positive selection in these five clinical experience was 7.16 ± 4.8 for total NCs, 9.5 ± 8.9 for CD34+ cells, and 6.9 ± 5.5 for CFU-GM. Clinical results have been recently published ***(14)*** (*see* **Note 13**).

4. Notes

1. CD34-positive selection from cryopreserved samples: First, we defined the best starting cells for an optimized expansion. CB stored in CB banks, ready-to-use, is peripheral blood of fetal origin consisting of 50% granulocytes, 40% lymphocytes, 5% monocytes, and 5% erythroblasts. The mean concentration of CD34+ cells is 0.30%. After thawing, the major part of granulocytes die, and the cells recovered consist of a mixture of 80% lymphocytes, 10% monocytes, and 10% erythroblasts, with a higher concentration of CD34+ cells (0.50%). Expansion condition directly using these cells in static cultures show a subopti-

mal expansion rate. At 6 d, after culturing with SCF, IL-3, and IL-6, only a maintenance in the ability to generate colonies in semisolid media is achieved using total cells. However, if we perform a positive selection of CD34+ cells, the expansion of colony-forming units is as high as 10 times comparing the starting population. As the overall recovery of CD34+ cells after positive selection in this situation is around 50%, the overall gain of this procedure is five times the net generation of committed progenitors (CD34+ cells or CFU-GM). The reason for the best control of expansion in this situation is a better definition of culture conditions (after CD34+ selection, only external cytokines are responsible for the culture kinetics), in contrast to the presence of high quantities of mature cells able to secret inhibitory cytokines affecting the development of the culture. Moreover, cell concentration in bags can affect the expansion kinetics by an inhibition for cell-to-cell contact.

2. Standardization of a positive selection protocol using thawed CB cells is difficult. We have recently published our approach in solving this problem. The method recommended is that based in a direct immunomagnetic protocol using commercially-available product such as Isolex-300i from Baxter. Using this method, the recovery and purity of CD34+ cells is 69 ± 16% and 52 ± 12%, respectively ***(15)***.
3. Stroma-free, serum-free, static culture: Culturing in external, fully controlled conditions make it possible to adapt culture to GMP rules for cell production. In our case, we have previously defined the objective and conditions of our expansion culture. The main objective of expansion is to decrease both the neutropenia period and the neutrophil nadir after the conditioning regimen in CB transplantation. Myeloid progenitors and precursors are obtained ex vivo from CD34+-selected cells. The culture is based in a stroma-free strategy, using media and cytokines with proteins of human origin. Cells are cultured in semipermeable bags in an incubator with gas and temperature controlled.
4. Reagents used in our laboratory are supplied by CellGenix (Friburg, Germany). The CellGro product line comprised a standardized serum-free medium for optimal growth of human hematopoietic progenitor cells (SCGM) and GMP-grade cytokines for the optimal growth support. These reagents can be obtained from other companies (i.e., Amgen, Thousand Oaks, CA).
5. The use of serum-free medium results in a lower coefficient of variation, warranting similar progenitor amplification. Quality controls include sterility, pH, osmolarity, endotoxin testings, as well as cell proliferation, viability, and colony-forming capacity determined under experimental conditions on CD34+ primary cultures.
6. Cytokine combination: Cytokines used in expansion cultures are selected following their action at hemopoiesis regulation. Cocktail design must consider the use of early-acting cytokines with function at the stem cell level, promoting the survival of cells in vitro and activating their proliferative responses. The most important cytokines at this level are Flt3-ligand and TPO ***(16,17)***. Synergy between cytokines at this level increases the effect substantially. The addition

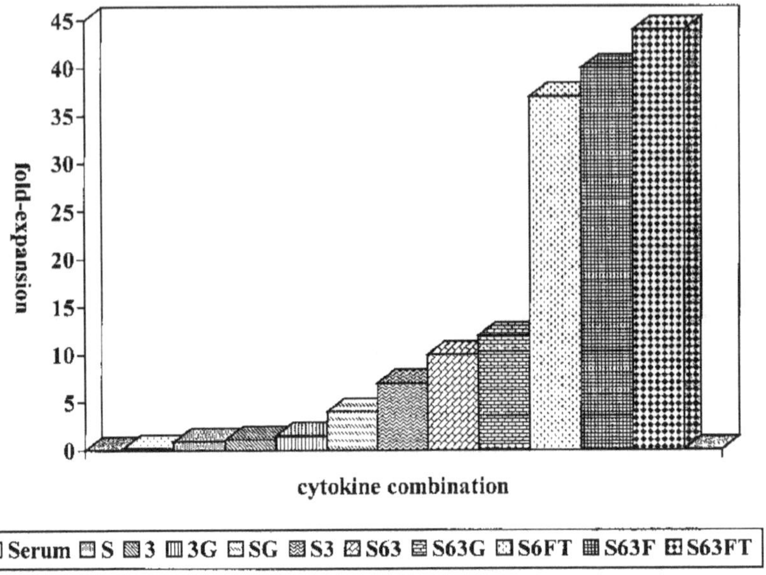

Fig. 1. Expansion potential of different cytokine combinations. The figure shows the overall amplification of CD34+ cells after a 6-d expansion of CB CD34+ selected cells using different cytokine combinations or serum alone. S (SCF), 3 (IL-3), 6 (IL-6), G (G-CSF), F (Flt3-L), and T (TPO).

of SCF increases even more the amplification of CD34+ cells. On the other hand, productive capacity of multilineage progenitors of our cytokine cocktail should be considered. Erythroid, myeloid, and megakaryocytic progenitors can be produced using a cocktail consisting of SCF, Flt3-L, TPO, and IL-6. **Figure 1** shows different expansion potentials depending on the cytokines used. The role of late-acting cytokines such as G-CSF or EPO is centered in their effect on the cellular maturation (higher viability after more than 6 d of culture) and in the antiapoptotic signals on specific lineage progenitors (granulocytic and erythroid).

7. Environmental conditions: After CD34+ cell selection, expansion culture starts fixing the initial cell concentration around 10^4/mL. The media used contains inorganic salts, amino acids, vitamins, glucose, and buffers similar to enriched hemopietic media (i.e., IMDM) supplemented with human albumin and serum substitutes, basically transferrin, low-density lipoproteins, and insulin. Regulatory agents used are early-acting and late-acting cytokine cocktails. Usually, we use cytokines at more than 10 ng/mL of concentration (saturating concentration) and feed with new cytokines every 3 d, in order to maintain a continuous presence of active cytokines avoiding apoptosis by deprivation. Media are added as necessary (expanding the culture volume) in order to maintain the cell concentration below 10^6 cells/mL.

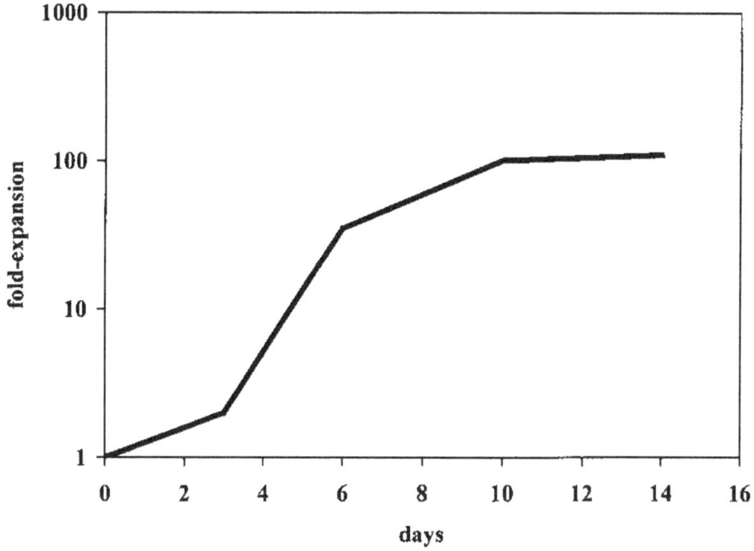

Fig. 2. Expansion kinetics using SCF, IL-6, Flt3-L, and TPO. Typical curve after an expansion culture of CB CD34+ selected cells using SCF, IL-6, Flt3-L, and TPO. During the first 10 d, an exponential curve is observed (induction phase). After this period, a maintenance phase starts reflecting the stem cell expansion within the culture.

8. Physical support to perform culture are semipermeable bags. Teflon bags can be used to this finality because they prevent dehydration and support a excellent gas interchange between the incubator environment (regulate for CO_2 at 5.5%) and the internal part of the bag. These bags can be adapted to different tube systems, and sterile connect with bags and the selection system performs all operation in a closed system, increasing the safety of the procedure.
9. Expansion kinetics: In the expansion culture, two different phases can be distinguished depending on the amplification of cell observed. In the first phase, a net amplification is observed because of the proliferation wave from the stem cell compartment and the differentiation in the culture of the cell belonging to all compartments: progenitor, precursors, and mature cells. After 2 wk, all compartments are fully represented in the culture and the capacity to generate new cells depends on the stem cell expansion. Thus, in a model of stem cell maintenance, the predictive curve to be observed is the maintenance of the number of CD34+ cells or nucleated cells. This is observed using this cocktail of cytokines. As shown in **Fig. 2**, after a phase of amplification, a maintenance curve is drawn and the expansion is only achieved as an accumulation of expansion and not as an expression of a higher content of progenitors in a specific control point. Using SCF, Flt3-L, TPO, and IL-6, amplification of CFU and CD34+ cells can increase up to 2 logs.

Table 2
Phenotypic Characteristics of Cells Generated Using SCF, Flt3-L, IL-6, and TPO

	Day			
	0	6	10	14
% CD34	95	60	20	6
% CD61	15	11	18	18
% GlycoA	3	1	7	7
% CD15 and/or CD11b	17	34	45	42

10. Colony-forming-unit assessment shows a similar rate of amplification. Multilineage expansion can be demonstrated for different colony types. CFU-GM can be expanded up to 117 times at 2 wk culture. The megakaryocytic progenitor assessed in semisolid media has been expanded at a similar rate (139 ± 23). However BFU-Es reach the highest amplification at 6 d (around 20 times), then decreases unless specific lineage cytokines are added.
11. Cell phenotyping: Cells generated during culture have different characteristics. At d 6, 50% of cells express the CD34 antigen and only 5% of them express CD34+ at d 14. Nevertheless, in absolute number, the culture continues producing new CD34+ cells. The antigen distribution of nucleated cells representing the maturation of different lineages is showed in **Table 2**. Differentiation to granulocytic and monocytic lineages is predominant. Erythroid cells are a minority in a cocktail without EPO, but a consistent generation of cells expressing CD61 antigen is demonstrated, indicating the presence of megakaryocytes.
12. The viability of CD34+ cells is maintained during an expansion culture higher than 90%. However, on the overall population, viability decreases substantially after 10 d of culture, mainly by the generation of mature cells that die by apoptosis as their natural way or as a consequence of the absence of late-acting cytokines.
13. For clinical use, several points must be taken into account:
    - Use of an optimized protocol for CD34+ selection from thawed CB cells adapted to small volumes in order to diminish the nonspecific losses of cells because of tube length.
    - Improve CD34+ cell recovery using a direct immunomagnetic approach, high ratio of immunobeads to target cells, and buffers containing DNase and citrate to prevent clumping.
    - Culture in tested serum-free media manufactured using GMP conditions and accepted for ex vivo use by regulatory agencies.
    - Define the environmental condition of the culture in terms of physical support, cell concentration, and feeding strategy: We use Teflon bags, with low cell

concentration (below 10^6 cells/mL) and feeding every 3 d to avoid cytokine deprivation.

- Define the optimal time for expansion to achieve our objective. In terms of progenitor cell amplification, 2 wk of expansion seems to be the maximum recommended time evaluating the cost–risk–benefit of all procedures. At that point, 2 logs of overall progenitor cells expansion is achieved.
- Develop the control test able to monitor the functional state of cells at every point of the culture. These analyses are based in flow cytometry antigen determination of CD34+ cells and subpopulation, and clonogenic assessment in semisolid media. Other assays such as long-term cultures or in vivo stem cells assays (non-obese diabetes severe-combined immunodeficiency [NOD–SCID] mice) are not useful in the scale-up phase because the high coefficient of variation and the difficulties to monitor all possible modification of the protocol.
- Work in appropriate controlled rooms, using closed systems and human-approved disposable reagents. A validated scale-up process must be available before starting the clinical assay. This validation needs to evaluate the reproducibility of the selected method and to identify the process critical points.

Acknowledgment

This work was supported by a grant from Fondo de Investigaciones Sanitarias (FIS) (01/0066-02).

References

1. Kurtzberg, J., Martin, P., Chao, N., Stevens, C., and Rubinstein, P. (2000) Unrelated placental blood in marrow transplantation. *Stem Cells* **18,** 153–154.
2. Rocha, V., Wagner, J. E., Jr., Sobocinski, K. A., et al. (2000) Graft-versus-host disease in children who have received a cord-blood or bone marrow transplant from an HLA-identical sibling. Eurocord and International Bone Marrow Transplant Registry Working Committee on Alternative Donor and Stem Cell Sources. *N. Engl. J. Med.* **342,** 1846–1854.
3. Barker, J. N., Davies, S. M., DeFor, T., et al. (2001) Survival after transplantation of unrelated donor umbilical cord blood is comparable to that of human leukocyte antigen-matched unrelated donor bone marrow: results of a matched-pair analysis. *Blood* **97,** 2957–2961.
4. Rocha, V., Cornish, T., Sievers, E. L., et al. (2001) Comparison of outcomes of unrelated bone marrow and umbilical cord blood transplants in children with acute leukemia. *Blood* **97,** 2962–71.
5. McNiece, I., Kubegov, D., Kerzic, P., Shpall, E. J., and Gross, S. (2000) Increased expansion and differentiation of cord blood products using a two-step expansion culture. *Exp. Hematol.* **28,** 1181–1186.

6. McNiece, I., Jones, R., Bearman, S. I., et al. (2000) Ex vivo expanded peripheral blood progenitor cells provide rapid neutrophil recovery after high-dose chemotherapy in patients with breast cancer. *Blood* **96,** 3001–3007.
7. Reiffers, J., Cailliot, C., Dazey, B., et al. (1999) Abrogation of post-myeloablative chemotherapy neutropenia by ex-vivo expanded autologous CD34-positive cells. [letter]. *Lancet* **354,** 1092–1093.
8. Dexter, T. M., Allen, T. D., and Lajtha, L. G. (1977) Conditions controlling the proliferation of haemopoietic stem cells in vitro. *J. Cell Physiol.* **91,** 335–344.
9. Brandt, J., Srour, E. F., van Besien, K., Briddell, R. A., and Hoffman, R. (1990) Cytokine-dependent long-term culture of highly enriched precursors of hematopoietic progenitor cells from human bone marrow. *J. Clin. Invest.* **86,** 932–941.
10. Srour, E. F., Abonour, R., Cornetta, K., and Traycoff, C. M. (1999) Ex vivo expansion of hematopoietic stem and progenitor cells: are we there yet?. *J. Hematother.* **8,** 93–102.
11. Aglietta, M., Bertolini, F., Carlo-Stella, C., et al. (1998) Ex vivo expansion of hematopoietic cells and their clinical use. *Haematologica* **83,** 824–848.
12. Querol, S., Gabarro, M., Amat, L., et al. (1998) The placental blood program of the Barcelona Cord Blood Bank. *Bone Marrow Transplant.* **22 (Suppl. 1),** S3–S5.
13. Martin-Henao, G. A., Picon, M., Amill, B., et al. (2000) Isolation of CD34+ progenitor cells from peripheral blood by use of an automated immunomagnetic selection system: factors affecting the results. *Transfusion* **40,** 35–43.
14. Fernandez, M. N., Regidor, C., Cabrera, R., et al. (2001) Cord blood transplants: early recovery of neutrophils from co-transplanted sibling haploidentical progenitor cells and lack of engraftment of cultured cord blood cells, as ascertained by analysis of DNA polymorphisms. *Bone Marrow Transplant.* **28,** 355–263.
15. Querol, S., Capmany, G., Azqueta, C., et al. (2000) Direct immunomagnetic method for CD34+ cell selection from cryopreserved cord blood grafts for ex vivo expansion protocols. *Transfusion* **40,** 625–631.
16. Capmany, G., Querol, S., Cancelas, J. A., and Garcia, J. (1999) Short-term, serum-free, static culture of cord blood-derived CD34+ cells: effects of FLT3-L and MIP-1alpha on in vitro expansion of hematopoietic progenitor cells. *Haematologica* **84,** 675–682.
17. Piacibello, W., Sanavio, F., Severino, A., et al. (1999) Engraftment in nonobese diabetic severe combined immunodeficient mice of human CD34(+) cord blood cells after ex vivo expansion: evidence for the amplification and self-renewal of repopulating stem cells. *Blood* **93,** 3736–3749.

27

Ex Vivo Expansion of Hematopoietic Stem Cells

Beatriz Albella, José Carlos Segovia, Guillermo Guenechea, and Juan Antonio Bueren

1. Introduction

A number of studies have demonstrated the feasibility of amplifying the short-term repopulating progenitors in vitro and have shown the relevance of ex vivo expansion approaches in hematopoietic transplantation *(1–3)* and gene therapy protocols *(4,5)*. In spite of the beneficial effects generally associated with ex vivo expansion strategies, a number of observations suggest that further studies on the biology of ex vivo expansion are required to facilitate the optimal implantation of this strategy in the clinics *(6,7)*. In this respect, although in some instances data showing an impairment in the long-term repopulating capacity of ex vivo expanded grafts *(8)*, ex vivo amplifications in other cases of very primitive progenitors have been observed *(9)*. To prevent the prompt differentiation of the hematopoietic stem cells (HSCs) during the ex vivo expansion process, new combinations of early-acting cytokines have been used to facilitate the self-renewal divisions in the HSC compartment *(1,10–12)*.

Regarding the experimental assays used for evaluating the effects of ex vivo expansion, several in vitro colony techniques have been developed for studying the committed progenitors *(13)*. However, the validity of in vitro assays for assessing the functionality of the true HSCs is still open to discussion. Both the CAFC (Cobblestone-area forming cell *(14)* and the LTC-IC (long-term culture-initiating cell *(15)* aimed at identifying the HSCs. However, a number of studies suggest that these assays are not capable of reflecting the functionality of the self-renewing HSCs *(16)*. Currently, the most reliable procedures used for assessing the primitive hematopoietic repopulating cells are based on in vivo reconstitution assays. In the mouse model, the competitive repopulation

From: *Methods in Molecular Biology, vol. 215:*
Cytokines and Colony Stimulating Factors: Methods and Protocols
Edited by: D. Körholz and W. Kiess © Humana Press Inc., Totowa, NJ

assay (CRA) is accepted as the most accurate assay for assessing the primitive HSCs *(17)*. In the human system, the transplantation of human hematopoietic grafts into immunodeficient mice is broadly accepted for this purpose *(16)*. Both assays have been successfully used in a wide range of studies, including ex vivo expansion, hematotoxicity testing, and gene therapy *(18–20)*.

The present chapter is focused on the ex vivo expansion of hematopoietic grafts, both from mouse and humans, as well as on the in vivo repopulating assays used for testing the expansion rate of the different repopulating cells.

2. Materials

2.1. Ex Vivo Expansion of Murine Bone Marrow Samples

1. Mice: P3D2F1 (B6.SJL-Ptprc^{a}Pep3^{b}/BoyJ-CD45.1 × DBA/2-CD45.2) and B6D2F1 (C57BL/6J-CD45.2 × DBA/2-CD45.2). Breeding pairs, originally obtained from Jackson Laboratory (Bar Harbor, ME), were maintained under high hygienic standards in well-ventilated conditions, with a regulated temperature of 22°C, a light/dark cycle of 7 AM to 7 PM, with food and water *ad libitum*. Animals were routinely screened for pathogens according to FELASA recommendations.
2. Iscove's modified Dulbecco's medium (IMDM) (Gibco Laboratories, Grand Island, NY).
3. Fetal bovine serum (FBS) (Gibco Laboratories).
4. Turk solution (2% acetic acid plus 0.01% methylene blue in water) and trypan blue (Sigma Chemical, St. Louis, MO).
5. 25-cm^2 Tissue culture flasks (Nunc, Roskilde, Denmark).
6. Hematopoietic growth factors (HGFs): recombinant murine stem cell factor (rmSCF) (4 U/mL), human interleukin-11 (rhIL-11) (100 ng/mL), human macrophage inhibitory protein-1α (rhMIP-1α) (100 ng/mL), and human flt-3 ligand (rhFlt3L) (50 ng/mL). All are commercially available.

2.2. Ex Vivo Expansion of Human Cord Blood Cells

1. Cord blood (CB) cells obtained from umbilical cord bloods scheduled for discard and processed during the next 12 h postpartum. Samples were collected in heparin.
2. Ficoll–Paque (1.077 g/mL; Pharmacia Biotech, Uppsala, Sweden).
3. PBE: phosphate-buffered saline without Ca^{2+} and Mg^{2+} (Sigma Chemical; PBS^{-}) plus 0.5% (w/v) bovine serum albumin (BSA) (Fraction V; Sigma Chemical) plus 5 m*M* EDTA.
4. Hematopoietic growth factors: rhSCF, rhFlt3L, and rhIL-11 at a final concentration of 100 ng/mL.
5. VarioMACS CD34 progenitor cell isolation kit (Myltenyi Biotech, Auburn, CA).
6. DMSO (dimethyl sulfoxide; Sigma, St Louis, MO).
7. Human albumin (Behring, Hoechst Iberica, Spain) and dextran-40 (Rheomacrodex 10%; Pharmacia Biotech).

8. PBA: PBS^- 1X + 0.1% BSA (w/v) + 0.02% NaN_3 (w/v).
7. 25-cm^2 Tissue culture flasks (Nunc, Roskilde, Denmark).
9. FBS (Gibco Laboratories).
10. Flow cytometer EPICS XL (Coulter, Hialeah, FL).

2.3. Competitive Repopulating Assay in Mice

1. Mice P3D2F1 (B6.SJL-$Ptprc^aPep3^b$/BoyJ-CD45.1 × DBA/2-CD45.2) and B6D2F1 (C57BL/6J-CD45.2 × DBA/2-CD45.2) (Jackson Laboratory, Bar Harbor, ME).
2. Philips MG 324 X-ray equipment (Philips, Hamburg, Germany).
3. Monoclonal antibodies (MAb): phytoerythrin (PE)-conjugated antimouse CD45.1 (clone A20; Pharmingen, San Diego, CA) and fluorceine isothiocyanate (FITC)-conjugated antimouse specific for T-cell (CD3), B-cell (B220), and myeloid (Gr-1) lineages (all from Pharmingen).
4. Ammonium chloride lysing solution (0.155 m*M* NH_4Cl + 0.01 m*M* $KHCO_3$ + 10^{-4} m*M* EDTA).
5. PBA: PBS^- 1X + 0.1% BSA (w/v) + 0.02% NaN_3 (w/v).
6. Flow cytometer EPICS XL (Coulter, Hialeah, FL).
7. Propidium iodide (PI), 2 µg/mL (Sigma).

2.4. Evaluation of Human Repopulating Cells

1. NOD/LtSz-scid/scid (NOD-SCID) mice (deficient in Fc receptors, complement function, natural killer cell B- and and T-cell function) were used as recipients of the human hematopoietic cells (Jackson Laboratory, Bar Harbor, ME).
2. Philips MG 324 X-ray equipment.
3. Monoclonal antibodies: antihuman-CD45–PECy5 (Clone J33, Immunotech, Marseille, France), anti-human-CD34–PE (anti-HPCA-2; Becton Dickinson Immunocytometry, San Jose, CA), anti-human-CD19–PE (anti-Leu-12, Becton Dickinson), and anti-human-CD33–PE (anti-Leu-M9, Becton Dickinson).
4. PBA: PBS^- 1X + 0.1% BSA (w/v) + 0.02% NaN_3 (w/v).
5. Propidium iodide 100X (PI), 200 µg/mL in PBS^- 1X (Sigma).

3. Methods

3.1. Ex Vivo Expansion of Murine Bone Marrow Samples

1. Flush the bone marrow (BM) cells from the femora of at least three 8- to 10-wk-old donor mice P3D2F1 (B6.SJL-$Ptprc^aPep3^b$/BoyJ-CD45.1 × DBA/2-CD45.2) and resuspend them in IMDM supplemented with 20% FBS.
2. Adjust cell density to 2×10^6 cells/mL by counting the cell suspension in an hemocytometer by 1/10 dilution with Turk solution.
3. Incubate the cells with HGFs and FBS-supplemented IMDM in 25-cm^2 tissue culture flasks (Nunc, Roskilde, Denmark). Suspension cultures are maintained in incubators at 37°C and 5% CO_2 in air. If cultures are maintained for more

than 1 wk, the medium must be changed and the cells diluted weekly to the initial cell density.

4. After the incubation period, wash the cells, determine the viability with trypan blue, and mix an appropriate number of cells with equivalent number of normal BM cells from B6D2F1.
5. Transplant the cells into irradiated recipients B6D2F1. We usually perform clonogenic assays for CFU-GM progenitors as a control of the capacity of expansion of the BM (*see* **Note 1**). Although in vitro expansion cultures of murine BM cells can be maintained through at least 1 mo, our observations indicate that the best repopulating results are obtained during the first week.

3.2. Ex Vivo Expansion of Human Cord Blood Cells

1. Obtain mononuclear cells (MNCs) from CB cells by layering the blood (2 volumes) onto Ficoll-Paque (1 vol) and centrifuge at 400*g* for 30 min.
2. Remove the layer in the interface, dilute it 1:5 with PBE and centrifuge at 400*g* for 15 min.
3. Discard supernatant, wash the cells again, and resuspend in PBE at a final concentration of 1×10^9 cells/mL.
4. To purify $CD34^+$ cells, subject the mononuclear fraction to immunomagnetic separation, using the VarioMACS CD34 progenitor cell isolation kit, according to the manufacturer's recommendations. Evaluate the purity of the $CD34^+$ population by flow cytometry.
5. Cryopreserve $CD34^+$ cells in liquid nitrogen using IMDM supplemented with 10% of DMSO and 20% of FBS until use.
6. Thaw $CD34^+$ cells according to an optimized protocol *(21)*. Briefly, dilute cells 1:1 with IMDM containing 2.5% human albumin and 5% dextran-40 and maintain at room temperature for 10 min.
7. Dilute cells with IMDM plus 10% FBS and centrifuge at 800*g* for 15 min.
8. For expansion protocols, incubate thawed $CD34^+$ cells in 25-cm^2 tissue culture flasks at a maximum of 2.5×10^4 cells/mL (5×10^3 to 2.5×10^4 cells/mL) in IMDM containing 20% FBS, rhSCF, rhFlt3L, and rhIL-11, all at a final concentration of 100 ng/mL (*see* **Note 2**). After 6 d of incubation, collect cells for assessing the content in CFC (colony-forming cells) progenitors (*see* **Note 3**) and the ability to reconstitute NOD-SCID mice.

3.3. Competitive Repopulating Assay in Mice

1. Recipients (B6D2F1 mice) of bone marrow grafts are total-body irradiated with a fractionated dose of 10 Gy (two doses of 5.0 Gy spaced 4 h apart; dose rate 1.03 Gy/min) using Philips MG 324 X-ray equipment at 300 kV and 10 mA. Earlier studies have revealed the myeloablative properties of this irradiation regimen *(22)*.
2. Transplant the chimeric population of BM expanded cells (test population from P3D2F1 mice) and fresh nonmanipulated cells (reference population

from B6D2F1 mice) into recipients. Transplantation is performed thorough intravenously injection into lateral vein. We routinely mix equivalent fractions of femoral BM from both sources, regardless of the ex vivo expansion values obtained after culture. We generally transplant each recipient mouse with 2×10^6 fresh cells from the reference population (CD45.2/CD45.2) together with the product generated by the expansion of 2×10^6 cells from the test population (CD45.1/CD45.2). The use of F1 animals instead of the original congenic strains increases the survival after the radiation conditioning regimen (*see* **Fig. 1A**).

3. At periodic times up to 1 yr after transplantation, obtain 200-μL samples of peripheral blood (PB) from the tail vein of recipients and mix with 20 μL of 0.5 *M* EDTA pH 8.0 to avoid coagulation. The analysis of PB samples at 20 d, 1 mo, 3 mo, 6 mo, and 12 mo posttransplantation generates very direct information about the kinetics of the different repopulating cells during the ex vivo expansion (*see* **Note 4**).
4. The competitive repopulating ability (CRA) of test samples is deduced from FACS analysis of CD45.1-positive PB cells using the anti-CD45.1 MAb. For multilineage reconstitution analysis, dual-color immunofluorescence using a PE-conjugated antibody for CD45.1 antigen and FITC-conjugated antibodies specific for T-cell (CD3), B-cell (B220), and myeloid (Gr-1) lineages can be employed.
5. Incubate 50 μL of PB with the corresponding monoclonal antibodies for 30 min at 4°C in the dark.
6. Lyse erythrocytes by addition of 2.5 mL of ammonium chloride lysing solution and incubation at room temperature for 10 min.
7. Wash cells with PBA, resuspend in PBA plus 2 μg/mL PI stain to exclude death cells, and analyze in a flow cytometer. For analyzing the chymerism of the BM inocula, resuspend 5×10^5 BM cells into 50 μL of PBA and then stain, lyse, and analyze as in the PB samples.
8. Perform flow-cytometric analyses by a serial gating strategy (*see* **Fig. 1B**). Briefly, a forward size scatter (FS) vs side size scatter (SSC) dot plot is set up and a gate is drawn to select only the mononuclear cells. This gate is applied to a second FL4 (675 ± 10-nm bandpass filter) vs FL3 (610 ± 10-nm bandpass filter) dot plot in which a new gate is drawn avoiding cells that appear as a diagonal, which are considered dead cells. Finally, both gates are applied to a third FL1 (525 ± 15 nm) vs FL2 (575 ± 10 nm) dot plot to analyze FITC and PE fluorescences. To establish autofluorescence signals, include samples from normal B6D2F1 and P3D2F1 mice as a routine in the analysis. Collect at least 10,000 events within the 2 consecutive gates per sample analyses. Off-line analysis can be done with the WinMDI free software package (http://www.cyto.purdue.edu/flowcyt/software/Winmdi.htm).
9. When 1:1 mixtures of the test and reference BM populations are used, data on CRA after 1 yr posttransplantation are deduced from the formula $CRA_{CD45.1}$ = (%CD45.1/100 – %CD45.1.) × 100 (*see* **Fig. 1C**).

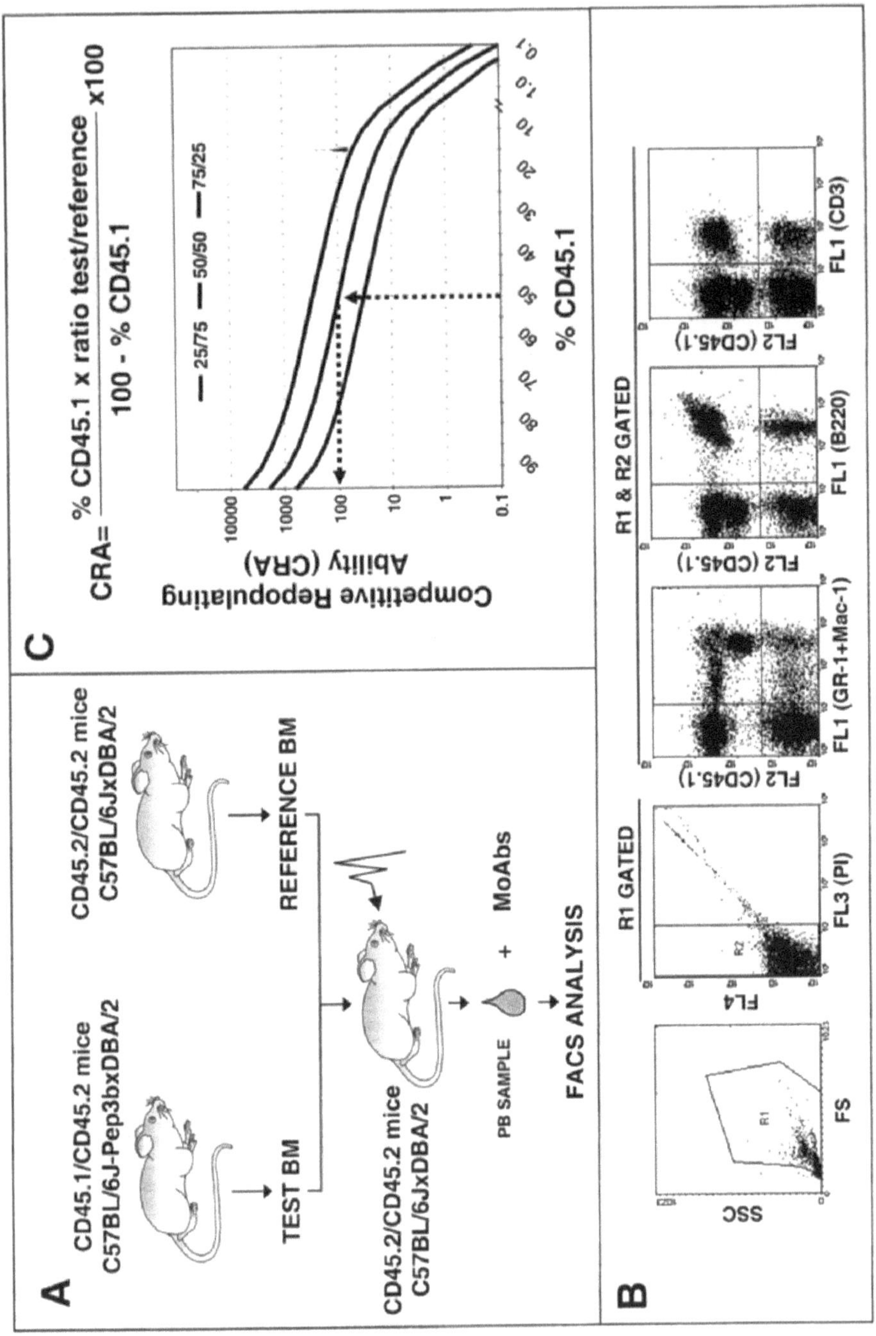
A
CD45.1/CD45.2 mice
C57BL/6J-Pep3bxDBA/2
TEST BM
CD45.2/CD45.2 mice
C57BL/6JxDBA/2
REFERENCE BM
CD45.2/CD45.2 mice
C57BL/6JxDBA/2
PB SAMPLE
+
MoAbs
FACS ANALYSIS
B
SSC
FS
R1
R1 GATED
FL4
FL3 (PI)
R2
R1 & R2 GATED
FL2 (CD45.1)
FL1 (GR-1+Mac-1)
FL1 (B220)
FL1 (CD3)
C
CRA= % CD45.1 x ratio test/reference / 100 - % CD45.1 x100
Competitive Repopulating Ability (CRA)
25/75
50/50
75/25
% CD45.1

3.4. Evaluation of Human Repopulating Cells

1. NOD-SCID mice 6–8 wk old must be handled under sterile conditions using laminar-flow hoods and maintained under microisolators with autoclaved food, soft wood pellets, and cages. Recipients of $CD34^+$ cord blood cells are total-body irradiated with a single sublethal dose of 2.5 Gy of X-rays.
2. Transplant aliquots of $\geq 10^4$ fresh $CD34^+$ cells or the corresponding product generated after ex vivo expansion into irradiated NOD-SCID mice. Optimal results are obtained by injecting 10^5 $CD34^+$ cells. In our experience, the capacity of reconstitution of NOD-SCID mice is markedly impaired if less than 10^4 $CD34^+$ cells are transplanted.
3. Test the presence of the human hematopoietic cells in NOD-SCID recipients in the BM of the recipient animals after 20, 40, 90, and 120 d posttransplantation by conducting small aspirates of femoral BM (*see* **Fig. 2A**) *(19)*.
4. Incubate aliquots of $(2–5) \times 10^5$ cells/tube with antibodies directed to the human common leukocyte antigen CD45 conjugated with PE/Cy5 tandem for 25 min at 4°C. Stain additional aliquots with anti-human-CD45–PECy5 in combination with anti-human-CD34–PE, anti-human-CD19–PE, or anti-human-CD33–PE. Different subpopulations of human primitive ($CD45^+CD34^+$), B-lymphoid ($CD45^+CD19^+$), and myeloid ($CD45^+CD33^+$) cells are determined with those antibody combinations. After staining, lyse red blood cells, wash in PBA, and resuspend in PBA + 2 μg/mL PI. Bone marrow cells from transplanted NOD-SCID mice and labeled with conjugated nonspecific isotype antibodies, and BM cells from untransplanted NOD-SCID mice stained with anti-human-CD45–PECy5 and CD34–PE antibodies are used as controls of nonspecific staining of human and mouse cells. For each FACS analysis, collect a total of $(10–15) \times 10^3$ PI^- cells.
5. Use serial gating strategy as described for mouse PB analyses for the flow-cytometric analysis (*see* **Fig. 2B**).
6. At the end of the experiment, kill the mice and analyze PB, spleen, thymus, and BM by flow cytometry for the presence of human cells. Perform cytometric analysis as described for bone marrow aspirations (*see* **Note 5**). In some cases, it is of interest to estimate the potential of the sample to repopulate secondary recipients *(23,24)* (*see* **Note 6**).

Fig. 1. *(see opposite page)* Competitive repopulating ability assay. **(A)** Scheme of the protocol used for the transplantation of chimeric grafts into irradiated recipients. **(B)** FACS analysis showing the gating strategy together with a repopulation analysis in three hematopoietic lineages. **(C)** Computer-generated graph showing the CRA values that can be deduced according to the percentages of CD45.1 obtained in the cytometer. The lines represent three different mixtures of test and reference BM.

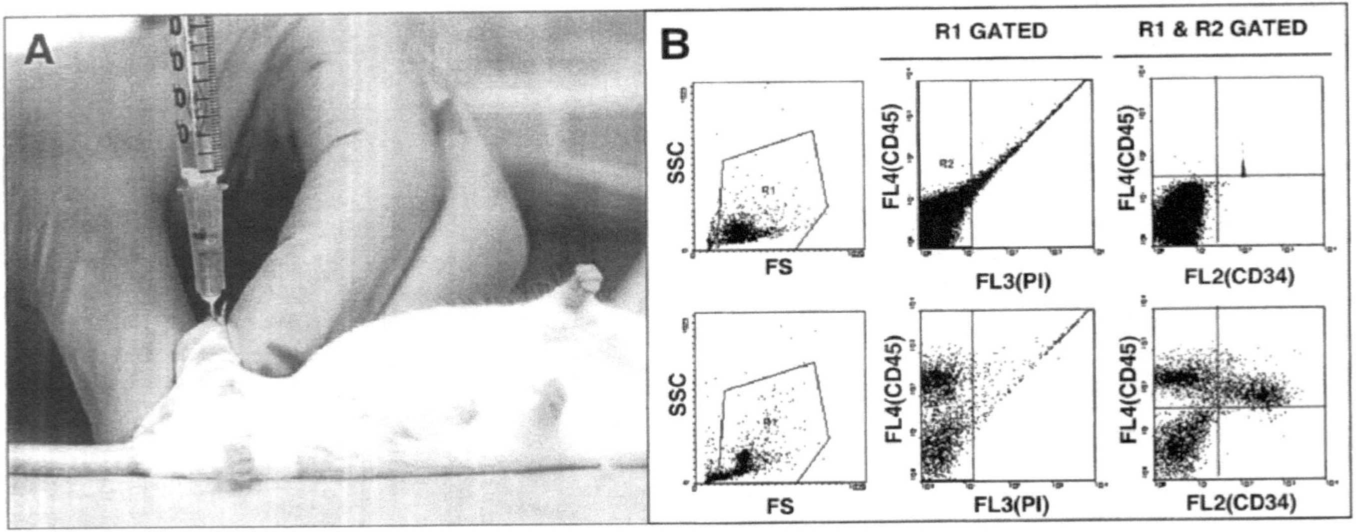

Fig. 2. Analysis of human hematopoiesis in immunodefficient NOD-SCID mice. (**A**) Picture of a femoral aspiration. A 25-gage needle is inserted through the knee into the femur of the recipient mouse. Bone marrow cells are aspirated with a 1-mL syringe containing 100 μL of PBS^-. (**B**) Evaluation by FACS of the human hematopoiesis in the BM of NOD-SCID mice. Upper row: analyses of a control nontransplanted mouse; lower row: analysis of a mouse transplanted 120 d earlier with 10^5 human $CD34^+$ cells.

4. Notes

1. Murine CFU-GM assay: Resuspend an appropriate number of BM cells (50,000 cells/mL for fresh BM and 2500 cells/mL for expanded BM cells) in IMDM supplemented with 25% horse serum (Gibco) and 10% Wehi-3b conditioned medium, mix with Bacto-agar (0.3% final concentration; Difco, Detroit, MI), and seed into 35-mm plastic tissue culture dishes (Nunc, Roskide, Denmark). Score colonies after 7 d of incubation at 37°C in a 95% humidified atmosphere with 5% CO_2 in air. Currently, CFU-GM assays can be easily performed in methylcellulose medium with recombinant cytokines from Stem Cell Technologies (GF M3434; Vancouver, BC, Canada) according to the manufacturer's instructions.
2. Some expansions are performed with other media and hematopoietic growth factor combinations. X-vivo 10 medium (BioWhittaker, Waldersville, MD) supplemented with 1% bovine serum albumin (Boehringer-Mannheim, Germany) and 2 m*M* L-glutamine (Gibco-BRL, Burlington, ON) and StemSpan (Stem Cell Technologies) are specially designed serum-free media for culture and expansion of hematopoietic cells. Hematopoietic growth factor combinations and concentrations are multiple; it is recommended to use of thrombopoietin (TPO) in most of the expansion procedures at doses ranging from 10 to 100 ng/mL.
3. Human CFC assay: MethoCult GF H4534 (Stem Cell Technologies) can be used for growing CFC progenitors. Cultures are kept under the same conditions as in murine CFU-GM, but colonies are scored after 14 d.
4. When the behavior of the most primitive murine HSCs has to be confirmed, secondary recipients can be transplanted with the BM from primary recipients. In these cases, after 3 or 6 mo posttransplantation, harvest the BM of primary recipients and transplant the total BM cells from one femur into one secondary recipient.
5. Mice with a low percentage of human engraftment, as detected by flow cytometry (<0.5%), can be also analyzed by Southern blotting using a human chromosome 17 specific α-satellite probe ***(25)***.
6. When BM cells from highly engrafted mice (60–90% $CD45^+$) are obtained, it is possible to transplant the cells into secondary recipients in order to estimate the self-renewal potential of the human repopulation cell. In these cases, BM cells from primary recipients are incubated with X-vivo 10 medium supplemented with IL-6 (20 ng/mL) and SCF (50 ng/mL). After a 2-d incubation period, cells are analyzed for human CD45-expressing cells and injected intravenously into irradiated secondary NOD-SCID recipients. BM aspirations are done 20–90 d after transplantation. Animals are killed 3 mo after transplantation and the presence of human cells is analyzed by flow cytometry or Southern blot.

References

1. Holyoake, T. L., Freshney, M. G., McNair, L., Parker, A. N., McKay, P. J., Steward, W. P., et al. (1996) Ex vivo expansion with stem cell factor and interleukin-11

augments both shot-term recovery posttransplant and the ability to serially transplant marrow. *Blood* **87,** 4589–4595.
2. Serrano, F., Varas, F., Bernad, A., and Bueren, J. A. (1994) Accelerated and long-term hematopoietic engraftment in mice transplanted with ex vivo expanded bone marrow. *Bone Marrow Transplant.* **14,** 855–862.
3. Reiffers, J., Cailliot, C., Dazey, B., Attal, M., Caraux, J., and Boiron, J. M. (1999) Abrogation of post-myeloablative chemotherapy neutropenia by ex-vivo expanded autologous CD34-positive cells. *Lancet* **354,** 1092–1093.
4. Bernad, A., Varas, F., Gallego, J. M., Almendral, J. M., and Bueren, J. A. (1994) Ex vivo expansion and selection of retrovirally transduced bone marrow: and efficient methodology for gene-transfer to murine lympho-haemopoietic stem cells. *Br. J. Haematol.* **87,** 6–17.
5. Flasshove, M., Banerjee, D., Mineishi, S., Li, M.-X., Bertino, J. R., and Moore, M. A. S. (1995) Ex vivo expansion and selection of human $CD34^+$ peripheral blood progenitor cells after introduction of a mutated dihydrofolate reductase cDNA via retroviral gene therapy. *Blood* **85,** 566–574.
6. Albella, B., Segovia, J. C., and Bueren, J. A. (1997) Does the granulocyte–macrophage colony-forming unit content in ex vivo-expanded grafts predict the recovery of the recipient leukocytes? *Blood* **90,** 464–470.
7. Guenechea, G., Segovia, J. C., Albella, B., Lamana, M., Ramirez, M., Regidor, C., et al. (1999) Delayed engraftment of nonobese diabetic/severe combined immunodeficient mice transplanted with ex vivo-expanded human CD34(+) cord blood cells. *Blood* **93,** 1097–1105.
8. Peters, S. O., Kittler, L. W., Ramshaw, H. S., and Quesenberry, P. J. (1996) Ex vivo expansion of murine marrow cells with interleukin-3 (IL-3), IL-6, IL-11, and stem cell factor leads to impaired engraftment in irradiated hosts. *Blood* **87,** 30–37.
9. Piacibello, W., Sanavio, F., Garetto, L., Severino, A., Bergandi, D., Ferrario, J., et al. (1997) Extensive amplification and self-renewal of human primitive hematopoietic stem cells from cord blood. *Blood* **89,** 2644–2653.
10. Yonemura, Y., Ku, H., Lyman, S. D., and Ogawa, M. (1996) In vitro expansion of hematopoietic progenitors and maintenance of stem cells: Comparison between FLT3/FLK-2 ligand and KIT ligand. *Blood* **89,** 1915–1921.
11. Bhatia, M., Bonnet, D., Kapp, U., Wang, J. C., Murdoch, B., and Dick, J. E. (1997) Quantitative analysis reveals expansion of human hematopoietic repopulating cells after short-term ex vivo culture. *J. Exp. Med.* **186,** 619–624.
12. Conneally, E., Cashman, J., Petzer, A., and Eaves, C. (1997) Expansion in vitro of transplantable human cord blood stem cells demonstrated using a quantitative assay of their lympho-myeloid repopulating activity in nonobese diabetic-scid/scid mice. *Proc. Natl. Acad. Sci. USA* **94,** 9836–9841.
13. Testa, N. and Molineux, G. (1993) Haemopoiesis. A practical approach, in *The Practical Approach*, vol. 111 (Rickwood, D., ed.), Oxford University Press, New York, p. 293.

14. Ploemacher, R. E., van der Sluijs, J. P., Voerman, J. S., and Brons, N. H. (1989) An in vitro limiting-dilution assay of long-term repopulating hematopoietic stem cells in the mouse. *Blood* **74,** 2755–2763.
15. Lemieux, M. E., Rebel, V. I., Lansdorp, P. M., and Eaves, C. J. (1995) Characterization and purification of a primitive hematopoietic cell type in adult mouse marrow capable of lymphomyeloid differentiation in long-term marrow "switch" cultures. *Blood* **86,** 1339–1347.
16. Larochelle, A., Vormoor, J., Hanenberg, H., Wang, J. C., Bhatia, M., Lapidot, T., et al. (1996) Identification of primitive human hematopoietic cells capable of repopulating NOD/SCID mouse bone marrow: implications for gene therapy. *Nat. Med.* **2,** 1329–1337.
17. Harrison, D. E., Jordan, C. T., Zhong, R. K., and Astle, C. M. (1993) Primitive hemopoietic stem direct cells: assay of most productive populations by competitive repopulation with simple binomial, correlation and covariance calculations. *Exp. Hematol.* **21,** 206–219.
18. Albella, B., Segovia, J. C., Guenechea, G., Pragnell, I. B., and Bueren, J. A. (1999) Preserved long-term repopulation and differentiation properties of hematopoietic grafts subjected to ex vivo expansion with stem cell factor and interleukin 11. *Transplantation* **67,** 1348–1357.
19. Keyes, K. A., Segovia, J. C., Bueren, J. A., Parchment, R. E., and Albella, B. (2001) Latent hematopoietic stem cell toxicity associated with protracted drug administration. *Exp. Hematol.* **29,** 286–294.
20. Barquinero, J., Segovia, J. C., Ramirez, M., Limon, A., Guenechea, G., Puig, T., et al. (2000) Efficient transduction of human hematopoietic repopulating cells generating stable engraftment of transgene-expressing cells in NOD/SCID mice. *Blood* **95,** 3085–3093.
21. Rubinstein, P., Dobrila, P., Rosenfield, R. E., Adamson, J. W., Migliaccio, G., Migliaccio, A. R., Taylor, P. E., and Stevens, C. E. (1995) Processing and cryopreservation of placental/umbilical cord blood for unrelated bone marrow reconstitution. *Proc. Natl. Acad. Sci. USA* **92,** 10119–10122.
22. Varas, F., Bernad, A., and Bueren, J. A. (1996) Granulocyte colony-stimulating factor mobilizes into peripheral blood the complete clonal repertoire of hematopoietic precursors residing in the bone marrow of mice. *Blood* **88,** 2495–2501.
23. Peled, A., Petit, I., Kollet, O., Magid, M., Ponomaryov, T., Byk, T., et al. (1999) Dependence of human stem cell engraftment and repopulation of NOD/SCID mice on CXCR4. *Science* **283,** 845–848.
24. Guenechea, G., Gan, O. I., Dorrell, C., and Dick, J. E. (2001) Distinct classes of human stem cells that differ in proliferative and self-renewal potential. *Nat. Immunol.* **2,** 75–82.
25. Kamel-Reid, S. and Dick, J. E. (1988). Engraftment of immune-deficient mice with human hematopoietic stem cells. *Science* **242,** 1706–1709.

28

Large-Scale Ex Vivo Expansion of Human Megakaryocytes for Clinical Use

Phil Lefebvre and Isaac Cohen

1. Introduction

This chapter will describe a method used to produce human megakaryocytes in vitro (*see* **Fig. 1**), in high yield, for a clinical product to supplement progenitor cell transplants. Producing significant numbers of megakaryocytes from human progenitor cells requires specific combinations of cytokines to promote their proliferation and maturation in vitro. Thrombopoietin (TPO) is the primary regulator of megakaryocytopoiesis *(1–6)*. Other cytokines, including interleukin-3 (IL-3), IL-6, IL-11, stem cell factor (SCF; also known as *kit*-ligand), and granulocyte–macrophage colony stimulating factor (GM-CSF) are also capable of promoting megakaryocytopoiesis and will synergize with TPO to increase megakaryocyte proliferation (reviewed in **ref.** *7*).

Properly maintained (e.g., refed and split at regular intervals), static in vitro liquid cultures of purified human bone marrow CD34+ cells can take approx 2 wk for the progenitor cells to terminally differentiate and undergo apoptosis *(8)*. In the presence of TPO alone, these cultures can reach 90+% CD41+ (the primary surface marker for megakaryocytes), but the range of megakaryocyte production and maturation is highly variable and is dependent on culture conditions and the quality of starting material. Human umbilical cord blood CD34+ cells are capable of greater proliferation, but less capable of producing high-ploidy megakaryocytes. Human peripheral blood CD34+ cells perform similarly to bone marrow cells, although the quality of the patient's progenitor cells will ultimately determine productivity *(9)*. Adding other mitogenic cytokines, such as IL-3, will significantly decrease the frequency of CD41+

From: *Methods in Molecular Biology, vol. 215: Cytokines and Colony Stimulating Factors: Methods and Protocols*
Edited by: D. Körholz and W. Kiess © Humana Press Inc., Totowa, NJ

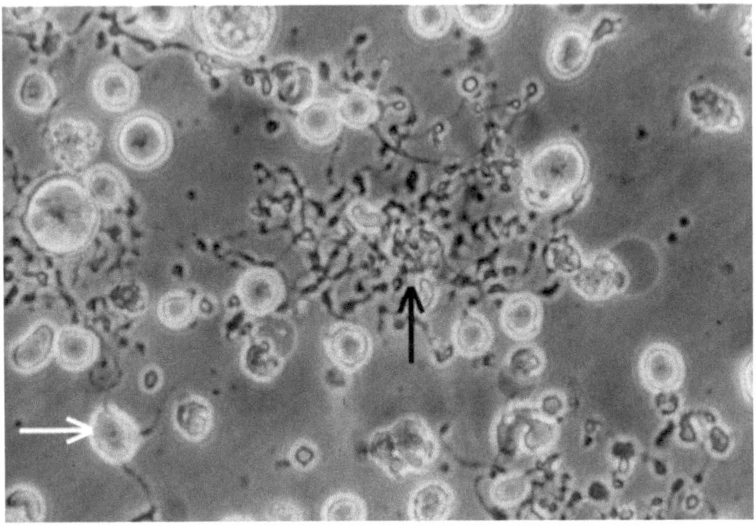

Fig. 1. Phase-contrast image of an 85% pure, 12-d human megakaryocyte culture. The white arrow points to a megakaryocyte that is beginning to form proplatelet extensions. The black arrow points to a megakaryocyte that has completed this process and shed platelets. Magnification = 400×.

cells, but will even more significantly increase the rate of cell expansion, resulting in a net increase in total megakaryocytes produced ***(10)***. Although numerous cytokines can be combined to promote ex vivo expansion of megakaryocytes ***(11)***, a significant expansion of megakaryocytes can be achieved with as little as two or three cytokines ***(10,12–14)***, which may prove more practical in a clinical setting.

Unlike murine megakaryocytes, no culture condition has yet been shown to promote normal polyploidal nuclear maturation of human megakaryocytes in vitro, defined as a modal ploidy of 16 *N* or 8 pairs of chromosome sets. Also, there is an inverse relationship between proliferation and maturation of cultured human megakaryocytes ***(12–15)***. Culturing with TPO alone, which is not a great promoter of cell proliferation, will generate a higher level of megakaryocyte maturation than when combined with more mitogenic cytokines, though a recent article has shown that stromal-cell-derived factor 1 can increase the number and ploidy of megakaryocytes derived in vitro from human CD34+ cells cultured in the presence of TPO at low cell concentrations under low oxygen tension ***(16)***, although a normal 16N modal ploidy was still not achieved.

We and others ***(10,11)*** have found that serum-free medium can be more productive for ex vivo expansion than medium containing plasma. Serum-free medium would also be preferable for ex vivo expansion of cells for transplant, because of the risks associated with allogeneic blood transfusions, and because it is a more consistent and well-defined product. Serum derived from clotted blood is not appropriate for promoting megakaryocyte expansion, as it contains platelet-derived factors inhibitory to megakaryocyte growth, such as transforming growth factor β, β-thromboglobulin, and platelet factor 4, all released from activated platelets ***(17–19)***. If cost is a factor, autologous, heparinized platelet-free plasma or autologous serum derived from calcium-clotted platelet-free plasma could also be suitable for culturing megakaryocytes, as it eliminates the risk associated with allogeneic blood transfusion.

2. Materials

2.1. Media, Buffers, and Cytokines

1. Culture medium: X-vivo 20 serum-free medium (BioWhittaker, Walkersville, MD) (*see* **Note 1**).
2. Thawing medium: X-vivo 20, containing 10 IU preservative-free heparin and 10 µg/mL of DNase.
3. Growth factors used (final concentration): TPO (100 ng/mL), IL-3 (10 ng/mL), Flt-3 ligand, (Flt-3L, 100 ng/mL). Aliquots of each reconstituted cytokine are stored at –70°C until needed. Thawed aliquots can be kept at 4°C for up to 2 wk.
4. Phosphate-buffered saline (PBS), without calcium or magnesium.
5. Hypotonic citrate; 1 mg/mL of sodium citrate in deionized water.
6. DNA dye; propidium iodide or 7-aminoactinomycin D (7-AAD).

2.2. CD34+ Cell Selection

Miltenyi MACS CD34+ selection system appropriate for the quantity of cells used.

2.3. Colony Assays

Needed if the primitive progenitor population is of interest. We have tested several different clonogenic assay protocols and have found the kits from Stem Cell Technologies (Vancouver, BC, Canada) for CFU-MK (MegaCult C) and CFU-GM (MethoCult) to be the most efficient.

2.4. Quantitative Cell Analysis

1. 1% Trypan blue (Sigma) in saline, for cell enumeration and determination of viability.
2. Anti-human CD34 and CD41 antibodies, coupled to different fluorochromes, for flow-cytometric analysis. A comparable isotypic antibody, labeled with the same fluorochrome, is used as a negative control.

2.5. Instrumentation

1. 37°C Water bath.
2. Class B sterile tissue culture hood.
3. Clinical bench centrifuge.
4. 37°C tissue culture incubator at 5% CO_2.
5. Variable adjustable electronic pipettor, EDP, Rainin.
6. Vacuum-driven automatic pipet device.

2.6. Supplies

1. Sterile pipet tips with aerosol barrier.
2. Sterile individually wrapped polypropylene transfer pipets.
3. 2-mL, 5-mL, 10-mL, 25-mL Sterile individually wrapped polystyrene pipets with an aerosol barrier.
4. Sterile centrifuge tubes.
5. 1-cm^3, 3-cm^3, 10-cm^3, 20-cm^3, 30-cm^3, 60-cm^3 sterile Luer-Lok™ syringes.
6. VueLife Teflon tissue culture bags, 1 L (270-mL nominal capacity) and 7-mL capacity, American Fluoroseal Corporation. Other sizes are also available, including custom sizes.
7. Luer-Lok connector tubing, available from American Fluoroseal Corp.
8. Sterile threaded Luer-Lok cannulas (BD Interlink System; Becton Dickinson product no. 303369).
9. Interlink Luer-Lok injection site (Baxter Healthcare product no. 2N3379).

3. Method

3.1. CD34 Cell Selection

For large-scale clinical use, the CliniMACS clinical-grade system is preferred. It is approved for clinical use in the European Community, but is presently under evaluation for clinical use in the United States. Smaller-scale systems are available for preclinical or basic research use. The manufacturer-supplied laboratory protocol is straightforward and easy to follow and should be followed closely to ensure successful CD34 cell purification. Purity of over 90% with high yield (70+%) is common with all MACS systems, when used before the expiration date.

3.2. Sample Thawing

If cells were frozen after purification for storage and use at a later date, then they will need to be carefully thawed, as freezing and thawing can damage cells.

1. Each frozen CD34+ cell unit is placed into a resealable plastic bag and the bag closed, to prevent any contaminating water seepage into the tube.

2. The sealed bag with the tube is then placed in a 37°C water bath and gently agitated until the cell suspension has thawed 90%, to a small ball of ice (approx 3–4 min for a 5-mL tube).
3. The bag is wiped dry and transferred to a Class B sterile tissue culture hood, where the tube is removed from the bag aseptically. (*See* **Notes 2–4**.)

3.3. Sample Washing

1. In the sterile tissue culture hood, the contents of a 5-mL tube of thawed cells is transferred by gentle addition into a sterile 50-mL centrifuge tube using a sterile polypropylene transfer pipet.
2. The cells are diluted 1 : 10 with 45 mL of cold thawing medium.
3. The tube is sealed and the cells gently mixed.
4. The suspension is centrifuged at 260*g* for 10 min at 4°C.
5. The cell pellet is resuspended in 3 mL of X-vivo 20 using a sterile polypropylene transfer pipet.
6. Dilute the cell suspension to 50 mL with X-vivo 20 (not thawing medium) by decanting fresh medium.
7. The cell suspension is centrifuged again at 260*g* for 10 min at 4°C.
8. The cell pellet is resuspended in 3 mL of X-vivo 20 using a sterile polypropylene transfer pipet.

3.4. Nucleated Cell Count

1. A 50-µL sample for cell concentration analysis is taken using a sterile pipet tip with an aerosol barrier.
2. The sample is diluted with trypan blue so that the cell concentration in the trypan blue is not more than 10^6 cells/mL.
3. Apply 10–12 µL of trypan blue cell suspension to a hemacytometer.
4. Determine the cell concentration and viability (cells that appear blue are not viable) using a phase-contrast microscope.
5. The final volume required to achieve a cell concentration of 400,000 cells/mL is calculated, based on the cell count.

3.5. Culture Preparation

1. The cells are injected using a syringe connected to a Luer-Lok cannula, through the Interlink injection port, into a Teflon bag of appropriate size (*see* **Fig. 2**).
2. The bag's nominal volume is adjusted with bag clip barriers so as to contain a volume sufficient to suspend the cells at 400,000 cells/mL at an average fluid height of 1 cm.
3. X-vivo 20 and sufficient cytokines to achieve a final concentration of 100 ng/mL Flt-3L, 100 ng/mL TPO, and 10 ng/mL IL-3 is then added using a 60-cm^3 syringe attached via Luer-Lok connections to the bag to reach the final volume. (*See* **Note 5**.)

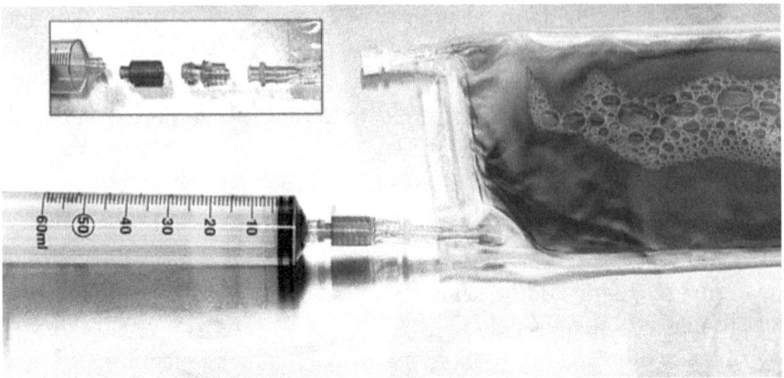

Fig. 2. Syringe–bag connection system. Inset: exploded view showing (from left to right) Luer-Lok 60-cm^3 syringe, Interlink threaded lock cannula, Interlink injection site, and Luer-Lok opening of Teflon bag.

3.6. Culture Maintenance

Cells should be assayed at least every other day to ensure that they have not proliferated beyond the capacity of the culture medium. In our experience, X-vivo 20 has a maximum cell capacity of up to 2×10^6 cells/mL, depending on the quality of the cells used.

1. After gently agitating the bag to resuspend the cells, a 5-mL aliquot is removed using a syringe and injected into a 7-mL Teflon bag for use as a test sample.
2. Both bags are stored in a 37°C tissue culture incubator at 5% CO_2 used solely for the purpose of clinical patient sample preparation. (*See* **Notes 6–8**.)

3.7. Refeeding the Culture

When cells have proliferated beyond the culture medium's limit, they will need to be split and refed.

1. Count the cells as in **Subheading 3.4.**
2. Calculate the amount of medium and cytokines required to readjust the cell concentration to 400,000 cells/mL.
3. Calculate the area required to maintain a 1-cm culture height and readjust the bag clips to that position.
4. Add that amount to the culture bag using the Luer-Lok syringes and cannula connectors.
5. If the culture medium required exceeds the capacity of the bag, additional bags are then attached to the culture bag using the Luer-Lok connection tubing, and an appropriate fraction of cell suspension is transferred to the new bag (*see* **Fig. 3**). (*See* **Note 9**.)

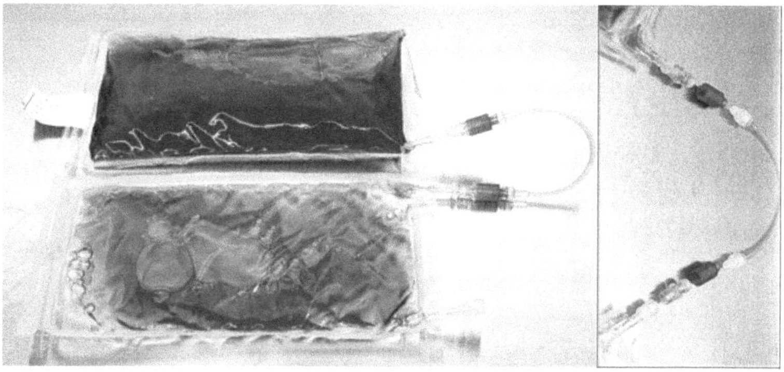

Fig. 3. Bag–bag connection system. Inset: Exploded view. A flexible tube with Luer-Lok connectors on each end is substituted for the syringe in **Fig. 2**.

3.8. Cell Harvest

When the designated culture period has passed, the sample is transferred to an appropriate vessel and gently washed, PBS with 1% albumin, or other clinically approved procedure. Cells can be washed in an automated cell washer or centrifuged using centrifuge tubes or bottles of sufficient capacity to contain the culture. The appropriate procedure needs to be determined empirically, based on culture size, equipment available, and whether a closed system is required for clinical use. Three wash steps should be used ensure that no residual cytokine remains in any product designated for transplant. (*See* **Note 10**.)

3.9. Cell Analysis

Flow cytometry is the most efficient way of measuring phenotype expression of cells in a liquid culture. Megakaryocytes can be difficult to assay, however, because of their fragility and great variation in size. Cells that may have been exposed to platelets (e.g., peripheral blood cells, umbilical cord cells, or cells in platelet-producing megakaryocyte cultures) should be incubated in EDTA-containing buffer to disassociate any adherent platelets ***(20)***. Then, standard flow-cytometric techniques can be used to assay megakaryocyte proliferation. (*See* **Note 11**.)

1. Place cell medium containing 100,000 cells into a 12 × 75-mm test tube, appropriate for use on a flow cytometer.
2. Add 2 m*M* EDTA and incubate for at least 15 min at room temperature. Longer incubations may be necessary if platelet contamination is high.
3. Fill the tube with PBS and centrifuge at 260*g* for 5 min at 4°C.

4. Remove all supernatant and add the recommended amount of fluorochrome-labeled antibody. Vortex gently.
5. Incubate for 15 min at 4°C in the dark.
6. Fill the tube with PBS and centrifuge at 260*g* for 5 min at 4°C.
7. Remove supernatant, resuspend the cells in 0.5–1 mL of PBS and assay by flow cytometry immediately.
8. If megakaryocyte ploidy measurement is desired, substitute 1 mL of hypotonic citrate buffer for the PBS. Add 1 μg/mL of propidium iodide or 7-AAD. Incubate an additional 30 min at room temperature before assaying.

4. Notes

1. We have tested numerous serum-free media and have found X-vivo 20 to be the most conducive to megakaryocyte proliferation. However, technology can continue to evolve, so it would be expected that improved media should become available in the future. Further testing would be warranted to find the medium optimal for each protocol.
2. It is our experience that mononuclear cells are more fragile during freezing and thawing than selected CD34 cells. In either case, DNase is recommended for the thawing medium to remove DNA released from lysed cells. However, work within the recommended levels, as too much DNase can be harmful to cells.
3. We and others ***(21)*** have found the Miltenyi CD34+ MACS selection systems to be the most efficient. The systems available range from a total cell capacity of 10^5–10^{11} cells. We have used several different Miltenyi systems, all with excellent results.
4. The Fluoroseal Teflon bags are impermeable to liquid, but permeable to gas. This allows adequate gas exchange while all ports remain fully sealed, so that vapor cannot escape the bag. What this means is that liquid medium loss is not a problem, so it is not necessary to keep water in the incubator for humidity, reducing contamination risk.
5. For a 1-L Teflon bag, only 270 mL is used so that the culture medium height remains at approx 1 cm, which has been determined to be optimal for culture performance. Lower heights reduce cell productivity, whereas higher media heights waste reagent. The use of bag clips allows for volumes less than 270 mL to be used in the 1-L bags, and then removed as the culture expands and is refed. American Fluoroseal supplies appropriate clips, but any plastic clip that would not damage the plastic bags would work.
6. It has been our experience that ex vivo cultured cells derived from oncology patients will often show a dramatic drop in viable cell numbers in the first few days of a culture. This phenomenon is true whether the cells were unselected mononuclear cells or selected CD34 cells and it is highly variable.
7. The use of a smaller, 7-mL bag is to be able to monitor culture progress without needing to access the main culture, reducing the risk of contamination. In our experience, the phenotypic development of the cells in either bag is identical.

8. We have found that megakaryocyte proliferation is enhanced with an initial seeding density of 400,000 CD34 cells/mL, compared to lower seeding densities. The culture conditions described have a total cell capacity in the range of up to 2×10^6 cells/mL (fivefold expansion) before having a significant negative impact on cell proliferation and viability. However, cell concentration may need to be readjusted after only a twofold to threefold expansion of viable cells to minimize any negative effects of large numbers of nonviable cells. This is most often in cases of less healthy or damaged cell samples that rapidly degenerate into nonviable states.
9. The easiest way to transfer expanded cells from on bag to another is by attaching the two bags together with a Luer-Lok connector tube and suspending the full bag so that medium runs down into the empty bag. When the cell culture appears to be equally distributed in each bag, they should be weighed and equilibrated by weight. This avoids excessive opening and taping of the bags, reducing contamination risk.
10. Once mature megakaryocytes are generated, care should be taken in handling the culture, as large high-ploidy megakaryocytes are fragile. Avoid extremes of handling and vortexing, if preservation of the larger, more mature megakaryocytes is important.
11. Although there are many procedures for fixing and permeabilizing megakaryocytes, the method using hypotonic citrate is the easiest and best preserves the large ploidy megakaryocytes, because of significantly less manipulation.

References

1. Bartley, T. D., Bogenberger, J., Hunt, P., Li, Y. S., Lu, H. S., Martin, F., et al. (1994) Identification and cloning of a megakaryocyte growth and development factor that is a ligand for the cytokine receptor Mpl. *Cell* **77(7)**, 1117–1124.
2. de Sauvage, F. J., Hass, P. E., Spencer, S. D., Malloy, B. E., Gurney, A. L., Spencer, S. A., et al. (1994) Stimulation of megakaryocytopoiesis and thrombopoiesis by the c-Mpl ligand. *Nature* **369(6481)**, 533–538.
3. Kaushansky, K., Lok, S., Holly, R. D., Broudy, V. C., Lin, N., Bailey, M. C., et al. (1994) Promotion of megakaryocyte progenitor expansion and differentiation by the c-Mpl ligand thrombopoietin. *Nature* **369(6481)**, 568–571.
4. Kuter, D. J., Beeler, D. L., and Rosenberg, R. D. (1994) The purification of megapoietin: a physiological regulator of megakaryocyte growth and platelet production. *Proc. Natl. Acad. Sci. USA* **91(23)**, 11,104–11,108.
5. Lok, S. and Foster, D. C. (1994) The structure, biology and potential therapeutic applications of recombinant thrombopoietin. *Stem Cells* **12(6)**, 586–598.
6. Wendling, F., Maraskovsky, E., Debili, N., Florindo, C., Teepe, M., Titeux, M., et al. (1994) cMpl ligand is a humoral regulator of megakaryocytopoiesis. *Nature* **369(6481)**, 571–574.
7. Hoffman, R. (1997) The role of other hemopoietic growth factors and the marrow microenvironment in megakaryocytopoiesis, in *Thrombopoiesis and Thrombopoi-*

etins: Molecular, Cellular, Preclinical, and Clinical Biology (Kuter, D. J., Hunt, P., Sheridan, W., and Zucker-Franklin, D., eds.), Humana, Totowa, NJ, pp. 165–178.

8. Zauli, G., Vitale, M., Falcieri, E., Gibellini, D., Bassini, A., Celeghini, C., et al. (1997) In vitro senescence and apoptotic cell death of human megakaryocytes. *Blood* **90(6),** 2234–2243.
9. Lefebvre, P., Winter, J. N., Rademaker, A. W., Goolsby, C., and Cohen, I. (1997) In vitro production of megakaryocytes from PIXY321 versus GM-CSF-mobilized peripheral blood progenitor cells. *Stem Cells* **15(2),** 112–118.
10. Lefebvre, P., Winter, J. N., Kahn, L. E., Giri, J. G., and Cohen, I. (1999) Megakaryocyte ex vivo expansion potential of three hematopoietic sources in serum and serum-free medium. *J. Hematother.* **8(2),** 199–208.
11. Bertolini, F., Battaglia, M., Pedrazzoli, P., Da Prada, G. A., Lanza, A., Soligo, D., et al. (1997) Megakaryocytic progenitors can be generated ex vivo and safely administered to autologous peripheral blood progenitor cell transplant recipients. *Blood* **89(8),** 2679–2688.
12. Schattner, M., Lefebvre, P., Mingolelli, S. S., Goolsby, C. L., Rademaker, A., White, J. G., et al. (1996) Thrombopoietin-stimulated ex vivo expansion of human bone marrow megakaryocytes. *Stem Cells* **14(2),** 207–214.
13. van den Oudenrijn, S., de Haas, M., Calafat, J., van der Schoot, C. E., and von dem Borne, A. E. G. K. (1999) A combination of megakaryocyte growth and development factor and interleukin-1 is sufficient to culture large numbers of megakaryocytic progenitors and megakaryocytes for 14. transfusion purposes. *Br. J. Haematol.* **106(2),** 553–563.
14. Lefebvre, P., Winter, J. N., Meng, Y. R., and Cohen, I. (2000) Ex vivo expansion of early and late megakaryocyte progenitors. *J. Hematother. Stem Cell Res.* **9(6),** 913–921.
15. Levin, J., Levin, F. C., Penington, D. G., and Metcalf, D. (1981) Measurement of ploidy distribution in megakaryocyte colonies obtained from culture: with studies of the effects of thrombocytopenia. *Blood* **57(2),** 287–297.
16. Guerriero, R., Mattia, G., Testa, U., Chelucci, C., Macioce, G., Casella, I., et al. (2001) Stromal cell-derived factor 1alpha increases polyploidization of megakaryocytes generated by human hematopoietic progenitor cells. *Blood* **97(9),** 2587–2595.
17. Mitjavila, M. T., Vinci, G., Villeval, J. L., Kieffer, N., Henri, A., Testa, U., et al. (1988) Human platelet alpha granules contain a nonspecific inhibitor of megakaryocyte colony formation: its relationship to type beta transforming growth factor (TGF-beta). *J. Cell. Physiol.* **134(1),** 93–100.
18. Gewirtz, A. M., Calabretta, B., Rucinski, B., Niewiarowski, S., and Xu, W. Y. (1989) Inhibition of human megakaryocytopoiesis in vitro by platelet factor 4 (PF4) and a synthetic COOH-terminal PF4 peptide. *J. Clin. Invest.* **83(5),** 1477–1486.
19. Abgrall, J. F., Han, Z. C., Sensebe, L., Le Gall, G., Guern, G., Renard, I., et al. (1991) Inhibitory effect of highly purified human platelet beta-thromboglobulin on in vitro human megakaryocyte colony formation. *Exp. Hematol.* **19(3),** 202–205.

20. Dercksen, M. W., Weimar, I. S., Richel, D. J., Breton-Gorius, J., Vainchenker, W., Slaper-Cortenbach, C. M., et al. (1995) The value of flow cytometric analysis of platelet glycoprotein expression of CD34+ cells measured under conditions that prevent P-selectin-mediated binding of platelets. *Blood* **86(10),** 3771–3782.
21. de Wynter, E. A., Coutinho, L. H., Pei, X., Marsh, J. C., Hows, J., Luft, T., et al. (1995) Comparison of purity and enrichment of CD34+ cells from bone marrow, umbilical cord and peripheral blood (primed for apheresis) using five separation systems. *Stem Cells* **13(5),** 524–532.

29

Cytokine-Mediated Expansion of Human NOD–SCID-Repopulating Cells

Kohichiro Tsuji, Takahiro Ueda, and Yasuhiro Ebihara

1. Introduction

There has been great interest in ex vivo expansion of human long-term repopulating hematopoietic stem cells (LTR-HSC) for a variety of developing clinical applications including HSC transplantation, gene therapy, and production of mature blood cells in manufacturing. Because the development of HSC is thought to be regulated, at least in part, by interactions of cytokine receptor signals, many investigators used various combinations of cytokines that have been shown to act on primitive hematopoietic cells to obtain the optimal culture condition for HSC expansion.

Transplantable human HSC should prove to have long-term repopulating ability. However, most of human HSC studies aimed at clinical application have used in vitro assay for CD34+ cells *(1)*, colony-forming cells in clonal culture *(2–4)*, cobblestone-area-forming cells (CAFC) *(1)* and long-term culture-initiating cells (LTC-IC) *(4)*, but these surrogate assays have been shown not to correctly reflect stem cell activity *(5–7)*. Recently, assays evaluating the stem cell activity have been developed using immunodeficient non-obese diabetes severe combined immunodeficient (NOD-SCID) mice. The NOD-SCID mice possess lack of mature lymphocytes, macrophage dysfunction, absence of circulating complements, and low natural killer (NK) cell activity *(8)*, resulting in efficient engraftment of human HSC in the mice *(9,10)*. The assay system has allowed us to exactly evaluate the stem cell activity of expanded hematopoietic cells.

From: *Methods in Molecular Biology, vol. 215: Cytokines and Colony Stimulating Factors: Methods and Protocols*
Edited by: D. Körholz and W. Kiess © Humana Press Inc., Totowa, NJ

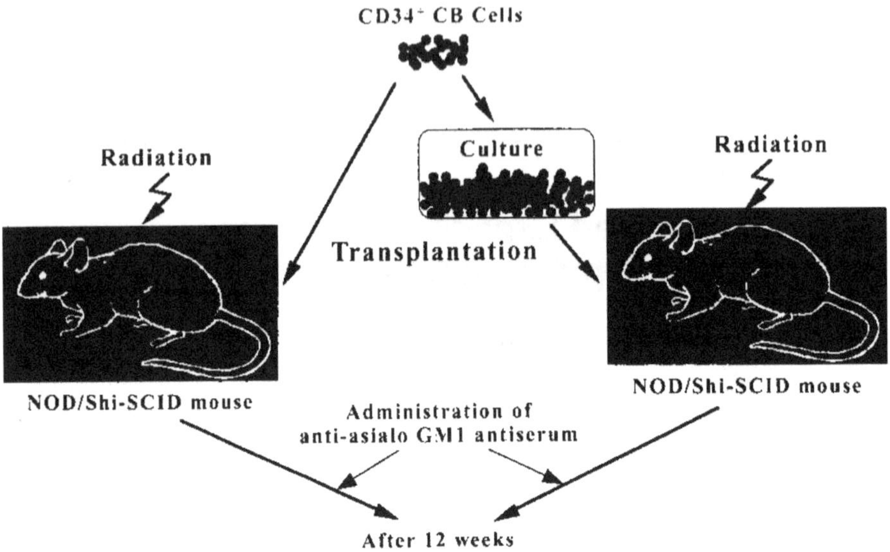

Fig. 1. Assay for HSC expansion by culture with SCF+FL+TPO+IL-6/sIL-6R. A total of (1–2) × 10^4 $CD34^+$ CB cells and their progenies cultured with SCF+FL+TPO+IL-6/sIL-6R were transplanted into irradiated NOD/Shi-*scid* mice, and anti-asialo GM1 antiserum was administered on d 0, 11, 22, and 33 of transplantation. Human blood cells repopulating in recipient BM and PB were analyzed by flow cytometry 12 wk after the transplantation.

We have developed a significant ex vivo expansion system of human HSC capable of repopulating in NOD-SCID mice, termed severe combined immune deficiency (SCID)-repopulating cells (SRC), using a combination of stem cell factor (SCF), Flk2/Flt3 ligand (FL), thrombopoietin (TPO) and a complex of interleukin (IL)-6 and soluble IL-6 receptor (IL-6/sIL-6R) ***(11)*** (*see* **Fig. 1**). SCF and FL have been used as key cytokines for previous HSC expansion systems, because c-Kit and Flk2/Flt3, tyrosine kinase receptors for SCF and FL, respectively, were shown to transduce signals crucial for HSC development. TPO, a ligand for c-Mpl, originally identified as a primary regulator for megakaryopoiesis, has been shown to stimulate the expansion of primitive hematopoietic cells ***(2,4)***. In addition, we have demonstrated that gp130 signal activated by IL-6/sIL-6R synergizes with the c-Kit or Flk2/Flt3 signal to expand primitive hematopoietic progenitor cells ***(3,12)***.

When (1–2) × 10^4 cord blood (CB) $CD34^+$ cells and their progenies cultured with SCF+FL+TPO+IL-6/sIL-6R for 7 d were transplanted into NOD-SCID mice, 6/25 (24%) and 13/16 (81%) recipients, respectively, revealed successful engraftment. Human $CD45^+$ blood cells in the recipient marrow engrafted

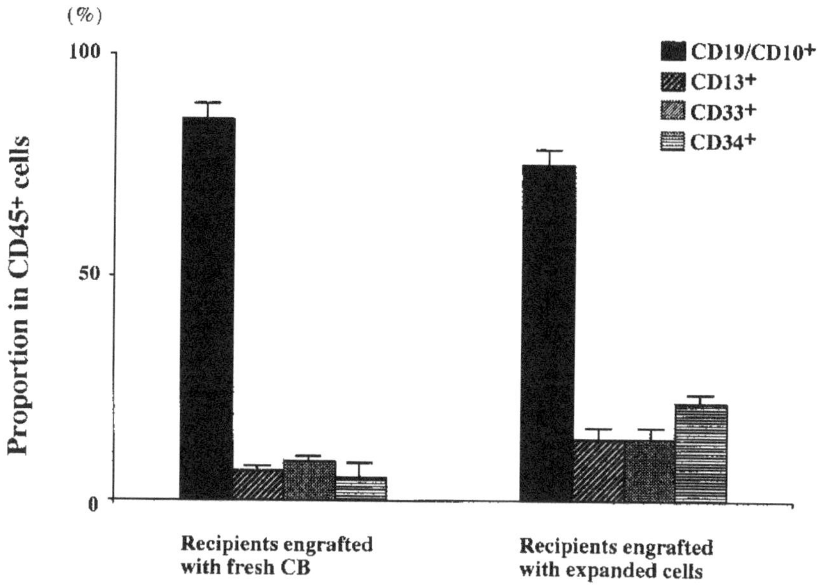

Fig. 2. Proportions of human CD 19/CD 10^+ cells, $CD13^+$ cells, $CD33^+$ cells and $CD34^+$ cells in human $CD45^+$ cells engrafted in BM of recipient NOD-SCID mice transplanted with fresh CB cells or the cells expanded with SCF+FL+TPO+IL-6/sIL-6R.

with the cells cultured with SCF+FL+TPO+IL-6/sIL-6R consisted of various lineages of cells 10–12 wk after the transplantation, and there was no difference in the proportion of each lineage of cells compared with that in the recipients engrafted with the initial CB $CD34^+$ cells (*see* **Fig. 2**). However, the proportion of human $CD45^+$ cells in the bone marrow (BM) was 10-fold higher in the recipients engrafted with the cultured cells than in those with initial CB cells, indicating the significant expansion of SRC. The expansion rate was estimated at 4.2-fold by a limiting dilution method. Similar HSC expansion was also observed in fetal bovine serum (FBS)-free culture, which, in clinical application, would exclude the possibility of unknown infections (*see* **Note 1**). In this article, the detail of the novel culture system is shown. Our culture system may pave the way for the clinical application of ex vivo expansion of human HSC.

2. Materials

2.1. Mice

1. Experimental NOD-SCID (NOD-Shi-*scid*) mice were obtained from a Central Institute for Experimental Animals (Kawasaki, Japan) (*see* **Note 2**). The mice were kept in microisolator cages on laminar flow racks in a clean experimental

room. They were maintained on an irradiated sterile diet and given autoclaved acidified water. Cages were exchanged once a week.

2.2. Cytokines

1. Recombinant human IL-6 and sIL-6R (gifts from Tosoh Corp., Ayase, Japan).
2. Recombinant human SCF (a gift from Amgen Biologicals, Thousand Oaks, CA).
3. Recombinant human TPO (a gift from Kirin Brewery, Tokyo, Japan).
4. Recombinant human FL (R&D Systems, Minneapolis, MN).

2.3. Purification of $CD34^+$ Cells

1. Ficoll–Paque (Pharmacia LKB, Uppsala, Sweden).
2. Silica (Immuno Biological Laboratories, Fujioka, Japan).
3. Deionized fraction V bovine serum albumin (BSA; Sigma, St. Louis, MO).
4. Dynabeads CD34 Progenitor Cell Selection System: Dynabeads M-450 CD34 and DETACHaBEAD CD34 (Dynal AS, Oslo, Norway).

2.4. Suspension Culture

1. α-Medium (Flow Laboratories, Rockville, MD).
2. Fetal bovine serum (FBS) (Hyclone, Logan, UT).
3. Deionized fraction V bovine serum albumin (BSA) (Sigma).
4. 12-Well tissue plates (Nunc, Roskilde, Denmark).
5. Deionized crystallized BSA (Sigma).
6. Human transferrin (Sigma).
7. Soybean lecithin (Sigma).
8. Cholesterol (Nakalai Tesque Inc, Kyoto, Japan).
9. Human recombinant insulin (Sigma).

2.5. Transplantation into NOD-SCID Mice

1. Anti-asialo GM1 antiserum (Wako, Osaka, Japan).
2. Deionized fraction V BSA (Sigma).
3. Neomycin sulfate (Gibco-BRL, Grand Island, NY).
4. 40-μm Cell Strainer (#2340; Falcon, Lincoln Park, NJ).

2.6. Flow-Cytometric Analysis

1. FACSCalibur (Becton Dickinson, Mountain View, CA).
2. Lysing solution (Nichirei Co, Tokyo, Japan).
3. Rabbit serum (Funakoshi, Tokyo, Japan).
4. Antibodies: anti-human CD45 conjugated with fluorescein isothiocyanate (CD45–FITC), CD34–FITC, CD10–FITC, CD3–FITC, CD33 conjugated with phycoerythrin (CD33–PE), CD19–PE, and CD13–PE (Becton Dickinson, San Jose, CA).
5. CD3–FITC and CD45 conjugated with PE-cyanine 5-succinimidylester (CD45–PE-Cy5) (Immunotech, Marseille, France).

2.7. Polymerase Chain Reaction (PCR) Analysis

1. DNA thermocycler (GeneAmp, PCR System 2400; Perkin-Elmer).
2. QIAGEN QIAamp DNA mini kit (Amersham, Uppsala, Sweden).
3. *Taq* polymerase (Perkin-Elmer Cetus, Foster City, CA).
4. Oligonucleotide primers: ALU-5, CACCTGTAATCCCAGCAGTTT-3; ALU-3, CGCGATCTCGGCTCACTGCA.

3. Methods

3.1. Purification of $CD34^+$ Cells

1. Human umbilical CB, collected according to institutional guidance, was obtained during normal full-term deliveries.
2. Mononuclear cells (MNCs) were separated from CB cells by Ficoll–Paque density gradient centrifugation after phagocytes were removed by silica on incubation at 37°C in a humidified atmosphere of 5% CO_2 in air for 30 min.
3. After washing MNCs with phosphate-buffered saline (PBS) containing 0.2% deionized fraction V BSA and 0.6% citrate (buffer B), cells were resuspended at $(3–5) \times 10^7$/mL in PBS containing 2% deionized fraction V BSA and 0.6% citrate (buffer A) and mixed with Dynabeads M-450 CD34, with a beads-to-cell ratio of 1:1.
4. The cell–beads suspension was resuspended and incubated at 4°C for 30 min with gentle rotation.
5. After incubation, the cell–beads volume was expanded with buffer B and placed in a DYNAL Magnetic Particle Concentrator (MPC) to collect the Dynabeads M-450 CD 34^+/rosetted cell.
6. The rosetted cells were washed four times with buffer B and then resuspended in buffer A.
7. One hundred microliters of DETACHaBEAD CD34 per 4×10^7/mL Dynabeads M-450 CD34 were added to the rosetted cell suspension and incubated at room temperature for 45 min to detach the Dynabeads M-450 CD34 from the positively selected cells.
8. The released cells ($CD34^+$) were collected by placing the tube in the MPC and were then washed twice with buffer B. When their purity was evaluated by flow-cytometric analysis, approx 95% of the separated cells were $CD34^+$.

3.2. Suspension Culture

1. One milliliter of culture mixture containing $(1–2) \times 10^4$ purified $CD34^+$ cells, α-medium, 20% FBS, 1% deionized fraction V BSA, 100 ng/mL of SCF, IL-6, and FL, 1000 ng/mL of sIL-6R, and 10 ng/mL of TPO was incubated for 1 wk in 12-well tissue plates at 37°C in a humidified atmosphere flushed with 5% CO_2 in air.
2. Fetal-bovine-serum-free suspension culture contained components identical to FBS-containing culture, except for 2% deionized crystallized BSA, 200 μg/mL

of human transferrin, 160 μg/mL of soybean lecithin, 96 μg/mL of cholesterol (*see* **Note 3**), and 10 μg/mL of human recombinant insulin replaced fraction V BSA and FBS. A one-milliliter aliquot of culture mixture was plated in 12-well tissue plates and incubated for 1 wk at 37°C in a humidified atmosphere flushed with 5% CO_2, 5% O_2, and 90% N_2 in air.

3.3. Transplantation into NOD-SCID Mice

1. Eight to 10-wk-old NOD-SCID mice were sublethally irradiated with 240 cGy as two divided doses by a ^{137}Cs source immediately before transplantation (*see* **Note 4**).
2. The freshly isolated CD34$^+$ cells and their progenies cultured with SCF+FL+TPO+IL-6/sIL-6R were resuspended in 0.5 mL PBS containing 1% deionized fraction V BSA and were injected into recipient mice via a tail vein with a 29- or 30-gage needle.
3. The recipient NOD-SCID mice were injected intraperitoneally with 400 μL of PBS containing 20 μL of anti-asialo GM1 antiserum immediately before the cell transplantation (*see* **Note 5**). Identical treatments were performed on d 11, 22, and 33 after the infusion of experimental cells (*see* **Note 6**).
4. After transplantation, mice received water containing prophylactic neomycin sulfate (1.1 g/1000 mL). The water was exchanged once a week.
5. Mice were killed in a CO_2 chamber 10–12 wk after the transplantation. Femurs and tibiae were collected and aspirated with 5% FBS-containing PBS to liberate BM cells. Cell suspensions were then filtered through a sterile 40-μm cell strainer to eliminate clumps and debris and then processed for flow-cytometric analysis.

3.4. Flow-Cytometric Analysis of Transplanted NOD-SCID Mice

1. Surface markers on human blood cells reconstituted in NOD-SCID mouse BM were analyzed by flow cytometry using the FACSCalibur.
2. After depletion of erythrocytes with lysing solution at room temperature for 5 min, 1×10^6 mouse BM cells suspended with 200 μL of PBS containing 0.1% deionized fraction V BSA were stained with antibodies at a concentration of 20 μL/mL. After the blocking of the surface Fc receptor by rabbit serum, all antibody incubations were carried out for 30 min on ice. Cells were then washed in PBS containing 0.1% deionized fraction V BSA and resuspended in 0.5 mL of the same buffer.
3. The presence of human blood cells was determined by detection of cells positively stained with CD45–FITC in flow-cytometric analysis. Successful engraftment by human HSC was defined by the presence of at least 1% of human CD45$^+$ cells in NOD-SCID mouse BM cells 10–12 wk after the transplantation (*see* **Note 7**). BM cells of untransplanted NOD-SCID mice were used as negative control.
4. Three-color flow cytometry was performed with BM cells in which human blood cells successfully repopulated. Specific subsets of human blood cells were quantified by gating on human CD45–PE-Cy5-positive cells and then assessing staining with anti-human CD34–FITC (immature cells), CD10–FITC (immature

B-cells), CD3–FITC (T-cells), CD33–PE (myeloid cells), CD19–PE (B-cells), and CD13–PE (myeloid cells) (*see* **Note 8**). For each mouse analyzed, an aliquot of cells was also stained with mouse IgG conjugated to FITC, PE, and PE-Cy5 as isotype controls.

3.5. PCR Analysis

To confirm the engraftment of human HSC in NOD-SCID mouse marrow, the detection of human ALU sequences in BM DNA was performed by PCR analysis (*see* **Note 9**).

1. DNA extracted from recipient mouse marrow by a QIAGEN QIAamp DNA mini-kit was subjected to PCR amplification using a pair of oligonucleotide primers. PCR was performed using 2 μg of total genomic DNA, 20 pmol of primers, and 2.5 U of *Taq* polymerase. Samples were denatured at 94°C for 4 min, then amplified by rounds consisting of 94°C for 1 min (denaturing), 55°C for 45 s (annealing), and 72°C for 1 min (extension) for 21 cycles, followed by 5 min at 68°C using a DNA thermocycler.
2. Products were separated on a 3.0% agarose gel, stained with ethidium bromide, and photographed. The sequence amplified by two primers is of 221 base pairs.

4. Notes

1. There have been some reports describing FBS-free expansion culture of human HSC whose long-term-repopulating ability was confirmed in vivo, using combinations of various cytokines. Bhatia et al. *(13)* and Conneally et al. *(14)* reported on twofold to fourfold expansion of human SRC in FBS-free culture containing SCF, FL, G-CSF, IL-3, and IL-6 for 4–7 d.
2. Some investigators are using, as recipient NOD-SCID mice, NOD-LtSz-*scid* mice, which was established in the animal facility of the British Columbia Cancer Research Center from breeders originally provided by The Jackson Laboratories *(5–9,13,14)*.
3. Twenty milligrams of soybean lecithin and 12 mg of cholesterol were placed on the bottom of a 25-mL glass breaker. A few drops of chloroform were added until the lipids were dissolved. When the dissolved lipids were completely evaporated, 10 mL of bicarbonate-free α-medium containing 1% deionized crystallized BSA was added. The beaker was immersed in ice water on a Measuring and Scientific Equipment sonicator for 10 min. The resulting solution was filtered and stored at 4°C.
4. Initial experiments were performed by irradiating mice with 300 cGy as two divided dose. At this dose, 70% of NOD-SCID mice died 1–2 wk after the transplantation.
5. The NK cell activity of NOD/Shi-*scid* mice we used was two-thirds to one-half of the activity of *scid* mice *(15)*.
6. Because it was reported that NK cell activity was markedly reduced on d 3 after the one-shot injection of anti-asialo GM1 antiserum, began to recover on d 7,

and rose to more than 50% of the original value on d 14 *(16)*, the injection was performed on d 0, 11, 22, and 33 of transplantation *(15)*.

7. In some recipient NOD-SCID mice, peripheral blood (PB) collected by puncture of the tail vein was analyzed for the presence of human $CD45^+$ blood cells. Although the proportion of human $CD45^+$ cells in PB was always lower than in BM, a stable number of $CD45^+$ cells were detected at least for 6 mo in all the engrafted recipients examined *(10)*.
8. All of the recipients transplanted with the cultured cells possessed $CD13^+$ and $CD33^+$ myeloid cells, $CD19^+$ and/or $CD10^+$ B-cells, and $CD34^+$ immature cells, but not $CD3^+$ T-cells, in BM $CD45^+$ cells. Although $CD41^+$ megakaryocytes and platelets were present, few glycophorin A^+CD71^+ or hemoglobin α-containing erythroid cells were detected *(17)*.
9. In all recipient NOD-SCID mice whose BM cells contained more than 1% $CD45^+$ cells, human ALU sequences were detected in BM DNA.

References

1. Luens, K. M., Travis, M. A., Chen, B. P., Hill, B. L., Scolly, R., and Murray, L. J. (1998) Thrombopoietin, kit ligand, and flk2/flt3 ligand together induce increased numbers of primitive hematopoietic progenitors from human $CD34^+Thy\text{-}1^+Lin^-$ cells with preserved ability to engraft SCID-hu bone. *Blood* **91,** 1206–1215.
2. Kobayashi, M., Laver, J. H., Kato, T., Miyazaki, H., and Ogawa, M. (1996) Thrombopoietin supports proliferation of human primitive hematopoietic cells in synergy with steel factor and/or interleukin-3. *Blood* **88,** 429–436.
3. Sui, X., Tsuji, K., Tanaka, R., Tajima, S., Muraoka, K., Ebihara, Y., et al. (1995) gp130 and c-Kit signallings synergize for ex vivo expansion of human primitive hematopoietic progenitor cells. *Proc. Natl. Acad. Sci. USA* **92,** 2859–2863.
4. Yoshida, M., Tsuji, K., Ebihara, Y., Muraoka, K., Tanaka, R., Miyazaki, H., et al. (1996) Thrombopoietin alone stimulates the early proliferation and survival of human eryhtroid, myeloid and multipotential progenitors in serum-free culture. *Br. J. Haematol.* **98,** 254–264.
5. Larochelle, A., Vormoor, J., Helmut, H., Wang, J. C., Bhatia, M., Lapidot, T., Moritz, T., et al. (1996) Identification of primitive human hematopoieitc cells capable of repopulating NOD/SCID mouse bone marrow: implications of gene therapy. *Nat. Med.* **2,** 1329–1337.
6. Bhatia, M., Wang, J.C., Kapp, U., Bonnet, D., and Dick, J. E. (1997) Purification of primitive human hematopoietic cells capable of repopulating immune-deficient mice. *Proc. Natl. Acad. Sci. USA* **94,** 5320–5325.
7. Gan, O. I., Murdoch, B., Larochelle, A., and Dick, J. E. (1997) Differential maintenance of primitive human SCID-repopulating cells, clonogenic progenitors, and long-term culture-initiating cells after incubation on human bone marrow stromal cells. *Blood* **90,** 641–650.
8. Shultz, L. D., Schweitzer, P. A., Chritianson, S. W., Gott, B., Schweitzer, I. B., Tennet, B., et. al. (1997) Multiple defects in innate and adaptive immunologic function in NOD/LtSz-scid mice. *J. Immunol.* **154,** 180–191.

9. Pflumio, F., Izac, B., Katz, A., Shultz, L. D., Vainchenker, W., and Coulombel, L. (1996) Phenotype and function of human hematopoietic cells engrafting immune-deficient CB17-severe combined immunodeficiency mice and nonobese diabetic-severe combined immnodeficiency mice after transplantation of human cord blood mononuclear cells. *Blood* **88,** 3731–3740.
10. Ueda, T., Yoshida, M., Yoshino, H., Kobayashi, K., Kawahata, M., Ebihara, Y., et al. (2001) Hematopoietic capability of $CD34^+$ cord blood cells: a comparison with $CD34^+$ adult bone marrow cells. *Int. J. Hematol.* **73,** 457–462.
11. Ueda, T., Tsuji, K., Yoshino, H., Ebihara, Y., Yagasaki, H., Hisakawa, H., et al. (2000) Expansion of human NOD/SCID-repopulating cells by stem cell factor, Flk2/Flt3 ligand, thrombopoietin, IL-6 and soluble IL-6 receptor. *J. Clin. Invest.* **105,** 1013–1021.
12. Ebihara, Y., Tsuji, K., Lyman, S. D., Sui, X., Yoshida, M., Muraoka, K., et al. (1997) Synergistic action of Flt3 and gp130 signalings in human hematopoiesis. *Blood* **90,** 4363–4368.
13. Bhatia, M., Bonnet, D., Kapp, U., Wang, J. C. Y., Murdoch, B., and Dick, J. E. (1997) Quantitative analysis reveals expansion of human hematopoietic repopulating cells after short-term ex vivo culture. *J. Exp. Med.* **186,** 619–624.
14. Conneally, E., Cashman, J., Petzer, A., and Eaves, C. (1997) Expansion in vitro of transplantable human cord blood stem cells demonstrated using a quantitative assay of their lympho-myeloid repopulating activity in nonobese diabetic-scid/scid mice. *Proc. Natl. Acad. Sci. USA* **94,** 9836–9841.
15. Yoshino, H., Ueda, T., Kawahata, M., Kobayashi, K., Ebihara, Y., Manabe, A., et al. (2000) Natural killer cell depletion by anti-asialo GM1 antiserum treatment enhances human hematopoietic stem cell engraftment in NOD/Shi-*scid* mice. *Bone Marrow Transplant* **26,** 1211–1216.
16. Habu, S., Fukui, H., Shimamura, K., Kasai, M., Nagai, Y., Okumura, K., et al. (1981) In vitro effects of anti-asialo GM1. I. Reduction of NK activity and enhancement of transplanted tumor growth in nude mice. *J. Immunol.* **127,** 34–38.
17. Ueda, T., Yoshino, H., Kobayashi, K., Kawahata, M., Ebihara, Y., Ito, M., et al. (2000) Hematopoietic repopulating ability of cord blood CD34+ cells in NOD/Shi-*scid* mice. *Stem Cells* **18,** 204–213.

V

Maturation and Differentiation of Dendritic Cells

30

Differentiation of Human Antigen-Presenting Dendritic Cells from CD34+ Hematopoietic Stem Cells In Vitro

Xin-Sheng Ju and Martin Zenke

1. Introduction

Antigen-presenting dendritic cells (DCs) originate from hematopoietic stem cells in bone marrow through successive steps of differentiation *(1–5)*. In peripheral tissues (e.g., skin), DCs are exposed to and capture antigens like viruses, bacteria, and other pathogens through endocytosis and phagocytosis. In the presence of maturation-inducing stimuli, such as inflammatory cytokines, DCs undergo a process of maturation and migrate via the lymphatic vessels to the secondary lymphoid organs. In the lymphoid tissue, they present the processed antigens to T cells in the context of major histocompatibility complex (MHC) class I and class II molecules and initiate potent antigen specific immune responses. Giving their unique properties as professional antigen-presenting cells, DCs are currently being assessed for medical therapy, such as immunotherapy of cancer *(6,7)*.

Human DCs can be generated in vitro from hematopoietic progenitor cells or monocytes by employing different protocols and cytokines *(1–5)*. For example, CD14+ peripheral blood monocytes develop into DCs by treatment with granulocyte–macrophage colony stimulating factor (GM-CSF) and interleukin-4 (IL-4) *(8,9)* and this protocol yields large numbers of DCs for application in DC-based immunotherapy *(10)*. DCs are also generated from CD34+ hematopoietic progenitor cells (HPCs) of bone marrow or cord blood with GM-CSF and tumor necrosis factor-α (TNF-α) *(11)* and including stem cell factor (SCF) and/or Flt-3 ligand (FL) to enhance DC production *(11,12)*. However, the total number of DCs that can be obtained in vitro remains

From: *Methods in Molecular Biology, vol. 215: Cytokines and Colony Stimulating Factors: Methods and Protocols*
Edited by: D. Körholz and W. Kiess © Humana Press Inc., Totowa, NJ

limited, which severely hampers their further biochemical and molecular characterization and their clinical application.

A variety of cytokines have been reported to efficiently expand CD34+ hematopoietic progenitors from human bone marrow and cord blood in vitro, including different combinations of SCF, FL, IL-3, thrombopoietin (TPO), IL-6, and soluble IL-6 receptor fusion protein (hyper-IL-6) ***(13–17)***. Such in vitro expanded progenitors reconstitute long-term hematopoiesis in immunodeficient mice and contain both CD14- and CD14+ DC precursors ***(15)***.

Here, we describe a two-step culture system that produces large numbers of fully functional DCs from CD34+ hematopoietic progenitors of cord blood or bone marrow. CD34+ cells are first grown with the cytokine combination of SCF, FL, TPO, and hyper-IL-6 in serum-free medium under culture conditions that preserve the progenitor phenotype and yield large cell numbers. Then, these progenitor cells are induced to undergo cell cycle arrest and to differentiate into DCs by exchanging the growth-promoting cytokines with the differentiation-inducing cytokines GM-CSF and IL-4; TNF-α is applied for 1 d to promote terminal maturation of DCs. This two-step culture system (growth→differentiation) follows the natural pathway of DC differentiation from hematopoietic progenitor cells and can be used to study growth and differentiation of DCs as separate genetic programs. It also provides a useful approach for generating large numbers of fully functional DCs for clinical application.

2. Materials

1. Cord blood harvested from full-term deliveries according to institutional guidelines is stored at 4°C in a 50-mL syringe containing 1000 U sodium heparin (Sodium Heparin Braun 5000 I.U./0.5 mL; Braun Melsungen AG, Melsungen, Germany) for less than 10 h before CD34+ stem cell isolation (*see* **Note 1**).
2. Cell strainer (70 µm; Becton Dickinson, Heidelberg, Germany).
3. Phosphate-buffered saline (PBS): Dissolve 8 g NaCl (137 m*M*), 0.2 g KCl (2.7 m*M*), 1.16 g Na_2HPO_4 (7.9 m*M*), and 0.2 g KH_2PO_4 (0.01 m*M*) in 1 L of H_2O, adjust pH to 7.0, and autoclave.
4. Ficoll–Hypaque (density 1.077 g/m; Eurobio, Paris, France).
5. StemSpan serum-free culture medium (Stem Cell Technologies, Vancouver, BC, Canada) containing 1% bovine serum albumin (BSA), 10 µg/mL bovine pancreatic insulin, 200 µg/mL human transferrin, 10^{-4} *M* of 2-mercaptoethanol, 2 m*M* of L-glutamine.
6. Eagle's medium (Gibco-BRL, Gaithersburg, MD).
7. CD34 Progenitor Cell Isolation Kit (Miltenyi Biotec, Bergisch Gladbach, Germany), which includes 2 mL FcR blocking reagent (reagent A1); 2 mL monoclonal hapten-conjugated anti-CD34 antibody, clone QBEND/10, isotype mouse IgG1 (reagent A2); 2 mL colloidal superparamagnetic MACS MicroBeads conjugated to an antihapten antibody (reagent B).

8. Magnetic cell separator unit MiniMACS, positive selection column MS+/RS+ (for $<2 \times 10^9$ mononuclear cells; Miltenyi Biotec, Bergisch Gladbach, Germany).
9. Isolation buffer: PBS supplemented with 0.5% BSA (Fraction V, Sigma, St. Louis, MO) and 2 m*M* EDTA (Merck, Darmstadt, Germany).
10. The following antibodies were used: CD1a (NA1/34; DAKO, Glostrup, Denmark), CD11a (HI111) and CD11c (B-Ly6; both BD Biosciences, San Diego, CA), CD14 (IOM2, clone RM052, Immunotech, Marseille, France), CD29 (MAR4, BD Biosciences, San Diego, CA), CD33 (WM53, Cymbus Biotechnology, Chandlers Ford, Hants, UK), CD34 (Anti-HPCA-2, clone 8G12), CD40 (5C3), CD49d (9F10) and CD54 (HA58; all BD Biosciences, San Diego, CA), CD71 (Ber-T8; DAKO, Glostrup, Denmark), CD80 (MAB104) and CD83 (HB15A, both Immunotech, Marseille, France), CD86 (B70/B7-2, clone 2331), CD117 (YB5.B8), mannose receptor (Clone 19.2) and CCR7 (2H4; all BD Biosciences, San Diego, CA) and HLA-DR (clone CR3/43; DAKO, Glostrup, Denmark).
11. Flow cytometry (FACSCalibur; Becton Dickinson, Heidelberg, Germany). Data are analyzed in CELLQuest software (Becton Dickinson).
12. RPMI 1640: complete RPMI 1640 medium containing 10% fetal calf serum (FCS), 100 U/mL penicillin–streptomycin, 2 m*M* of L-glutamine (all Gibco-BRL, Gaithersburg, MD), 10^{-4} *M* of 2-mercaptoethanol (Sigma, St. Louis, MO).
13. The following recombinant factors are used: human SCF and human TPO (both AMGEN, Thousand Oaks, CA); human FL (PeproTech, London, UK); hyper-IL-6 was produced in yeast as described *(17)* and was a gift from S. Rose-John (Kiel, Germany); human IL-4 (Schering-Plough, Kenilworth, NJ); human GM-CSF (Leukomax, Novartis, Nürnberg, Germany); human TNF-α was a gift from G. R. Adolf (Boehringer Ingelheim Austria, Vienna, Austria).
14. Centrifuge (Heraeus, Berlin, Germany).
15. 5% CO_2 incubator (Steri-Cult 200; Forma Scientific Inc., Marietta, OH).
16. Tissue culture dishes or flasks.
17. Aminoethyiouronium bromide (AET; Sigma, St. Louis, MO).
18. Dynabeads M-450 for Pan Human MHC-II+ cells (Dynal A.S., Oslo, Norway).
19. ^{3}H-thymidine (29 Ci/mmol; Amersham Pharmacia Biotech).
20. Liquid scintillation counting (Micobeta counter; Wallac, Turku, Finland).
21. Transwell inserts (5-μm pore size, Costar, Cambridge, UK).
22. ELC chemokine (Peprotech, London, UK).
23. Dynabeads (15-μm diameter; Nynal PolymersAS, Lillestrom, Norway).
24. Fluorescein-di-Acetate (FDA, Sigma, St. Louis).

3. Methods

Note: Human cord blood or bone marrow are biohazardous and therefore must be handled in a proper biohazard containment facility. All work must be performed in a sterile biohazard hood and Latex gloves must be worn when handling these samples. All waste must be autoclaved prior to disposal.

3.1. Isolation of Mononuclear Cells from Cord Blood

1. Cord blood samples are passed through a 70-μm cell strainer, diluted with 4 volumes of PBS containing 2 m*M* EDTA; then, 35 mL of diluted cell suspension are carefully layered onto 15 mL Ficoll–Hypaque in a 50-mL conical tube and centrifuged at 400*g* for 30 min at 20°C without setting brake (*see* **Note 1**).
2. Aspirate the upper layer leaving the mononuclear cell layer at the interface. Carefully transfer the interphase cells to a new 50-mL conical tube, fill the tube with PBS containing 2 m*M* EDTA, and centrifuge at 300*g* for 10 min at 20°C; remove the supernatant, resuspend the cell pellet in 50 mL PBS with 2 m*M* EDTA, and centrifuge 300*g* for 5 min at 20°C.
3. Resuspend the cell pellet in a final volume of 300 μL isolation buffer (PBS containing 0.5% BSA, 2 m*M* EDTA) per 10^8 cells (for less than 10^8 cells, adjust to 300 μL with isolation buffer).

3.2. Magnetic Labeling of CD34+ Cells

1. Add 100 μL FcR blocking reagent (reagent A1; *see* CD34 Progenitor Cell Isolation Kit) per 10^8 cells and mix gently; then, add 100 μL hapten-conjugated anti-CD34 antibody (reagent A2), mix again, and incubate cells for 15 min at 4°C (*see* **Note 2**).
2. Cells are suspended with 5 mL PBS containing 2 m*M* EDTA (washing buffer) and centrifuged at 300*g* for 10 min; remove the buffer completely. Resuspend the pellet in 400 μL isolation buffer per 10^8 cells, add 100 μL of paramagnetic MACS MicroBeads (reagent B), mix well, and incubate for 15 min at 4°C. Wash cells with washing buffer and resuspend cell pellet in 500 μL isolation buffer per 10^8 cells.

3.3. Magnetic Separation of Mononuclear Cells

1. Connect the MS+/RS+ column with a 21-gage needle; assemble MS+/RS+ column in the magnetic field of MACS separator.
2. Fill column and rinse with 500 μL isolation buffer.
3. Apply cells and pass through the column; wash three times with 500 μL isolation buffer (*see* **Note 2**).
4. Remove column from MACS separator, place column in 10-mL tube, and elute retained cells with 1 mL isolation buffer using the plunger supplied with the column.
5. Eluted cells are then applied to a new prefilled MS+/RS+ column; the column is washed three times and cells are eluted in isolation buffer as in **step 3**.

3.4. Evaluation of the Purity of the CD34+ Stem Cell Preparation by Flow Cytometry

1. Take an aliquot of eluted CD34+ cells and stain with FITC-conjugated anti-CD34 antibody (clone AC136, which recognizes an epitope different from that recog-

nized by the monoclonal antibody QBEND/10), 1:50 dilution; incubate for 15 min at 4°C.

2. After staining cells are washed with PBS containing 0.5% BSA (FACS buffer) and resuspended in 300 µL FACS buffer, propidium iodide (2 µg/mL) is added for gating on viable cells. CD34 expression is analyzed by flow cytometry using a FACSCalibur device and CELLQuest software (Becton Dickenson, Heidelberg, Germany).

3.5. Culture of CD34+ Stem/Progenitor Cells

1. Resuspend CD34+ cells at (0.3–0.5) × 10^6 cells/mL cell density in serum-free StemSpan medium containing 100 ng/mL SCF, 20 ng/mL TPO, 50 ng/mL FL, and 5 ng/mL hyper-IL-6. The cell number is determined in regular time intervals and cytokines are added every second day.
2. After 4 d of culture, cell density is kept at 1 × 10^6 cells/mL.

3.6. Differentiation of DCs from CD34+ Stem/Progenitor Cells

1. After 12–14 d of culture with SCF, TPO, FL, and hyper-IL-6, cell numbers have increased by more than 100-fold and cells are then induced to differentiate into DCs.
2. Then, cells are harvested by centrifugation (300*g*, 10 min), washed with Eagle's medium and centrifuged again (300*g*, 10 min). Cells are then resuspended at 0.5 × 10^6 cells/mL cell density in complete RPMI 1640 medium containing 10% FCS, GM-CSF (500 U/mL), and IL-4 (500 U/mL).
3. Every second day, half of the medium is replaced by fresh culture medium, and GM-CSF and IL-4 are added to the final concentration as in **step 2**.
4. Routinely, cells resemble immature DCs at d 5–6 of culture with prominent dendritic processes and hairlike cytoplasmic projection and express CD1a, CD11c, and CD80.
5. To induce maturation, cell numbers are adjusted to 0.5 × 10^6 cell/mL and cells are cultured for an additional 24 h with TNF-α (10 ng/mL). Mature DCs show long veils and high expression of MHC class I and II, CD40, CD80, CD83 CD86, and CCR7 (*see* **Table 1**).

3.7. Mixed Leukocyte Reaction (MLR)

1. Dendritic cells efficiently stimulate T-cell proliferation in mixed-leukocyte reaction (MLR) assay. Mixed-leukocyte reaction is done as described elsewhere *(18)*.
2. Briefly, HPCs and DCs are recovered, irradiated (5000 rad), and used as stimulator cells. T cells are obtained from peripheral blood and depleted of erythrocytes by rosetting with AET-treated sheep red blood cells.
3. The MHC-II+ cells are depleted by using Dynabeads. Increasing numbers of stimulator cells are incubated with 1 × 10^5 responder T cells/well in RPMI, 10% FCS for 5 d.

Table 1
Cell Surface Markers of Hematopoietic Progenitors (HPCs, d 14 of Growth) and Differentiated DCs (d 6 of Differentiation Culture)

Surface marker	Characteristic	HPCs (%)	DCs (%)	DCs+TNF (%)
CD1a	DC, Langerhans cells	3.1	45.6	75.7
CD11a	Integrin α_L	91.8	82.7	78.9
CD11c	Integrin $\alpha_x\beta_2$	4.2	77	82
CD14	Monocyte	20.1	7.8	1.8
CD29	Integrin β1	94.9	85.1	68.4
CD33	Myeloid cells	84.8	80.5	60.3
CD40	DCs, B cells	1.2	74	85
CD49d	Integrin α4	95.1	72.6	75
CD54	ICAM-1	60.7	77	94.9
CD71	Transferrin receptor	58	37.4	33.8
CD80	Costimulatory molecule	4.3	41.5	62
CD83	DCs, Langerhans cells	2.6	8.8	54
CD86	Cstimulatory molecule	4.5	16.5	59.5
HLA-DR	MHC class II	20.3	69	73
Mannose receptor	DC, macrophage	1.4	52.2	44
CD117	c-kit/SCF receptor	14	0	0
CCR7	Chemokine receptor	0	15	46.7

Results are mean of 3–6 independent experiments with standard deviation less than 20%.

4. Cells are then incubated with 1 µCi ^{3}H–thymidine (29 Ci/mmol) per well (6 h), harvested, and subjected to liquid scintillation counting.
5. T-cell proliferation is particularly high when TNF-α is applied to induce full maturation of DC, whereas hematopoietic progenitor cells grown with SCF, TPO, FL, and hyper-IL-6 are inactive in MLR.

3.8. Chemotaxis Assay

Finally, DCs are active in chemotaxis assay and migrate toward CCR7 ligand ELC gradients. Chemotaxis assay is performed as described with minor modifications ***(19)***.

1. Briefly, DCs are incubated for 2–4 h in serum-free RPMI medium with GM-CSF and IL-4.
2. Prior to analysis, Transwell inserts are preincubated in RPMI plus 1% BSA and 2 × 10^5 cells are seeded in the upper compartment in 100 µm RPMI plus 1% BSA.

3. ELC chemokine (100 ng/mL) is added to the upper or lower compartment to analyze migration against or toward gradient, respectively.
4. Ninety minutes later, 1×10^4 Dynabeads are added to the lower compartment to normalize for variations in the experimental procedure.
5. Cells are stained with Fluorescein-di-Acetate (FDA), and cells and beads are recovered and analyzed by flow cytometry.
6. By gating on beads, the ratio beads : FDA positive cells is determined and allows a precise calculation of transmigrated cells.
7. The chemotactic activity of DCs is particularly high after TNF-α treatment.

4. Notes

1. Do not use cord blood samples that were stored for more than 10 h, and cord blood should be kept at 4°C prior to cell fractionation, because these factors affect the quantity and quality of the isolated CD34+ cells.
2. During isolation of CD34+ cells, use cold solution (4°C) and work quickly during labeling with reagents A1, A2, and B, because increased temperature and prolonged incubation time may lead to unspecific labeling. Wash cells carefully and remove buffer completely from cell pellet after labeling. Avoid bubbles in the matrix of the column during separation, as this may lead to clogging of the column and a decreased quality of separation.
3. Isolation of CD34+ cells from bone marrow follows similar protocols, with the modification that bone marrow cells should be diluted 10 times with PBS containing 2 m*M* EDTA and then passed through 30-μm nylon mesh prior to Ficoll–Hypaque centrifugation.
4. Hematopoietic progenitor cells proliferate very fast with the culture condition described above. If apoptotic cells occur after about 10 d of culture, cells are harvested and diluted with an equal volume of Eagle's medium and subjected to Ficoll–Hypaque centrifugation as above. Recover old culture medium and resuspend cells at 10^6 cells/mL in 50% StemSpan medium and 50% old culture medium and add cytokines (SCF, TPO, FL, hyper-IL-6) to the final concentration as in **Subheading 3.5., step 1**.
5. CD34+ cells can also be expanded in RPMI 1640 medium with 10% FCS instead of StemSpan medium and differentiated into DCs as in **Subheading 3.5., step 1**. However, in RPMI 1640 culture medium, cumulative cell numbers are lower than those in StemSpan medium.
6. Differentiation of DCs from stem/progenitor cells is strongly influenced by the cell density in culture. If cells are kept at low cell density (less than 1×10^6 cells/mL), differentiation will occur faster than at high density, as monitored, e.g., by downregulation of CD14 expression. To effectively induce DC maturation, TNF-α is added to cells at low cell density (less than 0.5×10^6 cells/mL).
7. The concentration of GM-CSF and IL-4 can also affect DC differentiation. If DC differentiate with slow kinetics (e.g., CD14 expression is not downregulated), it is strongly suggested to increase the IL-4 concentration to 1000 U/mL.

8. During differentiation, some cells will adhere to the culture dish, whereas others will stay nonadherent or loosely adherent. We note, however, that the nonadherent cells do not significantly differ from the adherent or loosely adherent cells (e.g., in their surface marker expression [CD14, MHC class II, CD80, CD86, etc.]). Additionally, if the nonadherent cells are transferred to a new dish, some of them will become adherent again after a short time of culture. Thus, every transfer of cells during culture will result in some cell loss. Therefore, it is recommended that cells be cultured in the same dish during differentiation.
9. Dendritic cells can be generated in culture medium containing FCS or human serum. It is recommended to compare various batches of FCS, because FCS quality may influence DC differentiation. Generation of DCs under serum-free medium normally requires supplementation with transforming growth factor-β *(20)*. For clinical application, it is imperative to generate DCs in autologous human serum or under serum-free conditions.

References

1. Banchereau, J. and Steinman, R. M. (1998) Dendritic cells and the control of immunity. *Nature* **392,** 245–252.
2. Hart, D. N. (1997) Dendritic cells: unique leukocyte populations which control the primary immune response. *Blood* **90,** 3245–3287.
3. Cella, M., Sallusto, F., and Lanzavecchia, A. (1997) Origin, maturation and antigen presenting function of dendritic cells. *Curr. Opin. Immunol.* **9,** 10–16.
4. Grabbe, S., Kämpgen, E., and Schuler, G. (2000) Dendritic cells: multi-lineal and multi-functional. *Immunol. Today* **21,** 431–433.
5. Banchereau, J., Briere, F., Caux, C., Davoust, J., Lebecque, S., Liu, Y. J., et al. (2000) Immunobiology of dendritic cells. *Annu. Rev. Immunol.* **18,** 767–811.
6. Schuler, G. and Steinman, R. M. (1997) Dendritic cells as adjuvants for immune-mediated resistance to tumors. *J. Exp. Med.* **186,** 1183–1187.
7. Colaco, C. A. (1999) Why are dendritic cells central to cancer immunotherapy? *Mol. Med. Today* **5,** 14–17.
8. Sallusto, F. and Lanzavecchia, A. (1994) Efficient presentation of soluble antigen by cultured human dendritic cells is maintained by granulocyte/macrophage colony-stimulating factor plus interleukin 4 and downregulated by tumor necrosis factor alpha. *J. Exp. Med.* **179,** 1109–1118.
9. Romani, N., Gruner, S., Brang, D., Kämpgen, E., Lenz, A., Trockenbacher, B., et al. (1994) Proliferating dendritic cell progenitors in human blood. *J. Exp. Med.* **180,** 83–93.
10. Thurner, B., Haendle, I., Roder, C., Dieckmann, D., Keikavoussi, P., Jonuleit, H., et al. (1999) Vaccination with Mage-3A1 peptide-pulsed mature, monocyte-derived dendritic cells expands specific cytotoxic T cells and induces regression of some metastases in advanced stage IV melanoma. *J. Exp. Med.* **190,** 1669–1678.
11. Caux, C., Vanbervliet, B., Massacrier, C., Dezutter-Dambuyant, C., de Saint-Vis, B., Jacquet, C., et al. (1996) CD34+ hematopoietic progenitors from human cord

blood differentiate along two independent dendritic cell pathways in response to GM-CSF+TNF alpha. *J. Exp. Med.* **184,** 695–706.
12. Rosenzwajg, M., Camus, S., Guigon, M., and Gluckman, J. C. (1998) The influence of interleukin (IL)-4, IL-13, and Flt3 ligand on human dendritic cell differentiation from cord blood CD34+ progenitor cells. *Exp. Hematol.* **26,** 63–72.
13. Peters, M., Muller, A. M., and Rose-John, S. (1998) Interleukin-6 and soluble interleukin-6 receptor: direct stimulation of gp130 and hematopoiesis. *Blood* **92,** 3495–3504.
14. Piacibello, W., Sanavio, F., Garetto, L., Severino, A., Bergandi, D., Ferrario, J., et al. (1997) Extensive amplification and self-renewal of human primitive hematopoietic stem cells from cord blood. *Blood* **89,** 2644–2653.
15. Arrighi, J. F., Hauser, C., Chapuis, B., Zubler, R. H., and Kindler, V. (1999) Long-term culture of human CD34(+) progenitors with FLT3-ligand, thrombopoietin, and stem cell factor induces extensive amplification of a CD34(–)CD14(–) and a CD34(–)CD14(+) dendritic cell precursor. *Blood* **93,** 2244–2252.
16. Kollet, O., Aviram, R., Chebath, J., ben-Hur, H., Nagler, A., Shultz, L., et al. (1999) The soluble interleukin-6 (IL-6) receptor/IL-6 fusion protein enhances in vitro maintenance and proliferation of human CD34(+)CD38(–/low) cells capable of repopulating severe combined immunodeficiency mice. *Blood* **94,** 923–931.
17. Fischer, M., Goldschmitt, J., Peschel, C., Brakenhoff, J. P., Kallen, K. J., Wollmer, A., et al. (1997) A bioactive designer cytokine for human hematopoietic progenitor cell expansion. *Nat. Biotechnol.* **15,** 142–145.
18. Diebold, S. S., Lehrmann, H., Kursa, M., Wagner, E., Cotten, M., and Zenke, M. (1999) Efficient gene delivery into human dendritic cells by adenovirus polyethylenimine and mannose polyethylenimine transfection. *Hum. Gene Ther.* **10,** 775–786.
19. Kellermann, S. A., Hudak, S., Oldham, E. R., Liu, Y. J., and McEvoy, L. M. (1999) The CC chemokine receptor-7 ligands 6Ckine and macrophage inflammatory protein-3 beta are potent chemoattractants for in vitro- and in vivo-derived dendritic cells. *J. Immunol.* **162,** 3859–3864.
20. Strobl, H., Riedl, E., Scheinecker, C., Bello-Fernandez, C., Pickl, W. F., Rappersberger, K., et al. (1996) TGF-beta 1 promotes in vitro development of dendritic cells from CD34+ hemopoietic progenitors. *J. Immunol.* **157,** 1499–1507.

31

Dendritic Cell Development from Mobilized Peripheral Blood CD34+ Cells

Deanne M. R. Lathers, Nicholas Achille, and M. Rita I. Young

1. Introduction

Dendritic cells are not a homogenous population, but develop through different differentiation pathways, whose prominence is dependent on the cytokine milieu. Although dendritic cells can be readily derived from monocytes, dendritic cells that develop from $CD34^+$ progenitor cells have a more efficient antigen-presenting capability *(1)*. However, even the development of dendritic cells from $CD34^+$ progenitor cells can occur through several different pathways *(2,3)*. In one differentiation pathway, $CD34^+$ cells mature into committed CD14–$CD1a^+$ precursor cells that give rise to $CD1a^+$ dendritic cells. In a second differentiation pathway, $CD34^+$ cells first develop into bipotential $CD14^+CD1a^-$ precursor cells, which mature into bipotential $CD14^+CD1a^+$ intermediate precursor cells. These intermediate cells can differentiate into monocytic cells in the presence of colony stimulating factor 1 (CSF-1), or into dendritic cells in the absence of CSF-1 *(3,4)*.

The methods described herein can be used for differentiating mobilized human peripheral blood $CD34^+$ progenitor cells into dendritic cells and for identifying the pathway by which this differentiation occurs. The major procedures described for these protocols include the isolation of $CD34^+$ cells from the peripheral blood, culture conditions for the isolated $CD34^+$ cells, flow-cytometric detection or fluorescence-activated cell sorting of intermediate precursor cells, and identification of resultant dendritic cells.

From: *Methods in Molecular Biology, vol. 215: Cytokines and Colony Stimulating Factors: Methods and Protocols*
Edited by: D. Körholz and W. Kiess © Humana Press Inc., Totowa, NJ

2. Materials

2.1. Isolation of Peripheral Blood Mononuclear Leukocytes (PBMLs)

1. Fresh human blood collected in an anticoagulant such as heparin or EDTA (*see* **Note 1**).
2. Hank's balanced salt solution (HBSS; Sigma Chemical Co., St. Louis, MO), room temperature.
3. Histo-Paque (Sigma), room temperature.

2.2. Isolation of CD34+ Cells from the PBMLs

1. CD34 Progenitor Cell Isolation Kit (Miltenyi Biotec, Auburn, CA) (*see* **Note 2**):
   Human Ig (FcR blocking reagent; reagent A1);
   Hapten-conjugated CD34 antibody (reagent A2; clone QBEND/10);
   Colloidal superparamagnetic MACS MicroBeads conjugated to antihapten antibody (reagent B).
2. Separation buffer: phosphate-buffered saline (PBS) containing 0.5% bovine serum albumin and 2 m*M* EDTA; degassed.
3. Magnetic cell separator (MiniMACS; Miltenyi Biotec).
4. MACS-positive selection column(s) type MS+/RS+ (Miltenyi Biotec).

2.3. Culture Conditions for CD34+ Cells to Differentiate into Dendritic Cells

1. 24-Well Costar tissue culture plates.
2. Phenol-free RPMI 1640 culture medium (*see* **Note 3**) (Sigma) with the following:
   Endotoxin-free defined fetal bovine serum (Sigma);
   100 U/mL Penicillin (Sigma);
   0.25 µg/mL Amphotericin B (Sigma);
   100 µg/mL streptomycin (Sigma);
   0.02 *M* HEPES buffer (Sigma);
   5×10^{-5} *M* of 2-mercaptoethanol (Sigma);
   2 m*M* L-glutamine (Sigma).
3. Cytokines Supplements for Culture Medium
   100 U/mL Human recombinant granulocyte–macrophage colony-stimulating factor (GM-CSF; PharMingen, San Diego, CA);
   50 ng/mL Human stem cell factor (SCF; R&D Systems, Minneapolis, MN);
   25 U/mL Tumor necrosis factor-α (TNF-α; PharMingen).

2.4. Identification of Dendritic Precursor Cells in Cultures Established from CD34+ Cells

1. Human AB plasma (Sigma), inactivated by 30-min incubation at 56°C.
2. PBS.
3. Fluoresceine isothiocyanate (FITC)-conjugated CD1a (PharMingen, clone H149).

4. Phycoerythrin (PE)-conjugated CD14 antibodies (PharMingen, clone 3E2).
5. FITC- or PE-conjugated isotype control antibodies (PharMingen).
6. 1% Paraformaldehyde.
7. Flow cytometer such as FACS brand (Becton Dickinson, San Jose, CA).

2.5. Identification of Dendritic Cells in Cultures Established from CD34$^+$ Cells

1. Human AB plasma (Sigma), inactivated by 30-min incubation at 56°C.
2. 4-Color Dendritic Value Bundle (Becton-Dickinson):
   - Lineage 1 antibody cocktail:
     FITC–CD3 antibody (clone SK7);
     FITC–CD14 antibody (clone MϕP9);
     FITC–CD16 antibody (clone 3G8);
     FITC–CD19 antibody (clone SJ25C1);
     FITC–CD20 antibody (clone L27);
     FITC–CD56 antibody (clone NCAM 16.2);
   - PE–CD123 antibody (clone 9F5);
   - PerCP-HLA-DR antibody (clone L243);
   - APC–CD11c antibody (clone S-HCL-3);
   - Isotype control PE–IgG_1 (clone X40);
   - Isotype control APC–IgG_{2a} (clone X39).
3. Wash buffer: PBS.
4. 1% Paraformaldehyde.
5. Flow cytometer such as FACS brand (Becton Dickinson, San Jose, CA).

3. Method

3.1. Isolation of PBML

1. Pipet 10 mL Histo-Paque into 50-mL conical centrifuge tube.
2. Dilute freshly isolated peripheral blood with an equal volume of HBSS and layer diluted blood on top of the Histo-Paque within the 50-mL centrifuge tubes, making sure not to mix the two layers.
3. Centrifuge the gradient in swinging buckets at 400*g* for 15 min at 20°C with brake on low or off. After centrifugation, the PBMLs should form a band at the interface between the diluted plasma and the Histo-Paque; erythrocytes and granulocytes will be sedimented.
4. Use a pipet to remove liquid above the cell band and discard this liquid. Carefully collect the band of PBMLs and transfer into a new conical centrifuge tube.
5. Dilute isolated PBML cells by the addition of 30 mL HBSS.
6. Centrifuge at 300*g* for 10 min at 4°C, discard the liquid, and repeat cell washing.

3.2. Isolation of CD34$^+$ Cells from the PBMLs

1. The CD34$^+$ cells can be immunomagnetically isolated from the PBMLs with a high degree of progenitor cell purity *(5,6)*. This is accomplished by an indirect

magnetic labeling system using a hapten-conjugated primary monoclonal anti-CD34 antibody and an antihapten secondary antibody coupled to magnetic microbeads. Follow procedure exactly as described in the directions that accompany the CD34 Progenitor Cell Isolation Kit from Miltenyi Biotec. Keep all reagents cold. The described procedure is for a starting number of up to 10^8 cells.

2. Suspend the PBMLs at 300 µL separation buffer per 10^8 cells. For less than 10^8 PBMLs, resuspend to 300 µL.
3. Add 100 µL FcR blocking reagent (reagent A1) to the suspended cells and mix.
4. Add 100 µL of the anti-CD34 antibody mixture (reagent A2) and mix again.
5. Incubate cell suspension for 15 min at 4°C.
6. Wash cells by adding 2 mL separation buffer, centrifuging 300*g* for 10 min at 4°C, removing the supernatant and resuspending cells in 400 µL buffer.
7. Add 100 µL of the MACS MicroBead mixture (reagent B), mix, and incubate for 15 min at 4°C.
8. After washing cells, resuspend in 500 µL separation buffer. To separate the $CD34^+$ cells, place the positive selection column (MS+/RS+) in the magnetic field of the MACS separator.
9. Rinse the column with 500 µL separation buffer and discard the rinse.
10. Add the cells onto the column and allow cells to pass through the column, making sure that all liquid has passed through the column before initiating the next step.
11. Wash column two times, each with 500 µL buffer.
12. Remove the column from the MACS separator and place onto a tissue culture tube.
13. Add 500 µL separation buffer to the column to elute cells; repeat with another 500-µL elution. After each wash, use the column's plunger to remove remaining cells.
14. Add 1 mL culture medium to each tube.
15. Centrifuge cells at 300*g* for 10 min at 4°C and resuspend in cytokine-supplemented medium.

3.3. Culture Conditions for $CD34^+$ Cells to Differentiate into Dendritic Cells

1. After isolation, the $CD34^+$ cells are suspended in RPMI 1640 culture medium supplemented with the cytokines 100 U/mL GM-CSF, 50 ng/mL SCF, and 25 U/mL TNF-α at a density of 2.5×10^5 cells/mL, and seeded in a volume of 2 mL per well of 24-well plates.
2. Incubate the cells at 37°C in a humidified 5% CO_2 in air incubator for a total of up to 2 wk.
3. Gently aspirate medium and replace with fresh medium containing GM-CSF and TNF-α to the cultures every 4–5 d.

3.4. Identification of Dendritic Cell Precursors in Cultures Established from CD34⁺ Cells

1. After 5–6 and 8–9 d of culture, dendritic cell precursor populations can be detected phenotypically as $CD14^-CD1a^+$ cells, $CD14^+CD1a^-$ cells, and $CD14^+CD1a^+$ cells. Cultures can be immunostained and analyzed by flow cytometry for the presence of these cell populations. This procedure is described in **steps 2–9**. In addition, these cell populations can be isolated by fluorescent-activated cell sorting and further cultured to confirm their capacity to give rise to dendritic cells. This procedure will not be described in this chapter, but has been described previously *(7)*.
2. After 5–6 and 8–9 d of culture, collect cells that have developed from the $CD34^+$ cell cultures by pipetting the culture's medium several times to dislodge any loosely adhering cells. Residual adherent cells can be detached with a cell scraper.
3. Transfer cells to a 15-mL conical centrifuge tube and pellet the cells by centrifugation at 300*g* for 10 min at 4°C.
4. Discard the supernatant and resuspend cells in PBS at a concentration of 10^6 cells/mL.
5. Transfer 1 mL of cell suspension to a 3-mL polypropylene tube; pellet cells by centrifugation at 300*g* for 10 min at 4°C.
6. Resuspend cells in 100 µL PBS plus 50 µL human plasma to block nonspecific binding of antibody.
7. After 30 min of incubation on ice, wash the cells and resuspend in 20 µL FITC–CD1a plus 20 µL PE–CD14 antibodies or appropriate isotype control antibodies.
8. Incubate on ice in the dark for 30 min. After 30 min staining, wash cells twice, each time by dilution with 1 mL PBS, and centrifugation at 300*g* for 10 min at 4°C.
9. Resuspend cells in 0.5 mL of 1% paraformaldehyde and determine the percentage of positive staining cells using a flow cytometer such as a FACS 420 (Becton Dickinson, San Jose, CA).

3.5. Identification of Dendritic Cells in Cultures Established from CD34⁺ Cells

1. Two different subpopulations of dendritic cells can be detected to develop from the cultured $CD34^+$ cells. They can be detected by their lack of expression of lineage markers for monocytes, lymphocytes, and granulocytes ($lin1^-$) and by the expression of HLA-DR plus either interleukin (IL)-3Rα (CD123) or CD11c. This is accomplished with Becton Dickinson's 4-Color Dendritic Value Bundle kit. Follow procedure exactly as described in the directions that accompany the kit.
2. Resuspend cells in 100 µL PBS plus 50 µL human plasma to block nonspecific binding of antibody.

3. Combine antibodies as follows into each of two tubes for each sample being analyzed.
4. In the first tube, add 20 μL lin1 antibody mixture (antibodies to CD3, CD14, CD16, CD19, CD20, CD56), 10 μL anti-CD123, 10 μL anti-HLA-DR and 5 μL anti-CD11c.
5. In the second tube, combine 20 μL lin1 antibody mixture, 10 μL anti-HLA-DR, and isotype control antibodies at concentrations equivalent to the CD123 and CD11c antibodies.
6. To each tube, add 100 μL of cells (no more than 10^6 cells) collected from the CD34$^+$-derived cells that have been in culture for 2 wk.
7. Incubate 25 min on ice in the dark. After incubation, gently mix each tube and add 1 mL PBS.
8. Centrifuge both tubes at 300*g* for 5 min.
9. Remove supernatants and resuspend cells in 500 μL of 1% paraformaldehyde.
10. Analyze cells on a FACS brand flow cytometer within 24 h, acquiring at least 50,000 events.
11. To analyze samples, follow the instructions provided with the 4-Color Dendritic Value Bundle kit (*see* **Note 4**). Briefly, use forward and side scatter to exclude debris and dead cells (region 1). Live cells (gated region 1) that either stain dimly or negatively for lineage-specific markers are then identified and gated (region 2). Of this latter population (region 2), dendritic cells that are HLA-DR$^+$ plus either CD11c$^+$ or CD123$^+$ are then identified.

4. Notes

1. Most critical to successfully generating dendritic cells from CD34$^+$ cells is the freshness of the starting population of cells. The peripheral blood should be no older than 8 h. PBMLs can, however, be used after being stored frozen in liquid nitrogen in a mixture of 1 part dimethyl sulfonate and 9 parts fetal bovine serum.
2. The technical notes accompanying the CD34 Progenitor Cell Isolation Kit are superb in their description of both the underlying approach being used as well as the specific methods to be employed.
3. The rationale for recommending phenol-free culture medium is to reduce background fluorescence of cultured cells. To maintain the activity of working stocks of cytokines, dilute cytokines in culture medium and aliquot prior to freezing. Thaw cytokine aliquots just prior to addition to cultured cells. When working with a small number of cells, 96-well tissue culture plates have been used, but not successfully. Therefore, 24-well plates should be used, even if the number of CD34$^+$ cells is small.
4. Analysis of cells that have been stained with the dendritic cell bundle kit requires CellQuest, Attractors, or PAINT-A-GATE^{PRO} software. The fact sheet from Becton Dickinson that accompanies the kit very clearly demonstrates how these software applications are used to identify the stained dendritic cells.

References

1. Ferlazzo, G., Wesa, A., Wei, W.-Z., and Galy, A. (1999) Dendritic cells generated either from CD34$^+$ progenitor cells or from monocytes differ in their ability to activate antigen-specific CD8$^+$ T cells. *J. Immunol.* **163,** 3597–3604.
2. Lathers, D. M. R., Lubbers, E., Wright, M. A., and Young, M. R. I. (1999) Dendritic cell differentiation pathways of CD34$^+$ cells from the peripheral blood of head and neck cancer patients. *J. Leuk. Biol.* **65,** 623–628.
3. Caux, C., Massacrier, C., Vanbervliet, B., Dubois, B., Durand, I., Cella, M., et al. (1997) CD34$^+$ hematopoietic progenitors from human cord blood differentiate along two independent dendritic cell pathways in response to granulocyte–macrophage colony-stimulating factor plus tumor necrosis factor α: II. Functional analysis. *Blood* **90,** 1458–1470.
4. Caux, C., Vanbervliet, B., Massacrier, C., Dezutter-Dambuyant, C., de Saint-Vis, B., Jacquet, C., et al. (1996) CD34$^+$ hematopoietic progenitors from human cord blood differentiate along two independent dendritic cell pathways in response to GM-CSF + TNFα. *J. Exp. Med.* **184,** 695–706.
5. Garrity, T., Pandit, R., Wright, M. A., Benefield, J., and Young, M. R. I. (1997) Increased presence of CD34$^+$ cells in the peripheral blood of head and neck cancer patients and their differentiation into CD1a$^+$ cells. *Int. J. Cancer.* **73,** 663–669.
6. Young, M. R. I., Schmidt-Pak, A., Wright, M. A., Matthews, J. P., Collins, S. L., and Petruzzelli, G. (1995) Mechanisms of immune suppression in patients with head and neck cancer: Presence of immune suppressive CD34$^+$ cells in cancers that secrete granulocyte–macrophage colony-stimulating factor. *Clin. Cancer Res.* **1,** 95–103.
7. Lathers, D. M. R., Lubbers, E., Beal, N. M., Wright, M. A., and Young, M. R. I. (1999) Cultures derived from peripheral blood CD34+ progenitor cells of head and neck cancer patients and from cord blood are functionally different. *Hum. Immunol.* **60,** 1207–1215.

32

In Vitro Maturation of Dendritic Cells from Blood Progenitors

Lina Matera and Alessandra Galetto

1. Introduction

Dendritic cells (DCs) are the most potent antigen-presenting cells (APCs) because of their ability to elicit strong proliferative response to alloantigens and to recall antigens. Most importantly, DCs have the unique ability to initiate the immune response by capturing antigens in peripheral tissues and migrating to secondary lymphoid organs, where they sensitize naive CD4+ T cells to the antigen. DC migration is concomitant with maturation, during which the DCs lose their ability to acquire and process antigens *(1,2)*.

The difficulty of isolating DCs from tissues and blood has fueled the efforts toward unraveling the basic DC biology and later to development of methods to obtain DCs in large numbers in vitro. This has opened the possibility of using DCs to induce immunity in vivo.

In man, DCs can either be generated from rare, proliferating CD34+ precursors (1–2% in bone marrow, <0.1% in peripheral blood of healthy adults, <1% in cord blood) or from more frequent nonproliferating CD14+ monocytes (usually 5–10% of periferal blood mononuclear cells [PBMCs] in healthy adults). It should be stressed that although there are definite similarities with the "gold standard" of directly purified blood DC populations, there are also notable differences. The level of heterogeneity is dependent on the population of DC progenitors and the cytokines used to trigger progenitors to maturation. The CD34+ method uses granulocyte–macrophage colony stimulating factor and tumor necrosis factor-α (GM-CSF+TNF-α) as key cytokines, whereas the CD14+ approach requires GM-CSF+IL-4 (interleukin-4). For experimental

From: *Methods in Molecular Biology, vol. 215: Cytokines and Colony Stimulating Factors: Methods and Protocols*
Edited by: D. Körholz and W. Kiess © Humana Press Inc., Totowa, NJ

purposes, cord blood CD34+ cells and CD14+ monocytes are convenient sources for propagating DCs in vitro. For clinical use, adult blood is most accessible, but requires G-CSF pretreatment of the patient to increase the otherwise minimal percentage of CD34+ cells, whereas no such treatment is needed if CD14+ cells are used as the starting population. Indeed, in most of recent clinical trials, mature antigen-presenting DCs are typically generated in vitro from peripheral blood monocytes by means of a two-step culture analogous to the in vivo maturation process. Immature DCs (*i*DCs) develop after a 6-d incubation of monocytes with GM-CSF and IL-4 ***(3,4)***. Direct effects of GM-CSF and IL-4 on monocytes are upregulation of CD1a ***(5)*** and downregulation of CD14 ***(6–8)***, respectively. *i*DCs are characterized by a strong ability to capture antigens. The best-characterized mechanisms of antigen internalization are phagocytosis by macrophages ***(9)*** and Langerhans cells and receptor-mediated endocytosis by B cells ***(10)***. In monocyte-derived DCs and Langerhans cells, the most efficient routes of antigen internalization have been attributed to endocytosis, which utilizes both macropynocitosis and adsorptive uptake via the macrophage mannose receptor (MR) and Fc receptors. The latter can easily be measured by uptake of fluorosceine isothiocyanate (FITC)–dextran ***(11)***. At this stage, DCs are relatively weak in priming T cells but can process and present superantigens to sensitized specific T cell clones. Complete differentiation to mature DCs (*m*DCs) is induced by lipopolysaccharide (LPS), contact allergens, bacteria and virus, TNF-α, IL-1β, prostaglandin E_2 (PGE_2), interferon (IFN)-α, immunostimulatory unmetylated CpG oligonucleotides, poly(I:C) ***(12–17)***, and signaling molecules ***(18,19)***. In addition, factors released from activated monocytes (monocyte conditioned medium [MCM]) can support the final stage of DC maturation ***(20–22)***. *m*DCs are characterized by loss of Ag-uptake ability and acquisition of strong antigen presentation. The transport and turnover of major histocompatability complex (MHC) class II is, in fact, developmentally regulated. In addition, DCs have the unique ability to present exogenous proteins in the MHC class I molecules, thus directly priming both CD4 helper and CD8 cytotoxic lymphocytes (cross-priming) ***(23)***.

Mature DCs are defined in vitro by their enhanced ability to induce an allogeneic CD4+ T-cell mixed lymphocyte reaction, morphologic features and upregulation of MHC class II and of the costimulatory molecules CD40 and CD86. The DC-restricted antigens p55, CD25, and CD83 also come up after the addition of maturation stimuli ***(21,24)***. Upregulation of MHC class II mainly results from redistribution of molecules from intracellular vesicular compartments to the cell surface. Physical changes occur as a result of rearrangements in actin filaments and microtubules during DC maturation ***(25)***. *i*DCs appear loosely adherent to the surface of culture dishes, whereas *m*DCs detach from the plastic surface, appears as veiled cells, because of sheet-like lamellipodia and form large floating bright class II aggregates.

Maturation of DCs may continue until they interact with T cells ***(18)*** through crosslinking of CD40, which upregulates further the expression of the costimulatory molecules CD80 and CD86 ***(26)*** and activate IL-12 synthesis ***(19,27,28)***.

Because many applications of DCs for clinical treatments are being investigated, it is crucial that methods used in disparate laboratories to mature DCs from blood be compared, so that the most appropriate cell population to be used for each is identified. We will present here our experience on the requirements for the generation of monocyte-derived DC-optimized for CD4+ T-cell activation.

2. Materials

2.1. DC Cultures

1. The leukocyte fraction (buffy coat) from one unit of human blood.
2. Washing medium: RPMI 1640 (1X) (Gibco cat. no. 61870-028), penicillin–streptomycin (final concentrations 100 U/mL and 100 ng/mL, respectively, or gentamycin 50 μg/mL) HEPES (25 m*M*) (Gibco cat. no. 42401-018), 2% heat-inactivated fetal calf serum (FCS).
3. Complete medium (CM) (*see* **Note 1**): RPMI 1640, penicillin–streptomycin, L-glutamine (final concentration 2 m*M*), gentamicin (final concentration 50 μg/mL), 2-mercaptoethanol (2-ME) (final concentration 5×10^{-5} *M*) sodium piruvate (from 100X stock; Gibco cat. no. 11360-039), nonessential amino acids (from 100X stock, Gibco cat. no. 11140-035), 10% heat-inactivated human AB sera (HS) from blood of at least four different blood bank donors. Do not add HEPES to CM. The above ingredients are filter sterilized (0.2 μm) before use, stored at 4°C, and used within 30 d. HS must be filtered through a 0.4-μm filter and used within 7 d.
4. Ficoll (Lymphoprep, 1.077 g/mL; Nycomed Pharma, Oslo, Norway).
5. Granulocyte–Macrophage colony-stimulating factor (Leucomax) from Novartis (Basel, Switzerland) and IL-4 from R & D Systems (Space Import-Export s.r.l., Milan, Italy).
6. Nucleopore polycarbonate membrane filters 0.4-μm; 13-mm diameter.
7. 15 mL and 50 mL Plastic tubes (Falcon plastics).
8. 6- and 96-Well culture dishes (Corning).

2.2. DC Surface Phenotyping

1. FITC- or phycoerythrin (PE)-conjugated MAbs anti-human CD14, CD80, CD86, CD83, MHC class II (HLA-Dr), CD25, CD11c, and isotype controls (from Becton Dickinson).

2. IF buffer: phosphate-buffered saline (PBS): 12 m*M* Na_2HPO_4; 12 m*M* NaH_2PO_4, 150 m*M* NaCl, pH 7.2.

2.3. DC Antigen Uptake and Antigen-Presentation Activity

1. FITC dextran (Sigma cat. no. FD 40S).
2. Gamma irradiation source.
3. Tetanic toxoid (TT) (from NIBSC, South Mimms, Potters Bar, Hertfordshire, UK).
4. Methyl ^{3}H–TdR (specific activity 5 Ci/m*M*) (NEN-Dupont, Milan Italy).
5. Enzyme-linked immunosorbent assay (ELISA) kits (Duoset, R&D, Minneapolis, MN).

3. Methods

3.1. Preparation of iDCs from Monocytes

1. Diluite blood 1:1 with saline and stratify on Ficoll at the ratio of 3:2. Do not exceed 8 mL (if using 15-mL tubes) or 30 mL (if using 50-mL tubes) of diluted blood.
2. Centrifuge at 500*g* for 30 min at room temperature with the brake off.
3. Transfer cells at the interface in 30 mL washing medium.
4. Centrifuge at 150*g* for 10 min (brake on) (*see* **Note 2**).
5. Discard the supernatant and add washing medium. Repeat twice from **step 4**.
6. Suspend PBMCs in CM at 2×10^6/mL and plate 3 mL in a six-well plate.
7. Incubate for 2 h at 37°C in a 5% CO_2 humified incubator.
8. Wash the wells twice with 3 mL prewarmed (37°C) CM by gently rotating the plate and collect the nonadherent cells in a 50-mL tube. Freeze an aliquot of these cells (*see* **Subheading 3.5.2., step 1**).
9. Add to enriched monocytes (*see* **Note 3**) 3 mL CM (*see* **Note 4**) supplemented with 1000 U/mL of GM-CSF and 1000 U/mL of IL-4 (*see* **Note 5**).
10. Fresh CM containing cytokines must be replaced on d 3.
11. On d 6, the *i*DCs can be collected after one washing with cold nonenzymatic cell dissociation solution (Sigma–Aldrich, Germany; C-1544 Steinheim) and incubated with the same medium for 5 min at 4°C (*see* **Note 6**).
12. Freeze an aliquot of these cells (*see* **Subheading 3.5.2., step 6**).

3.2. Preparation of mDCs

Following the priming phase of 6 d, the second step or "maturation phase" is induced by adding to the umperturbed original GM-CSF+IL-4 medium 30% (v/v) MCM (*see* **Note 7**).

3.2.1. Preparation of MCM

1. Add 4 mL of human γ-globulin (10 mg/mL, Cappel Labs, Organon Teknika, Westchester, PA) to the plates for 1 min.
2. Remove residual γ-globulin.
3. Wash the plates three times with PBS.

4. Layer 5 × 10^7 PBMC onto Ig-coated plates for 1 h in volumes of 6–8 mL CM.
5. Wash off nonadherent cells with gentle aspiration.
6. Incubate γ-globulin adherent cells in CM with 1% of autologous plasma at 37°C for 24 h.
7. Collect and freeze at –20°C prior to use.

After 2 d, *m*DCs are collected as suspension cells by three quick washings with prewarmed culture medium (*see* **Note 8**) and frozen in 2 × 10^6 aliquots (*see* **Subheading 3.5.2., step 6**).

3.3. Phenotype of iDCs and mDCs

1. Prepare 2 × 10^5 cells in 200 μL IF buffer for each surface marker to be tested. Also include the isotype control for each MAb.
2. Add the appropriate experimentally predetermined amount of MAbs.
3. Incubate on ice for 30 min and wash twice in IF.
4. Fix the cells with 1% (w/v) paraformaldehyde in PBS for 30 min at room temperature.
5. Perform flow cytometry analysis (*see* **Note 9**).

3.4. Analysis of the Endocytic Activity of iDCs and mDCs

1. Prepare in wells of a 24-well plate three different sets (A, B, C) of DCs, each of 5 × 10^5 cells in 3 mL RPMI 10% AB serum.
2. Keep culture B 1 h at 4°C.
3. Add FITC–dextran (0.5 mg/mL) in cultures B and C.
4. Incubate culture B at 4°C for 30 min.
5. Incubate cultures A and C at 37°C in 5% CO_2 for 30 min.
6. Transfer cells to 15-mL tubes and wash three times in washing medium plus 2 m*M* NaN_3 at 250*g* for 10 min each.
7. Resuspend in 0.5 mL PBS and analyze FITC–dextran-positive cells by flow cytometry.

3.5. Analysis of the Antigen-Presentation Activity of iDCs and mDCs

3.5.1. Presentation of Alloantingen

One-way mixed lymphocyte reaction (MLR) is performed as follows:

1. CD3+ T-cells are isolated from PBMCs of allogeneic donors by positive selection with magnetically labeled CD3 microbeads on a column placed in the magnetic field of a MACS separator, according to the manufacturer's instructions.
2. Immature DCs and *m*DCs are irradiated (3000 rad, ^{137}Cs source). Decreasing numbers of DCs are added to 1 × 10^5 T-cells to make 1 : 10, 1 : 100, and 1 : 1000 ratios. Cultures are set in triplicate wells of a 96-well cell culture plate in a volume of 200 μL CM and are incubated at 37°C, in 5% CO_2 humified atmosphere.

3.5.2. Presentation of Soluble Antigens by iDCs and mDCs

1. CD4+ T cells are isolated from nonadherent cells (*see* **Subheading 3.1.**) by positive selection with magnetically labeled CD4 microbeads.
2. CD4+ lymphocytes are cultured in CM at 1.5×10^6/mL in the presence of 10.2 μg/mL TT.
3. After 2 d, cultures are supplemented with IL-2 (20 UI/mL).
4. After 5 d, cells are washed and resuspended in a fresh CM containing 5.1 μg/mL TT.
5. After 2 d, cultures are supplemented with IL-2 (20 UI/mL).
6. After 5 d, autologous *i*DCs and *m*DCs are thawed and incubated in CM containing 20 μg/mL TT.
7. After 16 h at 37°C, DCs are washed, irradiated, and mixed with CD4+ cells, as in **Subheading 3.5.1.**

3.5.3. Proliferation Assay

1. After 4 d (alloantigen) or 6 d (TT), cultures are admixed with 2 μCi/mL of ^{3}H–TdR and incubated for 16 h before harvesting onto glass fiber filters.
2. ^{3}H–TdR uptake is evaluated by β-scintillation counting and expressed as counts per minute (cpm)/10^6 cells.
3. The net cpm is calculated as the cpm of cultures with factors /cpm of cultures without factors (*see* **Note 10**).

3.5.4. IFN-γ Release Assay

Triplicate cultures are prepared as detailed in **Subheading 3.5.1.**, and after 48 h, the supernatant is collected and IFN-γ release is assessed by ELISA assay.

4. Notes

1. Media. For a clinical purpose special media are suggested. A medium studied for DC separation is CellGro DC Medium (Cell Genix. Freiburg, Germany).
2. The low speed favors depletion of contaminant platelets, which produce TGF-β_1. In the mouse, TGF-β_1 has been shown to block DC maturation in vitro ***(29)***.
3. Although plastic adherence of PBMC is presently the more diffused method to get high numbers of monocytes from blood, other methods have been suggested, namely adherence followed by metrizamide gradient centrifugation ***(30)***, T-cell-negative selection by E-rosetting ***(20)*** or T+B+NK depletion by immunomagnetic beads of PBMCs ***(21)***, and positive selection of CD14+ cells. The latter can easily be done by magnetic cell sorting (MACS) ***(31)***. We routinely use a kit available from Miltenyi Biotec GmbH (Bergisch Gladbach, Germany), following the manufacturer's protocol. In our experience, recovery of CD14+ cells is 10% of loaded PBMCs and the purity is higher than 85%.
4. Because the recovery is 12–19% of the plated cells, this will make a cell density of (2–4) $\times 10^5$/mL. (Cell density in this priming phase should range from 3×10^5 to 5×10^5/mL.)

5. We have also started cultures in serum-free medium (with bovine insulin [sodium, Zn-free, 5 μg/mL], human [holo] transferrin [5 μg/mL], and sodium selenite [5 ng/mL] [Redu-Ser™, UBI, Lake Placid, NY] added) to test the effect of prolactin (PRL) alone or in conjunction with either GM-CSF or IL-4. The study was limited to generation of *i*DCs. Increased expression of MHC class II, CD80 and CD86 as well stimulatory activity on allogeneic T-lymphocytes at a DC:T ratio of 1:10 were observed with the combination of GM-CSF and near-physiological (12–25 ng/mL) concentrations of human recombinant (Genzyme) PRL ***(32)***. In these serum-free conditions, the response induced by GM-CSF + PRL was of the same magnitute of that induced by GM-CSF (200 U/mL) + IL-4 (200 U/mL).
6. Most *i*DCs are still present as adherent cells.
7. Monocyte-conditioned medium can be replaced by a TNF-α + IL-1 + IL-6 + PGE_2 cocktail ***(33,34)***.
8. Mature DCs are characterized by the complete loss of adherence.
9. CD14 decreases from *i*DC to *m*DC (12% to 4%), whereas CD80 and CD86 almost double their expression on *m*DC (40–50% to 95% for both populations). CD83 is virtually absent on *i*DC and increases to 85% on *m*DC. Another striking upmodulation is observed with the CD25 marker (1.6% vs 92%).
10. Stimulation indexes of *m*DCs in allogeneic MLR should be higher than 20 at a DC:T cell ratio of 1:1000, compared to 10:1 of *i*DCs.

Acknowledgments

The Department of Internal Medicine of the University of Turin and the Department of Oncology of the Molinette Hospital of Turin are supported by Italian Ministry of Research, the Italian Ministry of Health, Regional Foundations, and private agencies.

References

1. Banchereau, J. and Steinman, R. (1998) Dendritic cells and the control of immunity. *Nature* **392,** 245–252.
2. Peters, J. H., Gieseler, R. K., Thiele, B., and Steinbach, F. (1996) Dendritic cells: from ontogenic orphans to myelomonocytic descendants. *Immunol. Today* **17,** 273–278.
3. Sallusto, F. and Lanzavecchia, A. (1994) Efficient presentation of soluble antigen by cultured human dendritic cells is maintained by granulocute–macrophage colony stimulating factor plus interleukin 4 and downregulated by tumor necrosis factor alpha. *J. Exp. Med.* **179,** 1109.
4. Romani, N., Gruner, S., Brang, D., Kampgen, E., Lenz, A., Trockenbacher, B., et al. (1994) Proliferating dendritic cell progenitors in human blood. *J. Exp. Med.* **180,** 83–93.
5. Kasinrerk, W., Baumruker, T., Majdic, O., Knapp, W., and Stockinger, H. (1993) CD1 molecule expression on human monocytes induced by granulocytes–macrophage colony-stimulating factor. *J. Immunol.* **150,** 579–584.

6. Peters, J. H., Rupper, J., Gieseler, R. K., Najar, H. M., and Xu, H. (1991) Differentiation of human monocytes into CD14 negative accessory cells: do dendritic cells derive from the monocytic line-age? *Pathobiology* **59,** 122–126.
7. Ruppert, J., Friedrichs, D., Xu, H., and Peters, J. H. (1991) IL-4 decreases the expression of monocyte differentiation marker CD14, paralleled by an increasing accessory potency. *Immunobiology* **182,** 449–464.
8. Lauener, R. P., Goyert, S. M., Geha, R. S., and Vercelli, D. (1990) Interleukin-4 downregulates the expression of CD14 in normal human monocytes. *Eur. J. Immunol.* **20,** 2375–2381.
9. Reis e Sousa, C., Stahl, P. D., and Austyn, J. M. (1993) Phagocytosis of antigen by Langerhans cells in vitro. *J. Exp. Med.* **178,** 509–519.
10. Lamaze, C. and Schmidt, S. L. (1985) The emergence of clathrin-independent pinocytic pathways. *Curr. Opin. Cell Biol.* **7,** 573–580.
11. Sallusto, F., Cella, M., Danili, C., and Lanzavecchia, A. (1995) Dendritic cells use macropinocytosis and the mannose receptor to concentrate macromolecules in the major histocompatibility complex class II compartment: downregulation by cytokine and bacterial products. *J. Exp. Med.* **182,** 389–400.
12. MacPherson, G. G., Jenkins, C. D., Stein, M. J., and Edwards, C. (1995) Endotoxin-mediated dendritic cell release from the intestine. Characterization of released dendritic cells and TNF dependence. *J. Immunol.* **154,** 1317–1322.
13. De Smedt, T., Pajak, B., Muraille, E., Lespagnard, L., Heinen, E., De Baetselier, P., et al. (1996) Regulation on dendritic cell numbers and maturation by lipopolysaccharide in vivo. *J. Exp. Med.* **184,** 1413–1424.
14. Rescigno, M., Citterio, S., Thery, C., Rittig, M., Medaglini, D., Pozzi, G., et al. (1998) Bacteria-induced neo-biosynthesis, stabilization, and surface expression of functional class I molecules in mouse dendritic cels. *Proc. Natl. Acad. Sci. USA* **95,** 5229–5234.
15. Sparwasser, T., Koch, E. S., Vabulas, R. M., Heeg, K., Lipford, G. B., Ellwart, J. W., et al. (1998) Bacterial DNA and immunostimulatory CpG oligonucleotides trigger maturation and activation of murine dendritic cells. *Eur. J. Immunol.* **28,** 2045–2054.
16. Cella, M., Salio, M., Sakakibara, Y., Langen, H., Julkunen, I., and Lanzavecchia, A. (1999) Maturation, activation and protection of dendritic cells induced by double-stranded RNA. *J. Exp. Med.* **189,** 821–829.
17. Verdijk, R. M., Mutis, T., Esendam, B., Kamp, J., Melief, C. J. M., Brand, A., et al. (1999) Polyriboinosinic polyribocytidylic acid (poly(I : C)) induces stable maturation of functionally active human dendritic cells. *J. Immunol.* **163,** 57–61.
18. Caux, C., Massacrier, C., Vanbervliet, B., Dubois, B., Van Kooten, C., Durand, I., et al. (1994) Activation of human dendritic cells through CD40 cross-linking. *J. Exp. Med.* **180,** 1263–1272.
19. Cella, M., Scheideffer, D., Palmer-Lehmann, K., Lane, P., Lanzavecchia, A., and Alber, G. (1996) Ligation of CD40 on dendritic cells triggers production of high levels of interleukin-12 and enhances T cell stimulatory capacity. *J. Exp. Med.* **184,** 747–752.

20. Bender, A., Sapp, M., Schuler, G., Steinman, R. M., and Bhardwaj, N. (1996) Improved methods for the generation of dendritic cells from nonproliferating progenitors in human blood. *J. Immunol. Methods* **196,** 121–135.
21. Romani, R., Reider, D., Heuer, M., Ebner, S., Kampgen, E., Eibl, B., et al. (1996) Generation of mature dendritic cells from human blood: an improved method with special regard to clinical applicability. *J. Immunol. Methods* **196,** 137–151.
22. Reddy, A., Sapp, M., Feldman, M., Subklewe, M., and Bhardwaj, N. (1997) A monocyte conditioned medium is more effective than defined cytokines in mediating the terminal maturation of human dendritic cells. *Blood* **90,** 3640–3636.
23. Ridge , J. P., Di Rosa, F., and Matzinger, P. (1998) A conditioned dendritic cell can be a temporal bridge between a CD4+ T-helper and a T-killer cell. *Nature* **393,** 474–478.
24. Zhou, L. J. and Tedder, T. F. (1996) CD14+ blood monocytes can differentiate into functionally mature CD83+ dendritic cells. *Proc. Natl. Acad. Sci. USA* **93,** 2588–2592.
25. Winzler, C., Rovere, P., Rescigno, M., Citterio, S., Granucci, F., Mutini, C., et al. (1997) Maturation steps of mouse dendritic cells in growth-factor dependent long-term cultures. *J. Exp. Med.* **185,** 317–328.
26. Lanier, L. L., O'Fallon, S., Somoza, C., Philips, J. H., Linsley, P. S., Okumura, K., et al. (1995) CD80 (B7) and CD86 (B70) provide similar costimulatory signals for T cell proliferation, cytokine production, and generation of CTL. *J. Immunol.* **154,** 97–105.
27. Macatonia, S. E., Hosken, N. A., Litton, M., Vieira, P., Hsieh, C. S., Culpepper, V. A., et al. (1995) Dendritic cells produce IL-12 and direct development of Th1 cells from naive CD4+ T cells. *J. Immunol.* **154,** 5071–5079.
28. Koch, F., Stanzl, U., Jennewien, P., Janke, K., Heufler, C., Kampgen, E., et al. (1996) High level IL-12 production by murine dendritic cells: upregulation via MHC class II and CD40 molecules and downregulation by IL-4 and IL-10. *J. Exp. Med.* **184,** 741–747.
29. Yasunori, Y., Tsumura, H., Miwa, M., and Inaba, K. (1997) Contrasting effects of TGF-β_1 and TNF-α on the development of dendritic cells from progenitors in mouse bone marrow. *Stem Cells* **15,** 144–153.
30. Tedder, T. F. and Jansen, P. J. (1998) Isolation and generation of human dendritic cells, Unit 7.32, in *Current Protocols of Immunology*, Wiley Interscience, New York.
31. Pickl, W. F., Majdic, O., Kohl, P., Stockl, J., Riedl, E., Scheinecker, C., et al. (1996) Molecular and functional characteristics of dendritic cells generated from highly purified CD14+ peripheral blood monocytes. *J. Immunol.* **157,** 3850–3859.
32. Matera, L., Galetto, A., Geuna, M., Vekemans, K., Ricotti, E., Contarini, M., et al. (2000) Individual and combined effect of GM-CSF and prolactin on maturation of dendritic cells from blood monocytes under serum-free conditions. *Immunology* **100,** 29–36.
33. Rieser, C., Bock, G., Klocker, H., Bartsch, G., and Thurnher, M. (1997) Prostaglandin E2 and tumor necrosis factor-α cooperate to activate human dendritic

cells: synergistic activation of interleukin 12 production. *J. Exp. Med.* **186,** 1603–1608.
34. Jonuleit, H., Kuhn, U., Muller, G., Steinbrink, K., Paragnik, L., Schmitt, E., et al. (1997) Pro-inflammatory cytokines and prostaglandins induce maturation of potent immunostimulatory dendritic cells under fetal calf serum-free conditions. *Eur. J. Immunol.* **27,** 3135–3142.

33

Generation of Human Type 1- and Type 2-Polarized Dendritic Cells from Peripheral Blood

Pawel Kalinski, Pedro Vieira, Joost H. N. Schuitemaker, Quan Cai, and Martien Kapsenberg

1. Introduction

Dendritic cells (DCs) are the most potent inducers of immune responses and potent regulators of immunity. They act as sentinel cells in the peripheral tissues, playing a key role in the development of effective immune responses to different types of pathogens *(1)*. At the same time, DC dysfunction may play a pathogenic role in different diseases, ranging from autoimmunity to chronic infections and cancer, and multiple pathogens developed ways to interfere with DC functions as a mean to avoid eradication by the immune system and to enhance their own survival within infected hosts *(2)*.

Both the efficiency of DC as an effective element of immune system and their susceptibility to pathogen-induced dysfunction result from an enormous plasticity of the DC system. Distinct DC subpopulations, such as mouse $CD8\alpha^-/CD11b^+$ vs $CD8\alpha^+/CD11b^-$ DCs, human plasmacytoid vs myeloid DCs, DCs isolated from different tissues, or DCs generated under different conditions, all show striking functional differences. One aspect of DC function that is subject to strict regulation is the ability of DCs to produce interleukin (IL)-12 and to induce Th1-polarized or Th2-polarized effector $CD4^+$ Th-cells *(3–6)*.

The ability of DCs to act both as the inducers of immune responses and as immunoregulatory cells led to the interest in their immunotherapeutic use in different disease types, ranging from cancer to autoimmunity, and as a tool to prevent the rejection of transplanted tissues and organs. Taking into account the plasticity of DCs and their ability to adopt different functions, it may

From: *Methods in Molecular Biology, vol. 215:*
Cytokines and Colony Stimulating Factors: Methods and Protocols
Edited by: D. Körholz and W. Kiess © Humana Press Inc., Totowa, NJ

be important to match the desired type of DC to the type of its clinical or laboratory application.

Recently, several groups, including ours, have demonstrated the feasibility of obtaining monocyte-derived DCs with different functions, by modulating the conditions of their early development ***(7)***, the conditions of their maturation ***(8–15)***, or the length of the maturation period of DCs ***(16,17)***. This chapter provides a detailed protocol of generating type-1- and type-2-polarized DCs, the protocols used to test the cytokine-producing capacity of these cells, and their ability to induce Th1-type responses in naive $CD4^+$ Th-cells (*see* **Fig. 1**).

2. Materials

2.1. Isolation of Peripheral Blood Monocytes and $CD4^+CD45RA^+$ Naive Th-Cells

1. Vacutainer blood collection tubes (sodium heparine; Becton-Dickinson, Franklin Lakes, NJ).
2. 50-mL Polypropylene tubes.
3. 10-mL Polypropylene tubes.
4. Lymphoprep (Nycomed, Torshov, Norway) (density (*d*) = 1.077).
5. Percoll (Pharmacia, Uppsala, Sweden) is aliquoted (30 mL) and stored at 4°C.
6. 10X Concentrated "acidic" (pH 4.6; 1.051 g/mL) phosphate-buffered saline (PBS), prepared from Sigma reagents:
   NaCl (13.5 g);
   Na_2HPO_4 (0.1 g [corresponding to 0.125 g of $Na_2HPO_4 \cdot 2H_2O$]);
   KH_2PO_4 (2.1 g);
   Distilled water: 200 mL.
   The PBS solution is sterilized by 0.22-µm filtration and stored at 4°C in 4-mL aliquots.
7. Medium for Percoll separation: Iscove's modified Dulbecco's medium (IMDM) (Biowhittaker, Walkersville, MD) with 10% fetal calf serum (FCS) (Hyclone, Logan, UT).
8. Medium for washing the cells: Hank's balanced salt solution (HBSS) (Gibco Life Technologies, Rockville, MD) with 2% FCS (Hyclone).
9. Isolation columns for human $CD4^+CD45RA^+$ naive Th-cells. We have been successfully using (1) a $CD4^+CD45R0^-$ negative isolation columns from R&D and (2) a customized StemSep system for the negative isolation of $CD4^+CD45R0^-$ cells (Stem Cell Technologies).

2.2. Generation of Immature DCs and DC Maturation in DC1- and DC2-Polarizing Conditions

1. Medium for DC culture: IMDM (Biowhittaker) with 10% FCS (Hyclone).
2. Medium for washing the cells: 2% FCS/HBSS.
3. rhuGM-CSF (a gift from Schering-Plough [currently Novartis], Uden, The Netherlands).

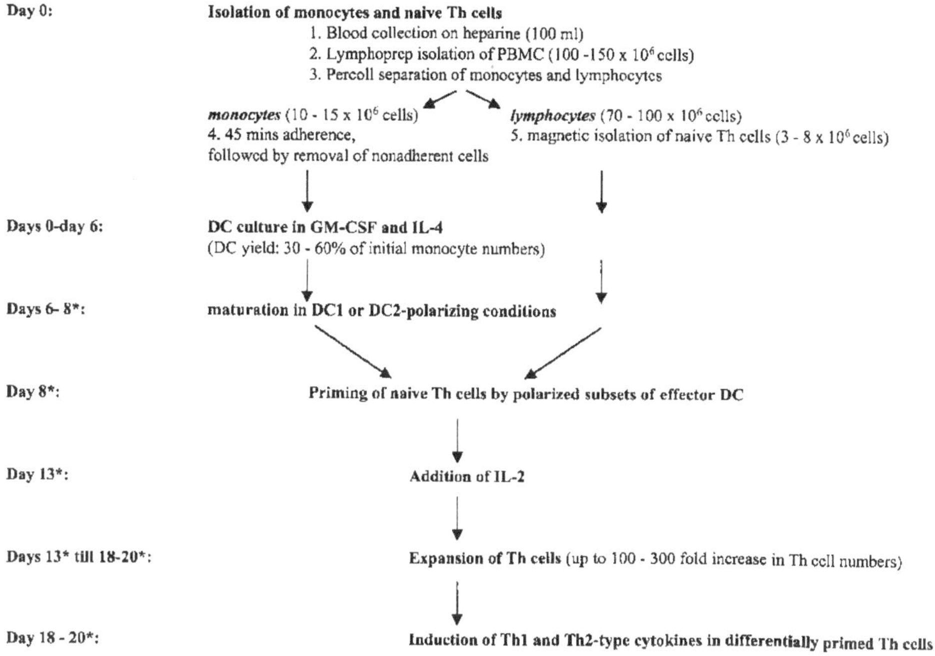

Fig. 1. The sequence of events during a typical experiment addressing the induction of Th1 and Th2 cells by polarized subsets of DCs. The numbers in brackets represent typical yields of the cells at different stages of an experiment utilizing 100 mL of fresh blood. *Typically, our experiments involve 48-h maturation of DCs, but the duration of this period may be modified, depending on the application.

4. Recombinant human interleukin-4 (rhuIL 4) (Strathmann Biotech Gmbh, Hannover, Germany).
5. Recombinant human tumor necrosis factor-α (rhuTNF-α) (Strathmann).
6. Recombinant human interleukin-1β (rhuIL-1β) (Strathmann).
7. Lipopolysaccharide (LPS) (from *Escherichia coli* 011:B4; Sigma, St. Louis, MO).
8. rhuIFN-γ (Strathmann).
9. Prostaglandin E_2 (PGE_2) (Sigma, St. Louis, MO).

2.3. Analysis of Cytokine Production by Differentially Polarized DCs

1. CD40L-transfected J558 cells were a kind gift from Dr. Peter Lane (University of Birmingham, Birmingham, UK). They express high levels of mouse CD40L that binds both mouse and human CD40.
2. sCD40LT (a gift of Immunex, Seattle, WA).
3. Human $CD4^+$ Th-cells (bulk population) are isolated by one of the two methods: (1) negative magnetic isolation StemSep columns (Stem Cell Technologies,

Vancouver, Canada); (2) CD4-specific Dynabeads/Detach-a-Bead system (Dynal AS, Oslo, Norway).
4. SEB (Sigma) is used as an Ag surrogate.

2.4. In Vitro Priming of $CD4^+CD45RA^+$ Naive Th-Cells with Polarized DC Subsets

1. Staphylococcal enterotoxin B (SEB) (Sigma).
2. rhuIL-2 (10 U/mL; a gift of Cetus Corp., Emeryville, CA).
3. CD3 MAb (CLB-T3/3; CLB, Amsterdam, The Netherlands) plus CD28 MAb (CLB-CD28/1; CLB), was used to induce the cytokine production in differentially primed populations of Th-cells.

3. Methods

3.1. Isolation of Peripheral Blood Monocytes and $CD4^+CD45RA^+$ Naive Th-Cells

3.1.1. Collection of Peripheral Blood

Collect blood into heparinized tubes and dilute 1 : 1 with HBSS.

3.1.2. Isolation of Peripheral Blood Monocyte Cells (PBMCs) on Lymphoprep

1. Overlay 30 mL of diluted blood over 15 mL of Lymphoprep in each 50-mL tube.
2. Centrifugate at 1000*g* for 30 min, at room temperature (RT; 21°C). Acceleration 1*g* to 1000*g* should take 60 s. Deceleration: 5 min.
3. Wash the cells twice at RT.

3.1.3. Isolation of the Light Fraction of PBMCs on Percoll Gradient

1. Prepare standard isotonic Percoll solution (SIP) by mixing nine parts of Percoll with one part of 10X concentrated "acidic" PBS.
2. Prepare three dilutions of SIP (v/v) in 10% FCS/IMDM (*see* **Notes 1–4**):
   a. 60% SIP (9 mL) + 40% FCS/IMDM (6 mL)
   b. 48% SIP (9.6 mL) + 52% FCS/IMDM (10.4 mL)
   c. 34% SIP (3.4 mL) + 66% FCS/IMDM (6.6 mL).
3. Suspend PBMCs (maximum 3×10^7 cells/mL) in 60% SIP. Layer 2–2.5 mL of cell suspension at the bottom of each 10-mL tube (max. 7.5×10^7 cells/tube) and overlay with 48% SIP (5 mL) and then with 34% SIP (2 mL).
4. Centrifugation: 2400*g*, 45 min at RT (21°C). Acceleration: 60 s; deceleration: 5 min.
5. Harvest monocytes from the upper interphase (the interphase corresponding to 48% [or 45%] SIP and 34% SIP) and lymphocytes from the lower interphase (60% SIP and to 48% [or 45%] SIP.
6. Wash the monocyte fraction three times and count the cells (*see* **Note 5**).

3.1.4. Adherence and Depletion of Nonadherent Cells

1. Seed the cells at 0.5×10^6/mL/well in a 24-well plate (or 2×10^6 in 4 mL in 6-well plate) and let them adhere for 45 min, 37°C, 5% CO_2 (*see* **Note 6**).
2. Remove nonadherent cells by washing the wells two to three times with a gentle stream of medium. This step requires eye control of the washing to assure high purity of monocytes and to prevent an excessive loss of the attached cells. Use washing medium at room temperature.

3.1.5. Isolation of $CD4^+CD45RA^+$ Naive Th-Cells from Peripheral Blood

1. Harvest the lymphocytes from the heavy fraction of PBMCs (*see* **Subheading 3.1.3.**) and wash two times.
2. Isolate naive Th-cells ($CD4^+CD45RA^+$ cells) by one of the negative selection systems (*see* **Subheading 2.1.**), according to the manufacturers' instructions.
3. Freeze the isolated naive Th-cells until use.

3.2. DC Culture and Maturation

1. After the last wash of adherent cells, add fresh IMDM with 10% FCS, containing 500 U/mL GM-CSF and 250 U/mL IL-4 (1 mL per well).
2. On d 3 of the culture, replace one-half of medium and add the same amount of fresh medium with the double-concentrated growth factors. At this time-point, a portion of the cells are already nonadherent, so it is necessary to let them sediment for 10 min, resting the plate at a certain angle, supported at one side. Gently, to avoid taking up the cells, take up 0.5 mL of medium with a 1-mL pipet from the lower side of each well. Add the new medium with double-concentrated GM-CSF and IL-4 (prewarmed) at the same spot, releasing the volume gently to reduce stirring up the cultures.
3. At d 6 (*see* **Note 7**), take out one-half of the spent medium and add new medium containing GM-CSF, a double-concentrated maturation-inducing factor without or with a polarizing factor. The standard maturation-inducing stimuli used in our studies are TNF-α (final conc. 50 ng/mL) in combination with with IL-1β (final conc. 25 ng/mL), or LPS (final conc. 250 ng/mL; *see* **Note 8**). As a type-1-polarizing factor, we use IFN-γ (final conc. 1000 U/mL, although even the concentrations of 1–10 U/mL have a distinct effect). PGE_2 (10^{-9}–10^{-6} *M*) shows a dose-dependent type-2 DC polarizing effect ***(10,14)***. Within 2 d, the expression of CD80 and CD86 will increase, and the cells will lose the ability to readhere, after moving to another well (*see* **Note 9**). At the very early stage of maturation (6–12 h), the cells become $CD83^+$ and lose the expression of CD115.

3.3. Analysis of Cytokine Production by Differentially Polarized DCs

1. Harvest DCs to polypropylene tubes and wash thoroughly to remove all of the cytokines.

2. Plate the cells at 2 × 10^4 cells/well in flat-bottomed 96-well plates.
3. Add the IL-12-inducing stimulus. We normally use three types of CD40L-based stimulus: J558–CD40L (5 × 10^4 cells/well), soluble rCD40LT (1 g/mL) in combination with rhuIFN (1000 U/mL), or $CD4^+$ T-cells (1 × 10^5 cells/well) in the presence of superantigen (SEB; 1 ng/mL). The stimulation is performed in a final volume of 200 µL/well (*see* **Note 10**).
4. Following either of the first two modes of stimulation, we harvest 24 h supernatants, whereas the T-cell-dependent IL-12p70 induction requires a longer 48-h stimulation.

3.4. In Vitro Priming of $CD4^+$ CD45RA Naive Th-Cells with Polarized DC Subsets

1. Harvest DCs to polypropylene tubes and wash thoroughly to remove all of the cytokines.
2. Plate the cells at 2 × 10^4 cells/well in flat-bottomed 96-well plates. Add SEB (1 ng/mL) and naive Th-cells (2 × 10^4/well; *see* **Subheading 3.1.** for the isolation procedure).
3. At d 5, add rhuIL-2 (final conc. 10 U/mL).
4. Starting from this point on the cells, proliferate rapidly over the period of next 4–6 d. The next day after the IL-2 addition, the cells usually need to be transferred to 1-mL wells. Subsequently, every 1–3 d, each well needs to be divided into two to three wells. At this point, the optimal culture density for the expansion of Th-cells is (1.5–3 × 10^6 cells/well (1 mL). The cultures reach quiescence about d 9–12 and need to be restimulated (*see* **Note 11**).
5. At 10–14 d after priming, induce the cytokine production in Th-cells by their restimulation for 24 h with CD3 MAb (1 µg/mL; Central Laboratory of the Netherlands' Blood Transfusion Service (CLB)–T3/3; CLB) plus CD28 MAb (1 µg/mL; CLB–CD28/1; CLB), as described *(7)*. The levels of IFN-γ, IL-4, and IL-5 in 24-h supernatants can be then analyzed by specific enzyme-linked immunosorbent assays (ELISAs). Alternatively, the differentially primed Th-cells can be stimulated for 6 h with ionomycine and phorbol myristate acetate (PMA) in the presence of Brefeldin A, followed by their fixation, permeabilization, and staining for the production of Th1- and Th2-type cytokines at the single-cell level *(18)*.

4. Notes

Our protocol of DC generation is based on the methods of Peters et al. ***(19)*** and Sallusto et al. ***(20)***. Our major modifications are the use of the double-step density-gradient isolation of monocytes, followed by a short-term adherence, the use of IMDM supplemented with 10% FSC, and the replacement of a half of the culture medium at d 3 and d 6 of DC culture. The described procedure of monocyte isolation allows a substantially higher yield and purity of the cells, compared to the isolation of monocytes by adherence alone. The rationale behind the use of IMDM is the superior performance of this medium in T-cell culture. We chose to use FCS-supplemented medium, rather than human serum, because, in our hands, DCs obtained in the presence of human serum do not express CD1a and they show a relative resistance to maturation. FCS/IMDM-generated DCs are also superior to the cells obtained in several types of serum-free medium, including AIM-V, which all result in a substantially lower yield of DCs, and allow only marginal production of IL-12p70. In addition, AIM-V medium is nonpermissive for the IFN-γ-induced type-1 DC polarization (Cai and Kalinski, manuscript in preparation) and carries the requirement for such factors as IL-6 and PGE_2 to achieve a complete maturation of DCs (***21***; Cai and Kalinski, unpublished).

The SEB-based model of naive Th cell priming was first described in **ref.** ***7***. It is based on the ability of SEB to activate a substantial proportion of naive T cells ***(22)***. This allows one to use it as a substitute of the T cell receptor (TCR)-transgenic models that are not available in the human system. In contrast, the traditional allogeneic mixed leukocyte reaction (MLR) model does not allow us to induce any detectable amounts of IL-12 within the first 3 d of DC–Th–cell interaction, most likely because of 100fold to 1000fold lower frequency of responsive T cells. The possible applications and the typical results obtained with use of the described protocols can be found in our previous publications ***(7,10–12,14,18,23–25)***.

Based on past experience with introducing the described protocol in several other labs, we would like to draw the attention of the reader to the following issues critical for its outcome.

1. Monocytes isolated from fresh blood give better results than monocytes isolated from buffy coats that often yield a lower percentage of $CD1a^+$ cells. In addition, DCs generated from buffy-coat-isolated monocytes frequently show signs of partial maturation (loss of CD115) and tend to produce lower amounts of IL-12p70. They are also less susceptible to polarization. The reason(s) for these differences is not completely clear to us, but the quality of DCs appears to inversely correlate with the level of platelet contamination that is substantially higher in the case of the monocytes isolated from the buffy coat, compared to fresh blood.
2. Isolation of monocytes should be performed at room temperature. Rapid changes of temperature increase the risk of monocyte activation and clumping. We advise the use of polypropylene to reduce cell attachment.
3. We also recommend the use of heparine as anticoagulant to avoid activation of monocytes in the course of decalcification/recalcification. Use Ca^{2+}/Mg^{2+}-containing media at all stages of the monocyte isolation.

4. A 48% layer of SIP is designed for freshly drawn blood. A lower-density layer of SIP (45%) should be used for the isolation of monocytes from buffy coats.
5. At this stage, the monocytes should be 80–90% pure (judged by CD14 expression). Higher contamination with $CD14^-$ cells indicates the need to reduce the concentration of SIP in the middle layer.
6. Do not exceed the starting cell density of 0.5×10^6 cells/1 mL of culture medium. Consider reducing it to 0.4×10^6 if the CD1a expression is poor. Generally, the lower the starting density of the monocytes, the higher the purity of resulting $CD1a^+$ DCs, although very low-density cultures result in a poor recovery of DCs (as a percentage of the plated monocytes).
7. At d 6, the cultures contain up to 90% $CD1a^+CD115^+$ immature DCs. They are expressing low to intermediate levels of CD80 and CD86 and lack CD83 expression. Poor CD1a expression may indicate (1) too high initial density of monocytes at the beginning, (2) poor batch of serum/medium (*see* **Notes 12–14**), and (3) poor MAb (in our hands, OKT6 proved superior to several other CD1a MAbs). It may also suggest poor activity of the IL-4 used and the need to increase its concentration.
8. Optimal type-1 polarization of DC requires full maturation of DC and is less pronounced in DC preparations that do not undergo full CD83 conversion. IFN-γ and the maturation-inducing factor should be administered simultaneously. Pretreatment of DC with either of the factors alone reduces the ability of DC to produce IL-12p70 after subsequent stimulation.
9. Although our standard protocol of generation of polarized effector DCs involves a 48-h maturation stage, the reader may consider reducing the maturation/polarization time, depending of the DC application.
10. Bacterial products, such as LPS or *Staphylococcus aureus* Cowan (SAC) (alone or in combination with IFN-γ), are effective inducers of IL-12p70 production in immature ($CD83^-$) DCs, but not in mature DCs. CD40L stimulation remains effective in mature DCs, although mature DCs show impaired responsiveness to the IL-12p70-enhancing action of IFN-γ ***(16)***.
11. The proliferation of Th-cells is very susceptible to the temperature changes, especially within the first 5 d of culture. To optimize the yield of the differentially primed Th-cells, try to minimize the length of time when the cells are outside of the incubator and use prewarmed medium to dilute the cultures.
12. A batch of FCS is important. We observed strong differences between several different batches of FCS in their ability to support the generation of $CD1a^+$ DCs.
13. To our surprise, the source of IMDM can make a difference as well, affecting the expression of CD1a.
14. We advocate using disposable plastic tubes, media flasks, and pipets to reduce the chance of endotoxin contamination at the onset of cultures.
15. Although difficult to avoid for some applications, γ-irradiation impairs the ability of DC to produce IL-12. Typically, the IL-12p70 production by 2500 R-irradiated DCs is only 15–25% compared to nonirradiated DCs.

Acknowledgments

This work was supported by NCI grant 1RO1CA82016, by the 2000 Kathy Salling Memorial Melanoma Research Foundation Award, and by the 2001 Pittsburgh Foundation Award (to PK).

References

1. Banchereau, J. and Steinman, R. M. (1998) Dendritic cells and the control of immunity. *Nature* **392,** 245–252.
2. Bhardwaj, N. (1997) Interactions of viruses with dendritic cells: a double-edged sword. *J. Exp. Med.* **186,** 795–799.
3. Kalinski, P., Hilkens, C. M., Wierenga, E. A., and Kapsenberg, M. L. (1999) T-cell priming by type-1 and type-2 polarized dendritic cells: the concept of a third signal. *Immunol. Today* **20,** 561–567.
4. Moser, M. and Murphy, K. M. (2000) Dendritic cell regulation of TH1–TH2 development. *Nat. Immunol.* **1,** 199–205.
5. Pulendran, B., Palucka, K., and Banchereau, J. (2001) Sensing pathogens and tuning immune responses. *Science* **293,** 253–256.
6. Liu, Y. J., Kanzler, H., Soumelis, V., and Gilliet, M. (2001) Dendritic cell lineage, plasticity and cross-regulation. *Nat. Immunol.* **2,** 585–589.
7. Kalinski, P., Hilkens, C. M., Snijders, A., Snijdewint, F. G., and Kapsenberg, M. L. (1997) IL-12-deficient dendritic cells, generated in the presence of prostaglandin E2, promote type 2 cytokine production in maturing human naive T-helper cells. *J. Immunol.* **159,** 28–35.
8. De Smedt, T., Van Mechelen, M., De Becker, G., Urbain, J., Leo, O., and Moser, M. (1997) Effect of interleukin-10 on dendritic cell maturation and function. *Eur. J. Immunol.* **27,** 1229–1235.
9. McRae, B. L., Semnani, R. T., Hayes, M. P., and van Seventer, G. A. (1998) Type I IFNs inhibit human dendritic cell IL-12 production and Th1 cell development. *J. Immunol.* **160,** 4298–4304.
10. Kalinski, P., Schuitemaker, J. H., Hilkens, C. M., and Kapsenberg, M. L. (1998) Prostaglandin E2 induces the final maturation of IL-12-deficient CD1a+ CD83+ dendritic cells: the levels of IL-12 are determined during the final dendritic cell maturation and are resistant to further modulation. *J. Immunol.* **161,** 2804–2809.
11. Vieira, P. L., Kalinski, P., Wierenga, E. A., Kapsenberg, M. L., and de Jong, E. C. (1998) Glucocorticoids inhibit bioactive IL-12p70 production by in vitro-generated human dendritic cells without affecting their T cell stimulatory potential. *J. Immunol.* **161,** 5245–5251.
12. de Jong, E. C., Vieira, P. L., Kalinski, P., and Kapsenberg, M. L. (1999) Corticosteroids inhibit the production of inflammatory mediators in immature monocyte-derived DC and induce the development of tolerogenic DC3. *J. Leukoc. Biol.* **66,** 201–204.

13. Vanderheyde, N., Verhasselt, V., Goldman, M., and Willems, F. (1999) Inhibition of human dendritic cell functions by methylprednisolone. *Transplantation* **67,** 1342–1347
14. Vieira, P. L., de Jong, E. C., Wierenga, E. A., Kapsenberg, M. L., and Kalinski, P. (2000) Development of Th1-inducing capacity in myeloid dendritic cells requires environmental instruction. *J. Immunol.* **164,** 4507–4512.
15. Gagliardi, M. C., Sallusto, F., Marinaro, M., Langenkamp, A., Lanzavecchia, A., and De Magistris, M. T. (2000) Cholera toxin induces maturation of human dendritic cells and licences them for Th2 priming. *Eur. J. Immunol.* **30,** 2394–2403.
16. Kalinski, P., Schuitemaker, J. H., Hilkens, C. M., Wierenga, E. A., and Kapsenberg, M. L. (1999) Final maturation of dendritic cells is associated with impaired responsiveness to IFN-gamma and to bacterial IL-12 inducers: decreased ability of mature dendritic cells to produce IL-12 during the interaction with Th cells. *J. Immunol.* **162,** 3231–3236.
17. Langenkamp, A., Messi, M., Lanzavecchia, A., and Sallusto, F. (2001) Kinetics of dendritic cell activation: impact on priming of Th1, Th2 and nonpolarized T cells. *Nat. Immunol.* **1,** 311–316.
18. Kalinski, P, Smits, H. H., Schuitemaker, J. H., Vieira, P. L., van Eijk, M., de Jong, E. C., et al. (2000) IL-4 is a mediator of IL-12p70 induction by human Th2 cells: reversal of polarized Th2 phenotype by dendritic cells. *J. Immunol.* **165,** 1877–1881.
19. Peters, J. H., Xu, H., Ruppert, J., Ostermeier, D., Friedrichs, D., and Gieseler, R. K. (1993) Signals required for differentiating dendritic cells from human monocytes in vitro. *Adv. Exp. Med. Biol.* **329,** 275–280.
20. Sallusto, F. and Lanzavecchia, A. (1994) Efficient presentation of soluble antigen by cultured human dendritic cells is maintained by granulocyte/macrophage colony-stimulating factor plus interleukin 4 and downregulated by tumor necrosis factor alpha. *J. Exp. Med.* **179,** 1109–1118.
21. Jonuleit, H., Kuhn, U., Muller, G., Steinbrink, K., Paragnik, L., Schmitt, E., et al. (1997) Pro-inflammatory cytokines and prostaglandins induce maturation of potent immunostimulatory dendritic cells under fetal calf serum-free conditions. *Eur. J. Immunol.* **27,** 3135–3142.
22. Fraser, J. D. (1989) High-affinity binding of staphylococcal enterotoxins A and B to HLA-DR. *Nature* **339,** 221–223.
23. Hilkens, C. M., Kalinski, P., de Boer, M., and Kapsenberg, M. L. (1997) Human dendritic cells require exogenous interleukin-12-inducing factors to direct the development of naive T-helper cells toward the Th1 phenotype. *Blood* **90,** 1920–1926.
24. Snijders, A., Kalinski, P., Hilkens, C. M., and Kapsenberg, M. L. (1998) High-level IL-12 production by human dendritic cells requires two signals. *Int. Immunol.* **10,** 1593–1598.
25. Kalinski, P., Vieira, P. L., Schuitemaker, J. H., de Jong, E. C., and Kapsenberg, M. L. (2001) Prostaglandin E2 is a selective inducer of IL-12p40 production and an inhibitor of bioactive IL-12p70 heterodimer. *Blood* **97,** 3466–3569.

34

Preparation of Human Dendritic Cells for Tumor Vaccination

Michael R. Shurin

1. Introduction

Naive T-lymphocytes specific for a given primary antigen, including tumor antigen, occur in very low frequencies and require the relevant antigen to be presented by specialized antigen-presenting cells (APCs). Dendritic cells (DCs), along with B-lymphocytes and mononuclear phagocytes, are professional APCs and are found in all tissues and organs of the body. These are cells with a highly efficient capability to present antigens in the context of major histocompatibility complex (MHC) class I and class II molecules. What makes DCs THE most potent APC, however, is their unique ability to present antigen to naive T-lymphocytes. This puts DCs in a key role in the initiation of specific immune responses and creates possibilities for their utilization in the development and improvement of immunotherapeutic approaches for the treatment of tumor and other diseases. Adoptive transfer of host defense cells, including DCs and T-cells, may be able to correct an otherwise defective generation of competent immune cells in patients with cancer. Most of the DC-based therapeutic strategies are designed to induce or burst weak T-cell responses to tumor-associated antigens (TAAs) in cancer patient. A number of in vitro and in vivo studies have demonstrated that DCs presenting TAA could efficiently stimulate cytotoxic T-lymphocytes (CTLs) and the development of TAA-specific protective and therapeutic immune responses. However, the initiation of human DC-based clinical trials in cancer patients was limited for many years by the absence of culture methods allowing one to obtain a sufficient number of DCs. Recent discoveries in this field markedly increase the yield of cultured

From: *Methods in Molecular Biology, vol. 215: Cytokines and Colony Stimulating Factors: Methods and Protocols*
Edited by: D. Körholz and W. Kiess © Humana Press Inc., Totowa, NJ

or ex vivo-expanded DCs and accelerated evaluation of DC capacity to induce antitumor immunity in clinical trials. In this chapter, we describe several methods of DC generation from CD34+ precursor cells and adherent macrophages. A variety of modifications of several basic procedures, used in preclinical and clinical research, are also discussed. In addition, we have illustrated several common and novel strategies how to load DCs with TAAs.

2. Materials

2.1. Isolation of Mononuclear Cells (MNC)

1. LymphoPrep or NycoPrep (Nycomed, Oslo, Norway).
2. Lymphocyte separation medium (Organon Teknika Co., Durham, NC).
3. Ficoll–Paque (Pharmacia, Uppsala, Sweden).
4. Histopaque-1077 (Sigma, St. Louis, MO).

Any of these solutions could be used if the density is adjusted to 1.077 g/mL.

2.2. Cytokines

1. rhGM-CSF.
2. rhIL-4.
3. rhCSF (kit ligand, Steel Factor).
4. rhFLT3 ligand (FLT3L).
5. rhTNF-α.

These growth factors and cytokines can be purchased from Genzyme (Cambridge, MA), Endogen (Woburn, MA), R&D Systems (Minneapolis, MN), PeproTech (Norwood, MA), or other sources.

2.3. Media

1. RPMI-1640 (Gibco-BRL, Gaithersburg, MD) supplemented with growth factors 9RPMI 1640, 100 U/mL penicillin, 100 μg/mL streptomycin, 2 m*M* L-glutamine, 10 m*M* HEPES, 0.1 m*M* nonessential amino acids, 1 m*M* sodium pyruvate, 10% fetal calf serum [FCS]) and cytokines 91000 U/mL GM-CSF [1.3×10^7 U/mg; Schering Plough, Kenilworth, NJ], 1000 U/mL IL-4 [2.975×10^7 U/mL; Schering Plough], 2.5 ng/mL TNF-α [Knoll Pharmaceuticals, Whippany, NY]).
2. Iscove's modified Dulbecco's medium (IMDM) supplemented with growth factors: IMDM with 100 U/mL penicillin, 100 μg/mL streptomycin (or 50 μg/mL gentamicin instead of penicillin + streptomycin), 10% heat-inactivated FCS, 50–100 ng/mL GM-CSF, and 10 ng/mL TNF-α.
3. AIM-V (Gibco-BRL).
4. Complete medium: RPMI 1640 with 10 m*M* HEPES, 2 m*M* L-glutamine, 50 μg/mL gentamicin, 0.1 m*M* nonessential amino acids, and 0.1 m*M* sodium pyruvate, with >200 U/mL GM-CSF and >500 U/mL IL-4.
5. X-vivo 15 (BioWhitaker, Walkersville, MD).

6. FCS (Gibco-BRL) treated with dextran-coated charcoal: add 5% dextran-coated charcoal (DCC, Sigma) + 0.5% dextran in water to FCS at 1:5 (v/v) ratio, stir overnight at 4°C, spin at 10,000*g* for 30 min, collect and sterilize the supernatant.
7. Human AB serum (Sigma).

2.4. Antibodies for FACScan Analysis and Immunocytochemistry

CD1a, CD11c, CD83, CD80, CD86, CD14, HLA-DR, CD34. Antibodies can be purchased from PharMingen (San Diego, CA), Serotec (Oxford, England), and Immunotech (Marseille, France).

2.5. Other Reagents

1. LeukoStat staining kit (Fisher Scientific, Pittsburgh, PA).
2. Hanks' balanced salt solution (HBSS) (Gibco-BRL).
3. Phosphate-buffered saline (PBS) (Gibco-BRL) with 1% bovine serum albumin (BSA) (Sigma).
4. Cell-labeling buffer (PBS with 0.5% BSA and 2 m*M* EDTA, pH 7.2).
5. Human serum albumin (HSA) (Sigma).
6. Collagenase (Sigma).
7. DNase (Sigma).
8. Hyaluronidase (Sigma).
9. Lipopolysaccharide (LPS) (Sigma).
10. sCD40L (Immunex, Seattle, WA).
11. EDTA disodium salt (Sigma).
12. Red blood cell sysing buffer (Sigma).

3. Methods

3.1. Generation of DCs from the Precursor Cells

The development of hematopoietic stem and progenitor cell isolation methods in combination with specific selection of growth factors and cytokines facilitated the development of ex vivo expansion techniques. The use of lineage-specific cytokines such as GM-CSF significantly simplify the generation of a large number of myeloid postprogenitor cells *(1)*. The generation of DCs (dendropoiesis) from CD34+ peripheral blood or cord blood progenitor cells is a good example of recent development of this field.

1. Use human peripheral blood, umbilical cord blood, leukopack (cells collected after leukapheresis), or bone marrow cells to isolate CD34+ precursors.
2. Dilute each sample 1:3–1:4 (v/v) with HBSS or PBS.
3. Isolate mononuclear cells (MNCs) by Ficoll–Hypaque density gradient centrifugation using one of the 1.077-g/mL media listed in **Subheading 2.3.** and corresponding conditions of centrifugation (usually 400*g* for 20 min at room temperature).

4. Wash cells twice in medium supplemented with 10% FCS.
5. Incubate cells ([1–2] × 10^6 cells/mL) in a Petri dish to deplete adherent cells.

Elimination of adherent cells is recommended if Dynobeads or cell sorting are used for the CD34+ cell separation and can be omitted with other CD34+ cell isolation procedures described in **Subheadings 3.1.1.–3.1.5.** It is important to remember that all cell isolation procedures might be divided into two groups: one group that utilizes positive selection (isolation of CD34+ target cells) protocols and the second group in which the concentration of CD34+ cells in suspension is significantly enriched by the depletion of nontarget cells (negative selection). The choice of a specific procedure depends on an individual application and both types of procedure are equally suitable for human DC cultures.

3.1.1. Isolation of CD34+ Cells Using CellPro CEPRATE Columns

The CEPRATE kit (CellPro, Bothell, WA) is a disposable cell-separation system that utilizes a proprietary avidin–biotin immunoaffinity process.

1. Wash peripheral blood mononuclear cells (PBMC) twice with 1% bovine serum albumin (BSA) in PBS.
2. Incubate for 25 min with biotin-labeled anti-CD34 antibody at room temperature, followed by an additional wash step.
3. Load cells onto an assembled continuous-flow column where labeled CD34+ cells are captured on avidin-coated beads.
4. Wash avidin column with PBS to remove unadsorbed (nontarget) cells.
5. Release and collect bound CD34+ cells by gentle squeezing of avidin column 5–10 times along its length.

After a short practice, this method will give you more than 95% purity and more than 80% recovery. Viability of cells are usually >98%. Note that CellPro columns may no longer be available; however, the principle and procedures with similar devices will be identical.

3.1.2. Isolation of CD34+ Cells Using Stem Cell Magnetic Beads

Stem Cell Technologies (Vancouver, Canada) produces all necessary antibodies, beads, columns, equipment, and media for the successful isolation and culture of human CD34+ cells from different origins. CD34+-enriched cell suspension can be prepared by negative selection from bone marrow or peripheral blood using the StemSep progenitor enrichment cocktail.

1. Mix cells with antibody cocktail.
2. Incubate for 15 min at room temperature or 30 min on ice.
3. Add magnetic colloid, mix well, and incubate for the same time at the same conditions.

4. Load sample into the column in magnet.
5. Add BSA solution.
6. Collect enriched CD34+ cell suspension as follow-through.

This simple and efficient method of negative selection of CD34+ progenitors always gave us high levels of purity and recovery similar to those described for CellPro columns (positive selection).

3.1.3. Isolation of CD34+ Cells Using MACS Cell Sorting

Highly pure CD34+ cells can be relatively easily isolated using MACS CD34 Progenitor Cell Isolation Kit (Miltenyi Biotec, Bergisch Gladbach, Germany), which contains all necessary antibodies, superparamagnetic beads, and cell labeling buffer.

1. Mix cells with blocking reagent (human IgG) to avoid nonspecific binding to Fc receptor.
2. Add hapten-conjugated CD34 Ab.
3. Incubate 15 min at 6–12°C and wash with included buffer.
4. Incubate with MACS MicroBeads recognizing the CD34 antibody for 15 min at 6–12°C.
5. Wash cells and apply to prefilled column in MACS separator followed by an additional wash.
6. Remove column from separator and elute remaining cells.
7. Repeat magnetic separation step.

The viability and purity of harvested CD34+ cells are usually >95%.

3.1.4. Isolation of CD34+ Cells Using Dynobeads or BioMag Beads

CD34-conjugated immunomagnetic beads (Dynobeads M-450 CD34) from Dynal A.S. (Oslo, Norway) is also a common technique used for the isolation of CD34+ precursors.

1. Incubate MNCs with magnetic beads in medium containing 10% FCS at 4°C with gentle rotation for 60 min.
2. Collect CD34+ cells with a magnet (MPC-1, Danal).
3. Release (if necessary) CD34-conjugated immunomagnetic beads by incubation with a goat anti-mouse-Fab polyclonal antibody (DETACHaBEADS, Danal) at room temperature for 90 min with gentle rotation.
4. Remove (if necessary) contaminated T- and B-cells by using CD2-conjugated and CD19-conjugated magnetic beads (Dynobeads M-450 PanT and PanB).

This method usually results in >95% purification of CD34+ cells, as assessed by flow cytometry.

A similar protocol of CD34+ cell isolation should be used with BioMag magnetic particles coated with anti-CD34 antibody (PerSeptive Diagnostics, Cambridge, MA or PE Biosystems, Foster City, CA). However, for detaching magnetic particles from separated cells two protocols are recommended:

1. Use a protease, such as chymopapain, to break the antigen–antibody bond and remove the particle with a magnet.
2. Culture cells for 48 h, during which magnetic particles fall away from cells because of cell surface changeover and may be removed magnetically.

Each of these procedures has limitations and requires special consideration depending on the application.

3.1.5. Isolation of CD34+ Cells Using Cell Sorter

1. Stain MNCs on ice for 20 min with fluorescein isothiocyanate (FITC)-conjugated anti-CD34 monoclonal antibody.
2. Wash and resuspend cells in PBS supplemented with 1% human serum albumin.
3. Separate by flow cytometry sorting immediately.

The viability and purity of sorted viable cells usually exceed 96–98%.

3.1.6. Cultures of Isolated CD34+ Cells

There are several common culture conditions, which are used for the ex vivo generation of DCs from CD34+ progenitors. Depending on the application and availability of reagents, the investigator should choose the most appropriate conditions by paying attention to the (1) type of medium, (2) supplementation with serum, (3) combination of cytokines, and (4) duration of the culture. We provide several examples with short comments.

3.1.6.1. Protocol A

1. Resuspend CD34+ cells ([0.5–1] × 10^5/mL) in IMDM supplemented with growth factors (*see* **Subheading 3.2.**).
2. Maintain culture at 37°C in humidified atmosphere with 5% CO_2 for 16–18 d for a maximum yield.
3. Replace medium (one-half of total volume) every 4–5 d and split the culture if the cells become confluent.

The addition of 50 ng/mL stem cell factor (SCF) and 50 ng/mL FLT3 ligand (FLT3L) increases the yield of DCs up to five times ***(2)***. It is important to note that using human pooled AB serum instead of FCS or the combination of GM-CSF + IL-4 supports very poor DC growth in this medium ***(2)***.

3.1.6.2. Protocol B

1. Resuspend CD34+ cells in serum-free AIM-V or X-vivo 15 medium, which could be additionally supplemented with 2–2.5% AB serum if required.
2. Add 500–1000 U/mL of GM-CSF, 500–1000 U/mL IL-4, 10 ng/mL TNF-α, and 100 ng/mL FLT3L and culture for 14 d in six-well plates (4 mL/well).
3. Exchange 1 mL of culture medium with fresh medium and cytokines every 4 d and split cells if necessary.

3.1.6.3. Protocol C

The following basic protocol is usually used in our laboratory:

1. Resuspend 5×10^4 cells/mL CD34+ cells in RPMI 1640 medium supplemented with growth factors.
2. Add 20 ng/mL SCF (R&D) at d 1 and 3 only.
3. Incubate 3 wk in 24-well plates (Costar) at 1 mL/well.
4. Replace 0.4 mL of culture medium twice per week and split cultures when cells are confluent.
5. To increase the yield of cultured DCs, add 50–100 ng/mL FLT3 ligand (Immunex or R&D) and treat FCS with dextran-coated charcoal.

Similar yields of functionally active cultured DC can be obtained if RPMI 1640 is replaced with AIM-V medium and 5–10% FCS. It is recommended that cultures undergo half-refeeding two times per week with complete medium.

3.1.6.4. Protocol D

An interesting modification of CD34+ culture condition was recently reported *(3)*. Isolated CD34+ cells were cultured in a new gas-permeable PL2417 culture containers in serum-free medium supplemented with GM-CSF and TNF-α. Cultured DCs were not overall different from control DCs generated under the other conditions.

3.2. Conversion of Adherent MNCs into DCs

Similar to CD34-related protocols, there are several commonly used modifications of DC propagation from adherent mononuclear cells (MNC). Below we present different protocols, including protocols used in our laboratory, which are different mostly in culture conditions.

3.2.1. Protocol A

1. Prepare PBMCs as described in **Subheading 3.1.**
2. Resuspend cells in RPMI 1640 supplemented with 5% FCS at $(4\text{–}6) \times 10^6$ cells/mL.

3. Incubate in six-well plates or flasks for 60–90 min at 37°C.
4. Remove the nonadherent cells by gentle washing with warm (!) PBS.
5. Culture adherent CD14+ cells in complete medium.
6. TNF-α can be also added (10 ng/mL) to support the dendropoiesis *(4)*.

We recommend using 500 U/mL of both GM-CSF and IL-4. As a source of serum, use 10% FCS, although human AB serum (up to 10%) or defibrinated plasma (up to 10%) can also be used. It has been recently shown that neither human serum nor autologous plasma inhibits DC generation from monocytes. Furthermore, autologous plasma is as efficient or even more efficient than homologous serum *(5)*. However, be aware that autologous serum or plasma from cancer patients might contain factors which may inhibit dendropoiesis in vitro *(6,7)*.

3.2.2. Protocol B

1. Prepare suspension of PBMCs in AIM-V medium.
2. Transport cells into plastic flask and incubate at 37°C for 2–3 h.
3. Collect nonadherent cells.
4. Replace medium with AIM-V supplemented with 500–1000 U/mL GM-CSF and 500–1000 U/mL IL-4.
5. Incubate for 7 d and harvest nonadherent and loosely adherent cells by vigorous washing.

Addition of heat-inactivated human AB serum or autologous serum up to 2% usually increases the yield of collected DCs.

3.2.3. Protocol C

CD14+ monocytes can also be isolated from PBMCs using FACScan cell sorting or elutriation by counterflow centrifugation using a JE-6B rotor (Beckman, Palo Alto, CA). For elutriation, use constant rotor frequency (1000*g*), pump speed from 18.5 to 50 mL/min, and collect cells in 100-mL fractions. Note that monocytes are elutriated at a pump speed of >32 mL/min.

3.2.4. Preparation of Mature Cultured DCs

3.2.4.1. Cytokines and Cytokine Combinations

To obtain mature (activated) cultured DCs, one of the following treatments is recommended: (1) stimulate the cells on d 7–9 with 2000 U/mL IL-1β and 1000 U/mL TNF-α; (2) TNF-α can also be used alone (200 U/mL and up or 10–50 ng/mL); (3) cocktail of cytokines: 10 ng/mL TNF-α + 1000 U/mL IL-6 + 10 ng/mL IL-1β + 1 μg/mL prostaglandin E_2 (PGE_2) *(8)*; (4) LPS (up to 100 ng/mL); (5) LPS (1 ng/mL) + interferon (IFN)-γ (20 ng/mL); (6) CD40L

(CD154) (*see* **Subheading 3.2.4.2.**); (7) macrophage-conditioned supernatant (*see* **Subheading 3.2.4.3.**).

3.2.4.2. CD40/CD40L Interaction

CD40 ligation on DCs stimulates their differentiation and activity ***(9)***. DCs at different stages of maturation can be stimulated with CD40L (CD154) using the following technique.

1. Load DC (approx 10^6 cells/mL) over adherent hCD40L-transfected irradiated murine fibroblast. We recommend using 24-well or 6-well plates with CD40L-transfected fibroblast cultured for at least 16–18 h (start with $[0.3–0.5] \times 10^6$ cells/mL) prior to the addition of DCs.
2. Centrifuge plates briefly (100*g*, 3 min).
3. Incubate mixed cell cultures for 24 h.
4. Collect nonadherent DCs.
5. Incubate DCs for an additional 6–16 h, depending on the design of the experiment.

A second technique of CD40 ligation on DCs includes the addition of anti-CD40 or anti-CD40L antibodies directly to DC cultures (1–10 μg/mL). The third technique includes DC stimulation with soluble CD40L (1 μg/mL) for 18–24 h. However, this method is thought to be less effective.

3.2.4.3. Macrophage-Conditioned Supernatant

To prepare macrophage-conditioned medium:

1. Coat tissue culture flasks with human Ig (human γ-globulin, 10 mg/mL in AIM-V medium for 2 h at 37°C).
2. Wash and add PBMCs ($[10–20] \times 10^6$ cells/mL in AIM-V medium).
3. Incubate 1–2 h at 37°C.
4. Remove nonadherent cells and wash adherent monocytes with warm (!) medium.
5. Culture cells in AIM-V medium for 24 h.
6. Collect conditioned medium and store aliquots at –80°C.

To stimulate DC maturation, add macrophage-conditioned supernatant to 5-d PBMC-derived DC cultures to a final concentration of 25% (v/v) for 48 h. Harvest and analyze DCs as described in **Subheading 3.5.** This procedure was modified from **ref.** ***8***.

3.2.5. Generation of DCs in Teflon Bags

A new modification of the monocyte-derived DC generation technique was recently published ***(10)***. The sizable number of monocytes were purified by elutriation from blood leukocytes and cultured in Teflon bags. Developed DCs were defined by morphology and phenotype. Interestingly, the DCs generated under adherent-free conditions exhibited low levels of CD1a and lack of CD83 expression, although they expressed a high level of CD86 and were potent stimulators for T-cell proliferation.

In summary, it still remains to be determined whether the use of CD14+ monocyte-derived or CD34+ hematopoietic precursor-derived DCs is more appropriate for human clinical protocols. The data regarding direct comparison are controversial. In fact, Bui et al. ***(11)*** reported that both DC populations prepared from one donor functioned equally well as superior T-cell stimulators in allogeneic primary mixed-leukocyte reaction although the yield of monocyte-derived DCs was significantly lower compared to CD34-derived DC ***(11)***. Similar data were reported by Chen et al. ***(12)***. In contrast, Mortarini et al. ***(13)*** demonstrated that DCs generated from CD34+ cells were superior to DCs from monocytes, as was assessed by the time required to obtain Melan-A-specific CTLs with both DC populations. In addition, CD34-derived DCs, in contrast to CD14-derived DCs, are proliferating cell population, as determined by ^{3}H–thymidine uptake and Ki67 staining, and thus can be used for gene transduction techniques, which require cell proliferation, like the retrovirus-based technique. Of clinical importance, the yield of DCs derived from adherent CD14+ cells cannot be expanded beyond the number of starting monocytes, although the number of DCs generated from CD34+ precursors may exceed the initial cell number by 100 and more times. On other side, the generation of DCs from adherent monocytes is significantly easier and less expensive than DC production from precursor cells. Thus, these and other thoughts should be carefully considered when choosing the method of DC generation ex vivo.

3.3. Isolation of DCs from Different Tissues

3.3.1. Isolation of DCs from Peripheral Blood

1. Separate MNCs from the heparinized peripheral blood or leukopack by density gradient centrifugation, as described in **Subheading 3.1.**
2. Wash MNCs obtained from the interface and allow them to adhere into plastic culture plates or flasks at a concentration of (2–5) $\times 10^6$ cells/mL in Dulbecco's modified Eagle's medium (DMEM) for 2 h (37°C, 5% CO_2).
3. Remove nonadherent cells containing lymphocytes and gently wash flasks twice with warm PBS.
4. Add fresh medium with 10% FCS or autologous serum and incubate cells overnight at 37°C.
5. Collect nonadherent cells, which represent an enriched population of peripheral blood DCs.

6. For the further enrichment of peripheral blood DCs, we recommend the isolation of low-density cells by separation over a Nycodenz (NycoPrep, 1.068 g/mL) or 14.5% metrizamide (Sigma) at 650*g* for 15 min at room temperature. Additional purification procedure may involve the depletion of contaminating T-, B-, and natural killer (NK) cells by panning with anti-CD3, anti-CD21, and anti-CD56 antibodies. Collect cells from the intermediate phase, wash twice, and resuspend in the medium of choice depending on further analysis. Note that the yield of isolated DCs is less than 1% of the starting number of PBMCs.
7. In addition, peripheral blood DCs can be isolated using a magnetic cell sorting procedure with specific antibodies. For instance, Stem Cell Technologies (Vancouver, Canada) offers the StemSep DC enrichment cocktail, which allows, based on the negative selection, the collection of mature human DCs (HLA-DR+CD4+lin– (CD3–, CD19–, CD14–, CD56–) with 20- to 40-fold enrichment. Miltenyi Biotec (Bergisch Gladbach, Germany) has the Blood DC Isolation Kit. In the first MACS separation, T cells, monocytes, and NK cells are magnetically labeled and efficiently depleted by retention on a depletion column. From the nonmagnetic fraction, CD4+ blood DCs are then positively harvested with CD4 MicroBeads.

3.3.2. Isolation of DCs from Solid Tissues

3.3.2.1. Skin

Dendritic cells can be isolated from human skin samples. To prepare a suspension of enriched epidermal DCs (Langerhans cells, LCs).

1. Cut skin samples into pieces of about (3–4) × (10–20) mm.
2. Incubate pieces of skin with 0.5% trypsin or 50 U/mL dispase in PBS for 100–120 min at 37°C.
3. Remove epidermal sheets with a pair of fine forceps and stir them for 30 min in RPMI 1640 medium supplemented with 10% FCS, 2 m*M* L-glutamine, antibiotics (**Subheading 2.3.**), and 0.01% DNase.
4. Collect single-cell suspension by filtration through sterile nylon mesh (75-µm pore size).
5. Load onto 1.077 g/mL NycoPrep and centrifuge (400*g*, 20 min at room temperature).
6. Harvest interface cell layer, wash cells twice in medium, and resuspend in a complete medium.

Langerhans cell enrichment usually reaches up to 50%. For further purification of LCs, we recommend using the FACScan sorting procedure or magnetic beads using CD1a expression as a marker for LCs. To prepare magnetic beads, incubate precoated with anti-mouse antibody magnetic beads (e.g., Dynobeads M-280 coated by sheep anti-mouse antibody) with mouse anti-human CD1a, wash, and add cell suspension for 1 h (4°C). Wash bound cells and repeat the procedure three to five times.

Note that spontaneously migrating LCs can be collected from the culture medium where epidermal sheets were incubated for 2–3 d at 37°C in a complete medium. In addition, dermal DCs are commonly collected after digestion of the dermis sample with DNase + collagenase + hyaluronidase (triple enzyme cocktail, in **Subheading 3.6.2.**) followed by intensive washing and isolation with specific antibody-coated magnetic beads, FACS cell sorting, or differential adhesion procedure.

3.3.2.2. Tumor Tissue

A similar method (*see* **Subheading 3.3.2.1.**) is commonly utilized to isolate DCs from the tumor tissues. Use a triple enzyme cocktail (2–4 h at room temperature) for most tumors and dispase/collagenase mixture for the skin malignancies. After digestion, remove lineage positive cells (CD2, CD3, CD8, CD14, CD16, CD19, CD20, CD56) using magnetic beads and purify remaining cells by a density gradient centrifugation (1.077 g/mL). We recommend to use the MACS cell sorting procedure with paramagnetic beads for removing lineage positive cells. It is a fast, reliable, and reproducible procedure if one follows the manufacture procedures.

3.4. Stimulation of DC Generation In Vivo

It has been recently demonstrated that administration of FLT3L markedly increases the generation and redistribution of DCs in both mice ***(14,15)*** and humans ***(16)***. It has been found that the treatment of advanced cancer patients with FLT3L expanded DCs 20-fold in vivo ***(17)***. These data allowed initiating several clinical trials with FLT3L in cancer patients. For instance, one of the ongoing clinical trials is designed to evaluate the immunologic and biologic activity of FLT3L and melanoma peptide antigens (MART-1, gp100, and tyrosinase) versus FLT3L alone in patients with metastatic melanoma, renal cell cancer, colon cancer, prostate cancer, and so forth ***(17–19)***. Numerous successful results in preclinical murine models give hope for a certain success in cancer patient trials as well.

A further development of this idea resulted in a new approach in which FLT3L was used to expand DCs in vivo followed by their isolation and readministration after the pulsing with TAAs. For instance, an interesting report describes a phase I clinical trial combining TAA vaccination with an approach to the acquisition of large numbers of DCs: the systemic administration of FLT3L to expand DCs in colon and nonsmall-cell lung cancer patients in vivo (patients received 20 μg/kg of FLT3L up to a maximum dose of 1.5 mg via daily sc injections for the 10 d preceding their leukopheresis) followed by their isolation, antigen loading, and infusion as a vaccine ***(17)***.

In addition, we have recently demonstrated that administration of IL-12 in mice markedly increases accumulation of DCs in lymphoid and nonlymphoid tissues in mice *(20,21)*. It is possible that a similar effect can be expected in humans as well. Evaluation of clinical importance of this finding is in progress now.

3.5. Identification of Prepared DCs

Cultured or isolated DCs are identified by three sets of criteria: (1) distinctive morphological features, including villous cell surface with dendritic projections and veils, lobulated nucleus, large Golgi apparatus, and multivescicular bodies, occasionally in continuity with membrane complexes *(2)*; (2) phenotypic markers, including expression of CD1a, CD83, CD11c, HLA class I and II, CD80 (B7-1), CD86 (B7-2) molecules, and the lack of macrophage-restricted CD14 marker; (3) functional characteristics, including a high stimulatory capacity to activate proliferation of T-lymphocyte in an allogeneic mixed-leukocyte reaction (MLR), phagocytosis, and the ability to produce high levels of IL-12 upon stimulation. For detailed methodological description of these methods, *see*, for instance, **refs.** *5,22*, and *23*).

3.6. Loading of DCs with TAAs

Because naive T-cells specific for a given TAA occur in low frequencies and require APCs for activation, the in vitro induction of primary T-cell responses remains a significant technical challenge. Using highly functionally active DCs as APCs might represent a marking breakthrough for this problem. In addition, the capacity of DCs to traffic to the T-cell areas of the lymph nodes make TAA-loaded DCs the preferred cellular vehicle for the initiation of T-cell responses to TAAs. Thus, DCs can be used as a tumor vaccine in vivo, suggesting an additional therapeutic approach for the induction of antitumor immune responses in cancer patients.

It is important to note here that the efficiency of loading antigen into DCs depends on the purity of the DC population, medium content, antigen, and DC activity/maturation *(22)*. Here, we describe several protocols for DC pulsing with TAAs. The technique of loading may depend on the peptide/protein, and the protocol should be optimized regarding the specific conditions and purposes.

3.6.1. Tumor Antigenic Peptides

1. Resuspend DCs ([5–10] × 10^6 cells/mL in 24-well plates) in a complete medium supplemented with 10% FCS or autologous serum and antigens for 16 h. Protein antigens are usually used at 5–50 µg/mL and peptides at 10–100 µ*M* concentrations.

2. For a peptide-loading procedure, we recommend trying 2-h incubation in a serum-free medium. Note that the time of incubation of DCs with TAAs can be increased up to 16–18 h.

3.6.2. Tumor Lysates

Tumor cell lysates are often used as a TAA source. Prepare lysate by a repeated freeze–thaw procedure.

1. Mince tumor tissue in a glass Petri dish in a small volume (2–3 mL) of HBSS (to keep the tumor moist).
2. Prepare single-cell suspension by mechanical disaggregation (chopping into smaller fragment or mashing through sieves of decreasing size) or enzymatic digestion (10 μg/mL deoxyribonuclease [DNase], 2.5 U/mL hyaluronidase, and 1 mg/mL collagenase, 2–4 h in a sterile bottle on a magnetic stirring device at room temperature).
3. Let the tissue fragments to settle under 1*g*.
4. Isolate the dead cells by gradient centrifugation.
5. Wash remaining cells and lyse in a serum-free AIM-V medium by three to four freeze cycles (on liquid nitrogen or dry ice) and thaw cycles (at room temperature or 37°C).
6. Remove large particles and aggregates (200*g*, 10 min).
7. Sterilize the supernatant by irradiation (preferred) or filtration through a 0.22-μm filter.
8. Determine protein concentration and store aliquots at –80°C.

3.6.3. Tumor Extracts

Tumor extract can be prepared by sonicating the tumor cells in a serum-free medium or water (approx [10–50] × 10^6 cells/mL). An alternative method of TAA extraction, which was popular several years ago, is the stripping of peptides from the surface of viable tumor cells.

1. Wash tumor cells in HBSS.
2. Add citrate–phosphate buffer (0.131 *M* citric acid, 0.066 *M* Na_2HPO_4, 0.15 *M* NaCl, pH 3.0) ([5–10] × 10^6 cells/mL) for 5 min at room temperature.
3. Collect supernatant after centrifugation (3000*g* for 30 min).
4. Concentrate peptides using SepPak18 cartridge (Millipore, Bedford, MA) and the manufacturer's recommendations.

3.6.4. Tumor Cell Membranes

The plasma membrane fraction of tumor cells can also be used successfully as a source of tumor antigens for DC pulsing.

1. Wash tumor cells in HBSS.
2. Homogenize with Polytron or Ultra-Turrax for approx 1 min (approx 10^7 cells/mL).

3. Centrifuge the homogenate sequentially at 15,000g for 30 min and 100,000g for 60 min.
4. Use a pellet as a plasma membrane fraction.

3.6.5. Autologous Tumor

Autologous cultured tumor cells are another source of TAAs.

1. Collect samples of tumor tissues and place in a cold serum-free medium.
2. Mince tissue, digest with collagenase and DNase, and wash extensively.
3. Seed cells in the small 25-mm^2 flasks or six-well plates at a cell density of $(0.5–1) \times 10^5$ cells/mL in a complete culture medium supplemented with 2–5% autologous serum or 5–10% FCS.
4. Remove nonadherent cells and add fresh medium after 2–4 d. When confluent, collect cells and lyse by the repeated freeze–thaw method or under hypotonic conditions (0.3 mL of distilled water per each well of a six-well plate or 0.6 mL per 25-mm^2 flask).
5. Collect lysates, clarify by centrifugation, irradiate at 100 Gy, aliquot, and store at –80°C.

Note that intact γ-irradiated (200 Gy) autologous tumor cells can also be used to stimulate specific T-cell clones in the presence of autologous DCs in vitro.

3.6.6. Pulsing of DCs with TAA

1. To pulse DCs, carefully remove 0.5 volume of medium from DC cultures.
2. Add lysate (approx 1:5 [v/v] lysate:medium ratio) that was preliminary equilibrated for the right osmolarity (10 μL of 10X HBSS for each 100 μL of water if lysate was obtained by hypotonic lysis).
3. Incubate DCs with the lysates for 60–120 min at 37°C.
4. Reconstitute the original volume in DC cultures with fresh medium and culture for additional 18–24 h.
5. Harvest DCs, wash three times, and resuspend in an appropriate medium or solution.

In addition, see the brief descriptions and recommendations for other popular, although experimental, approaches of DC loading with TAAs in **Subheading 3.7.**

3.7. Overview

3.7.1. Transfection and Transduction of DCs

Transduction of DCs with defined TAA genes to be used as vaccines is a potentially powerful strategy in cancer gene therapy. For instance, evaluating the generation of CTLs and T-helper cells against TAA HER2 (human epidermal growth factor receptor 2) using autologous CD34-derived DCs that have been retrovirally transduced with the HER2 gene, it has been shown that HER2-transduced DCs elicited HER2-specific CD8+ CTLs that lyse HER2-overexpressing tumor cells in the context of distinct HLA class I alleles *(24)*. However, this strategy requires cloning of gene for TAAs, characterization of a specific promoter, and generation of adenovirus or retrovirus vector expressing the TAA gene, as well as the preparation and characterization of anti-TAA antibody. In spite of some technical and methodological difficulties, a growing body of evidences supports the use of TAA-engineered DCs in genetic immunotherapy.

Both CD34-derived and monocyte-derived DCs can be genetically engineered to express TAAs. There is no common protocol available for DC transduction, but the replication-defective adenoviruses are considered to be an efficient method for transgene expression in DCs. Examples of successful protocols for transfection of human DCs to express TAAs are the following: (1) adenovirus-based transduction *(25,26)* and (2) vaccinia viral vector-based protocol *(27)*. In addition, (3) a TAA-expressing retrovirus has been used to transduce CD34+ hemopoietic progenitor cells, which, after differentiation into DCs, could successfully generate anti-TAA CTLs *(28,29)*. In another report, DCs were transfected with TAA genes using cationic liposomes (lipofectin reagent) *(30,31)*. Although only low levels of transgene expression were detected, transfected DCs were able to induce specific T-cell activation and CTL responses.

When several methods of DC transfection were compared, including liposomes, electroporation, $CaPO_4$ precipitation, and adenovirus, only the last one has demonstrated a high level of transgene expression combined with sustained viability *(32)*. Other reports confirm these data. Therefore, the adenovirus-based transfection of DCs with TAAs should be considered as a method of choice.

1. Perform transfection in a small volume of RPMI 1640 or IMDM medium supplemented with 2% human AB serum or FCS for 2 h at 37°C.
2. Wash cells twice, resuspend in a complete medium, and return to culture for 24–48 h. If the efficacy of the transfection is low, omit the washing step after transfection and wash DCs after 24–48 h before use.

However, it is necessary to remember that the use of adenovirus-transfected DCs might lead to the initiation of antiadenoprotein immune responses as well. As a potential improvement, one may construct and test next-generation adenoviral vectors that improve the TAA-specific reactivity and reduce viral immunogenicity.

3.7.2. DC/Tumor Fusion

Fusion of DCs with tumor cells is a new experimental approach, which has been recently used in the treatment of cancer *(33,34)*. In this system, tumor cells would supply TAAs that would be presented to the host immune cells by DCs. Benefits for this approach are that TAAs do not need to be identified and the antigenic peptides would be physiologically processed and presented by the DCs. Only a few reports may serve as a basis for developing effective strategies for clinical trials. In one study, human ovarian cancer cells were fused to human DCs as an alternative strategy to induce immunity against known and unidentified TAAs. Fusions of ovarian cancer cells to autologous DCs resulted in the formation of heterokaryons that express the CA-125 antigen and DC-derived costimulatory and adhesion molecules *(35)*. Similarly, primary human breast carcinoma cells were fused with autologous DCs and the results showed that the fusion cells also retained the functional potency of DCs and stimulated autologous T-cell proliferation *(36)*. The recent study of 17 patients with metastatic renal carcinoma vaccinated with hybrid cells indicates that after vaccination and with a mean follow-up time of 13 mo, four patients completely rejected all metastatic tumor lesions, one presented a "mixed response," and two had a tumor mass reduction of greater than 50% *(37)*. Because the data are still limited, no specific recommendations can be given for different human tumor cells.

3.7.3. Tumor DNA/RNA as a Source of TAAs

The major limitation of using proteins or peptides isolated from the patient's tumor cells is the availability of substantial amounts of tumor material. The use of DNA or RNA, as the form of antigen loaded onto DCs, would overcome this major practical limitation. The mRNA or cDNA can be synthesized from patient tumor cells, suggesting that multiple epitopes are present. The use of DCs transfected with mRNA isolated from tumor cells may allow specific immunotherapy even in cancers for which potent rejection antigens have not been identified. The cDNA library or the mRNA-like product could be transduced into the DCs, generating transiently the functional equivalent of a fusion hybridoma between the DCs and tumor cells *(38)*. Protocols for the TAA RNA transfection of human DCs is described elsewhere *(39,40)*.

3.7.4. Apoptotic Bodies of Tumor Cells as a Source of TAAs

Tumor apoptotic bodies are also used to pulse DCs. Stimulation of apoptotic death of autologous tumor cells can be achieved by irradiation, ultraviolet (UV) light, chemotherapeutic drugs, deprivation of growth factors, Fas/FasL interaction, using NK cells, and so forth. For instance, pulse cultured DCs with melanoma apoptotic bodies obtained by the treatment of tumor cells with 0.5 µg/mL of actinomycin-D for 24 h, as described ***(41)***. Deprivation of growth factors in cultured tumor cells causes a high level of apoptotic death and generates an excellent source of TAAs for a highly efficient pulsing of DCs. To obtain tumor apoptotic bodies, remove complete medium from cell culture by a simple aspiration for the adherent cultures or centrifugation for the nonadherent tumor cells. Wash cells twice in tumor medium without FCS and culture then in this medium for 24 h. Collect floating cells and check the cell viability with trypan blue. Irradiate the suspension of dead cells, concentrate it by centrifugation, and add the aliquot to DC cultures. Remember that the feeding dose should be determined for each condition and may vary for different DC cultures.

3.7.5. Human DC-Based Clinical Trials with Cancer Patients

1. B-Cell lymphoma ***(42)***: Monocyte-derived DCs loaded with idiotype protein (2 µg/mL, 24 h in RPMI 1640 with 10% autologous serum) were infused three times with 2 wk intervals.
2. Metastatic melanoma ***(43)***: Monocyte-derived DCs were pulsed with tumor lysate or peptides and injected (106 cells) weekly into an uninvolved inguinal lymph node over 1 mo. Boosted immunizations were repeated after 2 wk and thereafter in monthly intervals.
3. Metastatic renal cell carcinoma ***(44)***: Adherent PBMC-derived DCs loaded with cell lysate from cultured autologous tumor cells and keyhole limpet hemocyanin (KLH) and activated with (TNF-α + PGE_2) were administered by three iv infusions at monthly intervals.
4. Prostate cancer ***(45)***: CD34-derived DCs loaded with PA2024, Prostatic acid phosphatase (PAP)–GM-CSF fusion protein.
5. Prostate cancer ***(46,47)***: Dendritic cells enriched from PBMCs by density gradient purification were cultured with recombinant mouse PAP and injected once per month two times by intravenous, intradermal, or intralymphatic administration.
6. Prostate cancer ***(48,49)***: Monocyte-derived DCs were pulsed with prostate-specific membrane antigen (PSMA)-derived peptides (2 h, 1, or 10 µg/mL) and infused four to six times at 6-wk intervals.
7. Breast, ovarian, and pancreatic cancer ***(50,51)***: The PBMC-derived DCs were transfected with mucin gene (MUC1) using lipofection or pulsed with MUC1-derived peptides and injected sc every 3 wk.
8. Multiple myeloma ***(52)***: The precursor DCs were isolated from PBMCs by a series of density gradient centrifugation, incubated with purified multiple myeloma idiotype proteins for 36–48 h and administered iv once per month for 2 mo, followed by five monthly sc boosts of idiotype/KLH injections.
9. Melanoma ***(53)***: Monocyte-derived DCs were pulsed with a mixture of peptide antigens (MART-1/Melan-A, gp100, tyrosinase) and injected iv in weekly interval for four or more times.

10. Melanoma ***(54,55)***: The PBMC-derived DCs were pulsed with autologous tumor cell lysate. Patients received four injections of DCs at 4-wk intervals followed by 3 d of low-dose IL-2.
11. Colorectal cancer ***(56)***: Monocyte-derived DCs were pulsed with CEA 652 peptide and injected sc every week.
12. Metastatic colon and nonsmall-cell lung cancer ***(17)***: Mobilized FLT3L and isolated DCs were pulsed with CEA-derived peptide and used for vaccination.
13. Metastatic breast cancer ***(57)***: Using autologous tumor lysate-pulsed DCs, the authors have demonstrated that four cycles of DCs injection into the right supraclavicular lymph nodes resulted in regression of bilateral supraclavicular lymphogenous metastasis in a 50-yr-old woman.
14. Chronic myeloid leukemia ***(58)***: Autologous bcr/abl expressing DCs were generated from monocyte precursors and used for sc vaccination (1×10^6 to 5×10^7 cells).

3.7.6. Summary

The preliminary data from various DC-based clinical trials in cancer patients indicate that DC-based immunotherapy is a feasible, well-tolerated, and promising approach. Although various clinical trials currently utilize in vivo-generated DCs of a different origin and prepared under different conditions, many methodological problems should be addressed in order to compare and standardize isolation of progenitors/precursors and culture conditions. However, in spite of certain discrepancies and variabilities, DCs represent an extremely powerful tool to induce and activate specific antitumor immune responses both in vitro and in vivo.

4. Notes

1. NycoPrep solutions are different when being used with human or murine cells even if the density is identical. NycoPrep Animal solution has lower osmolality (approx 265 mOsm) than NycoPrep designed for human cells (approx 335 mOsm).
2. The more diluted the blood sample, the better the purity of the PBMCs.
3. It has been reported that mobilized CD34+ cells give rise to more CFU-DC (colony-forming unit dendritic cells) than steady-state bone marrow counterparts and generate a significantly higher number of functional DCs ***(23)***.
4. To increase the number of circulating CD34+ cells, administration of G-CSF is commonly used ***(59)***. Mobilized peripheral blood could be collected from adult volunteers who received 5 μg/kg body weight G-CSF (Amgen, Thousand Oaks, CA or Dompè-Biotec, Milan, Italy) for 4 d (maximum of 480 μg/d) and were pheresed on d 5. A simple and highly reproducible procedure for flow-cytometric enumeration of CD34+ cells in peripheral blood was described by Sutherland et al. ***(60)***.

5. Be aware that isolated CD34+ cells are mixed populations of different precursors and might be further purified based on the expression of additional cell surface markers ***(61–63)***.
6. Note that CD34+ cells collected after isolation are usually recovered in PBS or a diluted BSA solution. If cell culturing will be performed, transfer cells to the enriched medium.
7. It is necessary to mention that additional cytokines might abrogate or alter the direction of dendropoiesis in cultures. For instance, in the presence of angiogenic growth factors such as VEGF, bFGF, and IGF-1, the cultured immature DCs developed into endothelial-like cells, characterized by increased expression of Willebrand factor, VEGF receptor-2 kinase domain region (KDR), and Flt-4 and a disappearance of CD1a and CD83 ***(64)***.
8. It is important to note that not only does the culture condition play an important role on DC generation, but also the source of CD34+ cells may determine the yield of cultured cells. For instance, it was shown that the addition of IL-4 or IL-13 to GM-CSF+SCF+TNF-α+FLT3L-supplemented cord blood CD34+ cell cultures significantly increases the generation of DCs but has no effect when IL-4 or IL-13 was added to cultures conducted with mobilized peripheral blood CD34+ cells and autologous human serum ***(65)***. However, replacement of SCF and FLT3L with IL-4 on d 7 in mobilized peripheral blood CD34-derived cultures supplemented with FCS has been reported to increase the yield of CD1a+ DC and their immunostimulatory capacity (*see* **ref. *23***).
9. To increase the numbers of propagated DCs, one can add inhibitor of inducible nitric oxide synthase (iNOS) (for instance, 100 μ*M* L-NAME or NMLA; Alexis, San Diego, CA), 1 μg/mL indomethacin (Sigma), and 5 m*M* of 2-mercaptoethanol (Gibco-BRL) to the culture medium. Follow these recommendations only if these conditions do not interact with the purposes of the study, like evaluating the effect of nitric oxide (NO), prostaglandin E, or peroxide-related molecules on DC generation or maturation.
10. Many blood banks now use whole-blood inline filtration to produce leukocyte-depleted blood products. As a result, a common source of large numbers of DCs for research purposes, namely standard buffy coats, has been lost. An interesting adaptation of a conventional method for growing DCs from CD14+ precursors in order to make use of these filter units has been recently reported ***(66)***. A dextran solution containing human serum albumin was used to flush back the filters. After pelleting, mononuclear cells were obtained by standard density gradient centrifugation (Lymphoprep). T cells were removed using rosetting with sheep red blood cells, and contaminating CD14+ granulocytes were depleted by separation on an additional Percoll gradient. Culture of the cells and the induction of maturation was identical to the previously described procedures, except that the culture time was reduced from 7 to 5 d and the maturation time from 3 to 2 d. Analyses of the major molecules indicative of DC maturation (CD83, CD86, CD208/DC-LAMP) and functional analyses of the T-cell-stimulatory capacity of the DC population (using the MLR assay with normal peripheral T cells

and naive T cells) revealed no major differences from buffy-coat-derived DC preparations ***(66)***.

11. To increase the efficacy of the DC pulsing procedure, add 5–10 ng/mL TNF-α because the stimulation of DC maturation with inflammatory factors may increase the half-life of the MHC class II/antigen complex up to 10 times and may increase DC survival as well. In addition, it has been recently demonstrated that mature cultured DCs pulsed with a peptide derived from tyrosinase, MelanA/MART-1, or MAGE-1 and a recall antigen induced stronger antitumor T-cell responses after administration every 2 wk in different lymph nodes of the same patient when compared with immature DCs ***(67)***. This recommendation, however, may not apply for all protocols and should be verified for each used condition.
12. When planning to inject DCs into humans, note that saline or Ringer solutions are usually used. These solutions can be supplemented with 1% autologous serum. Do not use FCS!
13. Note that the direct comparison of FCS and autologous serum for the higher yield of DCs in cultures resulted in contradictory data. Because FCS is usually contraindicated for human use, we recommend comparing both supplementations in a pilot study. It is especially important if the designed study plans are directly relevant for DC-mediated vaccination programs.
14. Note that the stimulation of DC maturation before or after pulsing with antigens depends on the form of TAA loading. For instance, optimal stimulation of CEA-specific CTLs by peptide-loaded DCs occurs when DCs from cancer patients are matured with CD40L before exposure to CEA peptide, whereas optimal stimulation by RNA-transfected DCs occurs when the DCs are loaded with CEA RNA before stimulation with CD40L ***(68)***.

Acknowledgments

The technical help of Mrs. Robbie Mailliard and Ms. Madeline Wahl is gratefully acknowledged. We thank Dr. Clemens Esche, Dr. Galina V. Shurin, and Dr. Svetlana Bykovskaia for many helpful discussions. This work was supported in part by NIH grants RO1 CA80126, RO1 84270, and DOD DAMD17-00-1-0099 to M.R.S.

References

1. Scheding, S., Kratz-Albers, K., Meister, B., Brugger, W., and Kanz, L. (1998) Ex vivo expansion of hematopoietic progenitor cells for clinical use. *Semin. Hematol.* **35,** 232–240.
2. Siena, S., Di Nicola, M., Bregni, M., Mortarini, R., Anichini, A., Lombardi, L., et al. (1995) Massive ex vivo generation of functional dendritic cells from mobilized CD34+ blood progenitors for anticancer therapy. *Exp. Hematol.* **23,** 1463–1471.
3. Kowalkowski, K. L., Alzona, M. T., Aono, F. M., Van Epps, D. E., and Vachula, M. (1998) Ex vivo generation of dendritic cells from CD34+ cells in gas-permeable containers under serum-free conditions. *J. Hematother.* **7,** 403–411.

4. Kleihauer, A., Grigoleit, U., Hebart, H., Moris, A., Brossart, P., Muhm, A., et al. (2001) Ex vivo generation of human cytomegalovirus-specific cytotoxic T cells by peptide-pulsed dendritic cells. *Br. J. Haematol.* **113,** 231–239.
5. Eljaafari, A., Duperrier, K., Mazet, S., Bardin, C., Bernaud, J., Durand, B., et al. (1998) Generation of stable monocyte-derived dendritic cells in the presence of high concentrations of homologous or autologous serum: influence of extracellular pH. *Hum. Immunol.* **59,** 625–634.
6. Shurin, M. R. and Gabrilovich, D. I. (2001) Regulation of dendritic cell system by tumor. *Cancer Res. Ther. Control* **11,** 65–78.
7. Shurin, G. V., Shurin, M. R., Bykovskaia, S., Shogan, J., Lotze, M. T., and Barksdale, E. M., Jr. (2001) Neuroblastoma-derived gangliosides inhibit dendritic cell generation and function. *Cancer Res.* **61,** 363–369.
8. Thurner, B., Roder, C., Dieckmann, D., Heuer, M., Kruse, M., Glaser, A., et al. (1999) Generation of large numbers of fully mature and stable dendritic cells from leukapheresis products for clinical application. *J. Immunol. Methods* **223,** 1–15.
9. Felzmann, T., Buchberger, M., Lehner, M., Printz, D., Kircheis, R., Wagner, E., et al. (2001) Functional maturation of dendritic cells by exposure to CD40L transgenic tumor cells, fibroblasts or keratinocytes. *Cancer Lett.* **168,** 145–154.
10. Bernard, J., Ittelet, D., Christoph, A., Potron, G., Adjizian, J. C., Kochman, S., et al. (1998) Adherent-free generation of functional dendritic cells from purified blood monocytes in view of potential clinical use. *Hematol. Cell Ther.* **40,** 17–26.
11. Bui, L. A., Butterfield, L. H., Roth, M. D., Kiertscher, S., Jilani, S., Ribas, A., et al. (1997) A direct comparison of DC propagated from different progenitor cell populations from one donor. *Proc. American Association for Cancer Research* **38,** 31.
12. Chen, B., Stiff, P., Sloan, G., Kash, J., Manjunath, R., Pathasarathy, M., et al. (2001) Replicative response, immunophenotype, and functional activity of monocyte-derived versus CD34(+)-derived dendritic cells following exposure to various expansion and maturational stimuli. *Clin. Immunol.* **98,** 280–292.
13. Mortarini, R., Di Nicola, M., Siena, S., Bregni, M., Anichini, A., Gianni, A. M., et al. (1996) Cytokine requirements for eliciting melanoma specific cytotoxic T lymphocytes by melanoma antigen peptides loaded on CD34+ cell-derived versus monocyte-derived autologous dendritic cells. *Haematologica* **81(Suppl.),** 83.
14. Maraskovsky, E., Brasel, K., Teepe, M., Roux, E. R., Lyman, S. D., Shortman, K., et al. (1996) Dramatic increase in the numbers of functionally mature dendritic cells in Flt3 ligand-treated mice: multiple dendritic cell subpopulations identified. *J. Exp. Med.* **184,** 1953–1962.
15. Shurin, M. R., Pandharipande, P. P., Zorina, T. D., Haluszczak, C., Subbotin, V. M., Hunter, O., et al. (1997) FLT3 ligand induces the generation of functionally active dendritic cells in mice. *Cell Immunol.* **179,** 174–184.
16. Williams, D. E. (1997) In vivo effects of flt3 ligand. *Blood* **90,** 5022.
17. Fong, L., Hou, Y., Rivas, A., Benike, C., Yuen, A., Fisher, G. A., et al. (2001) Altered peptide ligand vaccination with Flt3 ligand expanded dendritic cells for tumor immunotherapy. *Proc. Natl. Acad. Sci. USA* **98,** 8809–8814.

18. Phase II study of Flt3 ligand alone or in combination with melanoma peptide antigens in patients with metastatic melanoma or renal cell cancer. (1999) *Curr. Clin. Trials Oncol.* **6,** P-600.
19. Morse, M. A., Nair, S., Fernandez-Casal, M., Deng, Y., St Peter, M., Williams, R., et al. (2000) Preoperative mobilization of circulating dendritic cells by Flt3 ligand administration to patients with metastatic colon cancer. *J. Clin. Oncol.* **18,** 3883–3893.
20. Esche, C., Subbotin, V. M., Hunter, O., Peron, J. M., Maliszewski, C., Lotze, M. T., et al. (1999) Differential regulation of epidermal and dermal dendritic cells by IL-12 and flt3 ligand. [in process citation]. *J. Invest. Dermatol.* **113,** 1028–1032.
21. Esche, C., Cai, Q., Peron, J. M., Hunter, O., Subbotin, V. M., Lotze, M. T., et al. (2000) Interleukin-12 and Flt3 ligand differentially promote dendropoiesis in vivo. *Eur. J. Immunol.* **30,** 2565–2575.
22. Mannering, S. I., McKenzie, J. L., and Hart, D. N. (1998) Optimisation of the conditions for generating human DC initiated antigen specific T lymphocyte lines in vitro. *J. Immunol. Methods* **219,** 69–83.
23. Ratta, M., Rondelli, D., Fortuna, A., Curti, A., Fogli, M., Fagnoni, F., et al. (1998) Generation and functional characterization of human dendritic cells derived from CD34 cells mobilized into peripheral blood: comparison with bone marrow CD34+ cells. *Br. J. Haematol.* **101,** 756–765.
24. Meyer Zum Buschenfelde, C., Metzger, J., Hermann, C., Nicklisch, N., Peschel, C., and Bernhard, H. (2001) The generation of both t killer and th cell clones specific for the tumor-associated antigen her2 using retrovirally transduced dendritic cells. *J. Immunol.* **167,** 1712–1719.
25. Butterfield, L. H., Jilani, S. M., Chakraborty, N. G., Bui, L. A., Ribas, A., Dissette, V. B., et al. (1998) Generation of melanoma-specific cytotoxic T lymphocytes by dendritic cells transduced with a MART-1 adenovirus. *J. Immunol.* **161,** 5607–5613.
26. Perez-Diez, A., Butterfield, L. H., Li, L., Chakraborty, N. G., Economou, J. S., and Mukherji, B. (1998) Generation of CD8+ and CD4+ T-cell response to dendritic cells genetically engineered to express the MART-1/Melan-A gene. *Cancer Res.* **58,** 5305–5309.
27. Kim, C. J., Prevette, T., Cormier, J., Overwijk, W., Roden, M., Restifo, N. P., et al. (1997) Dendritic cells infected with poxviruses encoding MART-1/Melan A sensitize T lymphocytes in vitro. *.J Immunother.* **20,** 276–286.
28. Reeves, M. E., Royal, R. E., Lam, J. S., Rosenberg, S. A., and Hwu, P. (1996) Retroviral transduction of human dendritic cells with a tumor-associated antigen gene. *Cancer Res.* **56,** 5672–5677.
29. Henderson, R. A., Nimgaonkar, M. T., Watkins, S. C., Robbins, P. D., Ball, E. D., and Finn, O. J. (1996) Human dendritic cells genetically engineered to express high levels of the human epithelial tumor antigen mucin (MUC-1). *Cancer Res.* **56,** 3763–3770.
30. Alijagic, S., Moller, P., Artuc, M., Jurgovsky, K., Czarnetzki, B. M., and Schadendorf, D. (1995) Dendritic cells generated from peripheral blood transfected

with human tyrosinase induce specific T cell activation. *Eur. J. Immunol.* **25,** 3100–3107.

31. Philip, R., Brunette, E., Ashton, J., Alters, S., Gadea, J., Sorich, M., et al. (1998) Transgene expression in dendritic cells to induce antigen-specific cytotoxic T cells in healthy donors. *Cancer Gene Ther.* **5,** 236–246.
32. Arthur, J. F., Butterfield, L. H., Roth, M. D., Bui, L. A., Kiertscher, S. M., Lau, R., et al. (1997) A comparison of gene transfer methods in human dendritic cells. *Cancer Gene Ther.* **4,** 17–25.
33. Gong, J., Chen, D., Kashiwaba, M., and Kufe, D. (1997) Induction of antitumor activity by immunization with fusions of dendritic and carcinoma cells. *Nat. Med.* **3,** 558–561.
34. Akasaki, Y., Kikuchi, T., Homma, S., Abe, T., Kofe, D., and Ohno, T. (2001) Antitumor effect of immunizations with fusions of dendritic and glioma cells in a mouse brain tumor model. *J. Immunother.* **24,** 106–113.
35. Gong, J., Nikrui, N., Chen, D., Koido, S., Wu, Z., Tanaka, Y., et al. (2000) Fusions of human ovarian carcinoma cells with autologous or allogeneic dendritic cells induce antitumor immunity. *J. Immunol.* **165,** 1705–1711.
36. Gong, J., Avigan, D., Chen, D., Wu, Z., Koido, S., Kashiwaba, M., et al. (2000) Activation of antitumor cytotoxic T lymphocytes by fusions of human dendritic cells and breast carcinoma cells. *Proc. Natl. Acad. Sci. USA* **97,** 2715–2718.
37. Kugler, A., Stuhler, G., Walden, P., Zoller, G., Zobywalski, A., Brossart, P., et al. (2000) Regression of human metastatic renal cell carcinoma after vaccination with tumor cell–dendritic cell hybrids. *Nat. Med.* **6,** 332–336.
38. Gilboa, E., Nair, S. K., and Lyerly, H. K. (1998) Immunotherapy of cancer with dendritic-cell–based vaccines. *Cancer Immunol. Immunother.* **46,** 82–87.
39. Nair, S. K., Boczkowski, D., Morse, M., Cumming, R. I., Lyerly, H. K., and Gilboa, E. (1998) Induction of primary carcinoembryonic antigen (CEA)-specific cytotoxic T lymphocytes in vitro using human dendritic cells transfected with RNA. *Nat. Biotechnol.* **16,** 364–369.
40. Heiser, A., Maurice, M. A., Yancey, D. R., Coleman, D. M., Dahm, P., and Vieweg, J. (2001) Human dendritic cells transfected with renal tumor RNA stimulate polyclonal T-cell responses against antigens expressed by primary and metastatic tumors. *Cancer Res.* **61,** 3388–3393.
41. Chang, J. W. C., Peng, M., Zhou, Y. M., Vaquerano, J. E., Clinton, R. A., and Leong, S. P. L. (1999) Induction of Th1 response by dendritic cells pulsed by autologous melanoma apoptotic bodies. *Proc. American Association for Cancer Research* **40,** 86.
42. Hsu, F. J., Benike, C., Fagnoni, F., Liles, T. M., Czerwinski, D., Taidi, B., et al. (1996) Vaccination of patients with B-cell lymphoma using autologous antigen-pulsed dendritic cells. *Nat. Med.* **2,** 52–58.
43. Nestle, F. O., Alijagic, S., Gilliet, M., Sun, Y., Grabbe, S., Dummer, R., et al. (1998) Vaccination of melanoma patients with peptide- or tumor lysate-pulsed dendritic cells [see comments]. *Nat. Med.* **4,** 328–332.

44. Holtl, L., Rieser, C., Papesh, C., Ramoner, R., Herold, M., Klocker, H., et al. (1999) Cellular and humoral immune responses in patients with metastatic renal cell carcinoma after vaccination with antigen pulsed dendritic cells. *J. Urol.* **161,** 777–782.
45. Burch, P. A., Breen, J. K., Buckner, J. C., Gastineau, D. A., Kaur, J. A., Laus, R. L., et al. (2000) Priming tissue-specific cellular immunity in a phase I trial of autologous dendritic cells for prostate cancer. *Clin. Cancer Res.* **6,** 2175–2182.
46. Fong, L., Benike, C., Brockstredt, D., Strang, G., Hao, Z. M., Smits, B., et al. (1999) Immunization with dendritic cells pulsed with xenogeneic prostatic acid phosphatase (PAP) administered via different routs induces cellular immune responses in prostate cancer patients. *Proc. American Association for Cancer Research* **40,** 85.
47. Fong, L., Brockstedt, D., Benike, C., Wu, L., and Engleman, E. G. (2001) Dendritic cells injected via different routes induce immunity in cancer patients. *J. Immunol.* **166,** 4254–4259.
48. Salgaller, M. L., Tjoa, B. A., Lodge, P. A., Ragde, H., Kenny, G., Boynton, A., et al. (1998) Dendritic cell-based immunotherapy of prostate cancer. *Crit. Rev. Immunol.* **18,** 109–119.
49. Murphy, G. P., Tjoa, B. A., Simmons, S. J., Jarisch, J., Bowes, V. A., Ragde, H., et al. (1999) Infusion of dendritic cells pulsed with HLA-A2-specific prostate-specific membrane antigen peptides: a phase II prostate cancer vaccine trial involving patients with hormone-refractory metastatic disease. *Prostate* **38,** 73–78.
50. Pecher, G., Haring, A., Schirrmann, T., and Thiel, E. (1999) Vaccination with mucin gene (MUC1) transfected dendritic cells in patients with breast or pancreatic cancer. *Proc. American Association for Cancer Research* **40,** 85.
51. Brossart, P., Wirths, S., Stuhler, G., Reichardt, V. L., Kanz, L., and Brugger, W. (2000) Induction of cytotoxic T-lymphocyte responses in vivo after vaccinations with peptide-pulsed dendritic cells. *Blood* **96,** 3102–3108.
52. Reichardt, V. L., Okada, C. Y., Liso, A., Benike, C. J., Stockerl Goldstein, K. E., Engleman, E. G., et al. (1999) Idiotype vaccination using dendritic cells after autologous peripheral blood stem cell transplantation for multiple myeloma—a feasibility study. *Blood* **93,** 2411–2419.
53. Esche, C., Shurin, M. R., and Lotze, M. T. (1999) The use of dendritic cells for cancer vaccination. *Curr. Opin. Mol. Ther.* **1,** 72–81.
54. Darrow, T. L., Vervaert, C. E., Pickett, N. J., Slingluff, C. L., and Seigler, H. F. (1999) A randomized phase I/II study immunizing melanoma patients with autologous dendritic cells pulsed with TAA and administered in combination with IL-2 in the adjuvant setting. In vitro monitoring of TAA specific immunity. *Proc. American Association for Cancer Research* **40,** 598.
55. Yang, S., Kittlesen, D., Slingluff, C. L., Jr., Vervaert, C. E., Seigler, H. F., and Darrow, T. L. (2000) Dendritic cells infected with a vaccinia vector carrying the human gp100 gene simultaneously present multiple specificities and elicit high-affinity T cells reactive to multiple epitopes and restricted by HLA-A2 and -A3. *J. Immunol.* **164,** 4204–4211.

56. Tsunoda, T., Tanimura, H., Matsuda, K., Umano, Y., Shono, Y., Iwahushi, M., et al. (1999) Cytotoxic peptide can elicit HLA-restricted CEA specific CTL from colorectal cancer patients in vivo—development of colorectal cancer vaccine. *Proc. American Association for Cancer Research* **40,** 310.
57. Kobayashi, T., Shinohara, H., Toyoda, M., Iwamoto, S., and Tanigawa, N. (2001) Regression of lymph node metastases by immunotherapy using autologous breast tumor-lysate pulsed dendritic cells: report of a case. *Surg. Today* **31,** 513–516.
58. Westermann, J., Kopp, J., Korner, I., Richter, G., Qin, Z., Blankenstein, T., et al. (2000) Bcr/abl+ autologous dendritic cells for vaccination in chronic myeloid leukemia. *Bone Marrow Transplant.* **25(Suppl. 2),** S46–S49.
59. Lemoli, R. M., Tafuri, A., Fortuna, A., Catani, L., Rondelli, D., Ratta, M., et al. (1998) Biological characterization of CD34+ cells mobilized into peripheral blood. *Bone Marrow Transplant.* **22(Suppl. 5),** S47–S50.
60. Sutherland, D. R., Keating, A., Nayar, R., Anania, S., and Stewart, A. K. (1994) Sensitive detection and enumeration of CD34+ cells in peripheral and cord blood by flow cytometry. *Exp. Hematol.* **22,** 1003–1010.
61. Canque, B., Camus, S., Dalloul, A., Kahn, E., Yagello, M., Dezutter-Dambuyant, C., et al. (2000) Characterization of dendritic cell differentiation pathways from cord blood CD34(+)CD7(+)CD45RA(+) hematopoietic progenitor cells. *Blood* **96,** 3748–3756.
62. Storms, R. W., Goodell, M. A., Fisher, A., Mulligan, R. C., and Smith, C. (2000) Hoechst dye efflux reveals a novel CD7(+)CD34(–) lymphoid progenitor in human umbilical cord blood. *Blood* **96,** 2125–2133.
63. Hao, Q. L., Zhu, J., Price, M. A., Payne, K. J., Barsky, L. W., and Crooks, G. M. (2001) Identification of a novel, human multilymphoid progenitor in cord blood. *Blood* **97,** 3683–3690.
64. Fernandez Pujol, B., Lucibello, F. C., Zuzarte, M., Lutjens, P., Muller, R., and Havemann, K. (2001) Dendritic cells derived from peripheral monocytes express endothelial markers and in the presence of angiogenic growth factors differentiate into endothelial-like cells. *Eur. J. Cell Biol.* **80,** 99–110.
65. Rosenzwajg, M. and Gluckman, J. C. (1999) Problems raised by the in vitro generation of dendritic cells for use in immunotherapy. [letter]. *Exp. Hematol.* **27,** 387–388.
66. Ebner, S., Neyer, S., Hofer, S., Nussbaumer, W., Romani, N., and Heufler, C. (2001) Generation of large numbers of human dendritic cells from whole blood passaged through leukocyte removal filters: an alternative to standard buffy coats. *J. Immunol. Methods* **252,** 93–104.
67. Jonuleit, H., Giesecke-Tuettenberg, A., Tuting, T., Thurner-Schuler, B., Stuge, T. B., Paragnik, L., et al. (2001) A comparison of two types of dendritic cell as adjuvants for the induction of melanoma-specific T-cell responses in humans following intranodal injection. *Int. J. Cancer* **93,** 243–251.
68. Morse, M. A., Lyerly, H. K., Gilboa, E., Thomas, E., and Nair, S. K. (1998) Optimization of the sequence of antigen loading and CD40-ligand-induced maturation of dendritic cells. *Cancer Res.* **58,** 2965–2968.

35

Generation of Leukemic Dendritic Cells from Patients with Acute Myeloid Leukemia

Mohamad Mohty, Béatrice Gaugler, and Daniel Olive

1. Introduction

Allogeneic bone marrow transplantation proved to be an efficient treatment for patients with acute myeloid leukemia (AML). This is mainly attributed to the so-called graft-versus-leukemia effect mediated by the donor-derived immune system, especially T-cells *(1)*. Thus, modulation of the immune system appears to be an attractive modality for the treatment of AML patients, especially those patients at high risk of relapse and who can not benefit from allogeneic transplantation. More than 20 years ago, some patients with AML received pooled, irradiated allogeneic leukemic cells in order to enhance their immune system *(2)*. At present, donor lymphocyte infusions (DLIs) following allogeneic transplantation is the treatment of choice for relapse in chronic myeloid leukemia and AML patients *(3)*. Unfortunately, DLIs is less successful in the treatment of AML patients. One possible reason for this decreased efficiency of DLIs in AML patients could be related to the fact that AML blasts are poor antigen-presenting cells and fail to induce a potent and sustained antileukemic immune response *(4)*.

Dendritic cells (DCs) are bone-marrow-derived leukocytes that are responsible for the initiation of immune responses and exert a sentinellike function *(5)*. We and others reported previously that myeloid leukemic blasts were able to differentiate in vitro into cells with mature DC features *(6–12)*. These leukemic-derived DCs were shown to have a potent capacity to induce T-cell proliferation while still retaining the leukemic chromosomal abnormality of the original blasts. These observations constitute a unique model where the best

From: *Methods in Molecular Biology, vol. 215: Cytokines and Colony Stimulating Factors: Methods and Protocols*
Edited by: D. Körholz and W. Kiess © Humana Press Inc., Totowa, NJ

DCs necessary for tumor antigen presentation and the tumor cells themselves correspond to the same cell. Therefore, DCs generated from AML patients can facilitate an immune response that might help in the induction of effective antileukemic T-cell responses.

In this work, we give details concerning the optimal methods we use in our laboratory for the generation and functional characterization of mature leukemic DCs in patients with AML, especially those belonging to the M4 and M5 subtypes.

2. Materials

2.1. Culture Medium and Growth Factors

1. We use as complete medium RPMI 1640 (Biowhittaker, Verviers, Belgium) supplemented with 10% heat-inactivated (56°C for 30 min) fetal calf serum (FCS) (Biowhittaker, Verviers, Belgium).
2. Granulocyte–macrophage colony stimulating factor (GM-CSF) was kindly provided by Novartis (Berne, Switzerland). Aliquots of GM-CSF at 0.1 mg/mL are stored at –80°C until use and were used for up to 1 wk after thawing.
3. Interleukin (IL)-4 was a kind gift of Shering-Plough Research Institute (Kenilworth, NJ). IL-4 is stored at –80°C as small aliquots at a concentration of 10 µg/mL and was used avoiding repeated freeze–thaw cycles.
4. T-Cell culture medium (defined as mixed-leukocyte reaction [MLR] medium) is RPMI 1640 supplemented with 10% FCS, 1 m*M* sodium pyruvate, 1.5 m*M* L-glutamine, 50 mg/mL streptomycin, 50 U/mL penicillin (Gibco-BRL, Life Technologies, Paisler, Scotland), and 5×10^{-5} *M* β_2-mercaptoethanol (Sigma, St. Louis, MO).

2.2. Reagents and Monoclonal Antibodies

1. Monoclonal antibodies: CD1a (BL6), CD13 (Immu103.44), CD14 (RMO52), CD33 (D3HL60.251), CD40 (MAB89), CD54 (84H10), CD58 (AICD58), CD80 (MAB104), CD83 (HB15a), anti-HLA-DR (Immu-357), anti-HLA-A,B,C (B9.12.1); all purchased from Beckman-Coulter (Marseille, France).
2. CD86 (IT2.2) purchased from Pharmingen (San Diego, CA).
3. Isotypic controls MAb used in our studies are IgG1(679.1Mc7), IgG2a (U7.27), and IgG2b (MOPC-195); all purchased from Beckman-Coulter.
4. All MAbs are used as fluoresceine isothiocyanate (FITC)-, phycoerthrin (PE), Cy-5-conjugated MAbs.

2.3. Cell Lines

Murine L-cells transfected with human CD40L were kindly provided by Schering-Plough Research Institute (Laboratory for Immunological Research, Dardilly, France). L-Cells are cultured in complete medium in 25-mL cell culture flasks (Costar, Cambridge, MA).

3. Methods

3.1. Patients' Blood Samples and Cell Preparation

1. Acute myeloid leukemia heparinized peripheral blood samples should be obtained after informed consent at diagnosis and before any chemotherapy.
2. Acute myeloid leukemia peripheral blood mononuclear cells (PBMCs) are separated on a density gradient by careful layering of 35 mL of blood over 15 mL of Ficoll–Hypaque (Pharmacia, Uppsala, Sweden) in a 50-mL conical tube (Falcon, Becton Dickinson Labware, Franklin Lakes, NJ) followed by 20 min of centrifugation at 400*g* at 20°C in a swinging-bucket rotor without brake.
3. The mononuclear cell (MNC) layer containing the malignant cells is carefully harvested and transferred to a new conical tube, washed twice with 40 mL of RPMI 1640 medium (Biowhittaker, Verviers, Belgium) containing 10% FCS and centrifuged at 300*g* for 10 min at 20°C.
4. After carefully removing the supernatant, the cell pellet is resuspended in 50 mL of RPMI/10% FCS; viable cells are counted by trypan blue exclusion.
5. For cryopreservation, 1-mL aliquots of cell suspension ([10–20] × 10^6/mL) in RPMI medium containing 20% FCS and 10% dimethyl sulfoxide (Sigma, St. Louis, MO) is rapidly transferred in 1.8-mL cryotubes (cryotube vials; NUNC Brand Products, Roskilde, Denmark), frozen to –80°C, and, finally, transferred into the gas phase of liquid nitrogen until use.

3.2. Culture of Leukemic Blasts

1. Leukemic blasts are thawed in a 37°C water bath, then subsequently transferred in a 50-mL conical tube containing complete medium for washing.
2. After 10 min of centrifugation at 300*g*, cell pellets are resuspended in complete medium and cell viability evaluated by trypan blue exclusion.
3. Viable cells are suspended at 0.5 × 10^6 cells/mL in complete medium supplemented with 100 ng/mL GM-CSF and 20 ng/mL IL-4.
4. Cell culture is performed by plating 4 mL of cell suspension in six-well tissue culture plates (Costar, Cambridge, MA) and incubating at 37°C and 5% CO_2.
5. On d 3 of culture, medium is replenished with both cytokines GM-CSF (400 ng/well) and IL-4 (80 ng/well).
6. On d 5, final maturation of DCs is induced by adding 75-Gy-irradiated CD40L-transfected cells (2 × 10^5/well) for 48 h.
7. Differentiation of leukemic blasts into DCs can be observed as soon as d 3 (*see* **Note 1**) with various proportions of cells displaying DC features among patients studied (*see* **Fig. 1**).

3.3. Flow Cytometry Analysis

1. Cells are harvested after 7 d of culture and washed.
2. To minimize FcR-mediated MAb binding, cells are incubated for 1 h at 4°C before staining in phosphate-buffered saline (PBS) supplemented with 30%

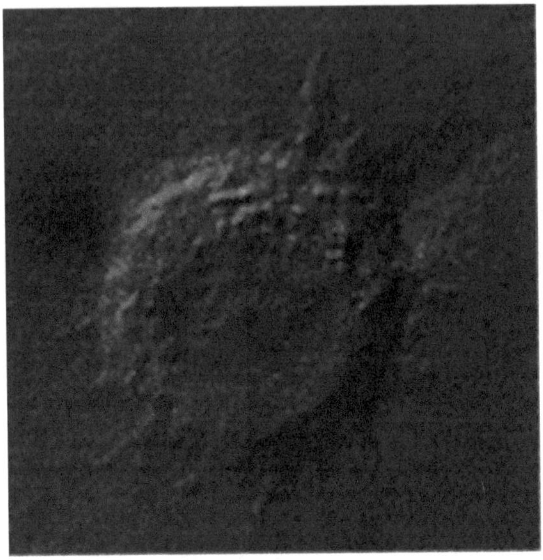

Fig. 1. Morphology of leukemic DCs. Leukemic DCs display dendritic morphology, as shown by interferential contrast transmission microscopy (100×).

human AB serum (AbCys, Paris, France), containing a high concentration of human IgG.

3. Cells are then transferred into V-bottomed 96-well microplates at a density of (2–4) × 10^5 cells/well in 100 μL and labeled by incubating with the mAb for 30 min.
4. After washing, cells are fixed in PBS +2% formaldehyde and analyzed using a FACSCalibur flow cytometer.
5. Dead cells and debris are gated out on the basis of their light-scatter properties. Data for 5000–10000 gated events are acquired and analyzed using CellQuest software (BD Biosciences).
6. This immunophenotypic analysis can indicate the percentage of leukemic blasts that can acquire the expression of CD83, which is a marker of mature DCs ***(13)*** and, to some extent, CD1a molecules. CD14 is downregulated on leukemic DCs. Like normal monocyte-derived DCs, leukemic DCs express the costimulatory molecules CD40, CD80, and CD86. The expression of HLA-ABC and HLA-DR is also upregulated. When examined, the expression of the adhesion molecules CD54 and CD58 is also upregulated. The expression of CD83 and costimulatory molecules can be considered as the major phenotypic features that can help to

define the maturation pattern of the leukemic blasts toward the DC lineage (*see* **Fig. 2**). (*See* **Note 2**.)

3.4. MLR Experiments

The ability to stimulate a T-cell primary response is one of the key features of DC function. Thus, leukemic DCs generated in vitro should be assessed for their ability to stimulate naive CD4+ T-cells in an allogeneic MLR:

1. To evaluate T cell proliferation capacity, graded numbers of irradiated (50 Gy) leukemic DCs are cocultured with 1×10^5 allogeneic naive CD4+ T cells in 96-well flat-bottomed plates for 6 d in T-cell culture medium (MLR medium), in a final volume of 200 µL in triplicates.
2. Proliferation of T cells is monitored by measuring methyl-(^{3}H)thymidine incorporation during the last 12 h of culture on a gas-phase β-counter.
3. Naive CD4+ T cells can be prepared from adult donor PBMCs, negatively depleted of CD8, CD14, CD19, CD56, and CD45RO+ cells using goat anti-mouse Ig-coated magnetic beads (Beckman-Coulter) according to the manufacturer's instructions. Ninety-eight percent of the resulting cells are usually CD4+CD45RA+ as controlled by FACS analysis.
4. Normally, leukemic DCs can efficiently stimulate the proliferation of naive CD4+ T cells. In the same experiments, autologous fresh leukemic blasts and normal PBMC from healthy volunteers induce very weak, if any, proliferation of naive allogeneic T cells.

3.5. Cytokine Production Assay

It has been previously shown that ligation of CD40 on DCs triggers IL-12 production. IL-12 stimulates natural killer (NK) cells, mediates T-cell development, and fosters cytotoxic T-lymphocyte (CTL) differentiation. Thus, for optimal immunotherapeutic strategies, DCs derived from leukemic blasts should produce the largest amounts of IL-12. We usually measure IL-12p70 in DC supernatants within 48 h after CD40L activation:

1. Supernatants of leukemic DC cultures are harvested at d 7 and cryopreserved for further analysis.
2. The IL-12p70 content is evaluated by standard enzyme-linked immunosorbent (ELISA) assays using reagents set OptEIA™ (Pharmingen, San Diego, CA).
3. The ELISA assays are performed according to the manufacturer's instructions.

4. Notes

1. After culture, leukemic cells displayed dramatic morphological changes with variable size increase, cytoplasmic protrusion and irregular cellular membrane

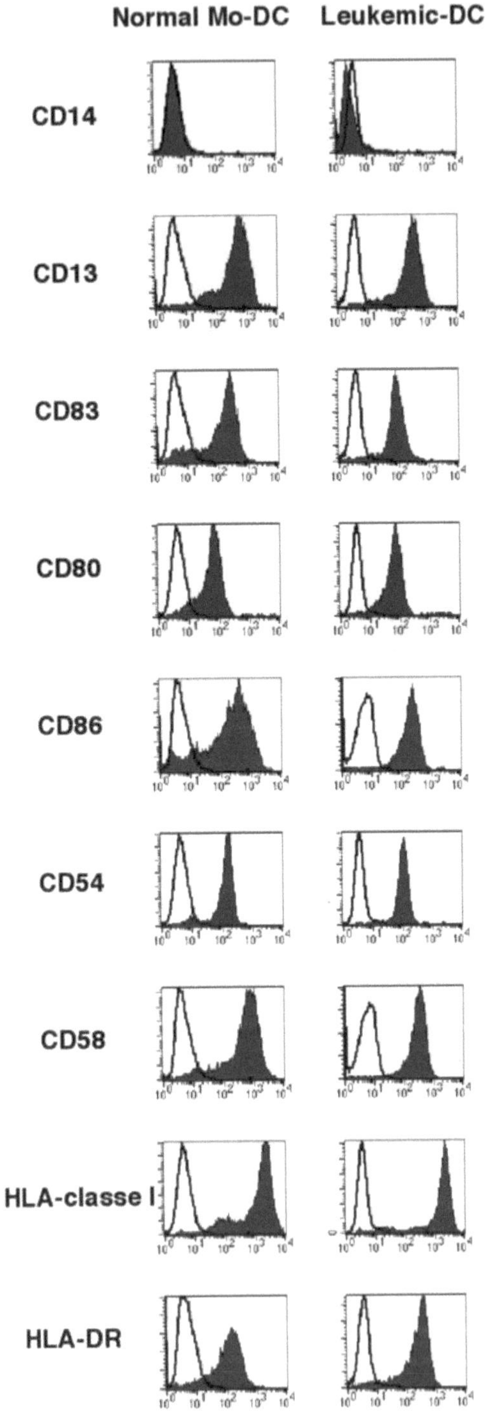
Normal Mo-DC
Leukemic-DC
CD14
CD13
CD83
CD80
CD86
CD54
CD58
HLA-classe I
HLA-DR

Fig. 2.

as soon as d 3 in some cases. As previously reported ***(6–12)***, various percentages of cells with DC features can be obtained using the widely used combination of GM-CSF and IL-4. However, in many reports, some authors added tumor necrosis factor (TNF)-α in their cytokine combination ***(7,11,14,15)*** because normal CD34+ stem cells can be differentiated into DCs by the use of GM-CSF and TNF-α ***(16)***. All of these cases concerned patients with chronic myeloid leukemia where the leukemic counterpart of the CD34+ stem cells can be found. In our experiments, we prefer the combination of GM-CSF and IL-4, used for the generation of DCs from circulating monocytes ***(17)***. For many reasons, it is unlikely in AML patients that CD34+ blasts might contribute to the generation of a major fraction of the leukemic DCs. Combined analyses of stem cell phenotypic antigen expression, genetic lesions and biological features provide evidence that major differences exist between normal and malignant progenitors ***(18)***. Furthermore, GM-CSF and IL-4 combination did not prove in vitro to induce DC differentiation from normal stem cells. Finally, differentiation of efficient DCs from normal stem cells requires a longer period of culture ***(16)***. This is not the case in our culture system, where fully mature leukemic DCs could be obtained after a maximum of 7 d, and in much shorter periods in the majority of cases.

In summary, the goal of our work is to induce leukemic blasts that differentiate into functional cells with antigen-presenting and DC features. Our culture system proved that leukemic DCs can be generated in a wide range of AML patients in the presence of GM-CSF, IL-4, and CD40L. Leukemic DCs exhibit a DC morphology, had a phenotype of mature DCs, and can induce a potent proliferative response in naive CD4+ T cells. Because leukemic DCs are an attractive tool for the design of clinical adoptive T cell or vaccine immunotherapy protocols for the adjuvant treatment of AML patients, the experimental approach described should be considered as a prerequisite before any immunotherapeutic applications.

2. The leukemic origin of these leukemic-derived DCs from AML patients was clearly established in previous studies by confirming the presence of cytogenetic abnormalities in fluorescence *in situ* hybridization (FISH) experiments. Thus, we do not feel that it is necessary to perform FISH experiments systematically.

References

1. Antin, J. H. (1993) Graft-versus-leukemia: no longer an epiphenomenon. *Blood* **82,** 2273–2277.

Fig. 2. *(see opposite page)* Phenotype of leukemic DCs. Leukemic DCs were analyzed by flow cytometry after 7 d of culture and compared to normal monocyte-derived DCs, for the expression of the indicated markers (black histograms). Open histogams represent cells stained with isotype-matched control MAbs.

2. Powles, R. L., Russell, J., Lister, T. A., Oliver, T., Whitehouse, J. M., Malpas, J., et al. (1977) Immunotherapy for acute myelogenous leukaemia: a controlled clinical study 2-1/2 years after entry of the last patient. *Br. J. Cancer* **35,** 265–272.
3. Kolb, H. J., Schattenberg, A., Goldman, J. M., Hertenstein, B., Jacobsen, N., Arcese, W., et al. (1995) Graft-versus-leukemia effect of donor lymphocyte transfusions in marrow grafted patients. European Group for Blood and Marrow Transplantation Working Party Chronic Leukemia. [see comments]. *Blood* **86,** 2041–2050.
4. Falkenburg, J. H., Smit, W. M., and Willemze, R. (1997) Cytotoxic T-lymphocyte (CTL) responses against acute or chronic myeloid leukemia. *Immunol. Rev.* **157,** 223–230.
5. Banchereau, J. and Steinman, R. M. (1998) Dendritic cells and the control of immunity. *Nature* **392,** 245–252.
6. Charbonnier, A., Gaugler, B., Sainty, D., Lafage-Pochitaloff, M., and Olive, D. (1999) Human acute myeloblastic leukemia cells differentiate in vitro into mature dendritic cells and induce the differentiation of cytotoxic T cells against autologous leukemias. *Eur. J. Immunol.* **29,** 2567–2578.
7. Choudhury, A., Gajewski, J. L., Liang, J. C., Popat, U., Claxton, D. F., Kliche, K. O., et al. (1997) Use of leukemic dendritic cells for the generation of antileukemic cellular cytotoxicity against Philadelphia chromosome-positive chronic myelogenous leukemia. *Blood* **89,** 1133–1142.
8. Choudhury, B. A., Liang, J. C., Thomas, E. K., Flores-Romo, L., Xie, Q. S., et al. (1999) Dendritic cells derived in vitro from acute myelogenous leukemia cells stimulate autologous, antileukemic T-cell responses. *Blood* **93,** 780–786.
9. Cignetti, A., Bryant, E., Allione, B., Vitale, A., Foa, R., and Cheever, M. A. (1999) CD34(+) acute myeloid and lymphoid leukemic blasts can be induced to differentiate into dendritic cells. *Blood* **94,** 2048–2055.
10. Robinson, S. P., English, N., Jaju, R., Kearney, L., Knight, S. C., and Reid, C. D. (1998) The in-vitro generation of dendritic cells from blast cells in acute leukaemia. *Br. J. Haematol.* **103,** 763–771.
11. Smit, W. M., Rijnbeek, M., van Bergen, C. A., de Paus, R. A., Vervenne, H. A., van de Keur, M., et al. (1997) Generation of dendritic cells expressing bcr-abl from CD34-positive chronic myeloid leukemia precursor cells. *Hum. Immunol.* **53,** 216–223.
12. Harrison, B. D., Adams, J. A., Briggs, M., Brereton, M. L., and Yin, J. A. (2001) Stimulation of autologous proliferative and cytotoxic T-cell responses by "leukemic dendritic cells" derived from blast cells in acute myeloid leukemia. *Blood* **97,** 2764–2771.
13. Zhou, L. J. and Tedder, T. F. (1995) Human blood dendritic cells selectively express CD83, a member of the immunoglobulin superfamily. *J. Immunol.* **154,** 3821–3835.
14. Eibl, B., Ebner, S., Duba, C., Bock, G., Romani, N., Erdel, M., et al. (1997) Dendritic cells generated from blood precursors of chronic myelogenous leukemia patients carry the Philadelphia translocation and can induce a CML-specific primary cytotoxic T-cell response. *Genes Chromosomes Cancer* **20,** 215–223.

15. Coleman, S., Throp, D., Fisher, J., Bailey-Wood, R., and Lim, S. H. (1997) Cytokine enhancement of immunogenicity in chronic myeloid leukaemia. *Leukemia* **11,** 2055–2059.
16. Caux, C., Dezutter-Dambuyant, C., Schmitt, D., and Banchereau, J. (1992) GM-CSF and TNF-alpha cooperate in the generation of dendritic Langerhans cells. *Nature* **360,** 258–261.
17. Sallusto, F. and Lanzavecchia, A. (1994) Efficient presentation of soluble antigen by cultured human dendritic cells is maintained by granulocyte/macrophage colony-stimulating factor plus interleukin 4 and downregulated by tumor necrosis factor alpha. *J. Exp. Med.* **179,** 1109–1118.
18. Brendel, C. and Neubauer, A. (2000) Characteristics and analysis of normal and leukemic stem cells: current concepts and future directions. *Leukemia* **14,** 1711–1717.

Index

F

G

H

I

R

S

FSC
www.fsc.org
MIX
Papier | Fördert gute Waldnutzung
FSC® C083411

Kolibri 360 GmbH
Ferdinand-Jühlke-Straße 7
99095 Erfurt, Deutschland
produktsicherheit@kolibri360.de